Inorganic Nanomaterials for Supercapacitor Design

Inorganic Nanomaterials for Supercapacitor Design

Edited by
Dr. Inamuddin
Rajender Boddula
Mohd Imran Ahamed
Abdullah Mohamed Asiri

CRC Press is an imprint of the
Taylor & Francis Group, an **informa** business

CRC Press
Taylor & Francis Group
6000 Broken Sound Parkway NW, Suite 300
Boca Raton, FL 33487-2742

First issued in paperback 2021

ISBN 13: 978-1-03-223816-6 (pbk)
ISBN 13: 978-0-367-23000-5 (hbk)

Publisher's Note
The publisher has gone to great lengths to ensure the quality of this reprint but points out that some imperfections in the original copies may be apparent.

Visit the Taylor & Francis Web site at
http://www.taylorandfrancis.com

and the CRC Press Web site at
http://www.crcpress.com

Contents

Preface..vii
Editors..ix
Contributors..xiii

Chapter 1 Niobium Based Materials for Supercapacitors..1

Prasun Banerjee, Adolfo Franco Junior, D. Baba Basha, K. Chandra Babu Naidu, and K. Srinivas

Chapter 2 Zinc-Based Materials for Supercapacitors..17

Prasun Banerjee, Adolfo Franco Junior, D. Baba Basha, K. Chandra Babu Naidu, and S. Ramesh

Chapter 3 Defect Engineered Inorganic Materials for Supercapacitors..33

Biswajit Choudhury and Anamika Kalita

Chapter 4 Vanadium-Based Compounds for Supercapacitors..59

Xuan Pan, Wenyue Li, and Zhaoyang Fan

Chapter 5 Future Prospects and Challenges of Inorganic-Based Supercapacitors..79

Vivian C. Akubude and Kevin N. Nwaigwe

Chapter 6 Tungsten Based Materials for Supercapacitors..89

Christelle Pau Ping Wong, Chin Wei Lai, and Kian Mun Lee

Chapter 7 Microwave-Assisted Inorganic Materials for Supercapacitors..101

M. Ramesh and Arivumani Ravanan

Chapter 8 Tin-Based Materials for Supercapacitor..119

Muhammad Mudassir Hassan, Nawshad Muhmmad, Muhammad Sabir, and Abdur Rahim

Chapter 9 Inorganic Materials-Based Next-Generation Supercapacitors..133

Muhammad Aamir, Arshad Farooq Butt, and Javeed Akhtar

Chapter 10 Synthesis Approaches of Inorganic Materials................................ 155

Bilal Akram, Arshad Farooq Butt, and Javeed Akhtar

Chapter 11 Metal-Organic Frameworks Derived Materials for Supercapacitors.. 187

E. Heydari-Soureshjani, Ali A. Ensafi, and Ahmad R. Taghipour-Jahromi

Chapter 12 Surface Morphology Induced Inorganic Materials for Supercapacitors 213

K. Chandra Babu Naidu, D. Baba Basha, S. Ramesh, N. Suresh Kumar, Prasun Banerjee, and K. Srinivas

Chapter 13 Molybdenum Based Materials for Supercapacitors Beyond TMDs...239

Swapnil S. Karade and Deepak P. Dubal

Chapter 14 Iron-Based Electrode Materials for an Efficient Supercapacitor 257

Leonardo Vivas, Rafael Friere, Javier Enriquez, Rajesh Kumar, and Dinesh Pratap Singh

Chapter 15 Metal-Organic Frameworks for Supercapacitors............................277

Mustapha Mohammed Bello, Anam Asghar, and Abdul Aziz Abdul Raman

Chapter 16 Amino Acid-Assisted Inorganic Materials for Supercapacitors 307

M. Hamed Misbah, Amr A. Essawy, Rawya Ramadan, E.F. ElAgammy, and Maged El-Kemary

Chapter 17 Co-Based Materials for Supercapacitors..325

N. Suresh Kumar, R. Padma Suvarna, S. Ramesh, D. Baba Basha, and K. Chandra Babu Naidu

Index...341

Preface

The crucial interest for clean energy and fast advancement of present-day electronics has driven the exciting research on novel energy storage devices, particularly supercapacitors. The most significant part of the supercapacitors is the designing of electrode materials with great capacitive performance. Among the various electrode materials used in supercapacitors, inorganic materials have received vast consideration owing to their redox chemistry, chemical stability, high electrochemical performance and high-power applications. The current developments in nanotechnology has opened up novel inorganic materials e.g. metal nitrides, metal carbides, perovskites, MXene and 2D single inorganic materials such as phosphorene, antimonene, etc. Various productive achievements have been accomplished toward creating inorganic materials-based supercapacitor electrodes. However, poor rate capability and energy density are the main limitations for inorganic materials-based supercapacitor devices. Therefore, exploring inorganic electrodes with high capacitance is becoming a key parameter to enable the high-energy density of supercapacitor devices.

This book provides an in-depth overview of the latest advances of the inorganic nanomaterials for supercapacitor energy storage devices. All the chapters discuss fundamental aspects, key factors, fabrication methods, designing of nanostructures, electrochemical performance of the inorganic-based materials designed for high-performance supercapacitor, and flexible supercapacitor applications. It also presents up-to-date literature coverage on a large, rapidly growing and complex inorganic supercapacitors. The chapters of the books provide indepth literature on the capacitors based on various inorganic materials such as zinc, niobium, vanadium, tungsten, tin molybdenum and iron cobalt. The chapters of the book are written by leading researchers throughout the world. This book will appeal to researchers, scientists and graduate and postgraduate students understudies from various disciplines, for example, material synthesis, inorganic chemistry, nanoscience and energy communities. The book content incorporates industrial applications and will fill the gap between the exploration works in the lab to viable applications in related ventures.

Dr. Inamuddin
King Abdulaziz University, Jeddah, Saudi Arabia
Aligarh Muslim University, Aligarh, India

Rajender Boddula
Chinese Academy of Sciences (CAS)
National Center for Nanoscience and Technology, Beijing, PR China

Mohd Imran Ahamed
Aligarh Muslim University, Aligarh, India

Abdullah Mohamed Asiri
King Abdulaziz University, Jeddah, Saudi Arabia

Editors

Dr. Inamuddin is currently working as Assistant Professor in the Chemistry Department, Faculty of Science, King Abdulaziz University, Jeddah, Saudi Arabia. He is a permanent faculty member (Assistant Professor) at the Department of Applied Chemistry, Aligarh Muslim University, Aligarh, India. He obtained Master of Science degree in Organic Chemistry from Chaudhary Charan Singh (CCS) University, Meerut, India, in 2002. He received his Master of Philosophy and Doctor of Philosophy degrees in Applied Chemistry from Aligarh Muslim University (AMU), India, in 2004 and 2007, respectively. He has extensive research experience in multidisciplinary fields of Analytical Chemistry, Materials Chemistry, and Electrochemistry and, more specifically, Renewable Energy and Environment. He has worked on different research projects as project fellow and senior research fellow funded by University Grants Commission (UGC), Government of India, and Council of Scientific and Industrial Research (CSIR), Government of India. He has received Fast Track Young Scientist Award from the Department of Science and Technology, India, to work in the area of bending actuators and artificial muscles. He has completed four major research projects sanctioned by University Grant Commission, Department of Science and Technology, Council of Scientific and Industrial Research, and Council of Science and Technology, India. He has published 147 research articles in international journals of repute and eighteen book chapters in knowledge-based book editions published by renowned international publishers. He has published 60 edited books with Springer (UK), Elsevier, Nova Science Publishers, Inc. (USA), CRC Press Taylor & Francis Asia Pacific, Trans Tech Publications Ltd. (Switzerland), IntechOpen Limited (UK), and Materials Research Forum LLC (USA). He is a member of various journals' editorial boards. He is also serving as Associate Editor for journals (Environmental Chemistry Letter, Applied Water Science and Euro-Mediterranean Journal for Environmental Integration, Springer-Nature), Frontiers Section Editor (Current Analytical Chemistry, Bentham Science Publishers), Editorial Board Member (Scientific Reports-Nature), Editor (Eurasian Journal of Analytical Chemistry), and Review Editor (Frontiers in Chemistry, Frontiers, U.K.). He is also guest-editing various special thematic special issues to the journals of Elsevier, Bentham Science Publishers, and John Wiley & Sons, Inc. He has attended as well as chaired sessions in various international and national conferences. He has worked as a Postdoctoral Fellow, leading a research team at the Creative Research Initiative Center for Bio-Artificial Muscle, Hanyang University, South Korea, in the field of renewable energy, especially biofuel cells. He has also worked as a Postdoctoral Fellow at the Center of Research Excellence in Renewable Energy, King Fahd University of Petroleum and Minerals, Saudi Arabia, in the field of polymer electrolyte membrane fuel cells and computational fluid dynamics of polymer electrolyte membrane fuel cells. He is a life member of the Journal of the Indian Chemical Society. His research interest includes ion exchange materials, a sensor for heavy metal ions, biofuel cells, supercapacitors and bending actuators.

Dr. Rajender Boddula is currently working for Chinese Academy of Sciences President's International Fellowship Initiative (CAS-PIFI) at National Center for Nanoscience and Technology (NCNST, Beijing). His academic honors includes University Grants Commission National Fellowship and many merit scholarships, and CAS-PIFI. He has published many scientific articles in international peer-reviewed journals and has authored twenty book chapters. He is also serving as an editorial board member and a referee for reputed international peer-reviewed journals. He has published edited books with Springer (UK), Elsevier, Materials Science Forum LLC (USA) and CRC Press Taylor & Francis Asia Pacific, Trans Tech Publications Ltd. (Switzerland). His specialized areas of research are energy conversion and storage, which include sustainable nanomaterials, graphene, polymer composites, heterogeneous catalysis for organic transformations, environmental remediation technologies, photoelectrochemical water-splitting devices, biofuel cells, batteries and supercapacitors.

Dr. Mohd Imran Ahamed received his PhD degree on the topic "Synthesis and characterization of inorganic-organic composite heavy metals selective cation-exchangers and their analytical applications," from Aligarh Muslim University, Aligarh, India in 2019. He has published several research and review articles in the journals of international recognition. Springer (UK), Elsevier, CRC Press Taylor & Francis Asia Pacific and Materials Research Forum LLC (USA). He has completed his BSc (Hons) Chemistry from Aligarh Muslim University, Aligarh, India, and MSc (Organic Chemistry) from Dr. Bhimrao Ambedkar University, Agra, India. His research work includes ion-exchange chromatography, wastewater treatment, and analysis, bending actuator and electrospinning.

Prof. Abdullah Mohamed Asiri is the Head of the Chemistry Department at King Abdulaziz University since October 2009 and he is the founder and the Director of the Center of Excellence for Advanced Materials Research (CEAMR) since 2010 till date. He is the Professor of Organic Photochemistry. He graduated from King Abdulaziz University (KAU) with BSc in Chemistry in 1990 and a PhD from University of Wales, College of Cardiff, UK, in 1995. His research interest covers color chemistry, synthesis of novel photochromic and thermochromic systems, synthesis of novel coloring matters and dyeing of textiles, materials chemistry, nanochemistry and nanotechnology, polymers and plastics. Prof. Asiri is the principal supervisors of more than 20 MSc and 6 PhD theses. He is the main author of ten books of different chemistry disciplines. Prof. Asiri is the Editor-in-Chief of *King Abdulaziz University Journal of Science*. A major achievement of Prof. Asiri is the research of tribochromic compounds, a new class of compounds which change from slightly or colorless to deep colored when subjected to small pressure or when grind. This discovery was introduced to the scientific community as a new terminology published by International Union of Pure and Applied Chemistry (IUPAC) in 2000. This discovery was awarded a patent from European Patent office and UK patent. Prof. Asiri involved in many committees at the KAU level and on the national level. He took a major role in the advanced materials committee working for King Abdulaziz City for Science and Technology (KACST) to identify the national plan for science and technology in 2007. Prof. Asiri played a major role in advancing the chemistry education and research in KAU. He has been awarded the best researchers from KAU for the past five years. He also awarded the Young Scientist Award from the Saudi Chemical Society in 2009 and also the first prize for the distinction in science from the Saudi Chemical Society in 2012. He also received a recognition certificate from the American Chemical Society (Gulf region Chapter) for the advancement of chemical science in the Kingdome. He received a Scopus certificate for the most publishing scientist in Saudi Arabia in chemistry in 2008. He is also a member of the editorial board of various journals of international repute. He is the Vice-President of Saudi Chemical Society (Western Province Branch). He holds four USA patents, more than one thousand publications in international journals, several book chapters and edited books.

Contributors

Muhammad Aamir
Materials Laboratory
Department of Chemistry
Mirpur University of Science and Technology (MUST)
Mirpur, Pakistan

Javeed Akhtar
Materials Laboratory
Department of Chemistry
Mirpur University of Science and Technology (MUST)
Mirpur, Pakistan

Bilal Akram
Department of Chemistry
Tsinghua University
Beijing, P.R. China

Vivian C. Akubude
Department of Agricultural and
Bioresource Engineering
Federal University of Technology
Owerri, Nigeria

Anam Asghar
Department of Chemical Engineering
School of Engineering
University of Mississippi
Chennai, India

Prasun Banerjee
Department of Physics
Gandhi Institute of Technology and Management (GITAM) University
Bangalore, India

and

Sri Krishnadevaraya University
Anantapuramu, India

D. Baba Basha
Department of Physics
College of Computer and Information Sciences
Majmaah University
Al Majmaah, Kingdom of Saudi Arabia

Mustapha Mohammed Bello
Department of Chemical Engineering
Faculty of Engineering
University of Malaya
Kuala Lumpur, Malaysia

Arshad Farooq Butt
Materials Laboratory
Department of Chemistry
Mirpur University of Science and Technology (MUST)
Mirpur, Pakistan

Biswajit Choudhury
Physical Sciences Division
Institute of Advanced Study in Science and Technology (An autonomous Institute under DST, Govt. of India)
Guwahati, India

Deepak P. Dubal
School of Chemistry
Physics and Mechanical Engineering
Queensland University of Technology (QUT)
Brisbane, Queensland, Australia

E.F. ElAgammy
Department of Physics
College of Science
Jouf University
Sakaka, Kingdom of Saudi Arabia

and

Glass Research Group
Physics Department
Faulty of Science
Mansoura University
Mansoura, Egypt

Maged El-Kemary
Nanoscience Department
Institute of Nanoscience & Nanotechnology
Kafrelshiekh University
Kafrelshiekh, Egypt

Javier Enriquez
Department of Physics
Millennium Institute for Research in Optics (MIRO)
University of Santiago Chile
Santiago, Chile

Ali A. Ensafi
Department of Chemistry
Isfahan University of Technology
Isfahan, Iran

Amr A. Essawy
Chemistry Department
Faculty of Science
Fayoum University
Fayoum, Egypt

and

Chemistry Department
College of Science
Jouf University
Sakaka, Kingdom of Saudi Arabia

Zhaoyang Fan
Department of Electrical and Computer Engineering
Texas Tech University
Lubbock, Texas

Adolfo Franco Junior
Instituto de Física
Universidade Federal de Goiás
Goiânia, Brazil

Rafael Friere
Department of Physics
CEDENNA, University of Santiago Chile
Santiago, Chile

Muhammad Mudassir Hassan
Department of Physics
COMSATS University Islamabad
Islamabad, Pakistan

E. Heydari-Soureshjani
Department of Chemistry
Isfahan University of Technology
Isfahan, Iran

Anamika Kalita
Physical Sciences Division
Institute of Advanced Study in Science and Technology (An autonomous Institute under DST, Govt. of India)
Guwahati, India

Swapnil S. Karade
Electrochemical Energy Laboratory
Department of Chemical and Biomolecular Engineering
Yonsei University
Seoul, Republic of Korea

N. Suresh Kumar
Department of Physics
Jawaharlal Nehru Technological University Anantapur
Anantapuramu, India

Rajesh Kumar
Department of Electrical and Electronic Information Engineering
Toyohashi University of Technology
Toyohashi, Japan

Chin Wei Lai
Nanotechnology & Catalysis Research Centre (NANOCAT)
Institute for Advanced Studies
University of Malaya
Kuala Lumpur, Malaysia

Kian Mun Lee
Nanotechnology & Catalysis Research Centre (NANOCAT)
Institute for Advanced Studies
University of Malaya
Kuala Lumpur, Malaysia

Wenyue Li
Department of Electrical and Computer Engineering
Texas Tech University
Lubbock, Texas

M. Hamed Misbah
Nanoscience Department
Institute of Nanoscience & Nanotechnology
Kafrelshiekh University
Kafrelshiekh, Egypt

Nawshad Muhmmad
Interdisciplinary Research Centre in Biomedical Materials
COMSATS University Islamabad
Islamabad, Pakistan

K. Chandra Babu Naidu
Department of Physics
Gandhi Institute of Technology and Management (GITAM) University
Bangalore, India

and

Sri Krishnadevaraya University
Anantapuramu, India

Kevin N. Nwaigwe
Department of Mechanical Engineering
University of Botswana
Gaborone, Botswana

Xuan Pan
Institutes of Science and Development
Chinese Academy of Sciences
Beijing, China

Abdur Rahim
Interdisciplinary Research Centre in Biomedical Materials
COMSATS University Islamabad
Islamabad, Pakistan

Rawya Ramadan
Physics Research Division
Microwave Physics and Dielectric Department
National Research Centre
Cairo, Egypt

Abdul Aziz Abdul Raman
Department of Chemical Engineering
Faculty of Engineering
University of Malaya
Kuala Lumpur, Malaysia

M. Ramesh
Department of Mechanical Engineering
KIT-Kalaignarkarunanidhi Institute of Technology
Coimbatore, India

and

Sri Krishnadevaraya University
Anantapuramu, India

S. Ramesh
Department of Physics
Gandhi Institute of Technology and Management (GITAM) University
Bangalore, India

Arivumani Ravanan
Department of Mechanical Engineering
KIT-Kalaignarkarunanidhi Institute of Technology
Coimbatore, India

Muhammad Sabir
Department of Physics
COMSATS University Islamabad
Islamabad, Pakistan

R. Padma Suvarna
Department of Physics
JNTUA
Anantapuramu, India

Dinesh Pratap Singh
Department of Physics
Millennium Institute for Research in Optics (MIRO)
University of Santiago Chile
Santiago, Chile

K. Srinivas
Department of Physics
Sri Krishnadevaraya University
Anantapuramu, India

and

Gandhi Institute of Technology and Management Deemed to be University
Bangalore, India

Ahmad R. Taghipour-Jahromi
Department of Chemistry
Isfahan University of Technology
Isfahan, Iran

Leonardo Vivas
Department of Physics
Millennium Institute for Research in Optics (MIRO)
University of Santiago Chile
Santiago, Chile

Christelle Pau Ping Wong
Nanotechnology & Catalysis Research Centre (NANOCAT)
Institute for Advanced Studies
University of Malaya
Kuala Lumpur, Malaysia

1 Niobium Based Materials for Supercapacitors

Prasun Banerjee, Adolfo Franco Junior, D. Baba Basha, K. Chandra Babu Naidu, and K. Srinivas

CONTENTS

1.1 Introduction 1
1.2 Historical Developments 3
1.3 Niobium as an Electrode Material in Supercapacitors Applications 3
 1.3.1 Nanoporous Niobium Pentoxide (Nb_2O_5) 4
 1.3.2 Core-Shell Structured Nb_2O_5 Nanoparticles 4
 1.3.3 Niobium Oxide Nanowires 6
 1.3.4 Niobium Oxide Nanorods 7
 1.3.5 Niobium Oxide Nanodots 8
 1.3.6 Titanium Niobium Oxide Nanotubes 9
 1.3.7 Titanium Niobium Oxide Nanofibers 9
1.4 Conclusions 10
Acknowledgements 11
References 12

1.1 INTRODUCTION

Unconventional energy sources and their need for growth are presently the most essential subjects in regard to rapid depletion of fossil fuels. These environmentally friendly and unconventional sources of energies, such as solar panels, wind turbines, and more, are mostly dependent on storage cell technologies. But on the other hand storage cells are mostly dependent on environmentally unfriendly components like lead acids, lithium, etc. Hence there is an urgent need for cleaner, greener methods that can replace the storage cell technologies to reduce environmental pollution. The present day alternative of the storage cell is definitely supercapacitors as they have the advantages to use environmentally friendly materials with high-power density along with high-withstanding limit with nearly half a million cycles [1–12]. In normal parallel plate capacitors, two metallic plates are separated from each other by an insulating medium of dielectrics. When an electric voltage is applied due to the insulating nature of the dielectric medium, static electric charges of opposite nature start building across the two metallic plates, which results in the storage of electrical energies. The charge that builds across the capacitor can be calculated by the formula $Q = C{\cdot}V$. The capacitance C of a parallel plate capacitor depends on

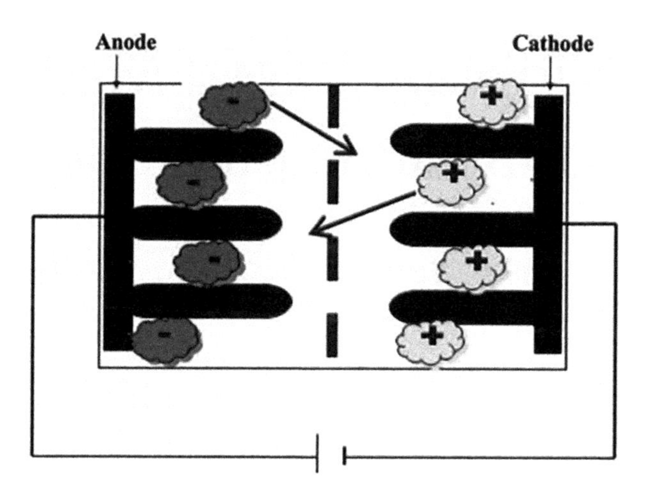

FIGURE 1.1 Schematic illustration of basic component of double-layer electrochemical capacitor. (Reprinted with the permission from Fisher, R.A. et al., *ECS J. Solid State Sci. Technol.*, 2, M3170–M3177, 2013. Copyright 2013, Electrochemical Society.)

dielectric constant k, area A and the distance d by the relation $C = \varepsilon_0 kA/d$, with ε_0 permittivity of free space [13].

Whereas the working principle of the supercapacitors or electrochemical capacitors (EC) are slightly different than that of the normal capacitors with somewhat special components, as shown in Figure 1.1, such as an "active electrode" material and a porous non-conductive separator [14]. In this particular arrangement in EC, the voltage at ion-electrode interface caused an electric double Helmholtz layer (order of 5–10 Å thickness) due to the attraction of electrodes and the electrolyte ions.

This specialty in the supercapacitor comes with great advantages, including with applications in hybrid vehicles, electronic devices and industrial energy/power management. The reasons behind the better performances in the supercapacitors are due to the small distance between the plates of the capacitors, with the high surface areas of the order of 1000–2000 m^2/g along with the nanometers level of thickness of the Helmholtz layer, which is in general microns level in a conventional capacitance [15]. It should also be noted that due to the absence of charge transfer between the plates of the capacitor through the dielectric medium, the degradation of the electrolyte and the electrodes are also minimal in case of the supercapacitors, which enhances the discharge-charge cycles of the supercapacitors [16] up to half a million cycles or more.

In contrary the development of the supercapacitors comes with major challenges along with added disadvantages due to its low power density and higher production cost. The most prominent approach to overcome these hurdles is to develop new kind of materials for the supercapacitor applications. The most popular material for the supercapacitor application is carbon particles due to their high surface areas. But carbon-based materials have some limitation due the physical storing of the charges in porous layers of electrodes known as the electrostatic double-layer supercapacitors (EDLS) with limited value of capacitance and low-power density. Due to these limitations the best approach is to replace the carbon-based material completely by a active material known as a faradic supercapacitors (FS), or it can be hybridized by mixing an active material with the carbon-based particles known as a hybrid double-layer supercapacitors (HDLC). It has already well-established that HDLC or FS comes with higher capacitance than that of the EDLC type. In this scenario niobium-based materials are the prominent candidates for the developments of the next generation supercapacitors. Hence in this chapter we concentrate on development of the supercapacitors by using niobium-based inorganic materials for the applications in supercapacitor electrodes.

1.2 HISTORICAL DEVELOPMENTS

The research on the topic of capacitors using electrolyte and metal with an accumulation of electrical charge at plates begins in the nineteenth centuries. In that direction of obtaining high surface area from porous-activated charcoal started by a group of electrical engineers in the mid 1950s a group of electrical engineers obtained high surface area from porous-activated charcoal. The first patent was filed on the supercapacitor field in 1957. Due to the interest in the hybrid vehicle field, the attention of researchers started growing in the field of supercapacitors from the early 1990s [17]. As it was then realized that to enhance the acceleration in the hybrid cars it is required to increase the energy of the batteries or the fuel cells by improving the supercapacitor technology [18]. Further it was found that the supercapacitors not only boost the performance of the energy associated with the fuel cells or the batteries, but also can replace them as well. It is the US Department of Energy that realized the supercapacitor can be an alternative for fuel cells and batteries for the future source of storage energy systems [19].

Thereafter, many enterprises and other governments departments started investing money and time to explore the possibilities gradually. Reports can be found with enhancement of the capacitance introduced with electrolyte along with redox additives with additional pseudo capacitive in the last decades [20–24]. Enhancement of the cell voltage can also be obtained in the asymmetry at anode and cathode in supercapacitors. Therefore, researches in this direction enlighten two directions in the overall improvement in the performances of the supercapacitors [25]. Firstly, electrodes with new advanced materials and the next one with the consideration of ion transport technology.

1.3 NIOBIUM AS AN ELECTRODE MATERIAL IN SUPERCAPACITORS APPLICATIONS

Ruthenium (IV) oxide is one of the most promising materials that have been studied extensively in the last decade as hybrid supercapacitor electrode because of its excellent cycle life, specific capacitance, and conductance [26]. The material application limitation as a supercapacitor was due to its expensive nature, but the study led by the Augustyn et al. indicates the fact that niobium-based inorganic materials have higher capacity at high rates, indicating a suitable material for the hybrid supercapacitor applications [27]. But the result also suggested that EC feature in niobium-based materials arises at the surface of the material rather than come as the bulk crystalline properties. The theoretical calculation of the Lubimtsev et al. [28] using the density functional theory method indicates that the nanoporous structure with niobium-based material can improve the performance of the pseudo capacitors by resulting high-rate pseudo capacitance. Hence a high-temperature synthesis of niobium-based materials with temperature more than 600°C can result in nanoclusters, which act as a deterrent factor between fast double-layer cathode and slow faradaic anode in the high-power electrode devices. Hence orthorhombic nanoparticles of niobium pentoxide (Nb_2O_5) can be considered one of the most suitable materials to improve the kinematics of the faradaic anode.

1.3.1 Nanoporous Niobium Pentoxide (Nb_2O_5)

Nanoporous Nb_2O_5 is the best alternative electrode material for the replacement of conventional ruthenium electrodes for its low price, higher specific capacitance, and energy density. Even the density functional theory calculated values of capacity 200 mA hg^{-1} for Nb_2O_5 is greater than that of the value of 175 mA hg^{-1} for $Li_4Ti_5O_{12}$ [29]. But the crystal structure of the Nb_2O_5 is highly dependent on the capacity, hence it is very difficult to apply Nb_2O_5 directly as electrode material [30]. It is first opted to improve the electron mobility in Nb_2O_5 crystal by applying carbon coating in it. But the obtained specific capacitance and energy density still is not suitable for most of the supercapacitor applications [31]. Hence the efforts have been made to verify supercapacitor applications on the various other structure of the Nb_2O_5 such as amorphous, H-Nb_2O_5 (pseudo hexagonal), O-Nb_2O_5 (orthorhombic), M-Nb_2O_5 (monoclinic) and T-Nb_2O_5 (tetragonal) phases as shown in Figure 1.2 [32]. Among all the available different phases O-Nb_2O_5 is the most promising one [30]. Hence the higher capacitance in Nb_2O_5 should be originated from the nanoporous structure of O-Nb_2O_5 phase which enhances the surface area as well as ionic conduction through these pores [33]. Lim et al. synthesized the nanoporous O-Nb_2O_5 structure by the heat treatment at a higher temperature with the evaporation of the added copolymer shown in Figure 1.3.

The as obtained power density 18,510 W kg^{-1} with the energy 74 Wh kg^{-1} and an great withstanding rate of 90% at 1 Ag^{-1} even after 1000 cycles indicates the suitability of nanoporous structured Nb_2O_5 ceramics as a suitable candidate for the supercapacitor system.

1.3.2 Core-Shell Structured Nb_2O_5 Nanoparticles

A core-shell structure of Nb_2O_5 and carbon is another alternative route for the application of the niobium-based material for the supercapacitor applications. Lim et al. utilized a facile one-pot technique by controlling the pH condition to synthesize pseudo hexagonal (TT) or orthorhombic (T) with carbon core-shell structure [34]. Here they show that the alkaline condition may lead to TT-Nb_2O_5@C NCs, whereas an acidic condition may be used to synthesize T-Nb_2O_5@C NCs. They systematically studied the EC properties and compared the results along with carbon shell-free Nb_2O_5, T-Nb_2O_5@C and TT-Nb_2O_5@C NCs. But due to the structural advantages and higher mobility of ions, T-Nb_2O_5@C NCs seems to be the best candidates among all these three types.

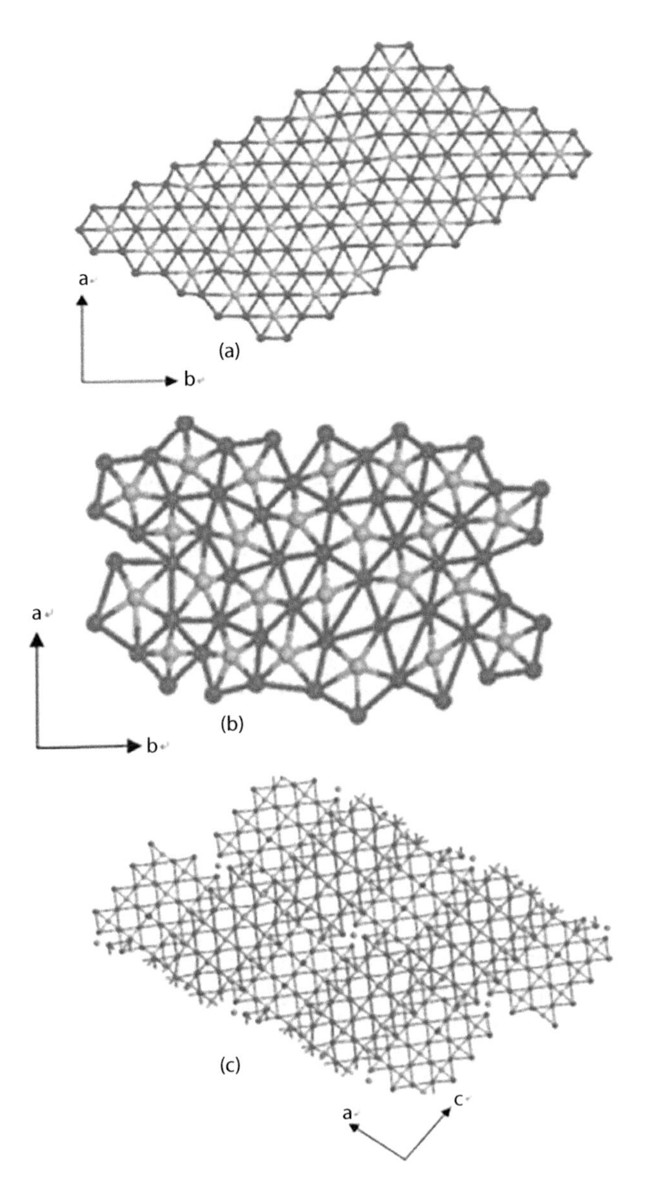

FIGURE 1.2 Structural schemes of Nb_2O_5 (a) TT-phase, (b) T-phase and (c) H-phase: Nb atom (blue); O atom (red). (With kind permission from Taylor & Francis: *Nano Rev.*, Nanostructured Nb_2O_5 catalysts 3, 2012, 17631, Zhao, Y. et al.)

HR-TEM images in Figure 1.4a. shows the core-shell structure of Nb_2O_5 and C. They observed the battery-type nature of the hybrid supercapacitor with redox peaks in the cyclic voltammetry graphs. The non-linear galvanostatic discharge/charge graphs shown in their work also proved the nature obtained from cyclic voltammetry results. It is also interesting to note that these hybrid supercapacitors retain 90% of its original specific capacitance magnitude even after a 1 k cycle of use. Another

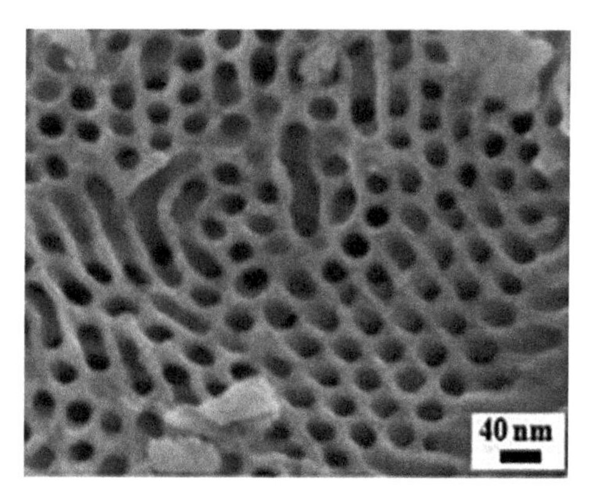

FIGURE 1.3 SEM images of m-Nb_2O_5-C. (Reprinted with permission from Lim, E. et al., *ACS Nano*, 8, 8968–8978, 2014. Copyright 2014 American Chemical Society.)

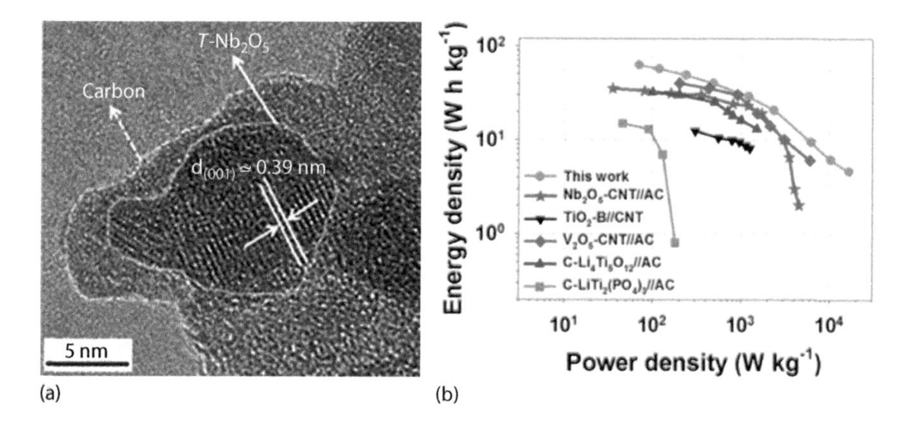

FIGURE 1.4 (a) HR-TEM image of T-Nb_2O_5@ C. (b) Ragone plots in comparison with other results. (Reprinted with permission from Lim, E. et al., *ACS Nano*, 9, 7497–7505, 2015. Copyright 2015 American Chemical Society.)

most important study with Ragone plot for this hybrid supercapacitor has been presented in Figure 1.4b. This establishes the superiority of the core-shell structured T-Nb_2O_5@C NCs with respect to the other materials. Hence the core-shell structured T-Nb_2O_5@C NCs show higher stability with improved performance in the design of hybrid supercapacitors.

1.3.3 Niobium Oxide Nanowires

The electron diffusion can greatly be benefited in EC from the 1-D nanomaterials like nanowires [35]. Hence it is inevitable to characterize the properties of 1-D nanowire of T-Nb_2O_5 for supercapacitor applications; however, it is quite difficult to synthesize the 1-D nanowire of T-Nb_2O_5. Because electro spinning method produced T-Nb_2O_5 along with other phases [36] where long-time etching at ammonium fluoride gives pseudo hexagonal phase [37], annealing under argon produces tetragonal

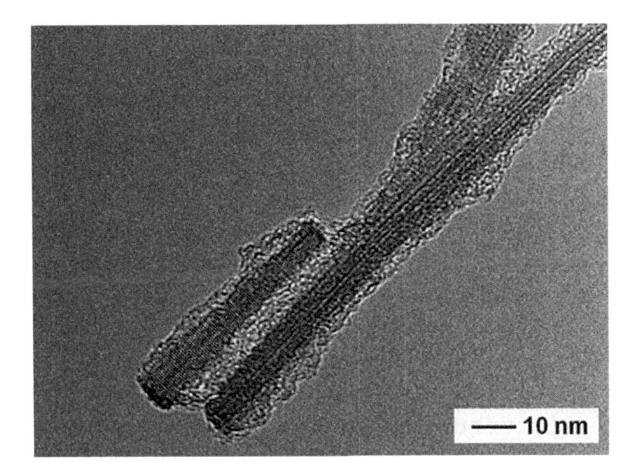

FIGURE 1.5 High magnification TEM image of sample C-T-Nb_2O_5. (Reprinted from *Nano Energ.*, 11, Wang, X. et al., Orthorhombic niobium oxide nanowires for next generation hybrid supercapacitor device, 765–772, Copyright 2014, with permission from Elsevier.)

phase [38] and under oxygen produces monoclinic phase [39]. Hence Wang et al. [40] synthesized carbon-coated 1-D nanowire of T-Nb_2O_5 using niobium oxalate coordination chemistry shown in Figure 1.5.

This 1-D nanowire of T-Nb_2O_5 can retain specific capacitance up to 140 mA hg^{-1} at 25°C and after the 1 k cycle it can retain 82% of specific capacitance at 5°C. Hence the hybrid supercapacitor builds with the 1-D nanowire of T-Nb_2O_5 shows high stability along with high power and energy density.

1.3.4 Niobium Oxide Nanorods

An optimal nanostructure to synthesize T-Nb_2O_5 NCs can be achieved by using a soft template of cellulose nanorod shown in Figure 1.6 [41]. This kind of morphology leads to a higher ion diffusion around electrodes and enhances effective

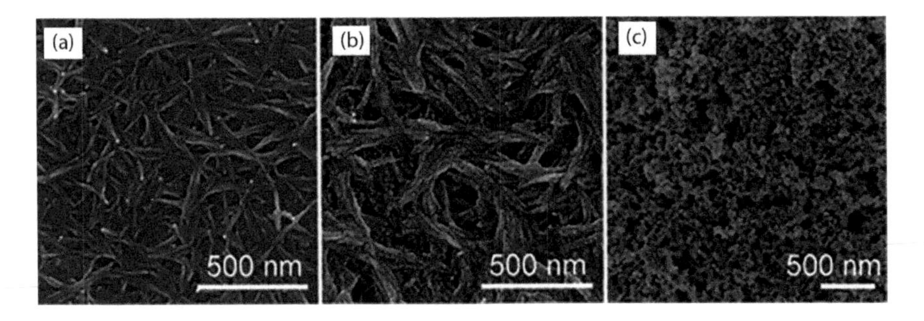

FIGURE 1.6 SEM image of Nb_2O_5 nanorods for (a) CNs film (b) Nb/CNs composite film and (c) mesoporous Nb_2O_5 film (Reprinted from *Mater. Chem. Phys.*, 149, Kong, Lingping et al., Ultrahigh intercalation pseudocapacitance of mesoporous orthorhombic niobium pentoxide from a novel cellulose nanocrystal template, 495–504, Copyright 2015, with permission from Elsevier.)

surface area by manifold [42]. They first prepared nanorods of cellulose fibers using acid hydrolysis with the generation of hydroxyl groups around the nanorods. This group further nucleates the niobium oxide phases very easily around the nanorods. Finally, annealing at higher temperature can easily evaporate the cellulose fibers from the structure, hence we can easily obtain a nanorod-like nanoporous T-Nb_2O_5 NCs. They observed a battery-type nature with redox peaks in the cyclic voltammetry graphs. The non-linear galvanostatic discharge/charge graphs in this nanorod-like nanoporous T-Nb_2O_5 NCs also proved the nature obtained from cyclic voltammetry results. It is also to note that the specific capacitance value remains constant even after 300 cycles, hence the T-Nb_2O_5 NCs nanorod show higher stability with improved performance in the design of hybrid supercapacitors as well. This unique behavior of the nanorod-like nanoporous T-Nb_2O_5 NCs with excellent energy and power density shows an interesting pathway to the research of the supercapacitors.

1.3.5 Niobium Oxide Nanodots

Electrode with materials consist of T-Nb_2O_5 nanodots dispersed around a thin sheet of 2-D graphite oxide can be a potential design in terms of binder-free, free-standing electrodes. This sort of design is the building block for the portable, wearable, and stretchable electronic devices. Here due to the high-energy density with high-specific capacitance the design of T-Nb_2O_5 nanodots with graphene composite meets the required criteria of providing the best performance in the wearable devices with limited space. Kong et al. synthesized the T-Nb_2O_5 nanodots around a thin sheet of 2-D graphite oxide films using simple hydrothermal reaction shown in Figure 1.7 [43].

They have loaded 74.2 wt% of T-Nb_2O_5 phase in the 2-D film with density 1.55 g cm^{-3} and obtained a very good conductivity of 2.5 S cm^{-1}. The volumetric and gravimetric analysis confirmed the value of 961.8 F cm^{-3} and 620.5 Fg^{-1} at 0.001 Vs^{-1}. The hybrid supercapacitor with T-Nb_2O_5 phase in the 2-D graphene film (anode) and AC (cathode) shows retention of 93% of the initial specific capacitance

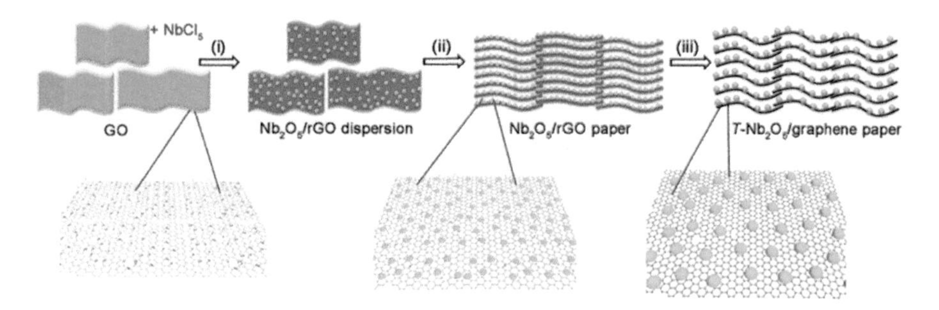

FIGURE 1.7 Schematic of the fabrication process for T-Nb_2O_5/graphene composite papers: (i) solvothermal process, (ii) vacuum filtration, (iii) heat treatment. (Reprinted with permission from Kong, L. et al., *ACS Nano*, 9, 11200–11208, 2015. Copyright 2015 American Chemical Society.)

after 2 k cycles with high power and energy density of value 18,000 W kg^{-1} and 47 Wh kg^{-1} respectively. This shows an effective way to design 3-D electrode structure by stacking 2-D layer of films with dispersed T-Nb_2O_5 nanodots phase as future flexible storage devices of energy.

1.3.6 Titanium Niobium Oxide Nanotubes

Titanium niobium oxide ($TiNb_2O_7$) is a potential material with theoretical capacity two times higher than lithium-based material with excellent cyclability and reversibility [44–51]. Design of hollow nanotube structures of $TiNb_2O_7$ can enhance the kinetics of the electrode to great extent. Li et al. synthesized nanoporous hollow 3-D structure of $TiNb_2O_7$ nanotubes for the supercapacitor applications [50] shown in Figure 1.8.

The hybrid supercapacitor with graphene as a cathode and nanoporous hollow 3-D $TiNb_2O_7$ nanotubes as an anode electrode have an extremely high retentivity of 93.8% even after 1k cycles with stable energy density at 1 Ag^{-1} with value 43.8 Wh kg^{-1}. It can maintain discharge/charge rate per 33.5 s at 34.5 Wh kg^{-1} and can deliver 74 Wh kg^{-1} optimal value of energy density ultimately. This indicates that the use of nanoporous 3-D $TiNb_2O_7$ nanotubes as an electrode is capable of electrochemical performance in supercapacitors applications.

1.3.7 Titanium Niobium Oxide Nanofibers

Easy ion diffusion, high surface area, and good contact made 1-D nanofiber material of titanium niobium oxide ($TiNb_2O_7$) shown in Figure 1.9 is one of the most promising material as an electrode of supercapacitor applications [48].

Madhavi et al. overcome the theoretical limitation of five moles by synthesizing 1-D $TiNb_2O_7$ nanofibers by using scalable electro spinning method [52]. They showed

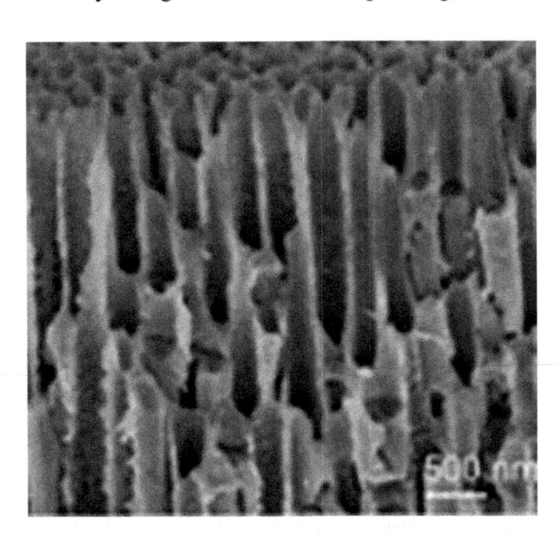

FIGURE 1.8 SEM image of 3D-O-P-$TiNb_2O_7$ nanotubes. (Li, H. et al., *J. Mat. Chem. A*, 3, 16785–16790, 2015. Reproduced by permission of The Royal Society of Chemistry.)

FIGURE 1.9 SEM image of 1-D nanofiber material of titanium niobium oxide ($TiNb_2O_7$). (Reprinted with permission from Jayaraman, S. et al., *ACS Appl. Mat. Interfaces*, 6, 8660–8666, 2014. Copyright 2014 American Chemical Society.)

that the hybrid supercapacitor with 1-D $TiNb_2O_7$ nanofibers as an electrode can retain 84% of initial specific capacitance till 3k cycles. Besides the power and energy density for 1-D $TiNb_2O_7$ nanofibers are also very high with the value of 3 kW kg^{-1} and 43 Wh kg^{-1} respectively. The new avenues of the growth of the high power capability with energy density supercapacitors are feasible with the use of niobium based 1-D $TiNb_2O_7$ nanofibers as an electrode material.

1.4 CONCLUSIONS

In summary, niobium based materials are one of the most important electrode materials for the developments of the next generation supercapacitors. Hence in this chapter we specifically concentrate on developments of the supercapacitors by using the niobium-based inorganic materials because of their high degree of thermal and chemical stability along with high-energy densities and excellent electrical conductivity. They are also eco friendly than lead-acid and lithium materials and due to their electro chemical properties niobium materials are in use as an active material in energy storage applications. Different structure of the niobium based nanomaterials can find application as a electrode material in supercapacitor such as nanoporous niobium pentoxide, core-shell structured Nb_2O_5 nanoparticles, niobium oxide nanowires, niobium oxide nanorods, niobium oxide nanodots, titanium niobium oxide nanotubes, titanium niobium oxide nanofibers etc. Nanoporous Nb_2O_5 is the best alternative electrode material for the replacement of conventional ruthenium electrodes because of its low cost, higher specific capacitance and energy density. Hence, the higher capacitance in Nb_2O_5 should be originated from the nanoporous structure of O-Nb_2O_5 phase which enhances the surface area as well as ionic conduction through these pores. The reported power density 18,510 W kg^{-1} with the energy 74 Wh kg^{-1} and a great withstanding rate of 90% at 1 Ag^{-1} even after 1000 cycles indicates the suitability of nanopores structured Nb_2O_5 ceramics as a suitable candidate for the supercapacitor system. A core-shell structure of Nb_2O_5 and carbon is another alternative route for the application of the niobium-based material for the supercapacitor applications. The battery-type nature of the hybrid supercapacitor with redox peaks has been observed in the cyclic voltammetry graphs. It is also interesting to note that this hybrid supercapacitor retains 90% of its original specific capacitance magnitude till 1 k cycle of use. Hence the core-shell structured T-Nb_2O_5@C NCs show higher stability with improved performance in the design of hybrid supercapacitors. Due to the better electron diffusion it is inevitable to synthesize the 1-D nanowire of T-Nb_2O_5 for supercapacitor applications. This 1-D nanowire of T-Nb_2O_5 can retain specific capacitance up to 140 mA hg^{-1} at 25°C and after the 1k cycle, it can retain 82% of specific capacitance at 5°C. Hence the hybrid supercapacitor builds with the 1-D nanowire of T-Nb_2O_5 shows high stability along with high power and energy density. The T-Nb_2O_5 nanorods synthesized with a soft template of cellulose leads to a higher ion diffusion around electrodes and enhances the effective surface area by manifold. The non-linear galvanostatic discharge/charge graphs in this nanorod like nanoporous T-Nb_2O_5 NCs also proved the nature obtained from cyclic voltammetry results. Hence this unique behavior of the nanorod like nanoporous T-Nb_2O_5 NCs with high energy and power density shows an interesting pathway to the research of the supercapacitors. On the other hand electrode materials consist of T-Nb_2O_5 nanodots dispersed around a thin sheet of 2-D graphite oxide can be a potential design in terms of binder-free, free-standing electrodes. The hybrid supercapacitor with T-Nb_2O_5 phase in the 2-D graphene film and AC shows retention of 93% of the initial capacitance till 2 k cycles with high power and energy density of value 18 kW kg^{-1} and 47 Wh kg^{-1} respectively. Design of hollow nanotube structures of $TiNb_2O_7$ can enhance the kinetics of the electrode in great extent in supercapacitor applications. The hybrid supercapacitor with graphene as a cathode and nanoporous hollow 3-D $TiNb_2O_7$ nanotubes as an anode electrode can retain 93.8% of the initial specific capacitance even after 1 k cycles with stable energy density at 1 Ag^{-1} with value 43.8 Wh kg^{-1}. Facile diffusion of ions, high surface area, and good contact has made 1-D nanofiber material of titanium niobium oxide ($TiNb_2O_7$) is one of the most promising material as an electrode of supercapacitor applications. Hence the niobium-based nanomaterials show higher stability with improved performance in the design of supercapacitors applications.

ACKNOWLEDGEMENTS

The author would like to thank UGC, India for the start-up financial grant no. F.30-457/2018 (BSR). One of us (D. Baba Basha) is also thankful to Deanship of Scientific Research at Majmaah University, K.S.A. for supporting this work. We also acknowledge the support provided to A. Franco Jr. by CNPq, Brazil with grant No. 307557/2015-4.

REFERENCES

1. Chowdhury, Geeta, William Adams, Brian Conway, and Srinivasa Sourirajan. "Thin film composite membrane as battery separator." U.S. Patent 5,910,366, issued June 8, 1999.
2. Burke, Lawrence W., Edward Bukowski, Colin Newnham, Neil Scholey, and William Hoge. *HSTSS Battery Development for Missile and Ballistic Telemetry Applications.* No. ARL-MR-477. Army Research Lab, Aberdeen Proving Ground MD, 2000.
3. Burke, Andrew, Marshall Miller, and Hengbing Zhao. "Lithium batteries and ultracapacitors alone and in combination in hybrid vehicles: Fuel economy and battery stress reduction advantages." *JSR* 21, no. 23 (2010): 15.
4. Frackowiak, Elzbieta, and Francois Beguin. "Carbon materials for the electrochemical storage of energy in capacitors." *Carbon* 39, no. 6 (2001): 937–950.
5. Pandolfo, A. G., and A. F. Hollenkamp. "Carbon properties and their role in supercapacitors." *Journal of Power Sources* 157, no. 1 (2006): 11–27.
6. Pushparaj, Victor L., Manikoth M. Shaijumon, Ashavani Kumar, Saravanababu Murugesan, Lijie Ci, Robert Vajtai, Robert J. Linhardt, Omkaram Nalamasu, and Pulickel M. Ajayan. "Flexible energy storage devices based on nanocomposite paper." *Proceedings of the National Academy of Sciences* 104, no. 34 (2007): 13574–13577.
7. Luo, Qingtao, Huamin Zhang, Jian Chen, Dongjiang You, Chenxi Sun, and Yu Zhang. "Preparation and characterization of Nafion/SPEEK layered composite membrane and its application in vanadium redox flow battery." *Journal of Membrane Science* 325, no. 2 (2008): 553–558.
8. Xu, Chunmei, Yishan Wu, Xuyang Zhao, Xiuli Wang, Gaohui Du, Jun Zhang, and Jiangping Tu. "Sulfur/three-dimensional graphene composite for high performance lithium–sulfur batteries." *Journal of Power Sources* 275 (2015): 22–25.
9. Einhorn, Markus, Valerio F. Conte, Christian Kral, Jürgen Fleig, and Robert Permann. "Parameterization of an electrical battery model for dynamic system simulation in electric vehicles." In *2010 IEEE Vehicle Power and Propulsion Conference*, pp. 1–7. IEEE, 2010.
10. Inagaki, Hiroki, and Norio Takami. "Nonaqueous electrolyte battery, battery pack and vehicle." U.S. Patent 7,662,515, issued February 16, 2010.
11. Chen, Shu-Ru, Yun-Pu Zhai, Gui-Liang Xu, Yan-Xia Jiang, Dong-Yuan Zhao, Jun-Tao Li, Ling Huang, and Shi-Gang Sun. "Ordered mesoporous carbon/sulfur nanocomposite of high performances as cathode for lithium–sulfur battery." *Electrochimica Acta* 56, no. 26 (2011): 9549–9555.
12. Li, Xianglong, and Linjie Zhi. "Managing voids of Si anodes in lithium ion batteries." *Nanoscale* 5, no. 19 (2013): 8864–8873.
13. Fang, B., Y. Z. Wei, K. Maruyama, and M. Kumagai. "High capacity supercapacitors based on modified activated carbon aerogel." *Journal of Applied Electrochemistry* 35, no. 3 (2005): 229–233.
14. Fisher, Robert A., Morgan R. Watt, and W. Jud Ready. "Functionalized carbon nanotube supercapacitor electrodes: A review on pseudocapacitive materials." *ECS Journal of Solid State Science and Technology* 2, no. 10 (2013): M3170–M3177.
15. Shukla, A. K., S. Sampath, and K. Vijayamohanan. "Electrochemical supercapacitors: Energy storage beyond batteries." *Current Science* 79, no. 12 (2000): 1656–1661.
16. Cericola, D., R. Kötz, and Alexander Wokaun. "Effect of electrode mass ratio on aging of activated carbon based supercapacitors utilizing organic electrolytes." *Journal of Power Sources* 196, no. 6 (2011): 3114–3118.
17. Husain, Iqbal. *Electric and Hybrid Vehicles: Design Fundamentals.* CRC press, London, UK, 2003.

18. Chu, Andrew, and Paul Braatz. "Comparison of commercial supercapacitors and high-power lithium-ion batteries for power-assist applications in hybrid electric vehicles: I. Initial characterization." *Journal of Power Sources* 112, no. 1 (2002): 236–246.
19. Barsukov, Igor V., Christopher S. Johnson, Joseph E. Doninger, and Vyacheslav Z. Barsukov, eds. *New Carbon Based Materials for Electrochemical Energy Storage Systems: Batteries, Supercapacitors and Fuel Cells*. Vol. 229, Springer Science & Business Media, Dordrecht, the Netherlands, 2006.
20. Su, Ling-Hao, Xiao-Gang Zhang, Chang-Huan Mi, Bo Gao, and Yan Liu. "Improvement of the capacitive performances for Co–Al layered double hydroxide by adding hexacyanoferrate into the electrolyte." *Physical Chemistry Chemical Physics* 11, no. 13 (2009): 2195–2202.
21. Roldán, Silvia, Zoraida González, Clara Blanco, Marcos Granda, Rosa Menéndez, and Ricardo Santamaría. "Redox-active electrolyte for carbon nanotube-based electric double layer capacitors." *Electrochimica Acta* 56, no. 9 (2011): 3401–3405.
22. Wu, Jihuai, Haijun Yu, Leqing Fan, Genggeng Luo, Jianming Lin, and Miaoliang Huang. "A simple and high-effective electrolyte mediated with p-phenylenediamine for supercapacitor." *Journal of Materials Chemistry* 22, no. 36 (2012): 19025–19030.
23. Chen, Wei, R. B. Rakhi, and Husam N. Alshareef. "Capacitance enhancement of polyaniline coated curved-graphene supercapacitors in a redox-active electrolyte." *Nanoscale* 5, no. 10 (2013): 4134–4138.
24. Senthilkumar, S. T., R. Kalai Selvan, and J. S. Melo. "Redox additive/active electrolytes: A novel approach to enhance the performance of supercapacitors." *Journal of Materials Chemistry A* 1, no. 40 (2013): 12386–12394.
25. Lu, Luyao, Tianyue Zheng, Qinghe Wu, Alexander M. Schneider, Donglin Zhao, and Luping Yu. "Recent advances in bulk heterojunction polymer solar cells." *Chemical Reviews* 115, no. 23 (2015): 12666–12731.
26. Zheng, J. P., P. J. Cygan, and T. R. Jow. "Hydrous ruthenium oxide as an electrode material for electrochemical capacitors." *Journal of the Electrochemical Society* 142, no. 8 (1995): 2699–2703.
27. Augustyn, Veronica, Jérémy Come, Michael A. Lowe, Jong Woung Kim, Pierre-Louis Taberna, Sarah H. Tolbert, Héctor D. Abruña, Patrice Simon, and Bruce Dunn. "High-rate electrochemical energy storage through Li^+ intercalation pseudocapacitance." *Nature Materials* 12, no. 6 (2013): 518.
28. Lubimtsev, Andrew A., Paul RC Kent, Bobby G. Sumpter, and P. Ganesh. "Understanding the origin of high-rate intercalation pseudocapacitance in Nb_2O_5 crystals." *Journal of Materials Chemistry A* 1, no. 47 (2013): 14951–14956.
29. Han, Jian-Tao, Dong-Qiang Liu, Sang-Hoon Song, Youngsik Kim, and John B. Goodenough. "Lithium ion intercalation performance of niobium oxides: KNb_5O_{13} and K_6Nb_{10}. 8O30." *Chemistry of Materials* 21, no. 20 (2009): 4753–4755.
30. Kim, Jong Woung, Veronica Augustyn, and Bruce Dunn. "The effect of crystallinity on the rapid pseudocapacitive response of Nb_2O_5." *Advanced Energy Materials* 2, no. 1 (2012): 141–148.
31. Wang, Xiaolei, Ge Li, Zheng Chen, Veronica Augustyn, Xueming Ma, Ge Wang, Bruce Dunn, and Yunfeng Lu. "High-performance supercapacitors based on nanocomposites of Nb_2O_5 nanocrystals and carbon nanotubes." *Advanced Energy Materials* 1, no. 6 (2011): 1089–1093.
32. Zhao, Yun, Xiwen Zhou, Lin Ye, and Shik Chi Edman Tsang. "Nanostructured Nb_2O_5 catalysts." *Nano Reviews* 3, no. 1 (2012): 17631.
33. Lim, Eunho, Haegyeom Kim, Changshin Jo, Jinyoung Chun, Kyojin Ku, Seongseop Kim, Hyung Ik Lee et al. "Advanced hybrid supercapacitor based on a mesoporous niobium pentoxide/carbon as high-performance anode." *ACS Nano* 8, no. 9 (2014): 8968–8978.

34. Lim, Eunho, Changshin Jo, Haegyeom Kim, Mok-Hwa Kim, Yeongdong Mun, Jinyoung Chun, Youngjin Ye et al. "Facile synthesis of Nb_2O_5@ carbon core–shell nanocrystals with controlled crystalline structure for high-power anodes in hybrid supercapacitors." *ACS Nano* 9, no. 7 (2015): 7497–7505.
35. Yan, Jian, Eugene Khoo, Afriyanti Sumboja, and Pooi See Lee. "Facile coating of manganese oxide on tin oxide nanowires with high-performance capacitive behavior." *ACS Nano* 4, no. 7 (2010): 4247–4255.
36. Viet, A. Le, M. V. Reddy, R. Jose, B. V. R. Chowdari, and S. Ramakrishna. "Nanostructured Nb_2O_5 polymorphs by electrospinning for rechargeable lithium batteries." *The Journal of Physical Chemistry C* 114, no. 1 (2009): 664–671.
37. Wen, Hao, Zhifu Liu, Jiao Wang, Qunbao Yang, Yongxiang Li, and Jerry Yu. "Facile synthesis of Nb_2O_5 nanorod array films and their electrochemical properties." *Applied Surface Science* 257, no. 23 (2011): 10084–10088.
38. Varghese, Binni, Sow Chorng Haur, and Chwee-Teck Lim. "Nb_2O_5 nanowires as efficient electron field emitters." *The Journal of Physical Chemistry C* 112, no. 27 (2008): 10008–10012.
39. Kang, Jun Ha, Yoon Myung, Jin Woong Choi, Dong Myung Jang, Chi Woo Lee, Jeunghee Park, and Eun Hee Cha. "Nb_2O_5 nanowire photoanode sensitized by a composition-tuned CdS x Se 1– x shell." *Journal of Materials Chemistry* 22, no. 17 (2012): 8413–8419.
40. Wang, Xu, Chaoyi Yan, Jian Yan, Afriyanti Sumboja, and Pooi See Lee. "Orthorhombic niobium oxide nanowires for next generation hybrid supercapacitor device." *Nano Energy* 11 (2015): 765–772.
41. Kong, Lingping, Chuangfang Zhang, Jitong Wang, Donghui Long, Wenming Qiao, and Licheng Ling. "Ultrahigh intercalation pseudocapacitance of mesoporous orthorhombic niobium pentoxide from a novel cellulose nanocrystal template." *Materials Chemistry and Physics* 149 (2015): 495–504.
42. Arico, Antonino Salvatore, Peter Bruce, Bruno Scrosati, Jean-Marie Tarascon, and Walter Van Schalkwijk. "Nanostructured materials for advanced energy conversion and storage devices." In *Materials For Sustainable Energy: A Collection of Peer-Reviewed Research and Review Articles from Nature Publishing Group*, World Scientific, Singapore, pp. 148–159, 2011.
43. Kong, Lingping, Chuanfang Zhang, Jitong Wang, Wenming Qiao, Licheng Ling, and Donghui Long. "Free-standing T-Nb_2O_5/graphene composite papers with ultrahigh gravimetric/volumetric capacitance for Li-ion intercalation pseudocapacitor." *ACS Nano* 9, no. 11 (2015): 11200–11208.
44. Guo, Bingkun, Xiqian Yu, Xiao-Guang Sun, Miaofang Chi, Zhen-An Qiao, Jue Liu, Yong-Sheng Hu, Xiao-Qing Yang, John B. Goodenough, and Sheng Dai. "A long-life lithium-ion battery with a highly porous $TiNb_2O_7$ anode for large-scale electrical energy storage." *Energy & Environmental Science* 7, no. 7 (2014): 2220–2226.
45. Jo, Changshin, Youngsik Kim, Jongkook Hwang, Jongmin Shim, Jinyoung Chun, and Jinwoo Lee. "Block copolymer directed ordered mesostructured $TiNb_2O_7$ multimetallic oxide constructed of nanocrystals as high power Li-ion battery anodes." *Chemistry of Materials* 26, no. 11 (2014): 3508–3514.
46. Han, Jian-Tao, Yun-Hui Huang, and John B. Goodenough. "New anode framework for rechargeable lithium batteries." *Chemistry of Materials* 23, no. 8 (2011): 2027–2029.
47. Fei, Ling, Yun Xu, Xiaofei Wu, Yuling Li, Pu Xie, Shuguang Deng, Sergei Smirnov, and Hongmei Luo. "SBA-15 confined synthesis of $TiNb_2O_7$ nanoparticles for lithium-ion batteries." *Nanoscale* 5, no. 22 (2013): 11102–11107.
48. Jayaraman, Sundaramurthy, Vanchiappan Aravindan, Palaniswamy Suresh Kumar, Wong Chui Ling, Seeram Ramakrishna, and Srinivasan Madhavi. "Exceptional performance of $TiNb_2O_7$ anode in all one-dimensional architecture by electrospinning." *ACS Applied Materials & Interfaces* 6, no. 11 (2014): 8660–8666.

49. Li, Hongsen, Laifa Shen, Gang Pang, Shan Fang, Haifeng Luo, Kai Yang, and Xiaogang Zhang. "$TiNb_2O_7$ nanoparticles assembled into hierarchical microspheres as high-rate capability and long-cycle-life anode materials for lithium ion batteries." *Nanoscale* 7, no. 2 (2015): 619–624.
50. Li, Hongsen, Laifa Shen, Jie Wang, Shan Fang, Yingxia Zhang, Hui Dou, and Xiaogang Zhang. "Three-dimensionally ordered porous $TiNb_2O_7$ nanotubes: A superior anode material for next generation hybrid supercapacitors." *Journal of Materials Chemistry A* 3, no. 32 (2015): 16785–16790.
51. Park, Hyunjung, Taeseup Song, and Ungyu Paik. "Porous $TiNb_2O_7$ nanofibers decorated with conductive Ti 1− x Nb x N bumps as a high power anode material for Li-ion batteries." *Journal of Materials Chemistry A* 3, no. 16 (2015): 8590–8596.
52. Aravindan, Vanchiappan, Jayaraman Sundaramurthy, Akshay Jain, Palaniswamy Suresh Kumar, Wong Chui Ling, Seeram Ramakrishna, Madapusi P. Srinivasan, and Srinivasan Madhavi. "Unveiling $TiNb_2O_7$ as an insertion anode for lithium ion capacitors with high energy and power density." *ChemSusChem* 7, no. 7 (2014): 1858–1863.

2 Zinc-Based Materials for Supercapacitors

Prasun Banerjee, Adolfo Franco Junior, D. Baba Basha, K. Chandra Babu Naidu, and S. Ramesh

CONTENTS

2.1 Introduction 17
2.2 Zinc as an Electrode Material in Supercapacitors Applications 18
2.3 Different Structural Nanomaterials Using for Supercapacitor Applications 20
2.3.1 Zinc Oxide Nanowire 20
2.3.2 Zinc Hydroxychloride Nanosheets 21
2.3.3 Zinc Sulphide Nanospheres 23
2.3.4 Flower-Like Zinc Molybdate ($ZnMoO_4$) 25
2.4 Conclusions 26
Acknowledgments 28
References 28

2.1 INTRODUCTION

The growth of the global economy and at the same time shortage of fossil fuels with degradation of the environment indicate towards the urgency not only for a greener, cleaner, and renewable source of energy, but also for some kind of advanced energy storage/conversion systems.

This energy storage system depends on fuel cells, batteries, and supercapacitors (ES). Present-day fuel cells/batteries with high-energy storage capabilities in general use potentially dangerous materials to the environment such as lead-acid, lithium, and so on at the same time with a short life cycle. On the other hand supercapacitors with half a million life cycle with high power density, discharge, and charge rate is an alternative to the battery and conventional capacitor with greener and cleaner materials [1–12]. In general, for the construction of a parallel plate capacitor, a pair of electrodes along with a dielectric medium in between the plates is needed. When a voltage applied along the two plates of the capacitor due to the accumulation of charges at the surface of each plate, the capacitor stores charges according to the formula $E = 1/2CV^2$. The capacitance of the parallel plate capacitor can be expressed as $C = \varepsilon_0 KA/d$ where K is the dielectric constant of the dielectric medium, d is the distance between the two plates, and A is the cross sectional area of the electrodes [13].

The principle of operation of modern day supercapacitors is similar to normal capacitors with somewhat special components as shown in Figure 1.1 [14], along with an "active electrode" material and a porous non-conductive separator. In this arrangement with the application of the voltage at ion-electrode interface an electric double Helmholtz layer of the order of 5–10 Å thickness formed due to the attraction of electrodes and the electrolyte ions. This specialty in the supercapacitor comes with a great advantage, along with a wide variety of applications in hybrid vehicles, consumer electronics, and industrial energy/power management. The reasons behind the better performances in the supercapacitors lie due to the small distance between the plates of the capacitors, with the high surface areas of the order of 1000–2000 m^2/g along with the nanometers level of thickness of the Helmholtz layer which is in general microns level in a conventional capacitance [15]. It also can be noted that due to the absence of charge transfer between the plates of the capacitor through the dielectric medium, the degradation of the electrolyte and the electrodes are also minimal in case of the supercapacitors, which enhances the discharge-charge cycles of the supercapacitors [16].

On the other hand, the development of the supercapacitors comes with major challenges along with added disadvantages due to its low power density and higher production cost. The most prominent approach to overcome these hurdles is to develop a new kind of materials for the supercapacitor applications. The most popular material for the supercapacitor application is carbon particles due their high surface areas. But carbon-based materials have some limitation due to the physical storing of the charges in porous layers of electrodes known as the electrostatic double-layer supercapacitors (EDLS) with limited value of capacitance and low power density. Due to these limitation the best approach is to replace the carbon-based material completely by an active material, which is known as a faradic supercapacitor (FS), or it can be hybridized by mixing a active material with the carbon-based particles, which is known as a hybrid double-layer supercapacitors (HDLC). It has already well established that HDLC or FS comes with higher capacitance than that of the EDLC type. In this scenario zinc-based materials are the prominent candidates for the developments of the next generation supercapacitors. Hence in this chapter we specifically concentrate on developments of the supercapacitors by using the zinc-based inorganic materials for the applications in supercapacitor electrodes.

2.2 ZINC AS AN ELECTRODE MATERIAL IN SUPERCAPACITORS APPLICATIONS

Recently zinc-based materials started receiving special attention as an electrode material in supercapacitor applications because of its high degree of thermal and chemical stability along with high-energy densities and excellent electrical conductivity [26–29]. The wurtzite crystal structure of ZnO is shown in Figure 2.1. Because of its eco friendly and electrochemical properties it is now in use as an active material in battery applications [30]. Its only from 2015, scientists start using ZnO as an electrode material in the supercapacitor applications [31]. In that design Ni foam has been used as a substrate on which zinc oxide deposited as an electrode with polymer hydrogel as an electrolyte shown in Figure 2.2a [32]. The enhancement in the catalytic activity can be noticed in the design from the obtained specific capacitance

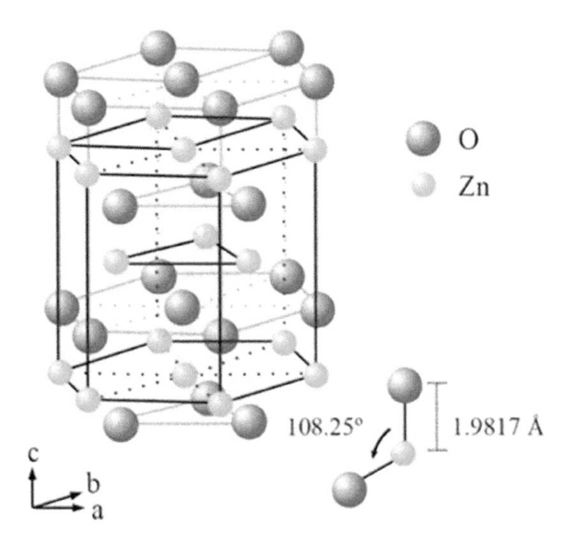

FIGURE 2.1 3-D representation of the unit cell structure of ZnO.

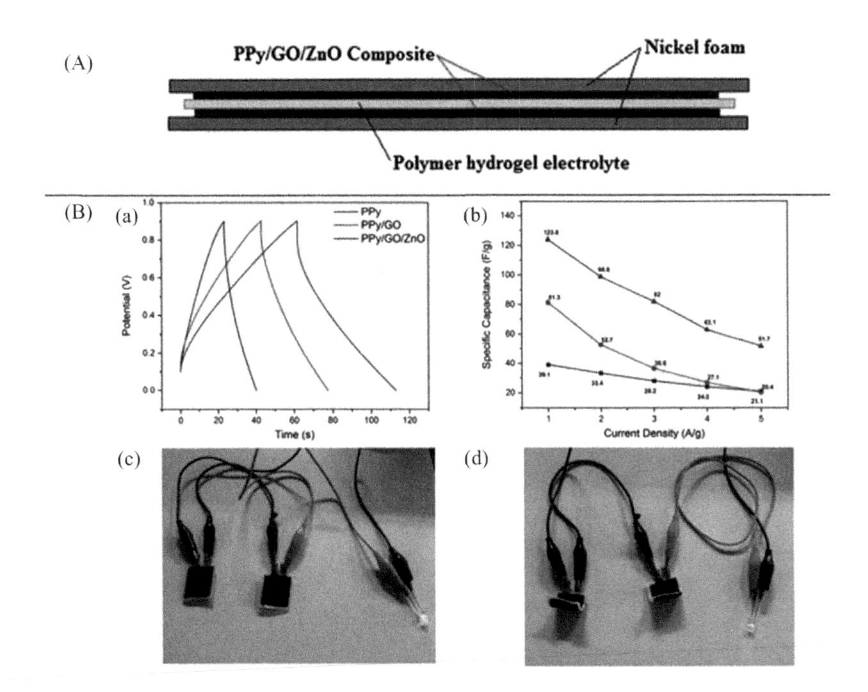

FIGURE 2.2 (A) Two-electrode configuration of PPy/GO/ZnO supercapacitor. (B): (a) Galvanostatic charge/discharge cycles, (b) specific capacitances at various current densities, (c) prepared supercapacitor connected in series lighting up an LED circuit, and (d) supercapacitor bent at 90 degree lighting up an LED circuit. (Reprinted from *J. Power Sources*, 296, Ng, C.H. et al., Potential active materials for photo-supercapacitor: A review, 169–185, Copyright 2015, with permission from Elsevier.)

values at current density 1 Ag^{-1} with value 123.8 Fg^{-1}. Figure 2.2b shows almost ideal capacitor behavior of the ZnO from the discharge/charge profile in the pseudo triangular galvanostatic cycle. The strong van der Waal force prevented the restacking here due to the ZnO. These supercapacitors can even be bent by 90 degrees during energy storage devise applications shown in Figure 2.2c.

2.3 DIFFERENT STRUCTURAL NANOMATERIALS USING FOR SUPERCAPACITOR APPLICATIONS

The different structure of the zinc-based nanomaterials can find application as an electrode material in supercapacitors due to their different surface areas and electrochemical activities. The range of nanomaterials, such as zinc oxide nanowire, zinc hydrochloride nanosheet, zinc sulfide nanosphere, and flower-like zinc molybdate, etc. can be used as an electrode materials in supercapacitor.

2.3.1 Zinc Oxide Nanowire

For decade's carbon-based materials have dominated as an electrode material conventionally, but with the use of ZnO nanowires the trends have been changing gradually with its added advantages with its peers. The added advantages of ZnO NWs are coming with its relatively high surface areas with its good electrical conductivity due to the presence of zinc ions [33]. Not only that, the ZnO NWs can be synthesized easily below 900°C and can be grown at any shape and sizes in any substrates. Figure 2.3 shows the X-ray diffraction of ZnO with fitting with Rietveld refinement techniques. Table 2.1 shows the Rietveld refinement parameters for the ZnO. There can be two approaches that have been adopted recently to use ZnO

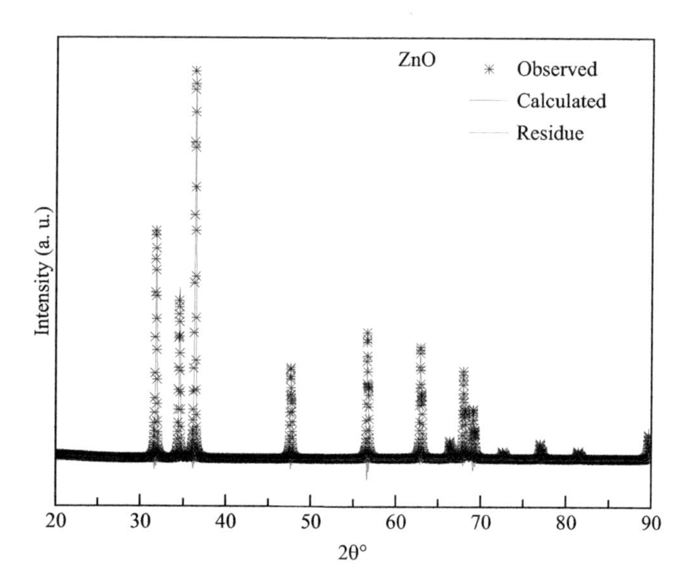

FIGURE 2.3 X-ray diffraction of ZnO with fitted with Rietveld refinement techniques.

TABLE 2.1
Rietveld Refinements Data for ZnO

Parameters	Refinement Values
a (Å)	3.249004
c (Å)	5.203241
V (Å^3)	47.567
ρ (g/cm^3)	5.682
u	0.38210
R_{wp} (%)	11.81
R_p (%)	8.37
GOF	2.10

NWs as supercapacitor applications. In the first approach the NWs can be grown in the shape of fibers and then can be entangling the same around a Kevlar fiber. Here poly (methyl methacrylate) (PMMA) can be used as insulation between both the fibers to act as electrodes [34]. In the second approach a core-shell structure can be used to use ZnO NWs as electrode materials. In this case highly conducting ZnO NWs can act as the core material whereas transitional metal hydroxides/oxides pseudo capacitors as shell materials [35].

Figure 1.1a in [36] shows the close views of the first type of highly flexible ZnO NWs discussed above. The NWs can be as long as 6 mm in length with a wide range of diameters from 220 to 700 nm shown in Figure 1.1b in [36]. The design of the NWs-based supercapacitor with ZnO NWs entangled around a gold-coated Kevlar fiber as shown in Figure 1.1c in [36]. A large-scale device is seemingly possible by forming a yarn with a bundle of such fibers. Even this device can also replace conventional micro supercapacitors, which in general builds with thin film technology that is very difficult to scale up. Combination of zinc oxide nanogenerators with NW supercapacitor can also recharge the supercapacitors very easily with day to day activity like footsteps, heartbeats, shaking, etc., and can be applied as wearable electronic devices. On the other hand a core-shell approach with ZnO NWs as conducting core and transitional metal hydroxide as the shell can improve the specific capacitance up to 10.678 F/cm^2. The typical SEM images of the ZnO ceramics can be seen in Figure 2.4.

2.3.2 Zinc Hydroxychloride Nanosheets

Nanosheets of zinc hydroxychloride ($Zn_5(OH)_8C_{12}$), popularly known as Simonkolleite (Simonk), have recently drawn the attention of the research community in the field of supercapacitors due to their unique chemical and physical properties associated with the electrode materials in the energy storage applications [37]. Simonk is a soft material by nature with specific gravity and Mohs hardness 3.2 and 1.5 respectively [38]. Nanomaterials of simonk can be grown along [001] plane direction with hexagonal shapes [39]. The presence of oxygen vacancies along the surface of such grown crystals shows a high degree of electrochemical activities in the materials [37]. It is

FIGURE 2.4 SEM image of ZnO ceramics.

already a well-established fact for other families of zinc materials of colossal permittivity ($\varepsilon' > 10^3$) due to the presence of oxygen vacancies [40] shown in Figure 2.5. There are numerous reports are available in the literature to improve the electrical conductivity of simonk materials with the addition of materials like graphene, active carbon (AC), Ni, Al, Co, etc. [37].

Chemical bath deposition technique can be adapted to grow simonk nanosheets on Ni foam as an electrode shown in Figure 2.6 [37]. The obtained surface area in this case can be as high as 119 m^2g^{-1} with a specific capacitance of 222 mF per square cm area. The absence of any polymer binder in the design maintains the performance up to 96% even after 5000 cycles of uses. This indicates the suitability of simonk nanosheets for the design of electrodes for supercapacitors. Not only that some other transitional metal ions are also can be incorporated with the simonk and nanosheets can be grown using chemical bath deposition techniques [41].

It is possible to fabricate the hybrid supercapacitors with active carbon and simonk-based nanosheets shown in Figure 2.7a [41]. The mass ratio of such hybrid

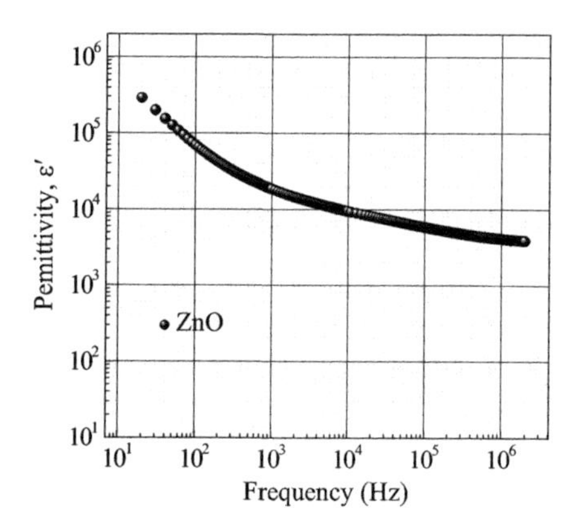

FIGURE 2.5 Colossal permittivity at room temperature for ZnO material.

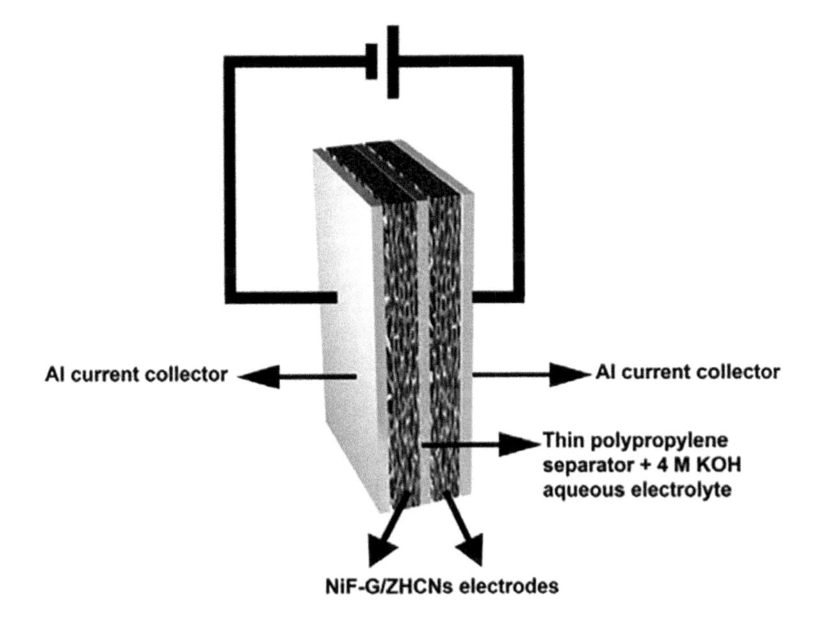

FIGURE 2.6 Schematic of NiF-G/ZHCNs composite electrode of a symmetric supercapacitor. (Reprinted from *Appl. Surf. Sci.*, 405, 37. Khamlich, S. et al., High performance symmetric supercapacitor based on zinc hydroxychloride nanosheets and 3D graphene-nickel foam composite, 329–336, Copyright 2017, with permission from Elsevier.)

supercapacitors can be made up to 0.312. The battery-type nature of the hybrid supercapacitor with redox peaks can be observed in the cyclic voltammetry graphs shown in Figure 2.7b [41]. The non-linear galvanostatic discharge/charge graphs shown in Figure 2.7c [41] also proved the nature obtained from cyclic voltammetry results. The variation of the specific capacitance with current densities of such hybrid supercapacitors has been shown in Figure 2.7d [41]. It is also interesting to note from Figure 2.7e [41] that this hybrid supercapacitor retains 90% of its original specific capacitance magnitude even after a 10k cycle of use. Another most important study with Ragone plot for these hybrid supercapacitor has been presented in Figure 2.7f [41]. Hence the simonk nanosheets show higher stability with improved performance in the design of hybrid supercapacitors as well.

2.3.3 Zinc Sulphide Nanospheres

Complex semiconductor nanoparticles such as zinc sulfide (ZnS) plays an important role in the design of supercapacitor applications due to the possibility to tune their band gaps very easily [42]. Figure 2.8 [42] shows such an application where the comparative results with pure and doped ZnS nanoparticles have been presented for the supercapacitor applications.

Numerous reports of ZnS nanoparticles with different shapes and sizes have been reported so far for the supercapacitor applications. Nanocables reported by Wang et al. [43] and nanoforests reported by Zhang et al. [31] are quite interesting

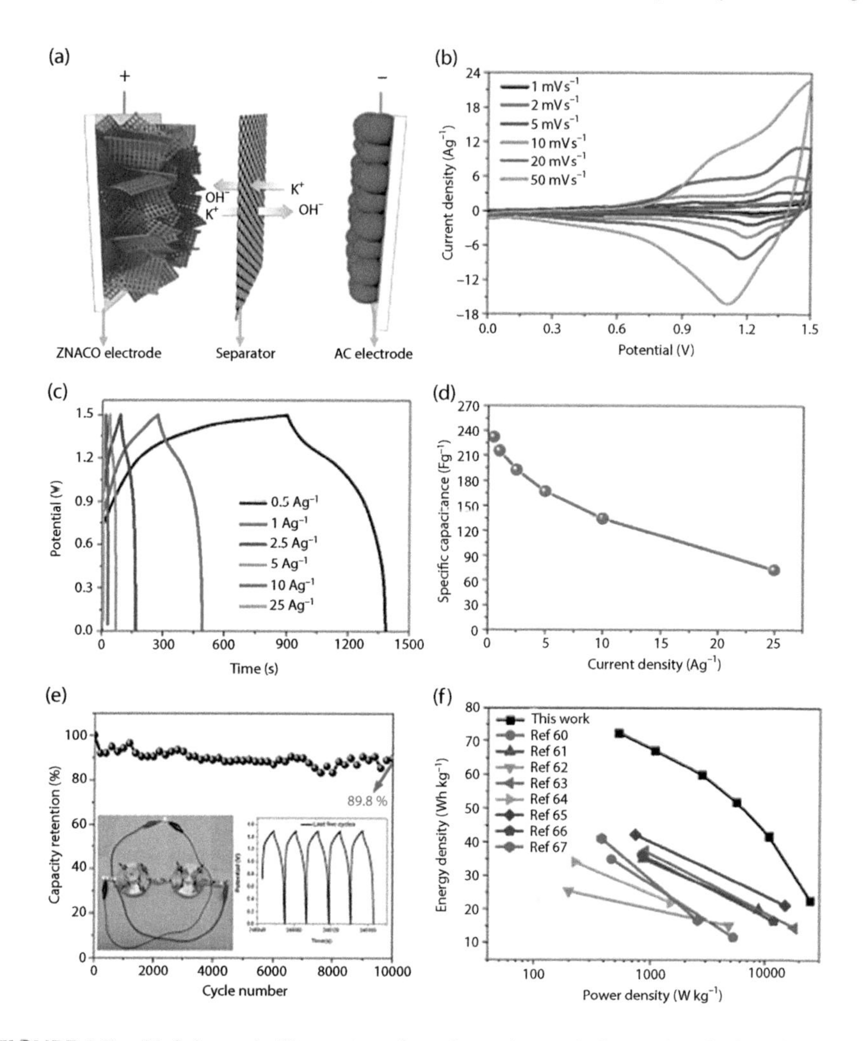

FIGURE 2.7 (a) Schematic illustration of configuration and electrochemical performance of the ZNACO//AC hybrid supercapacitor, (b) cyclic voltammetry curves, (c) galvanostatic charge/discharge curves, (d) cell capacitance vs current density, (e) cycling stability performance, inset is its corresponding charge/discharge curves of the last 5 cycles and photograph of a greed round LED indicator powered by two hybrid supercapacitors, and (f) Ragone plot of energy density vs power density. (Reprinted from *Nano Energy*, 28, Zhang, Q. et al., High-performance hybrid supercapacitors based on self-supported 3D ultrathin porous quaternary Zn-Ni-Al-Co oxide nanosheets, 475–485, Copyright 2016, with permission from Elsevier.)

so far. But the superior design with homogenous sizes and large surface area makes the ZnS nanospheres one of the best material for the supercapacitor applications. Fabrication of ZnS nanospheres on a conducting carbon textile is shown in Figure 2.9 [44]. Retention of 95% of its initial specific capacitance of 540 Fg^{-1} even after 5000 cycles indicates its stability with performance with energy density 51 Wh kg^{-1}.

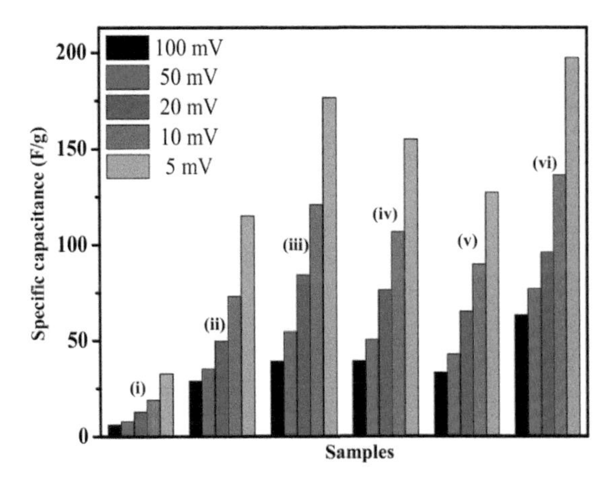

FIGURE 2.8 A graphical representation of specific capacitance at various scan rates for different compositions of the samples (i) Pure ZnS (ii) bare graphene (iii) ZnS/G-10 (iv) ZnS/G-20 (v) ZnS/G-40 and (vi) ZnS/G-60 nanocomposites. (Reprinted from *Electrochim. Acta*, 178, Ramachandran, R. et al., Solvothermal synthesis of Zinc sulfide decorated Graphene (ZnS/G) nanocomposites for novel Supercapacitor electrodes, 647–657, Copyright 2015, with permission from Elsevier.)

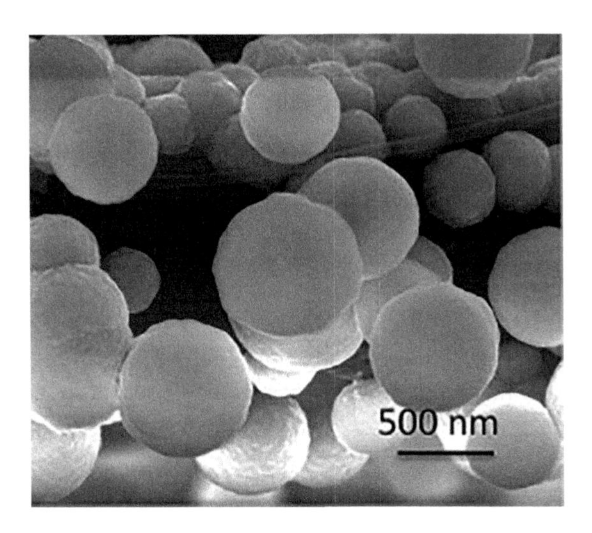

FIGURE 2.9 SEM images of the synthesized ZnS nanospheres grown on carbon textile. (Javed, M.S. et al., *J. Mat. Chem. A*, 4, 667–674, 2016. Reproduced by permission of The Royal Society of Chemistry.)

2.3.4 Flower-Like Zinc Molybdate ($ZnMoO_4$)

Due to the abundance, inexpensive, high electrochemical properties and presence of zero to +6 oxidation states, zinc molybdates attracted the attention of the scientific community for supercapacitor application [45]. Not only zinc molybdates but

TABLE 2.2
Specific Capacitances of Different Metal Molybdate Electrode

Compounds	Specific Capacitances in Fg^{-1}	References
$ZnMoO_4$	704.8 at 1 Ag^{-1}	[45]
$ZnFe_2O_4$	32.22 at 1 Ag^{-1}	[46]
$ZnCo_2O_4$	10.9 at 30 mV s^{-1}	[47]
$MnMoO_4$/polypyrrole	313 at 1 Ag^{-1}	[48]
$CoMoO_4/MnO_2$	152 at 1 Ag^{-1}	[49]
$MnMoO_4$/grapheme	364 at 2 Ag^{-1}	[50]

other metal molybdates are also potential members for energy storage applications. A comparative study [45–50] of their specific capacitance of such metal molybdates is presented in Table 2.2.

It can be observed that among all metal molybdates the zinc molybdates are the most promising. This is due to the special crystal structure of the zinc molybdates shown in Figure 2.10A(a) [45]. The flower-like zinc molybdates can be synthesized by using simple hydrothermal methods shown in Figure 2.10A(b) [45].

The hybrid supercapacitor can be designed by flower-like zinc molybdates as an anode material and activated carbon as a cathode material. The battery-type nature of the hybrid supercapacitor with redox peaks can be observed in the cyclic voltammetry graphs, shown in Figure 2.10B(a) [45]. The non-linear galvanostatic discharge/charge graphs shown in Figure 2.10B(b) [45] also proved the nature obtained from cyclic voltammetry results. The variation of the specific capacitance with current densities of such hybrid supercapacitors has been shown in Figure 2.10B(c) [45]. It is also interesting to note from Figure 2.10B(d) [45] that these hybrid supercapacitors retain 94% of its original specific capacitance magnitude even after a 10 k cycle of use. Another most important study with Ragone plot show a power density of 800 kW kg^{-1} with an energy density of 22 Wh kg^{-1} for these hybrid supercapacitors. Hence the flower-like zinc molybdates show higher stability with improved performance in the design of hybrid supercapacitors applications.

2.4 CONCLUSIONS

In summary, zinc-based materials are one of the prominent candidates for the developments of the next generation supercapacitors. In this chapter, we specifically concentrate on developments of the supercapacitors by using the zinc-based inorganic materials because of their high degree of thermal and chemical stability along with high energy densities and excellent electrical conductivity. They are also eco friendly than lead-acid and lithium materials and due to their electrochemical properties zinc materials are in use as an active material in battery applications. The different structure of the zinc-based nanomaterials can find application as an electrode material in supercapacitors such as zinc oxide

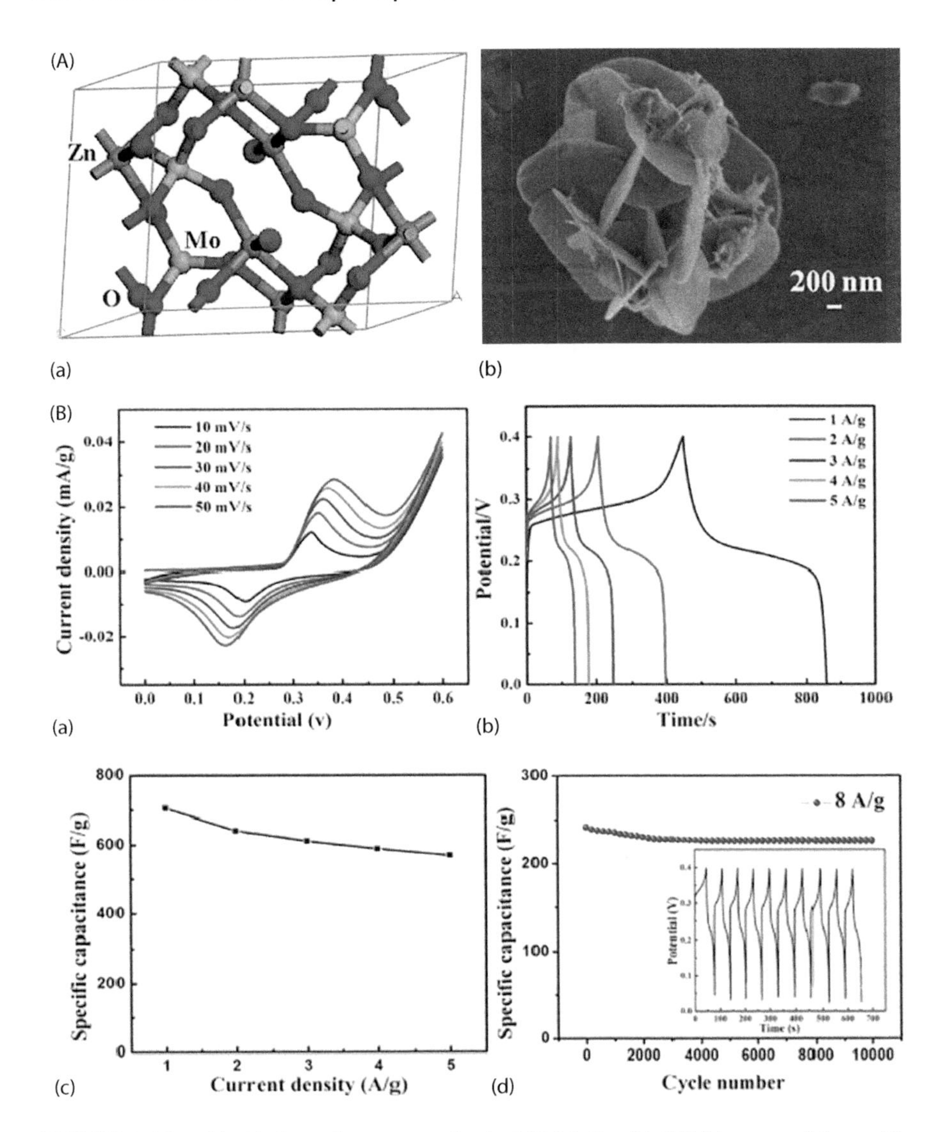

FIGURE 2.10 (A): (a) Crystal structure of spinel $ZnMoO_4$, (b) SEM images of flower-like $ZnMoO_4$. (B): (a) CVs of $ZnMoO_4$ nanoflowers, (b) GCDs of $ZnMoO_4$ nanoflowers, (c) the capacitance on the basis of current density, and (d) the cycling performance of $ZnMoO_4$ nanoflowers. (Reprinted from *J. Alloys Compd.*, 731, Gao, Y. et al., High-performance symmetric supercapacitor based on flower-like zinc molybdate, 1151–1158, Copyright 2017, with permission from Elsevier.)

nanowire, zinc hydrochloride nanosheet, zinc sulfide nanosphere, and flower-like zinc molybdate etc. due to their different surface areas and electrochemical activities. ZnO NWs comes with high surface areas, good electrical conductivity, and low synthesis temperature along with added advantage to grow at any shape and sizes in any substrates. Either the NWs can be grown in the shape of fibers and then entangling the same around a Kevlar fiber or a core-shell structure can be used to use ZnO NWs as electrode materials along with transitional metal hydroxides/oxides. Nanosheets of zinc hydroxychloride (simonk) also carry unique chemical and physical properties. Simonk is a soft material by nature with specific gravity and Mohs hardness 3.2 and 1.5 respectively and it can be grown along [001] plane direction with hexagonal shapes. The presence of oxygen vacancies along the surface of such grown crystals shows a high degree of electrochemical activities along with colossal permittivity in the materials. On the other hand complex semiconductor nanoparticles of zinc sulfide (ZnS) plays an important role in supercapacitors due to the possibility of tuning of their band gaps. Reports of ZnS nanocables, nanoforests, and nanospheres are quite interesting with different shape and sizes for the supercapacitor applications. Due to the abundance, inexpensive, high electrochemical properties, and presence of zero to +6 oxidation states zinc molybdates nanoparticles is another very important zinc-based material for supercapacitor applications. The hybrid supercapacitors of zinc molybdates along with activated carbons can retain 94% of its original specific capacitance magnitude even after a 10k cycle of use; therefore, the zinc-based nanomaterials show higher stability with improved performance in the design of supercapacitors applications.

ACKNOWLEDGMENTS

The author would like to thank UGC, India for the start-up financial grant no. F.30-457/2018 (BSR). One of us (D. Baba Basha) is also thankful to Deanship of Scientific Research at Majmaah University, K.S.A. for supporting this work. We also acknowledge the support provided to A. Franco Jr. by CNPq, Brazil with grant No. 307557/2015-4.

REFERENCES

1. Chowdhury, Geeta, William Adams, Brian Conway, and Srinivasa Sourirajan. "Thin film composite membrane as battery separator." U.S. Patent 5,910,366, issued June 8, 1999.
2. Burke, Lawrence W., Edward Bukowski, Colin Newnham, Neil Scholey, and William Hoge. *HSTSS Battery Development for Missile and Ballistic Telemetry Applications*. No. ARL-MR-477. Army Research Lab, Aberdeen Proving Ground, MD, 2000.
3. Burke, Andrew, Marshall Miller, and Hengbing Zhao. "Lithium batteries and ultracapacitors alone and in combination in hybrid vehicles: Fuel economy and battery stress reduction advantages." *JSR* 21, no. 23 (2010): 15.
4. Frackowiak, Elzbieta, and Francois Beguin. "Carbon materials for the electrochemical storage of energy in capacitors." *Carbon* 39, no. 6 (2001): 937–950.
5. Pandolfo, A. G., and A. F. Hollenkamp. "Carbon properties and their role in supercapacitors." *Journal of Power Sources* 157, no. 1 (2006): 11–27.
6. Pushparaj, Victor L., Manikoth M. Shaijumon, Ashavani Kumar, Saravanababu Murugesan, Lijie Ci, Robert Vajtai, Robert J. Linhardt, Omkaram Nalamasu, and Pulickel M. Ajayan. "Flexible energy storage devices based on nanocomposite paper." *Proceedings of the National Academy of Sciences* 104, no. 34 (2007): 13574–13577.

7. Luo, Qingtao, Huamin Zhang, Jian Chen, Dongjiang You, Chenxi Sun, and Yu Zhang. "Preparation and characterization of Nafion/SPEEK layered composite membrane and its application in vanadium redox flow battery." *Journal of Membrane Science* 325, no. 2 (2008): 553–558.
8. Xu, Chunmei, Yishan Wu, Xuyang Zhao, Xiuli Wang, Gaohui Du, Jun Zhang, and Jiangping Tu. "Sulfur/three-dimensional graphene composite for high performance lithium–sulfur batteries." *Journal of Power Sources* 275 (2015): 22–25.
9. Einhorn, Markus, Valerio F. Conte, Christian Kral, Jürgen Fleig, and Robert Permann. "Parameterization of an electrical battery model for dynamic system simulation in electric vehicles." In *2010 IEEE Vehicle Power and Propulsion Conference*, pp. 1–7. IEEE, 2010.
10. Inagaki, Hiroki, and Norio Takami. "Nonaqueous electrolyte battery, battery pack and vehicle." U.S. Patent 7,662,515, issued February 16, 2010.
11. Chen, Shu-Ru, Yun-Pu Zhai, Gui-Liang Xu, Yan-Xia Jiang, Dong-Yuan Zhao, Jun-Tao Li, Ling Huang, and Shi-Gang Sun. "Ordered mesoporous carbon/sulfur nanocomposite of high performances as cathode for lithium–sulfur battery." *Electrochimica Acta* 56, no. 26 (2011): 9549–9555.
12. Li, Xianglong, and Linjie Zhi. "Managing voids of Si anodes in lithium ion batteries." *Nanoscale* 5, no. 19 (2013): 8864–8873.
13. Fang, B., Y. Z. Wei, K. Maruyama, and M. Kumagai. "High capacity supercapacitors based on modified activated carbon aerogel." *Journal of Applied Electrochemistry* 35, no. 3 (2005): 229–233.
14. Fisher, Robert A., Morgan R. Watt, and W. Jud Ready. "Functionalized carbon nanotube supercapacitor electrodes: A review on pseudocapacitive materials." *ECS Journal of Solid State Science and Technology* 2, no. 10 (2013): M3170–M3177.
15. Shukla, A. K., S. Sampath, and K. Vijayamohanan. "Electrochemical supercapacitors: Energy storage beyond batteries." *Current Science* 79, no. 12 (2000): 1656–1661.
16. Cericola, D., R. Kötz, and Alexander Wokaun. "Effect of electrode mass ratio on aging of activated carbon based supercapacitors utilizing organic electrolytes." *Journal of Power Sources* 196, no. 6 (2011): 3114–3118.
17. Husain, Iqbal. *Electric and Hybrid Vehicles: Design Fundamentals.* CRC Press, London, UK, 2003.
18. Chu, Andrew, and Paul Braatz. "Comparison of commercial supercapacitors and high-power lithium-ion batteries for power-assist applications in hybrid electric vehicles: I. Initial characterization." *Journal of Power Sources* 112, no. 1 (2002): 236–246.
19. Barsukov, Igor V., Christopher S. Johnson, Joseph E. Doninger, and Vyacheslav Z. Barsukov, eds. *New Carbon Based Materials for Electrochemical Energy Storage Systems: Batteries, Supercapacitors and Fuel Cells.* Vol. 229. Springer Science & Business Media, Dordrecht, the Netherlands, 2006.
20. Su, Ling-Hao, Xiao-Gang Zhang, Chang-Huan Mi, Bo Gao, and Yan Liu. "Improvement of the capacitive performances for Co–Al layered double hydroxide by adding hexacyanoferrate into the electrolyte." *Physical Chemistry Chemical Physics* 11, no. 13 (2009): 2195–2202.
21. Roldán, Silvia, Zoraida González, Clara Blanco, Marcos Granda, Rosa Menéndez, and Ricardo Santamaría. "Redox-active electrolyte for carbon nanotube-based electric double layer capacitors." *Electrochimica Acta* 56, no. 9 (2011): 3401–3405.
22. Wu, Jihuai, Haijun Yu, Leqing Fan, Genggeng Luo, Jianming Lin, and Miaoliang Huang. "A simple and high-effective electrolyte mediated with p-phenylenediamine for supercapacitor." *Journal of Materials Chemistry* 22, no. 36 (2012): 19025–19030.
23. Chen, Wei, R. B. Rakhi, and Husam N. Alshareef. "Capacitance enhancement of polyaniline coated curved-graphene supercapacitors in a redox-active electrolyte." *Nanoscale* 5, no. 10 (2013): 4134–4138.

24. Senthilkumar, S. T., R. Kalai Selvan, and J. S. Melo. "Redox additive/active electrolytes: A novel approach to enhance the performance of supercapacitors." *Journal of Materials Chemistry A* 1, no. 40 (2013): 12386–12394.
25. Lu, Luyao, Tianyue Zheng, Qinghe Wu, Alexander M. Schneider, Donglin Zhao, and Luping Yu. "Recent advances in bulk heterojunction polymer solar cells." *Chemical Reviews* 115, no. 23 (2015): 12666–12731.
26. Chee, W. K., H. N. Lim, I. Harrison, K. F. Chong, Z. Zainal, C. H. Ng, and N. M. Huang. "Performance of flexible and binderless polypyrrole/graphene oxide/zinc oxide supercapacitor electrode in a symmetrical two-electrode configuration." *Electrochimica Acta* 157 (2015): 88–94.
27. Dong, Xiaochen, Yunfa Cao, Jing Wang, Mary B. Chan-Park, Lianhui Wang, Wei Huang, and Peng Chen. "Hybrid structure of zinc oxide nanorods and three dimensional graphene foam for supercapacitor and electrochemical sensor applications." *RSC Advances* 2, no. 10 (2012): 4364–4369.
28. Kim, Chang Hyo, and Bo-Hye Kim. "Zinc oxide/activated carbon nanofiber composites for high-performance supercapacitor electrodes." *Journal of Power Sources* 274 (2015): 512–520.
29. Wang, Xu, Afriyanti Sumboja, Mengfang Lin, Jian Yan, and Pooi See Lee. "Enhancing electrochemical reaction sites in nickel–cobalt layered double hydroxides on zinc tin oxide nanowires: A hybrid material for an asymmetric supercapacitor device." *Nanoscale* 4, no. 22 (2012): 7266–7272.
30. Hassanpour, Mohammad, Hossein Safardoust-Hojaghan, and Masoud Salavati-Niasari. "Rapid and eco-friendly synthesis of NiO/ZnO nanocomposite and its application in decolorization of dye." *Journal of Materials Science: Materials in Electronics* 28, no. 15 (2017): 10830–10837.
31. Zhang, Siwen, Bosi Yin, He Jiang, Fengyu Qu, Ahmad Umar, and Xiang Wu. "Hybrid ZnO/ZnS nanoforests as the electrode materials for high performance supercapacitor application." *Dalton Transactions* 44, no. 5 (2015): 2409–2415.
32. Ng, C. H., H. N. Lim, S. Hayase, I. Harrison, A. Pandikumar, and N. M. Huang. "Potential active materials for photo-supercapacitor: A review." *Journal of Power Sources* 296 (2015): 169–185.
33. Katwal, Giwan, Maggie Paulose, Irene A. Rusakova, James E. Martinez, and Oomman K. Varghese. "Rapid growth of zinc oxide nanotube–nanowire hybrid architectures and their use in breast cancer-related volatile organics detection." *Nano Letters* 16, no. 5 (2016): 3014–3021.
34. Meng, Fancheng, Qingwen Li, and Lianxi Zheng. "Flexible fiber-shaped supercapacitors: Design, fabrication, and multi-functionalities." *Energy Storage Materials* 8 (2017): 85–109.
35. Zeng, Wei, Lei Wang, Huimin Shi, Guanhua Zhang, Kang Zhang, Hang Zhang, Feilong Gong, Taihong Wang, and Huigao Duan. "Metal–organic-framework-derived ZnO@ C@ $NiCo_2O_4$ core–shell structures as an advanced electrode for high-performance supercapacitors." *Journal of Materials Chemistry A* 4, no. 21 (2016): 8233–8241.
36. Bae, Joonho, Min Kyu Song, Young Jun Park, Jong Min Kim, Meilin Liu, and Zhong Lin Wang. "Fiber supercapacitors made of nanowire-fiber hybrid structures for wearable/flexible energy storage." *Angewandte Chemie International Edition* 50, no. 7 (2011): 1683–1687.
37. Khamlich, S., Zhypargul Abdullaeva, J. V. Kennedy, and M. Maaza. "High performance symmetric supercapacitor based on zinc hydroxychloride nanosheets and 3D graphene-nickel foam composite." *Applied Surface Science* 405 (2017): 329–336.
38. Khamlich, S., T. Mokrani, M. S. Dhlamini, B. M. Mothudi, and M. Maaza. "Microwave-assisted synthesis of simonkolleite nanoplatelets on nickel foam–graphene with enhanced surface area for high-performance supercapacitors." *Journal of Colloid and Interface Science* 461 (2016): 154–161.

39. Shinagawa, Tsutomu, Mitsuru Watanabe, Jun-ichi Tani, and Masaya Chigane. "(0001)-oriented single-crystal-like porous ZnO on ITO substrates via quasi-topotactic transformation from (001)-oriented zinc hydroxychloride crystals." *Crystal Growth & Design* 17, no. 7 (2017): 3826–3833.
40. Pessoni, Herminia VS, Prasun Banerjee, and A. Franco. "Colossal dielectric permittivity in Co-doped ZnO ceramics prepared by a pressure-less sintering method." *Physical Chemistry Chemical Physics* 20, no. 45 (2018): 28712–28719.
41. Zhang, Qiaobao, Bote Zhao, Jiexi Wang, Chong Qu, Haibin Sun, Kaili Zhang, and Meilin Liu. "High-performance hybrid supercapacitors based on self-supported 3D ultrathin porous quaternary Zn-Ni-Al-Co oxide nanosheets." *Nano Energy* 28 (2016): 475–485.
42. Ramachandran, Rajendran, Murugan Saranya, Pratap Kollu, Bala PC Raghupathy, Soon Kwan Jeong, and Andrews Nirmala Grace. "Solvothermal synthesis of Zinc sulfide decorated Graphene (ZnS/G) nanocomposites for novel Supercapacitor electrodes." *Electrochimica Acta* 178 (2015): 647–657.
43. Rai, Satish C., Kai Wang, Yong Ding, Jason K. Marmon, Manish Bhatt, Yong Zhang, Weilie Zhou, and Zhong Lin Wang. "Piezo-phototronic effect enhanced UV/visible photodetector based on fully wide band gap type-II ZnO/ZnS core/shell nanowire array." *ACS Nano* 9, no. 6 (2015): 6419–6427.
44. Javed, Muhammad Sufyan, Jie Chen, Lin Chen, Yi Xi, Cuilin Zhang, Buyong Wan, and Chenguo Hu. "Flexible full-solid state supercapacitors based on zinc sulfide spheres growing on carbon textile with superior charge storage." *Journal of Materials Chemistry A* 4, no. 2 (2016): 667–674.
45. Gao, Yong-Ping, Ke-Jing Huang, Chen-Xi Zhang, Shuai-Shuai Song, and Xu Wu. "High-performance symmetric supercapacitor based on flower-like zinc molybdate." *Journal of Alloys and Compounds* 731 (2018): 1151–1158.
46. Kang Xiao, Lu Xia, Guoxue Liu, Suqing Wang, Liang-Xin Ding, Haihui Wang. "Honeycomb-like $NiMoO_4$ ultrathin nanosheet arrays for high-performance electrochemical energy storage." *Journal of Materials Chemistry A*. 3 (2015): 6128–6135.
47. Debasis Ghosh, Soumen Giri, Chapal Kumar Das. "Synthesis, characterization and electrochemical performance of graphene decorated with 1D $NiMoO_4 \cdot nH_2O$ nanorods." *Nanoscale* 5 (2013): 10428–10437.
48. Jianfeng Shen, Xianfu Li, Na Li, Mingxin Ye. "Facile synthesis of $NiCo_2O_4$-reduced graphene oxide nanocomposites with improved electrochemical properties." *Electrochimica Acta* 141 (2014): 126–133.
49. X. Xu, J. Shen, N. Li, M. Ye. "Microwave-assisted synthesis of graphene/$CoMoO_4$ nanocomposites with enhanced supercapacitor performance." *Journal of Alloy and Compounds* 616 (2014): 58–65.
50. L. Aleksandrov, T. Komatsu, R. Iordanova, Y. Dimitriev. "Structure study of MoO_3-ZnO-B_2O_3 glasses by Raman spectroscopy and formation of α-ZnMoO4 nanocrystals." *Optical Materials* 33 (2011): 839–845.

3 Defect Engineered Inorganic Materials for Supercapacitors

Biswajit Choudhury and Anamika Kalita

CONTENTS

3.1 Introduction 33
3.2 Defect Engineered Metal Oxides 34
3.3 Defect Engineered Metal Chalcogenides 47
3.4 Summary 54
References 55

3.1 INTRODUCTION

Supercapacitors are regarded as the next generation electrochemical energy storage devices which link the gap between batteries and usual capacitors. An efficient supercapacitor has an advantage of high power density, long cycle life, and short period for charging-discharging (Li et al. 2018; Lu et al. 2014). Materials for supercapacitor electrode must have a high surface area with large porosity for electrolyte accessibility and transportation, high electronic and ionic conductivity for minimizing capacitance losses at an increased scan/current densities (Yu et al. 2015). Inorganic nanomaterials comprising metal oxides and metal chalcogenides are widely explored electroactive materials for supercapacitor applications (Shi et al. 2014; Simon and Gogotsi 2008; Choi et al. 2017). MnO_2 has a high theoretical capacitance ~ 1370 Fg^{-1} but poor conductivity for electricity (10^{-5} to 10^{-6} S cm^{-1}). Although TiO_2 has a comparatively higher electrical conductivity (10^{-5} to 10^{-2} Scm^{-1}) than MnO_2, the conductivity is still less for designing efficient supercapacitor devices (Cao et al. 2015). Hematite α-Fe_2O_3 also serves as an excellent negative electrode but suffers from poor electrical conductivity (10^{-14} S cm^{-1}) (Lu et al. 2014). A 2D nanosheet provides a short diffusion path for the transportation of ions/electron during ion insertion/extraction process. The electron/ion diffusion distance on the surface of a 2D nanosheet is very short. Therefore, a strong redox reaction can lead to enhanced electrochemical performance (Xiang et al. 2017). The different ways for producing oxygen-deficient metal oxides are the electrochemical reduction, hydrogenation or vacuum calcination, and hydrogen plasma treatment. Incorporation of oxygen vacancies into a wide-band gap semiconductor introduce sufficient donor electrons and improves the electronic conductivity of the system (Zhou and Zhang 2014; Lu et al. 2012; Zhai et al. 2014; Wu et al. 2013). Because of the excess donor carriers, the semiconductor attains a semi-metallic behavior. Other advantages of defect engineering are to generate multiple oxidation states. For example, oxygen vacancies introduce multiple valence states in an oxide. These multiple valence states provide sufficient redox active sites (Zhai et al. 2014; Zhou and Zhang 2014).

Among two-dimensional (2D) nanostructured materials, transition metal dichalcogenides (TMDs) are considered a suitable replacement for graphene for a range of applications. These include photovoltaic devices, energy storage devices, and spintronics (Bissett et al. 2016; Choi et al. 2017; Heine 2015; Manzeli et al. 2017; Rasmussen and Thygesen 2015). They retain a layered, graphene-like van der Waals structures composed of a transition metal layer inserted between two chalcogenides. They have a general formula of MX2, where M refers to transition metal atom (e.g., molybdenum, tungsten, titanium, etc.) and X is a chalcogenide atom (e.g., sulphur, selenium, and tellurium) (Jacobs-Gedrim et al. 2014). As bulk layered materials, TMDs exist in 1T (tetragonal), 2H (hexagonal), and more rarely in 3R (rhombohedral) polymorphs (Heine 2015). The atomic-level defects in transition metal dichalcogenides enhances the specific surface area and provide sufficient exposed redox active sites at the electrode/electrolyte interface (Ding et al. 2019).

3.2 DEFECT ENGINEERED METAL OXIDES

Metal oxides contain abundant of defects including oxygen vacancies, cationic vacancies on the substitutional, interstitial positions. A series of metal oxides are available wherein with manipulation of intrinsic defects an enhanced supercapacitor performance can be generated. In this chapter, we will consider a number of metal oxides, including TiO_2, MnO_2, Co_3O_4, RuO_2, $NiCo_2O_4$. We will provide a discussion on each of these metal oxides.

We will start our discussion with TiO_2. This is a group IVB transition metal oxide and is abundantly found in metastable anatase and stable rutile phases (Choudhury and Choudhury 2014). The other existing phases of TiO_2 are brookite and (B) phase (Li et al. 2014; Hua et al. 2015). Unlike bulk TiO_2 which has a band gap of 3–3.2 eV, the defect engineered TiO_2 has a tunable band gap due to the formation of intermediate defect states providing excess donor electrons (Choudhury et al. 2016). Defect engineered TiO_2 has a high surface area, excellent electronic conductivity, and numerous redox active sites which makes it an efficient material for supercapacitor applications (Lu et al. 2012; Shi et al. 2014). Of the various nanostructure forms, TiO_2 nanotube arrays are better suited for supercapacitor applications because of the open channels for intercalation and migration of ions (Zhou and Zhang 2014). Moreover, the surface electric fields reduce carrier recombination. This is realized because of the confinement of the injected electrons in the central part of the nanotube (Salari et al. 2011). A widely adopted method for the production of titania nanotubes arrays is electrochemical anodization of titanium foil. Self-doping of Ti^{3+} with H^+ intercalation into titanium nanotubes arrays can be performed by electrochemical reduction of Ti^{4+} to Ti^{3+} at the potentials −1.2, −1.4, −1.6, and −1.8 V (Zhou and Zhang 2014) (Figure 3.1a). Proton intercalation leads to lattice expansion with a concomitant formation of oxygen

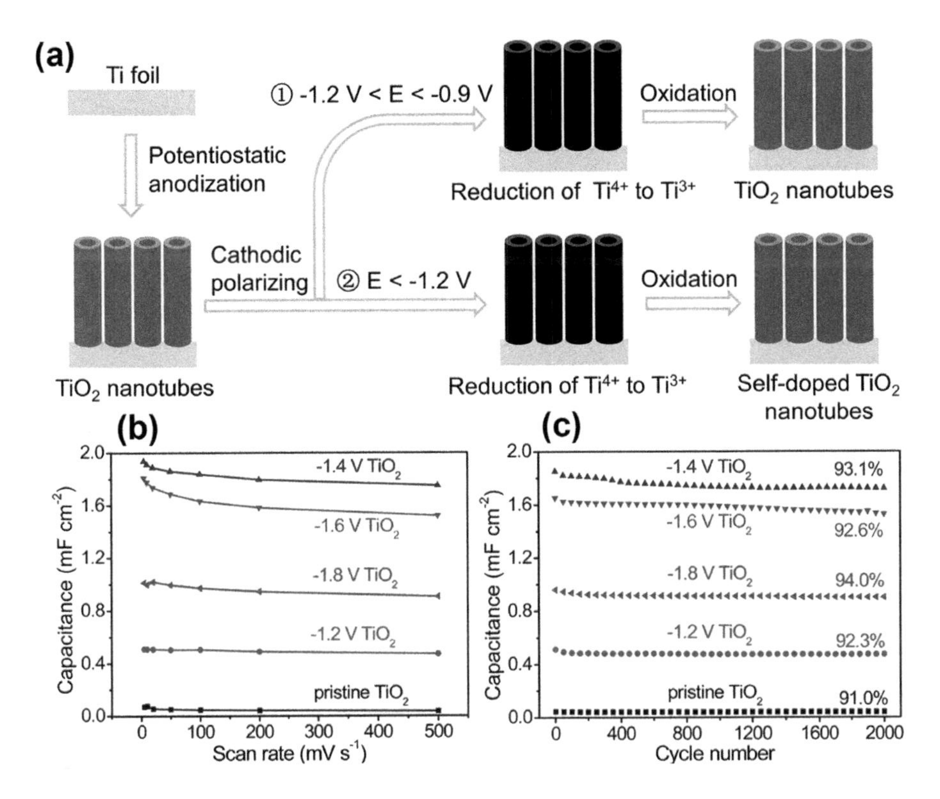

FIGURE 3.1 (a) Schematic shows the fabrication of self-doped TiO_2 nanotube arrays by electrochemical process, where E is the potential used during polarization process. (b) Areal capacitances of the samples measured as a function of scan rate. (c) Cycle performance is measured for 2000 cycles at a scan rate of 100 mV s^{-1}. (Reprinted with permission from Zhou and Zhang 2014. Copyright 2014 American Chemical Society.)

vacancies. This self-doping process induces semi-metallic behavior in semiconducting titania nanotubes arrays and generates high carrier densities. The carrier density as shown by the pristine titania nanotubes is 1.55×10^{16} cm^{-3}. Reduced titania nanotubes arrays at –1.4 V delivers a carrier density of 2.14×10^{21} cm^{-3}. It is estimated that the electrochemical reduction at moderate –1.4 V generates sufficient oxygen vacancies without bringing any nanotube disruption. However, reduction at a more negative potential (–1.6 and –1.8 V) leads to cracking in the nanotube (Zhou and Zhang 2014). The reduced –1.4 V TiO_2 delivers a capacitance of 1.84 mF cm^{-2} when the scan rate 5 mV s^{-1} within 5–500 mV s^{-1} scanning range (Figure 3.1b). In comparison to that of pure titanium nanotubes arrays (0.047 mF cm^{-2}) the resultant enhancement in capacitance in reduced-titanium nanotubes arrays is more than 39 times. Reduced TiO_2 at –1.4 V shows capacitance retention of 90.4% when the scan rate is increased from 5 to 500 mV s^{-1}. The stability of the samples is measured at a scan rate of 100 mV s^{-1} to 2000 cycles. The reduced TiO_2 show more than 90% capacitance retention after 2000 cycles indicating that there is no depletion in the oxygen defect content even after the successive cycle tests (Figure 3.1c) (Zhou and Zhang 2014).

Electrochemical reduction provide only 1% transformation of Ti^{4+} to Ti^{3+} with associated formation of oxygen vacancies. The Ti^{3+} generation can be enhanced to more than 22% if an accurate potential setting is done by performing the electrochemical reduction experiment in a three-electrode configuration (Li et al. 2015b). This is ascribed due to the generation of high-density of Ti^{3+} in the inner wall of titanium nanotubes arrays. The reduction starts at the bottom which slowly extends to the top of the nanotubes. By adopting this method the reduced titania nanotubes arrays display a carrier density of 9×10^{22} cm^{-3}, whereas the value is only 3.6×10^{20} cm^{-3} in untreated titania nanotubes arrays. The specific capacitance of reduced titania nanotubes arrays (24.07 mF cm^{-2}) is nearly 1094 times higher than that of pure titania nanotubes arrays (0.02 mF cm^{-2}) (Li et al. 2015a). Moreover, even after 2000 cycles of measurement about 98.1% capacitance is retained. Instead of an electrochemical reduction, annealing titania nanotubes arrays in a reduced atmosphere can lead to more than 31.7% conversion of Ti^{4+} to Ti^{3+} (Li et al. 2015a).

It is observed that titania nanotubes arrays with preferential C-axis orientation can be developed by controlling the water content during anodization step followed by crystallization in an oxygen-poor environment (Pan et al. 2014) (Figure 3.2). This leads to the generation of 31.7% Ti^{3+} in C-titania nanotubes arrays with high carrier density (2.62×10^{23} cm^{-3}) than that of randomly oriented titanium nanotubes arrays (2.7×10^{21} cm^{-3}). The cyclic voltammetry (CV) curves of C-titania nanotubes arrays and random-titania nanotubes arrays are collected at a scan rate of 100 mV s^{-1} (Figure 3.3a). The rectangular shape of the curves indicates electrical double-layer capacitive behavior. The measurement of areal capacitance as a function of scanning rate shows that C-titania nanotubes arrays electrode delivers 8.21 mF cm^{-2}. This is 25 times higher than that of random-titania nanotubes arrays (0.32 mF cm^{-2}). The capacitance of C-titania nanotubes arrays decreases from 11.44 at 10 mV s^{-1} to 5.83 mF cm^{-2} at 1000 mV s^{-1} (Figure 3.3b). Thus, C-titania nanotubes array show 51% of retention of initial capacitance, whereas random-titania nanotubes array retain only 13% of initial capacitance. The galvanostatic charge/discharge measurement reveals a symmetric and an extended curve indicating good capacitive behavior (Figure 3.3c). C-titania nanotubes arrays show long-term stability with only 16% loss in initial capacitance after 5000 cycles. In case of random-titania nanotubes arrays the capacitance loss is 49% after 5000 cycles (Pan et al. 2014) (Figure 3.3d). The preferential C-axis orientation of the nanotubes and the presence of oxygen defects enhance the electrical conductivity and fast H^+ diffusion in the electrode.

A report shows that annealing of titania nanotubes arrays at 450, 500, 550, 600, and 650°C for 5 h in Ar can lead to a significant alteration in their specific capacitance values (Salari et al. 2012). An increase in the annealing temperature from 450°C to 600°C could lead to an increment in the specific capacitance from 1.1 to 7.6 Fg^{-1} measured at the current density of 5 μA cm^{-2}.

The enhancement in the capacitance with annealing temperature is ascribed to the generation of oxygen vacancies and titanium interstitials. It is, however, observed that the capacitance reduces from 7.6 to 7.1 Fg^{-1} when the annealing temperature is set at 650°C. The lowering in the specific capacitance is owing to the collapse of the nanopores of titania nanotubes arrays leading to a poor ionic diffusion. It is further observed that two-step annealing of titania nanotubes arrays in Ar can lead to the

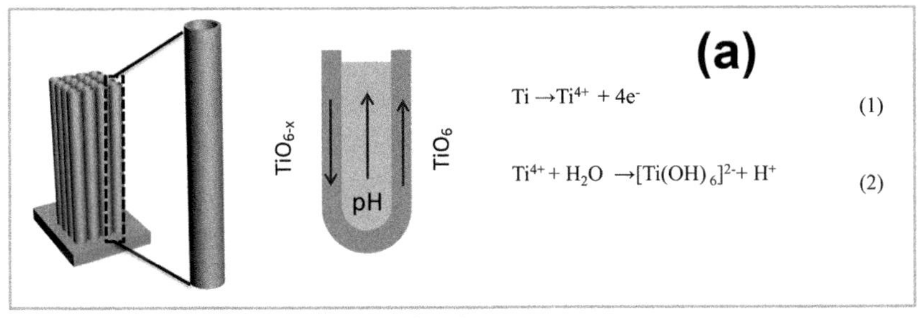

As-anodized TNA

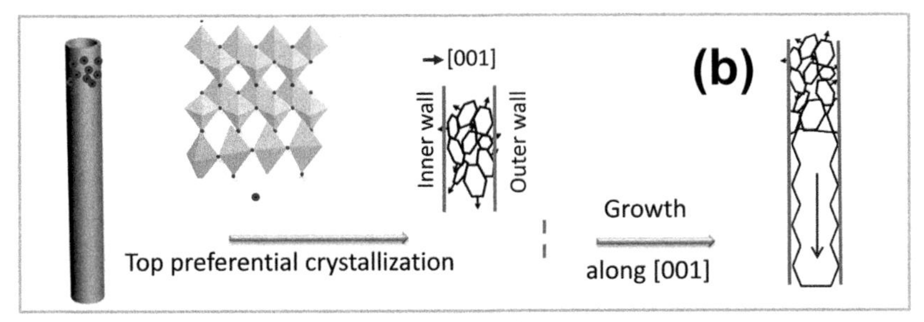

Highly [001] oriented TNA

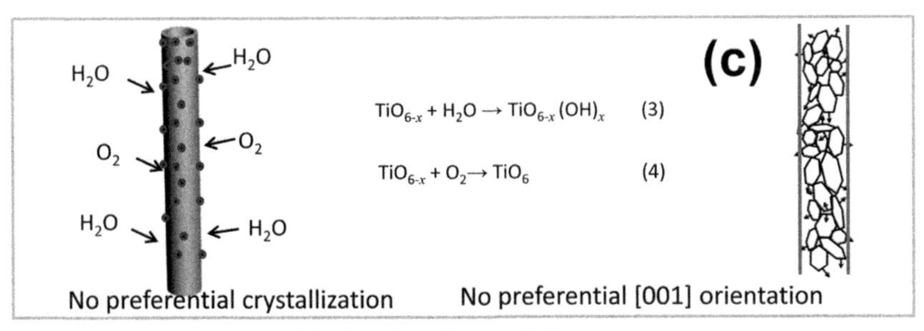

Randomly oriented TNA

FIGURE 3.2 Schematic showing fabrication of C-axis oriented (b) and randomly oriented TiO_2 (c) from as anodized TiO2 nanotube arrays (a). (Pan, D. et al., *J. Mater. Chem. A*, 2, 2014, 11454–11464, 2014. Reproduced by permission of The Royal Society of Chemistry.)

generation of higher numbers of oxygen vacancies than the one-step annealing process (Salari et al. 2011). The sample obtained after single step annealing at 580°C in Ar for 1 h is again calcined for another 1 h in Ar but at a higher temperature, i.e., 630°C. Similarly, another batch of samples is prepared first by air annealing followed by Ar annealing for the same time and at the temperatures 580°C and 630°C, respectively. The two-step calcination in Ar leads to the generation of higher density of oxygen vacancies and facilitate anatase to rutile phase transformation. The specific capacitance of the sample measured at a scan rate of 1 mV s^{-1} is 911 μF cm^{-2} for the two-step annealing process in Ar, whereas this value is only 521 μF cm^{-2} for a

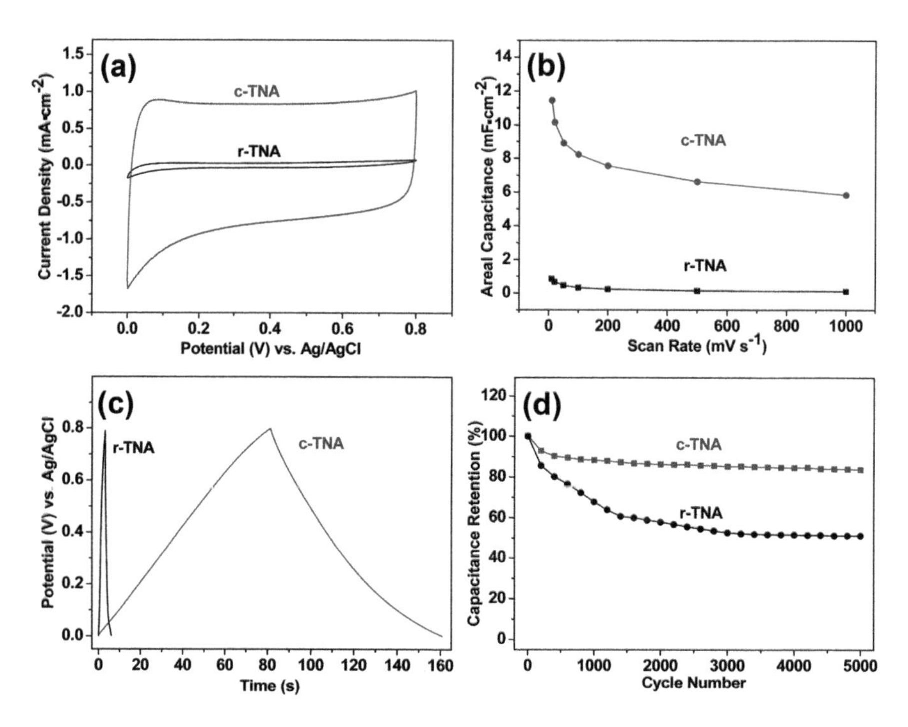

FIGURE 3.3 (a) CV curves of the C-axis oriented and randomly oriented titania nanotubes arrays at the scan rate of 100 mV s^{-1}. (b) Measured areal capacitance at various scan rates. (c) Galvanostatic charge/discharge curves obtained at a current density of 100 $\mu A\ cm^{-2}$. (d) Cycle performances for 5000 cycles at a scan rate of 100 mV s^{-1}. (Pan, D. et al., *J. Mater. Chem. A*, 2, 2014, 11454–11464, 2014. Reproduced by permission of The Royal Society of Chemistry.)

single step annealing in Ar. The recorded value of specific capacitance for single step air annealing of titania nanotubes arrays is only 30 $\mu F\ cm^{-2}$ (Salari et al. 2011).

Hydrogenation is another approach to fabricate reduced TiO_2 with a high density of free carriers (Lu et al. 2012; Zhang et al. 2018). Anodization of titanium foil results in vertically aligned ordered TiO_2 nanotube arrays on titanium fiber. The as-prepared nanotubes are crystallized through air and hydrogen annealing at 400°C for 60 min. The process is schematically shown in Figure 3.4a. The hydrogenated titania nanotubes (H-TNT) arrays exhibit carrier densities of 1.4×10^{23} cm^{-3}, whereas this value is only 3.4×10^{20} cm^{-3} and 3.6×10^{20} cm^{-3} for the untreated and air-calcined titania nanotubes arrays (Lu et al. 2012). The higher carrier density in air annealed titania nanotubes (A-TNT) arrays than that of untreated titania nanotubes arrays is due to the improved crystallinity in the former than the later. A further enhancement in carrier density on hydrogenation is due to the generation of sufficient numbers of oxygen vacancies. The recorded value of the areal capacitance of H-TNT is 3.8 mF cm^{-2} at a scan rate of 10 mV s^{-1} (Figure 3.4b). When the scan rate is 100 mV s^{-1}, the measured areal capacitance is 3.24 mF cm^{-2} for H-TiO_2, 0.08 mF cm^{-2} for A-TNT and 0.026 mF cm^{-2} for untreated TiO_2. Therefore, hydrogenation leads to an enhancement in capacitance by 40-fold to 124-fold in comparison to untreated TiO_2 and A-TNT.

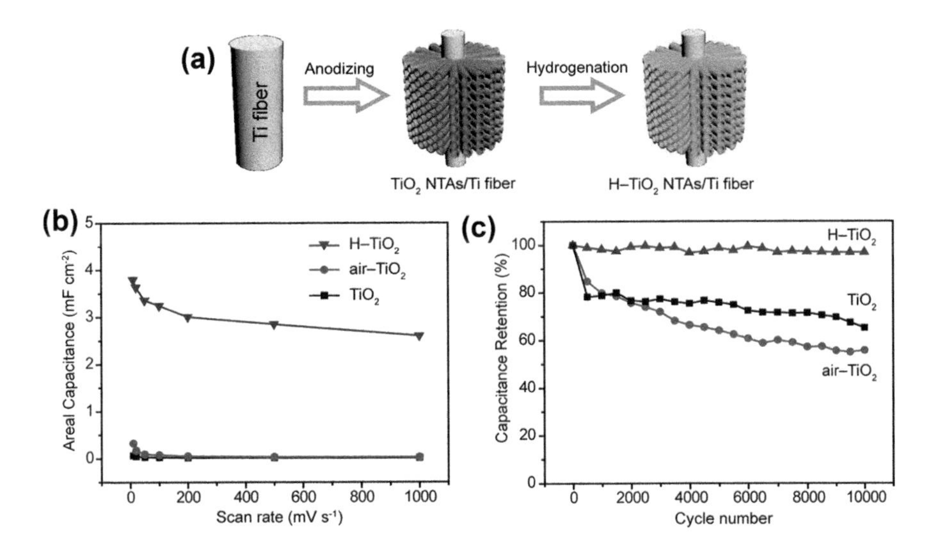

FIGURE 3.4 (a) Schematic shows the stepwise fabrication of hydrogenated TiO_2 nanotube arrays. (b) Measurement of areal capacitance versus scan rates. (c) Stability test of the samples for 10,000 cycles at a scan rate of 100 mV s^{-1} for 10,000 cycles. (Reprinted with permission from Lu, X. et al., *Nano Lett.*, 12, 1690–1696, 2012. Copyright 2012 American Chemical Society.)

Stability test shows that after 10,000 cycles, only 3.1% reduction in initial capacitance is recorded for H-TNT. However, there is a drop of 34.8% and 44.3% of the initial capacitance in untreated and A-TNT after 10,000 cycles (Figure 3.4c) (Lu et al. 2012). The lowest efficiency in untreated TiO_2 is because of their amorphous structure, the improvement in air treated TiO_2 is because of their improved crystallinity, and the best result in H-TNT is due to the hydrogenation induced generation of oxygen vacancies. These oxygen vacancies enhance carrier density facilitating easy charge transport, and sufficient density of –OH groups functionalized over TNT surface (Lu et al. 2012).

An important controlling parameter during hydrogenation is the heating rate during hydrogenation (Zhang et al. 2018). Defective titania nanotube is initially prepared by anodization of Ti foil at 400°C for 1 h under a hydrogen atmosphere. Afterward, the heating rate during calcination is 10°C, 30°C, 50°C, 75°C min^{-1}. It is reported that the measurement of the areal capacitance of H-TNT at a scan rate of 100 mV s^{-1} can lead to a significant enhancement in the areal capacitance from 27.36 mF cm^{-2} (10°C min^{-1}) to 52.4 mF cm^{-2} (50°C min^{-1}). The system shows remarkable stability after 5000 cycles with 95.2% retention of the initial capacitance. The enhancement is attributed to the generation of sufficient density of oxygen vacancies (large carrier density) and surface –OH groups (Zhang et al. 2018).

Hydrogen plasma treatment is considered to be an alternative to gas phase hydrogenation to introduce surface disorder over TiO_2 (Wu et al. 2013). The reason is the requirement of less time during plasma treatment as compared to long exposure under gas phase hydrogenation. Hydrogen plasma treatment can give surface roughness on nanotube walls and increase the specific surface area allowing sufficient active sites for ion adsorption (Xu et al. 2014). Anodic TiO_2 nanotubes are treated under hydrogen plasma for 1.5 h in a plasma-enhanced chemical vapor deposition (CVD) chamber. The plasma treatment generates an amorphous layer having a thickness of 1–2 nm and contains sufficient densities of Ti^{3+} and surface –OH groups. The specific capacitance is measured in a three-electrode configuration, and the measured value of specific capacitance for the untreated anodized titanium oxide (ATO) is 0.97 mF cm^{-2} at a discharge current density of 0.05 mA cm^{-2}. The specific capacitance recorded for the hydrogen plasma treated (H-ATO) is 7.22 mF cm^{-2} at 0.05 mA cm^{-2}. An increase in current density from 0.05 to 2 mA cm^{-2} could lead to a reduction of specific capacitance to 6.37 mF cm^{-2} due to the resistance to ion diffusion. The benefit of hydrogen plasma treatment is the generation of donor oxygen vacancies and reactive Ti^{3+} sites. Furthermore, plasma treatment leads to an enhancement in the surface area thus offering facile ion adsorption. The H-ATO electrode delivers excellent performance during stability test for 10,000 cycles measured at a current density of 0.1 mA cm^{-2}. The attributed enhancement is due to the excellent intercalation/deintercalation of ions on the active sites of the disordered surface of the hydrogenated nanotube (Wu et al. 2013).

Another widely used metal oxide for supercapacitor application is MnO_2. The theoretical specific capacitance of MnO_2 is 1233 Fg^{-1}. MnO_2 exists in different crystallographic phases such as α, δ, β, γ, and λ (Tanggarnjanavalukul et al. 2017). The 1D tunnel structure of MnO_2 is exhibited by α, β, and γ phases. On the other hand, the 2D tunnel structure and 3D spinel structure are exhibited by δ (birnessite) and λ phases (Thackeray 1997; Tanggarnjanavalukul et al. 2017). The different phases of MnO_2 deliver specific capacitance which follows an order $>\alpha = \delta > \gamma > \lambda > \beta$ (Devaraj and Munichandraiah 2008). It is stated that β-MnO_2 having narrow tunnels limit accommodation of cations during the charge-discharge process. On the other hand, the 2D layered structure of δ-MnO_2 and α-MnO_2 have larger tunnel size leading to maximum adsorption of ions and maximum diffusion of alkali cations during charge-discharge process (Devaraj and Munichandraiah 2008; Gao et al. 2017). Defect engineering of MnO_2 can enrich the system with excess oxygen vacancies and introduce multiple valence states of Mn for better redox active reactions (Zhai et al. 2014). Hydrogenated MnO_2 nanorods show excellent pseudocapacitive behavior due to the low valence Mn^{2+} and Mn^{3+} states on the surface of MnO_2 which is unlike in untreated MnO_2. CV measurement of H-MnO_2, and air annealed MnO_2 at a scan rate of 10 mV s^{-1} delivers an areal capacitance of 0.15 and 0.05 F cm^{-2} (Zhai et al. 2014). The H-MnO_2 shows 44% retention of initial capacitance when the scan rate is increased from 10 to 200 mV s^{-1}. The areal capacitance measured at a current density of 1 mA cm^{-2} is found to be 0.20 F cm^{-2} in H-MnO_2. It is also observed that an increase in the hydrogenation temperature from 250°C to 650°C can result in the change in the oxidation state from +3.44 to 2.15 with a concomitant decrease in the areal capacitance from 0.20 to 0.089 F cm^{-2}. Thus, Mn^{+3} is the favorable oxidation state of Mn at which maximum areal capacitance can be recorded. In its +3 state, the excess negative charge accumulated can be transferred to the neighboring O atoms. Hydrogenation results in the mixed valence MnO_2/Mn_2O_3 phases with sufficient interfacial oxygen vacancies generation (Liu et al. 2019). Hydrogenation leads to the formation of Mn^{4+}/Mn^{3+} redox couple. A higher scan rate offers resistance to ion and electron transfer. The areal capacitance of the electrode measured from the charge/discharge curves is 640 mF cm^{-2} at a current density of 1 mA cm^{-2}. Capacitance reduces to 573.3 mF cm^{-2} at the current density of 10 mA cm^{-2}. The measured values indicate that the H-MnO_2 nanosheets retain 89.6% of specific capacitance even after 10 times increase in current density. The excellent conductivity and good rate capability are because of the interconnected nanosheets of MnO_2 which provides facile electrons and ions transfer path. There is 84.6% of capacitance retention after 1000 cycles indicating long stability of the electrode (Liu et al. 2019).

2D δ-MnO_2 nanosheets are self-assembled into 3D porous MnO_2 nanosheets by varying pH during self-assembly (Gao et al. 2017). The pH variation results in the formation of "Frenkel defect pair" in the 3D nanostructures which comprise of a surface Mn vacancy and six-fold coordinated Mn^{3+} on another side of the surface. The Mn defect concentration increases from 19.9% to 26.5% on decreasing pH from 4 to 2. An acidic medium enhances proton sorption onto MnO_2 surface and facilitates expulsion of more numbers of in-plane Mn to the surface generating high numbers of Mn vacancies. The measured specific capacitance is 306 Fg^{-1} at a current density of 0.2 Ag^{-1} and pH = 2 and 209 Fg^{-1} at pH = 4. A high density of "Frenkel defect pair" provides higher intercalation sites for Na^+. Moreover, Mn^{3+} participates in the polaron hopping conduction and improves electrical conductivity (Gao et al. 2017).

The core-shell serves as an excellent electroactive material for supercapacitor applications (Fu et al. 2018). Oxygen vacancy (Ov) rich MnO_2@MnO_2 microspheres with a yolk@shell@shell configuration is prepared from MnO_2@MnO_2 by hydrogenation (Figure 3.5a). Hydrogenation of the microspheres generate sufficient oxygen vacancies into the system and enlarges the void thickness between yolk and shell. The specific capacitance values of MnO_2@MnO_2 measured at 1 Ag^{-1} is nearly 226.3 Fg^{-1}, whereas Ov-MnO_2@MnO_2 delivers 452.4 Fg^{-1} (Figure 3.5b). The oxygen-deficient core-shell still delivers 316.1 Fg^{-1} specific capacitance when the current density is at 50 Ag^{-1}. The cyclic performance measured at a current density of 10 Ag^{-1} shows that the oxygen-deficient microspheres retain 92.2% of its initial capacitance after 10,000 cycles (Figure 3.5c). The excellent supercapacitor performance of this material is because of its large surface

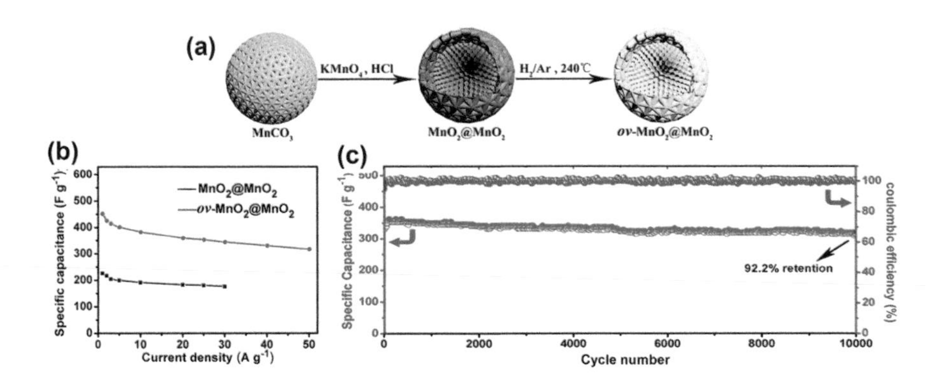

FIGURE 3.5 (a) Fabrication of oxygen vacancy rich MnO_2@MnO_2. (b) Measurement of specific capacitance values under different current densities (c) Stability test of Ov-MnO_2@MnO_2 for 10,000 cycles at a current density of 10 Ag^{-1}. (Fu, Y. et al., *J. Mater. Chem. A*, 6, 1601–1611, 2018. Reproduced by permission of The Royal Society of Chemistry.)

area providing sufficient electrode-electrolyte interface resulting in a shorter charge diffusion path; the yolk-shell structure provide more redox active sites; and lastly the void space between yolk and shell providing higher space for volume expansion/contraction during charge/discharge (Fu et al. 2018).

Co_3O_4 has the prominent oxidation state of +2 and +3. An increase in oxygen vacancy in the system can favor an increased Co^{2+}/Co^{3+} ratio. One simple method for fabrication of two-dimensional Co_3O_4 nanosheets is through the air calcination of $Co(OH)_2$ precursor at 300°C in air for 2 h. The reduction of the nanosheets is performed either through hydrogenation at 200°C for 2 h which results in H200-Co_3O_4 nanosheets, or by reducing with 1 M $NaBH_4$ (Xiang et al. 2017). Scanning electron microscope image shows the nanosheets of OVR-Co_3O_4 (Figure 3.6a).

The nanosheets of Co_3O_4 with oxygen vacancies display a higher electrical conductivity of 7.3×10^{-3} Sm^{-1} than that of pure Co_3O_4 (2.8×10^{-4} Sm^{-1}). X-ray photoelectron spectroscopy results demonstrate that OVR-Co_3O_4 has higher Co^{2+}/Co^{3+} ratio (1.70) than that of H200-Co_3O_4 (1.45) and pristine Co_3O_4 (1.07). The specific capacitance of OVR-Co_3O_4 at 1 Ag^{-1} is 2195 Fg^{-1}, which is 3.6 and 2.6 times higher than that of pristine Co_3O_4 and H200-Co_3O_4 (Figure 3.6b). There is good retention of 72.5% of

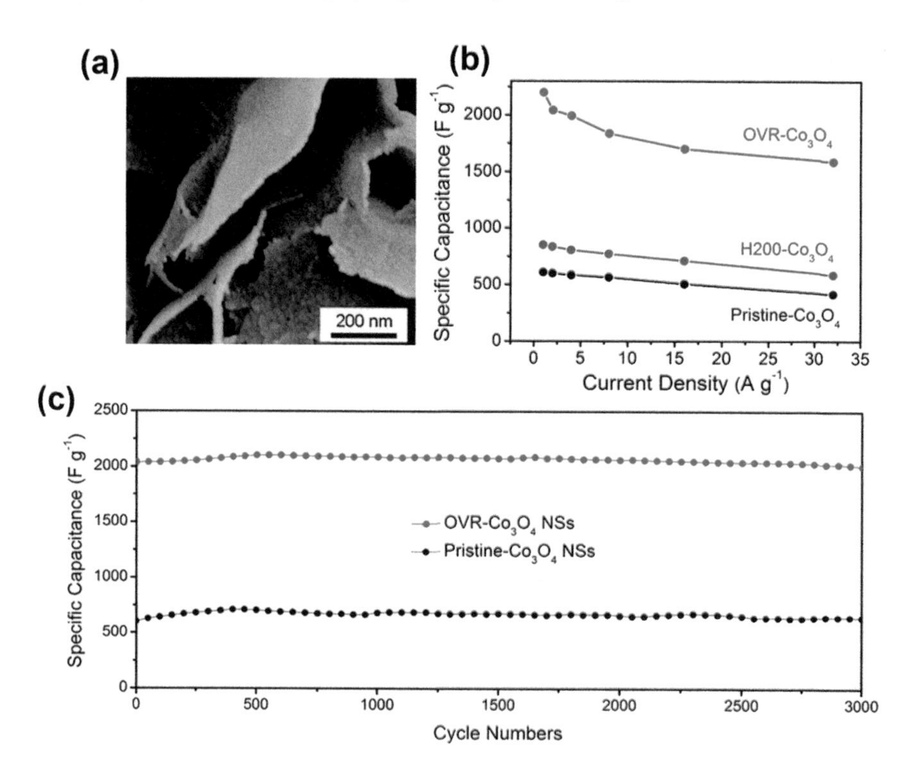

FIGURE 3.6 (a) Scanning electron microscope image of oxygen vacancy rich (OVR) Co_3O_4 nanosheet. (b) Specific capacitance values at an increasing current density. (c) Stability test of OVR-Co_3O_4 and pristine Co_3O_4 for 3000 cycles. (Xiang, K. et al., *Chem. Commun.*, 53, 12410–12413, 2017. Reproduced by permission of The Royal Society of Chemistry.)

initial capacitance as current density increases from 1 to 32 Ag^{-1} (Figure 3.6b). OVR-Co_3O_4 retains 95% of initial capacitance after 3000 cycles, whereas only 90% by pristine Co_3O_4, indicating long-term stability of OVR-Co_3O_4 (Figure 3.6c) (Xiang et al. 2017). These oxygen-deficient sites attract O_2^{2-}/O- species. Oxygen vacancies increase the Co^{2+} states near the oxygenated donors. The areal specific capacitance measured at a current density of 9 mA cm^{-2} is 1.58 F cm^{-2} (Cheng et al. 2017). This value drops to 1.36 F cm^{-2} at a current density of 60 mA cm^{-2}. There is 86% retention of initial capacitance indicating that the oxygen-deficient Co_3O_4 with surface functionalization provides less charge transfer resistance for ion diffusion. One study shows that an increase in grain boundary defects in Co_3O_4 can serve as redox active sites and provides diffusion paths for electrolytes leading to an enhanced supercapacitor performance (Hao et al. 2018). This is achieved by incorporating Pd in the reaction mixture of Co_3O_4 during the hydrothermal reaction. Incorporation of Pd hinders nanocrystals growth, increases surface disorder and oxygen vacancies on the grain boundary. The disordered structure has high electronic conductivity, abundant redox reaction sites, provides easy paths for the diffusion of electrolytes and ion/electron transfer. Cyclic voltammetry curves of Co_3O_4 and oxygen deficient-Co_3O_4 (Pd-Co_3O_4) are obtained at a scan rate of 10 mV s^{-1} in the potential range of 0–0.6 V (Figure 3.7a). The redox reaction peaks indicate the

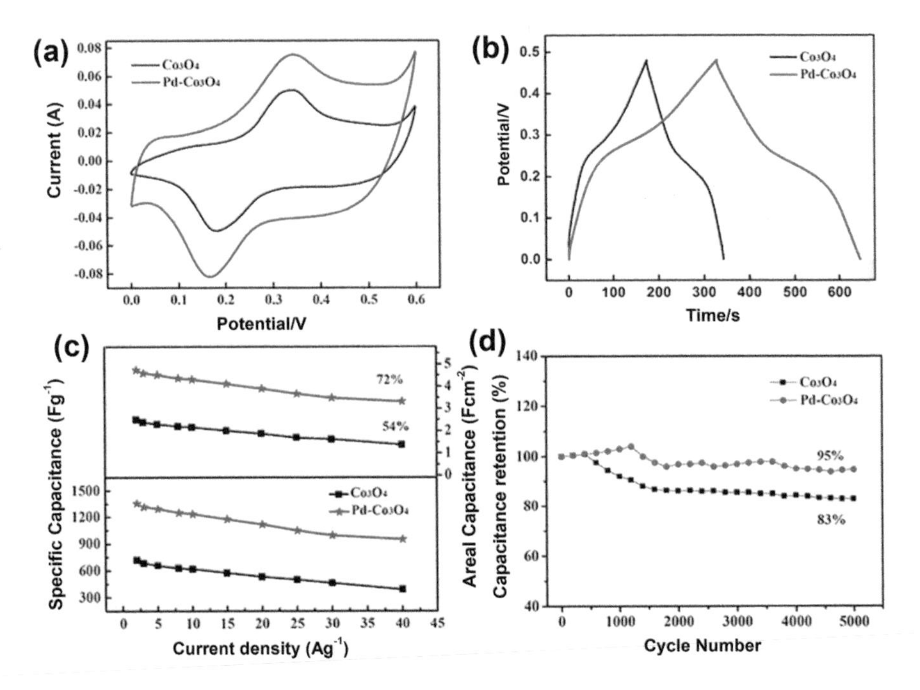

FIGURE 3.7 Comparative analysis of the electrochemical properties of Co_3O_4 and Ov-Co_3O_4 (Pd-Co_3O_4). (a) CV curves at a scan rate of 10 mV s^{-1}. (b) Galvanostatic charge/discharge spectra measured at a current density of 2 Ag^{-1} (7 mA cm^{-2}). (c) Specific and areal capacitance at different current densities. (d) Cycle performance recorded for 5000 cycles at a current density of 70 mA cm^{-2}. (Hao, J. et al., *J. Mater. Chem. A*, 6, 16094–16100, 2018. Reproduced by permission of The Royal Society of Chemistry.)

pseudocapacitive characteristic of the electrodes, which originates due to the Co^{2+}/Co^{3+} states in Co_3O_4. Oxygen deficient-Co_3O_4 has a higher current density and larger integrated area than pristine Co_3O_4 indicating good capacitive behavior in the former than the later.

The galvanostatic charge/discharge data of oxygen deficient-Co_3O_4 acquires symmetric shape and is much more extended than pristine Co_3O_4 (Figure 3.7b). The areal capacitance measurement at a current density of 10 Ag^{-1} shows that oxygen-deficient-Co_3O_4 delivers 4.31 F cm^{-2} and pure Co_3O_4 delivers only 2.2 F cm^{-2} (Figure 3.7c). There is 72% and 54% retention in the initial capacitance in oxygen-deficient-Co_3O_4 (or Pd-Co_3O_4) and Co_3O_4, respectively when the current density is 40 Ag^{-1}. The stability measurement shows that about 95% of initial capacitance is retained by oxygen-deficient-Co_3O_4, whereas only 83% is retained by pure Co_3O_4 after 5000 cycles measured at a current density of 70 mA cm^{-2} (Figure 3.7d) (Hao et al. 2018).

Thermal oxidation of corundum (c)-V_2O_3 in H_2/Ar at 300°C can generate a 3D core-shell structure of c-V_2O_3/r-VO_{2-x} with corundum (c)-V_2O_3 as the core and rutile (r) VO_{2-x} as the shell (Liu et al. 2018). The shell r-VO_{2-x} layer is sandwiched between conductive r-VO_2 slabs. The shell is composed of quasi-hexagonal oxygen vacancy tunnels improving the conductivity and facilitating Na^+ intercalation/deintercalation. The V-V bonds formation at the interface between V_2O_3 and r-VO_{2-x} provides an easy pathway for electron transport and provides a much lowered interfacial charge transfer resistance. The 3D core-shell films of c-V_2O_3/r-VO_{2-x} show excellent pseudocapacitive behavior in aqueous electrolyte. The volumetric capacitance delivered by c-V_2O_3/r-VO_{2-x} is 1933 F cm^{-3} at a scan rate of 5 mV s^{-1} and current density of 10.4 A cm^{-3}. With capacitance retention of 792 F cm^{-3} at 1000 mV s^{-1} scan rate, the core-shell structure outperforms the supercapacitor performance of r-VO_2 nanoparticle (37 F cm^{-3}). Long-term durability of the core-shell at current density 80 Ag^{-1} is confirmed by 90% capacitance retention after 10,000 cycles (Liu et al. 2018). Oxygen vacancies can induce semi-metallic characteristic in a semiconductor and semiconducting behavior in an insulator. SiO_2 behaves as an insulator and when excess oxygen defects are introduced it behave as a semiconductor. A high density of oxygen vacancies can reduce the effective distance by 2.5 nm between redox active sites for providing efficient charge transfer. These oxygen deficient SiO_2 shows supercapacitor performance of 337 Fg^{-1} at a current density of 1 Ag^{-1}. The bimodal porosity of SiO_2 with a high density of oxygen vacancies are responsible for the enhanced supercapacitor performance.

WO_3 shows excellent capacitive behavior due to the interconversion states of W^{6+} and W^{5+}. Hydrogenation can lead to the generation of oxygen vacancies into WO_{3-x} (Wang et al. 2018). Increase in the oxygen vacancies leads to a proportionate increase in the W^{5+}/W^{6+} ratio. These oxygen deficient WO_3 facilitates faster ion diffusion, improves conductivity, and increase charge transfer as well as reversible redox reaction rates. WO_{3-x} delivers 1.83 F cm^{-2} performance at a current density of 1 mA cm^{-2} while WO_3 shows only 1.15 F cm^{-2}. α-Fe_2O_3 is an excellent negative electrode material for asymmetric supercapacitor application (Wang et al. 2018). α-Fe_2O_3 is an excellent negative electrode material for asymmetric supercapacitor application. The poor electrical conductivity of α-Fe_2O_3 (10^{-14} S cm^{-1}) is much enhanced in the presence of oxygen vacancies. Oxygen deficient α-Fe_2O_3 nanorods is prepared by the thermal decomposition of FeOOH on a carbon cloth in a N_2 atmosphere (Lu et al. 2014). The oxygen vacancies act as shallow donors and increase the donor electron density from 5.5×10^{18} cm^{-3} in air annealed Fe_2O_3 to 8.1×10^{19} cm^{-3} in oxygen-deficient-Fe_2O_3. The areal capacitance of oxygen deficient-Fe_2O_3 is 382.7 mF cm^{-2} at a current density of 0.5 mA cm^{-2}. Oxygen deficient-Fe_2O_3 shows excellent stability with capacitance retention of 95.2% after 10000 cycles. One study demonstrates that amorphous α-Fe_2O_3 with structural disorder provides isotropic diffusion and percolation of ions. Thus, a coating of an amorphous layer over an α-Fe_2O_3 core could be an active supercapacitor material because of the faster ion diffusion through the open framework of the core-shell structure (Sun et al. 2018b). A crystalline core/amorphous shell interface with oxygen vacancies on the shell can improve electrical conductivity and provide faster ion transfer. The areal capacitance displayed by oxygen-deficient core/shell α-Fe_2O_3 and oxygen deficient α-Fe_2O_3 at a current density of 1 mA cm^{-2} are 350 and 168 mF cm^{-2}, respectively. This is because a disordered amorphous structure could provide more percolation paths for ion diffusion (Sun et al. 2018).

NiO is another electroactive material with a theoretical specific capacitance value of 2584 Fg^{-1} (Li et al. 2018). In order to establish an efficient electrical contact between electroactive material and a current collector, hexagonal NiO nanoplates enriched with oxygen vacancies are fabricated in an in-situ method over Ni foam which acts as a current collector. The method involves hydrothermal treatment of a Ni foam in an aqueous H_2O_2 solution followed by annealing in N_2 at 400°C for 1 h. Because of the generation of oxygen vacancies, the surface coordination of Ni atoms surrounding the vacancy decrease leading to enhanced surface reactivity. The recorded values of specific capacitance at a scan rate of 1 mV s^{-1} are 2495 Fg^{-1} (defective-NiO) and 1867 Fg^{-1} (defect free-NiO) (Li et al. 2018). The enhanced performance of defective-NiO is due to increased donor density, electrical conductivity, and presence of redox active sites.

In comparison to single-component metal oxides, bimetallic metal oxides such as $NiMoO_4$, $CoMoO_4$ show excellent supercapacitor applications. The specific capacitance values of various forms of $CoMoO_4$, such as nanorods, nanowires, and micro-rhombohedra are measured (Dam et al. 2017). Figure 3.8a shows the diameter of the semi-circular loop for nanorods, nanowires, and micro-rhombohedra, which are measured to be 1.7, 1.8, and 4.0 Ω, respectively. These values imply a fast redox reaction for the nanorods and accelerated charge transport property than that of the other nanostructures. The stability test of the samples at 6 Ag^{-1} current density for 2000 cycles has shown that nanorods show excellent retention of the initial capacitance followed by nanowires and micro-rhombohedron (Figure 3.8b). The specific capacitance of the various forms of $CoMoO_4$ is measured at different current densities (Figure 3.8c). A comparison of the specific capacitance at a current density of 1 Ag^{-1} reveals that the nanorods of $CoMoO_4$ deliver higher specific capacitance (420 Fg^{-1}) than that of nanowires (362 Fg^{-1}) and micro-rhombohedra (263 Fg^{-1}) at a current density of 1 Ag^{-1} (Figure 3.8c) (Dam et al. 2017). The reason is ascribed to the presence of defects and dislocated atoms on the nanorods which facilitate a fast redox reaction. The measured energy density for nanorods is 14.6 Wh kg^{-1} at a power density of 250 W kg^{-1} (Figure 3.8d).

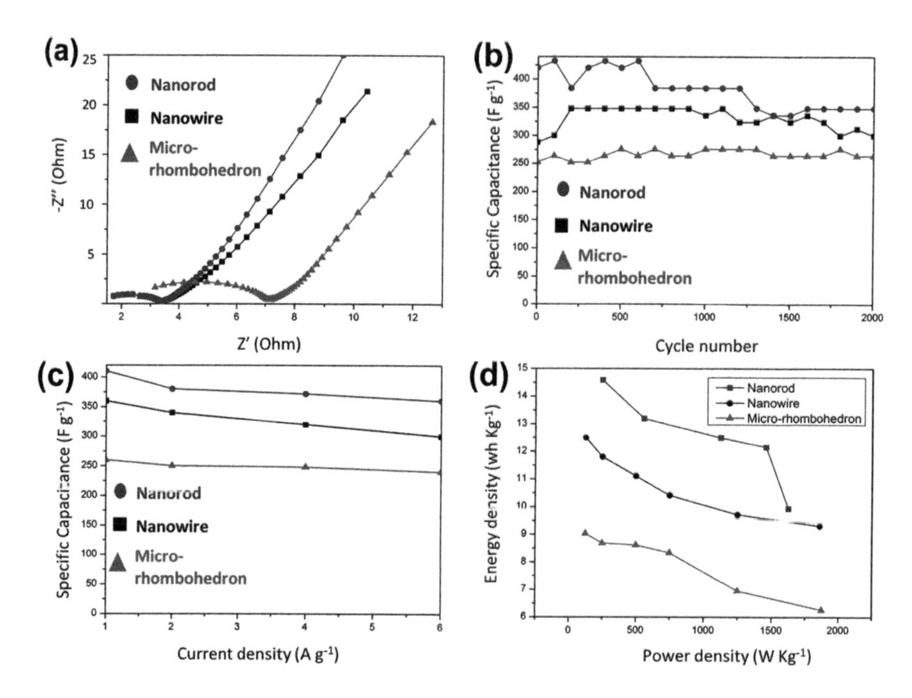

FIGURE 3.8 Comparison of the various electrochemical properties of nanorods, nanowires, and micro-rhombohedra of $CoMoO_4$. (a) Complex-plane impedance plots; (b) stability test for 2000 cycles at a current density of 6 Ag^{-1}; (c) specific capacitance as a function of current densities; (d) energy density *vs.* power density of the different forms of $CoMoO_4$ nanorods. (Dam, D.T. et al., *Sustainable Energy Fuels*, 1, 324–335, 2017. Reproduced by permission of The Royal Society of Chemistry.)

$NiMoO_4$ nanoflake and nanowires are synthesized by the hydrothermal method following hydrogenation process to enrich the nanostructures with oxygen defects (Qing et al. 2018). Defect assists the change in the oxidation state from Mo(+6) to Mo(+4). Pristine nanoflake and nanowires show areal capacitances of 1.075 and 0.909 F cm^{-2} at a current density of 1 mA cm^{-2}. On the other hand, vacancy rich nanoflake and nanowires show areal capacitances of 2.718 and 2.331 F cm^{-2}, respectively. The stability of the hydrogenated samples are measured at a current density of 5 mA cm^{-2}. It is revealed that nearly 94% of initial capacitance is retained after 6000 cycles. Oxygen vacancies lower the diffusion resistance for faster diffusion of electrolyte and improve the ionic conductivity (Qing et al. 2018).

Spinels such as $NiCo_2O_4$ are excellent materials for supercapacitor applications. A theoretical study shows that oxygen vacancy formation is easier near a Ni atom than near a Co atom (Shi et al. 2017). This leads to a conclusion that oxygen vacancy formation is easier over $NiCo_2O_4$ surface than over Co_3O_4. Moreover, the fractional change of valency Ni^{2+}/Ni^{3+} and Co^{2+}/Co^{3+} provides more redox active sites. Furthermore, it is observed that oxygen vacancies in $NiCo_2O_4$ nanosheets and $NiCo_2O_4$ nanowires can help the transformation of Co^{3+}/Co^{2+} and Ni^{3+}/Ni^{2+} (Yan et al. 2018). Oxygen defects can destroy parts of Co-O and Ni-O bonds and degrades the crystallinity of the nanosheets and nanowires. These oxygen vacancies boost the transportation of ions/electrons and provide redox active sites for acceleration of redox reactions (Sun et al. 2018). Pure nanosheets and nanowires exhibit a specific capacitance of 898 and 610 Fg^{-1} at the current density of 1 Ag^{-1}. The reduced nanosheets and nanowires show capacitance of 1590 and 1248 Fg^{-1}, respectively. An amorphous double-shell hollow spheres of $NiCo_2O_4$ show capacitance value much higher than single-shell $NiCo_2O_4$ (Li et al. 2015a). The single shell and double-shell show capacitance values of 445 and 568 Fg^{-1} at the current density of 1 Ag^{-1}. It is further stated that hydrogenation of the double-shell spheres can increase the specific capacitance from 568 to 718 Fg^{-1} at the current density of 1 Ag^{-1}. The oxygen defect density is likely to be much higher in the amorphous double-shell hollow spheres than that of the single shell spheres. The oxygen vacancies help in the multivalent states in the spheres and improves the electronic conductivity (Li et al. 2015a). Three-dimensional $NiCo_2O_4$ microspheres containing radial chain like $NiCo_2O_4$ nanowires are synthesized via hydrothermal method (Zou et al. 2013). These nanowires are constructed of small nanocrystals. These defect enriched small nanocrystals have different exposed reactive sites, which promote facile ion transport. The chain-like nanowires provide easy transport of ions and accelerate redox reaction. The measured specific capacitances of the microspheres at 2 and 20 Ag^{-1} are 1284 and 986 Fg^{-1}. Thus, the microspheres show excellent stability with only 23% capacitance loss at a large current density (Zou et al. 2013).

3.3 DEFECT ENGINEERED METAL CHALCOGENIDES

Transition metal dichalcogenides, including metal sulphides and metal selenides, have fascinated immense attention for several applications due to their high electrical conductivity, good thermal stability, earth abundance, etc. Among the large volume of works available on transition metal dichalcogenides, our discussion mainly focuses on MoS_2, WS_2, VS_2, highlighting their defect engineered features on the charge storage capabilities for supercapacitor applications.

Surface defects with unsaturated sulfur atoms in the MoS_2 ultrathin nanosheets provide redox active sites for the intercalation of ions for an improved supercapacitor performance. The structural defects and surface disorder helps in reducing the surface energy and improves the stability of the nanosheets. intercalate strongly with the small ions (Wu et al. 2015). Ultrathin nanosheets of defect-rich MoS_2 containing unsaturated sulfur atoms are synthesized using an appropriate ratio of Mo(VI) and L-cysteine with 1,6-hexanediamine as the chelating agent. A synergetic performance of L-cysteine and 1,6-hexanediamine help in the formation of the defect-rich MoS_2 ultrathin nanosheets. In the initiation of the reaction, a linkage is established between L-cysteine and Mo(VI). L-cysteine reduces Mo(VI) to Mo(IV) and produces primary MoS_2 nanocrystallites. 1,6-Hexanediamine attach to the surface of MoS_2 nanocrystallite and hinders its growth leading to a defect enriched MoS_2 nanosheets (Wu et al. 2015) (Figure 3.9).

The capacitive performances of the ultrathin defective MoS_2 nanosheets is investigated using cyclic voltammetry in 1 M KCl solution (Figure 3.10a).

The curves do not display any redox peak. This is an indication of the typical electrical double-layer capacitance behavior of the electrode. Figure 3.10b shows the variation

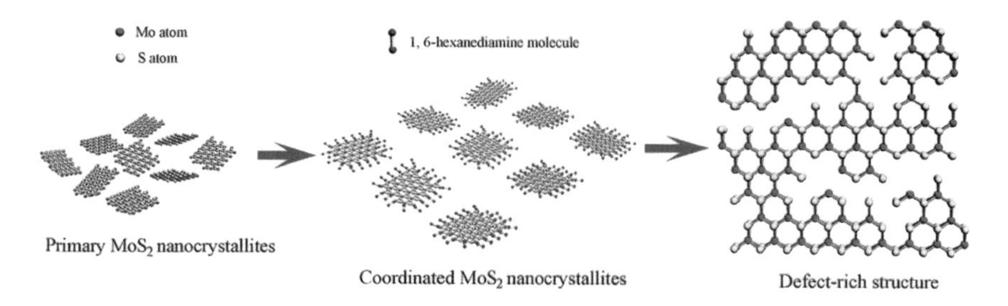

FIGURE 3.9 Schematic showing the reaction pathway for the fabrication of defective MoS_2 nanosheets. (Wu, Z. et al., *J. Mater. Chem. A*, 3, 19445–19454, 2015. Reproduced by permission of The Royal Society of Chemistry.

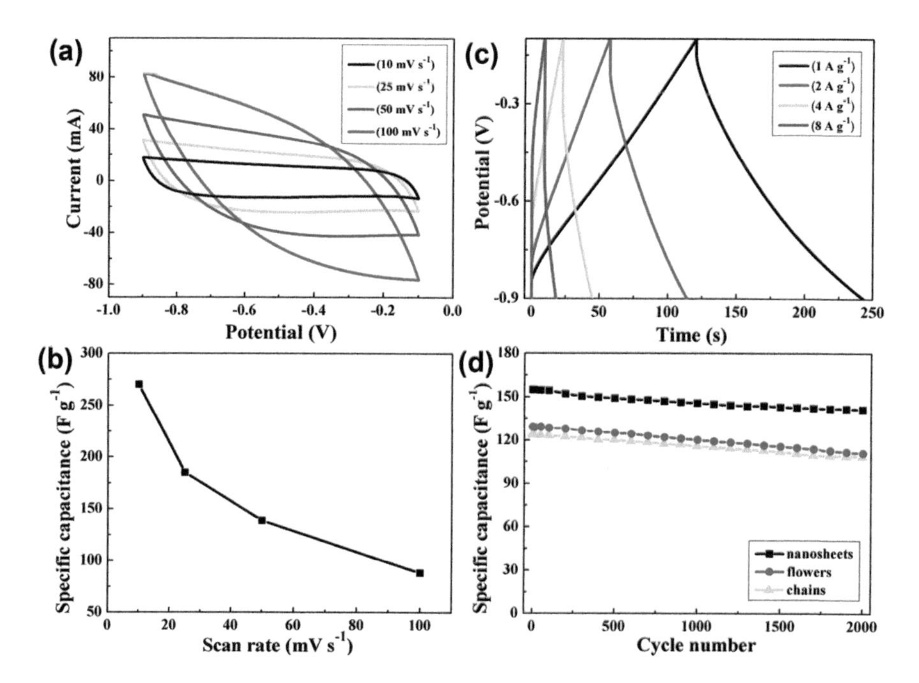

FIGURE 3.10 Supercapacitor performance of defect enriched MoS_2 nanostructures. (a) CVs of the electrode at various scan rates; (b) values of specific capacitance as a function of scan rate; (c) galvanostatic charge/discharge curves with variation of current densities; and (d) cyclic performance recorded at a current density of 1 Ag^{-1}. (Wu, Z. et al., *J. Mater. Chem. A*, 3, 19445–19454, 2015. Reproduced by permission of The Royal Society of Chemistry.)

of the specific capacitance at various scan rates. This suggest that the specific capacitance of the defective MoS_2 electrode increases as scan rate decreases. When the scan rate is low, the interlayer of MoS_2 provides less resistance for the ion migration and therefore, the ions reach more active sites for efficient charge-transfer reactions. At a scan rate of 10 mVs^{-1}, the maximum specific capacitance recorded for the electrode is 270.3 g^{-1}. Galvanostatic charge/discharge profile of the defective MoS_2 electrode is measured by varying the current densities (Figure 3.10c). The curves show a nearly symmetric and triangular shapes. The internal resistance of the electrode material is the reason for the pronounced potential drop at the beginning of the discharge process. The measured specific capacitances of the ultrathin defective MoS_2 nanosheet are 154.9, 143.5, 125.5, and 92.0 Fg^{-1}. The measurement is performed at current densities of 1, 2, 4 and 8 Ag^{-1}, respectively, with a relatively good retention ratio of the specific capacitances at current densities of 1 and 2 Ag^{-1}. The stability test of the electrode is performed for 2000 cycles at a current charge/discharge between −0.9 and −0.1 V and at a current density of 1 Ag^{-1}. After 2000 cycles, the specific capacitance remained as high as 141.3 $Fg^{-1}$with a retention up to 91.2% (Figure 3.10d). The measured value is 111.0 Fg^{-1} for the flower type electrode with a capacitance retention of 86.1%; for the chain electrode the capacitance and the retention value are 108.5 Fg^{-1} and 87.6%, respectively. The supercapacitor performance of the defective MoS_2 electrode was analyzed in various electrolytes (Na_2SO_4, LiOH, KCl and KOH). The activity test shows that the defective MoS_2 electrode can adsorb and intercalate different types of ions. Finally, the enhanced electrochemical performance of the defective MoS_2 nanosheet is due to the presence of defects on the basal planes of the ultrathin nanosheets. The defective structure provides enhanced structural stability and more active sites for fast electron transport (Wu et al. 2015).

A hydrothermal route of preparation can also result into a defect engineered metallic 1T MoS_2 (Figure 3.11).

Numerous active edge sites are exposed through this reaction pathway. To produce a defect-rich structure, excess thiourea was employed as a reductant to reduce Mo (VI) to Mo(IV). Thiourea provides sulphur source for the formation of MoS_2 during hydrothermal reaction and an effective additive to stabilize the ultrathin nanosheet morphology (Xie et al. 2013). An excess thiourea can hinder the oriented crystal growth by covering the surface of primary nanocrystallites leading to a defect enriched MoS_2 structures.

Defect enriched 1T MoS_2 nanosheets with active edges were utilized for high-performance supercapacitor electrodes. 1 T MoS_2 is highly conducting in nature and contains several active edges. This enables enhanced electrical conductivity, a smooth charge transfer process and an easy intercalation of ions with the accessible active sites. This defect enriched 1T MoS_2 can be a potential candidate for high performance supercapacitor electrode application (Joseph et al.2018). The cyclic voltammetry (CV) is conducted in a 1 M KOH electrolyte solution using a three electrode system. Figure 3.12a shows the comparative CV curves for both high-defect content-MoS_2 and low-defect content-MoS_2. The scan rate is 10 mV s^{-1} with a potential sweep of 0 to 0.55 V. The oxidation-reduction peaks indicate the pseudocapacitive nature of the electrode material. The increase in the current density for high defect content-MoS_2 is because of the presence of active exposed molybdenum edge atoms for the redox reaction. The galvanostatic charge/discharge (GCD) curves (Figure 3.12b) for high defect content-MoS_2 and low defect content-MoS_2 is measured at current density 1 Ag^{-1}. The specific capacitance recorded for high defect content MoS_2 and low defect content MoS_2 are 379 and 270 Fg^{-1}, respectively (Joseph et al. 2018).

The relatively lower conductivity and the specific capacitance of low-defect content-MoS_2 nanosheet is due to the presence of 2H phase and the limited access to the available active sites. The excellent electrochemical performance of highly defect content MoS_2 is because of the presence of molybdenum edges. The molybdenum edges improve the

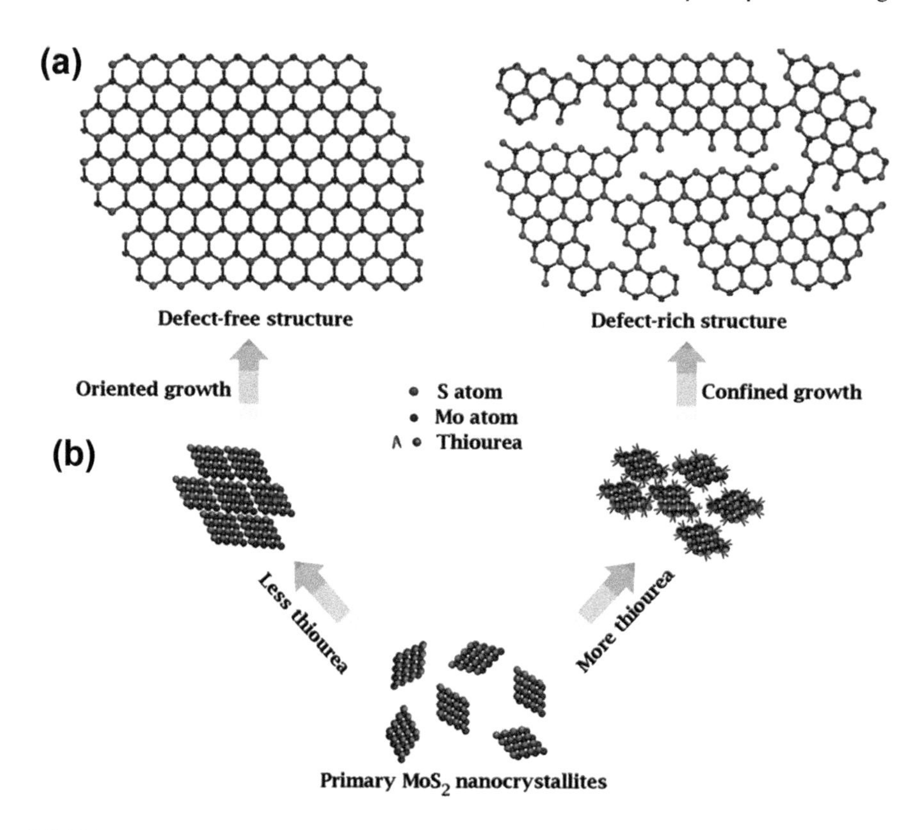

FIGURE 3.11 (a) Defect-free and defect-rich structural models of MoS_2 nanosheets. (b) The synthetic pathways to obtain the above two structures. (From Xie, J. et al.: Defect-rich MoS_2 ultrathin nanosheets with additional active edge sites for enhanced electrocatalytic hydrogen evolution. *Advanced Material*, 2013. 25. 5807–5813. Copyright Wiley-VCH Verlag GmbH & Co. KGaA. Reproduced with permission.)

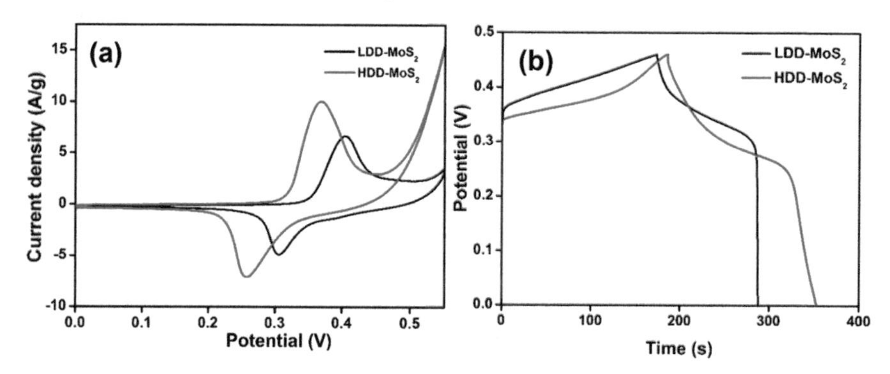

FIGURE 3.12 Electrochemical properties of high defect content-MoS_2 and low defect content-MoS_2 in the 1 M KOH electrolyte. (a) A comparison of the CV performances of the two types of electrodes at a scan rate of 10 mV s^{-1}. (b) Galvanostatic charge/discharge curves of the MoS_2 electrodes at 1 Ag^{-1} current density. (Joseph, N. et al., *New J. Chem.*, 42, 12082–12090, 2018. Reproduced by permission of The Royal Society of Chemistry.)

electrical conductivity and contribute redox active sites to the electrode. These results indicate that engineering the defect density on the 1T MoS_2 material can induce drastic improvement in the electrochemical energy storage mechanism (Joseph et al. 2018).

An electrode material with high surface area, high density of active sites, and available edges for the adsorption/desorption of ions is beneficial for designing excellent charge storage devices. Quantum dot-based devices can show an enhanced performance as supercapacitor devices due to the high surface areas, available edges and active sites for adsorption and desorption (Ghorai et al. 2018). Flexible solid-state supercapacitor devices were prepared using WS_2 nanosheets and quantum dots. Both the nanosheets and quantum dots of WS_2 were synthesized via a lithium bromide-assisted lithium intercalation and sonication method.

Defect-rich quantum dots of size range from 1 to 3 nm are formed by disruption of WS_2 nanosheets. The electrochemical performance of the electrode with WS_2 quantum dots and nanosheets as the active materials have been evaluated by cyclic voltammetry (CV). They were blended separately with acetylene black to enhance the conductivity of the energy storage materials. CV of the devices made of WS_2 quantum dots at different scan rates are depicted in Figure 3.13a. The performance is compared

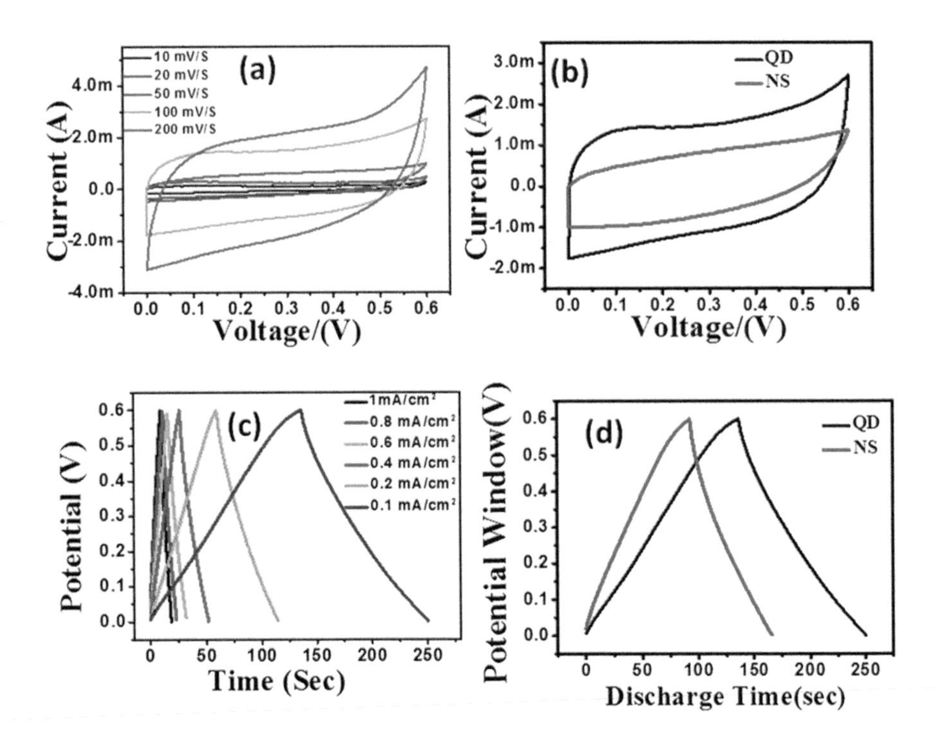

FIGURE 3.13 (a) Typical CVs of WS_2 quantum dots at different scan rates; (b) CVs of WS_2 quantum dots and nanosheets based supercapacitors at a scan rate of 100 mV s^{-1}; (c) cyclic test of WS_2 quantum dots at different current densities; (d) comparative charge–discharge profiles of WS_2 nanosheets and quantum dots. Quantum dot devices show a longer discharge time. (Ghorai, A. Et al., *New J. Chem.*, 42, 3609–3613, 2018. Reproduced by permission of The Royal Society of Chemistry.)

with the nanosheet-based device in Figure 3.13b. The specific capacitance value for the quantum dots is measured to be 22 and 13 mF cm^{-2} for nanosheet-based devices at a scan rate of 100 mV s^{-1}, respectively (Ghorai et al. 2018). The high specific surface area, available surface active sites and open edges provides numerous sites for strong adsorption-desorption of ions resulting in a higher specific capacitance in the quantum dots-based devices than a nanosheets-based devices. Cyclic charge-discharge curves at various current densities for quantum dots (Figure 3.13c), and a comparison of the charging-discharging comparison between dots and nanosheets are shown in Figure 3.13d, respectively. The nearly triangular shape of the charge-discharge measurement indicates an ideal capacitive behavior (Ghorai et al. 2018). The mechanical flexibility for a flexible solid-state supercapacitor is demonstrated at different bending angles, with no change in CV and capacitance is observed, as shown in Figure 3.14a (Ghorai et al. 2018). Cycle stability of the devices using WS_2 quantum dots and nanosheet-based materials is measured by running the experiments up to 10000 cycles (Figure 3.14b). The specific capacitance retention is nearly 80% after 10,000 cycles for the WS_2 quantum dots-based devices. Figure 3.14c shows the comparative Ragone plot of WS_2 quantum dots- and nanosheet-based devices

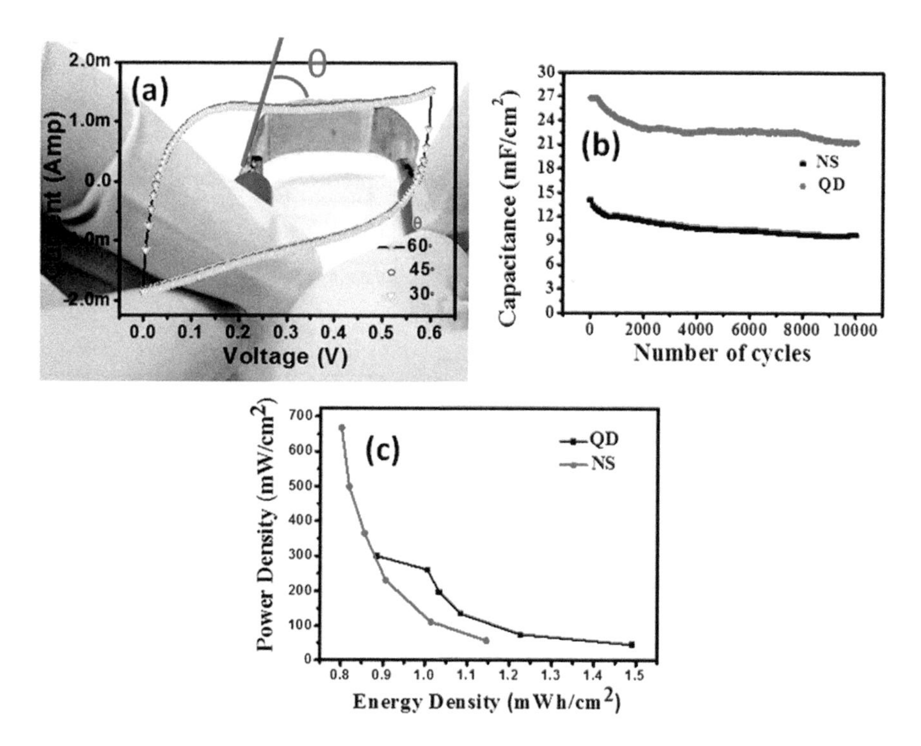

FIGURE 3.14 (a) CV of flexible WS_2 quantum dot based devices at different bending angles. The scan rate is 100 mV s^{-1}; (b) performance test of the devices after 10,000th cycle; (c) comparative Ragone plot of WS_2 quantum dot and nanosheet based devices. (Ghorai, A. et al., *New J. Chem.*, 42, 3609–3613, 2018. Reproduced by permission of The Royal Society of Chemistry.)

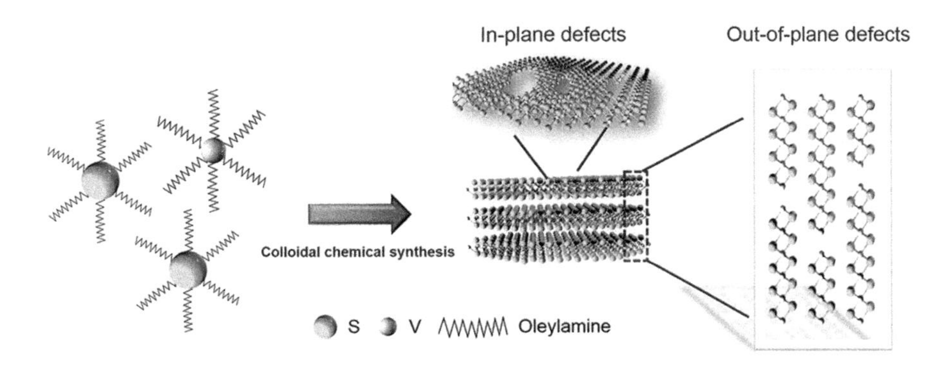

FIGURE 3.15 Schematic showing the formation of defect rich VS_2 nanoplate. (Guo, Z. et al., *J. Mater. Chem. A*, 6, 14681–14688, 2018. Reproduced by permission of The Royal Society of Chemistry.)

indicating a higher areal specific power density (47.05 mW cm^{-2}) in the quantum dots (Ghorai et al. 2018).

An ultrathin VS_2 nanoplate containing in-plane and out-of-plane defects were prepared by a simple colloidal method (Figure 3.15). The ultrathin structure and rich defect content with exposed active sites make VS_2 an ideal electrode material for enhanced redox reactions at the electrode/electrolyte interface (Guo et al. 2018). The prepared VS_2 nanoplates were used in a three-electrode geometry for supercapacitor performance.

Cyclic voltammetry (CV) curves (Figure 3.16a) of the VS_2 nanoplates is obtained at various scanning rates in 1M KOH aqueous solution. The CV exhibit a pair of redox peaks at the scan rate of 5 mV s^{-1}, indicating that the capacitance can be determined by faradaic redox reactions. Galvanostatic charge-discharge curves (GCD) of the defect-rich VS_2 nanoplates at various current density are shown in Figure 3.16b. The measured specific capacitances at the current densities of 1, 2, 5, and 10 Ag^{-1}, are 2200, 1945, 1660 and 1275 Fg^{-1}, respectively (Figure 3.16c) (Guo et al. 2018). The ultrahigh specific capacitance value for rich-defect VS_2 nanoplates (2200 Fg^{-1} at a current density of 1 Ag^{-1}) is recognized as one of the best among all reported value. The conductivity of the VS_2 electrode and the ion migration rate were measured by electrochemical impedance spectroscopy (EIS) in the frequency range from 0.01 to 100 KHz in a three-electrode configuration. From the Nyquist plots in Figure 3.16d, the equivalent series resistance is about 0.81 Ω at high frequency for the electrode and the electrolyte. The charge transfer resistance at the electrode material/electrolyte interface is about 0.49 Ω.

Compared to defect-free VS_2 nanoplates, improved conductivity of defect-rich VS_2 is due to the more number of defects and standing edges in the rich-defect nanoplates, facilitating the electron transfer process in the redox reaction. Subsequently, fabrication of asymmetric super-capacitor (ASC) using the as-prepared rich-defect VS_2 nanoplates as the anode material were also achieved with excellent energy density (66.54 Wh kg^{-1} at a power density of 0.75 kW kg^{-1}), power density (6.62 kW kg^{-1} at an energy density of 31.25 Wh kg^{-1}), and long-life cycling of (5000). These rich defects can expose number of active sites and provide the pathway of the electrolyte

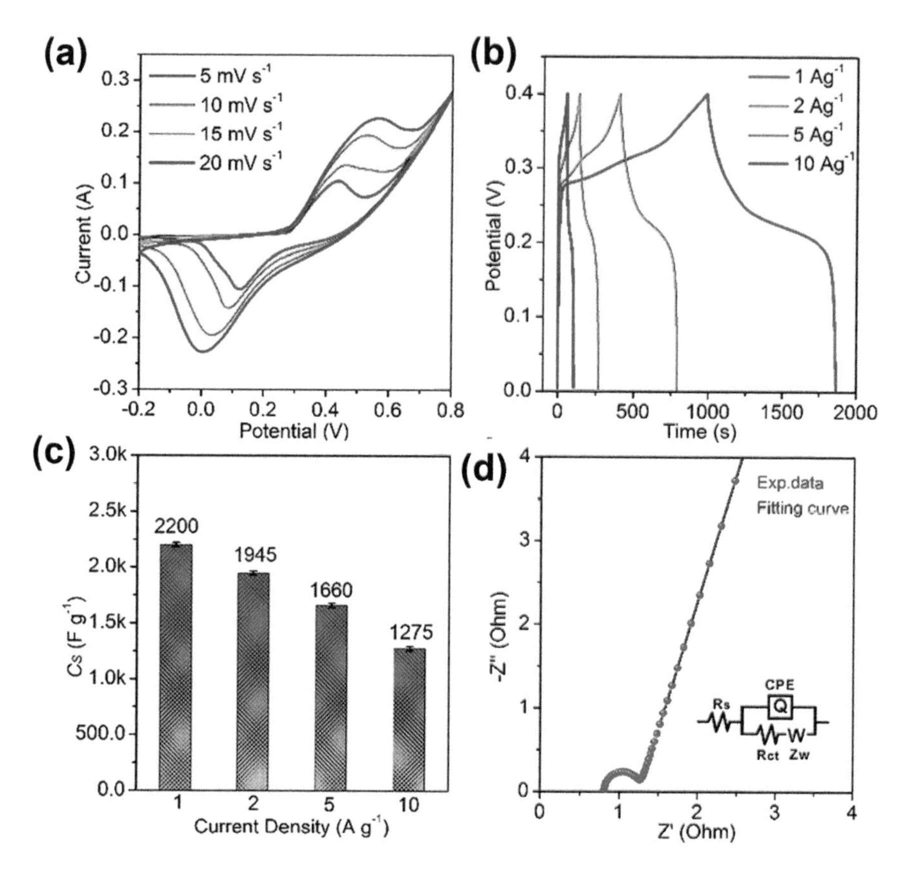

FIGURE 3.16 Supercapacitor tests of VS_2 nanoplates in a three electrode cell. (a) CV as a function of scan rates. (b) Galvanostatic charge–discharge curves as a function of current density. (c) Specific capacitance of VS_2 nanoplates at different current density. (d) Nyquist plots of VS_2 nanoplates (inset is the equivalent circuit). (Guo, Z. et al., *J. Mater. Chem. A*, 6, 14681–14688, 2018. Reproduced by permission of The Royal Society of Chemistry.)

ion, which confirms the excellent performance of VS_2 nanoplates as electrochemical energy storage anode materials (Guo et al. 2018).

3.4 SUMMARY

In this chapter, the importance of defects on the supercapacitor performance of metal oxides and metal chalcogenide have been thoroughly discussed. There are a variety of methods by which a pristine metal oxide can be enriched with defects. Some of the widely adopted strategies for defect engineering are electrochemical reduction, hydrogenation and vacuum annealing, and hydrogen plasma treatment. The benefit of defect engineering is the increase in the surface area, donor electron density, multiple valence states and numerous redox active sites, and easy transportation of ions. The morphology of the defect-engineered metal oxides also affect the supercapacitor performance. Nanotube arrays and two-dimensional nanosheets processes a high surface area and numerous redox active sites. The occurrence of defects in the nanostructures of transition metal dichalcogenides leads to the exposure of active edge sites for excellent performance in a supercapacitor. Presence of defects induces additional active edge sites, leading to an excellent supercapacitor activity in transition metal dichalcogenides. The success of defect engineering of supercapacitor electrode may offer an effective platform for an improvement of energy storage application.

REFERENCES

Bissett, M. A., S. D. Worrall, I. A. Kinloch, et al. 2016. Comparison of two-dimensional transition metal dichalcogenides for electrochemical supercapacitors. *Electrochim. Acta* 201:30–37.

Cao, X. Y., X. Xing, N. Zhang, et al. 2015. Quantitative investigation on the effect of hydrogenation on the performance of MnO_2/H-TiO_2 composite electrodes for supercapacitors. *J. Mater. Chem. A* 3:3785–93.

Cheng, G., T. Kou, J. Zhang, et al. 2017. O_2^{2-}/O^- functionalized oxygen-deficient Co_3O_4 nanorods as high performance supercapacitor electrodes and electrocatalysts towards water splitting. *Nano Energy* 38:155–66.

Choi, W., N. Choudhary, G. H. Han, et al. 2017. Recent development of two-dimensional transition metal dichalcogenides and their applications. *Mater. Today* 20:116–30.

Choudhury, B., and A. Choudhury. 2014. Oxygen defect dependent variation of band gap, Urbach energy and luminescence property of anatase, anatase–rutile mixed phase and of rutile phases of TiO_2 nanoparticles. *Physica E* 56:364–71.

Choudhury, B., S. Bayan, A. Choudhury, et al. 2016. Narrowing of band gap and effective charge carrier separation in oxygen deficient TiO_2 nanotubes with improved visible light photocatalytic activity. *J. Colloid Interface Sci.* 465:1–10.

Dam, D. T., T. Huang, and J.-M. Lee. 2017. Ultra-small and low crystalline $CoMoO_4$ nanorods for electrochemical capacitors. *Sustainable Energy Fuels* 1:324–35.

Devaraj, S., and N. Munichandraiah. 2008. Effect of crystallographic structure of MnO_2 on its electrochemical capacitance properties. *J. Phys. Chem. C* 112:4406–17.

Ding, X., F. Peng, J. Zhou, et al. 2019. Defect engineered bioactive transition metals dichalcogenides quantum dots. *Nat. Commun.* 10:41.

Fu, Y., X. Gao, D. Zha, et al. 2018. Yolk–shell-structured MnO_2 microspheres with oxygen vacancies for high-performance supercapacitors. *J. Mater. Chem. A* 6:1601–11.

Gao, P., P. Metz, T. Hey, et al. 2017. The critical role of point defects in improving the specific capacitance of δ-MnO_2 nanosheets. *Nat. Commun.* 8:14559.

Ghorai, A., A. Midya, and S. K. Ray. 2018. Superior charge storage performance of WS_2 quantum dots in a flexible solid state supercapacitor. *New J. Chem.* 42:3609–13.

Guo, Z., L. Yang, W. Wang, et al. 2018. Ultrathin VS_2 nanoplate with in-plane and out-of-plane defects for an electrochemical supercapacitor with ultrahigh specific capacitance. *J. Mater. Chem. A* 6:14681–88.

Hao, J., S. Peng, H. Li, et al. 2018. A low crystallinity oxygen-vacancy-rich Co_3O_4 cathode for high-performance flexible asymmetric supercapacitors. *J. Mater. Chem. A* 6:16094–100.

Heine, T. 2015. Transition metal chalcogenides: Ultrathin inorganic materials with tunable electronic properties. *Acc. Chem. Res.* 48:65–72.

Hua, X., Z. Liu, P. G. Bruce, et al. 2015. The morphology of TiO_2(B) nanoparticles. *J. Am. Chem. Soc.* 137:13612–23.

Jacobs-Gedrim, R. B., M. Shanmugam, N. Jain, et al. 2014. Extraordinary photoresponse in two-dimensional In_2Se_3 nanosheets. *ACS Nano* 8:514–21.

Joseph, N., P. Muhammed Shafi, and A. Chandra Bose. 2018. Metallic 1T-MoS_2 with defect induced additional active edges for high performance supercapacitor application. *New J. Chem.* 42:12082–90.

Li, L., J. Zhang, J. Lei, et al. 2018. O^- Vacancy-enriched NiO hexagonal platelets fabricated on Ni foam as a self-supported electrode for extraordinary pseudocapacitance. *J. Mater. Chem. A* 6:7099–106.

Li, X., L. Jiang, C. Zhou, et al. 2015a. Integrating large specific surface area and high conductivity in hydrogenated $NiCo_2O_4$ double-shell hollow spheres to improve supercapacitors. *Npg Asia Mater.* 7:e165.

Li, Z., S. Cong, and Y. Xu. 2014. Brookite vs anatase TiO_2 in the photocatalytic activity for organic degradation in water. *ACS Catal.* 4:3273–80.

Li, Z., Y. Ding, W. Kang, et al. 2015b. Reduction mechanism and capacitive properties of highly electrochemically reduced TiO_2 nanotube arrays. *Electrochim. Acta* 161:40–47.

Liu, B.-T., X.-M. Shi, X.-Y. Lang, et al. 2018. Extraordinary pseudocapacitive energy storage triggered by phase transformation in hierarchical vanadium oxides. *Nat. Commun.* 9:1375.

Liu, X. Y., T. C. Gong, J. Zhang, et al. 2019. Engineering hydrogenated manganese dioxide nanostructures for high-performance supercapacitors. *J. Colloid Interface Sci.* 537:661–70.

Lu, X., G. Wang, T. Zhai, et al. 2012. Hydrogenated TiO_2 Nanotube Arrays for Supercapacitors. *Nano Lett.* 12:1690–96.

Lu, X., Y. Zeng, M. Yu, et al. 2014. Oxygen-deficient hematite nanorods as high-performance and novel negative electrodes for flexible asymmetric supercapacitors. *Adv. Mater.* 26:3148–55.

Manzeli, S., D. Ovchinnikov, D. Pasquier, et al. 2017. 2D transition metal dichalcogenides. *Nat. Rev. Mater.* 2:17033.

Pan, D., H. Huang, X. Wang, et al. 2014. C-axis preferentially oriented and fully activated TiO_2 nanotube arrays for lithium ion batteries and supercapacitors. *J. Mater. Chem. A* 2:11454–64.

Qing, C., C. Yang, M. Chen, et al. 2018. Design of oxygen-deficient $NiMoO_4$ nanoflake and nanorod arrays with enhanced supercapacitive performance. *Chem. Eng. J.* 354:182–90.

Rasmussen, F. A., and K. S. Thygesen. 2015. Computational 2D materials database: Electronic structure of transition-metal dichalcogenides and oxides. *J. Phys. Chem. C* 119:13169–83.

Salari, M., K. Konstantinov, and H. K. Liu. 2011. Enhancement of the capacitance in TiO_2 nanotubes through controlled introduction of oxygen vacancies. *J. Mater. Chem.* 21:5128–33.

Salari, M., S. H. Aboutalebi, A. T. Chidembo, et al. 2012. Enhancement of the electrochemical capacitance of TiO_2 nanotube arrays through controlled phase transformation of anatase to rutile. *Phys. Chem. Chem. Phys.* 14:4770–79.

Shi, F., L. Li, X.-l. Wang, et al. 2014. Metal oxide/hydroxide-based materials for supercapacitors. *RSC Adv.* 4:41910–21.

Shi, X., S. L. Bernasek, and A. Selloni. 2017. Oxygen deficiency and reactivity of spinel $NiCo_2O_4$ (001) surfaces. *J. Phys. Chem. C* 121:3929–37.

Simon, P., and Y. Gogotsi. 2008. Materials for electrochemical capacitors. *Nat. Mater.* 7:845.

Sun, D., Y. Li, X. Cheng, et al. 2018a. Efficient utilization of oxygen-vacancies-enabled $NiCo_2O_4$ electrode for high-performance asymmetric supercapacitor. *Electrochim. Acta* 279:269–78.

Sun, S., T. Zhai, C. Liang, et al. 2018b. Boosted crystalline/amorphous $Fe_2O_{3-}\delta$ core/shell heterostructure for flexible solid-state pseudocapacitors in large scale. *Nano Energy* 45:390–97.

Tanggarnjanavalukul, C., N. Phattharasupakun, K. Kongpatpanich, et al. 2017. Charge storage performances and mechanisms of MnO_2 nanospheres, nanorods, nanotubes and nanosheets. *Nanoscale* 9:13630–39.

Thackeray, M. M. 1997. Manganese oxides for lithium batteries. *Prog. Solid State Chem.* 25:1–71.

Wang, R., Y. Lu, L. Zhou, et al. 2018. Oxygen-deficient tungsten oxide nanorods with high crystallinity: Promising stable anode for asymmetric supercapacitors. *Electrochim. Acta* 283:639–45.

Wu, H., C. Xu, J. Xu, et al. 2013. Enhanced supercapacitance in anodic TiO_2 nanotube films by hydrogen plasma treatment. *Nanotechnology* 24:455401.

Wu, Z., B. Li, Y. Xue, et al. 2015. Fabrication of defect-rich MoS_2 ultrathin nanosheets for application in lithium-ion batteries and supercapacitors. *J. Mater. Chem. A* 3:19445–54.

Xiang, K., Z. Xu, T. Qu, et al. 2017. Two dimensional oxygen-vacancy-rich Co_3O_4 nanosheets with excellent supercapacitor performances. *Chem. Commun.* 53:12410–13.

Xie, J., H. Zhang, S. Li, et al. 2013. Defect-rich MoS_2 ultrathin nanosheets with additional active edge sites for enhanced electrocatalytic hydrogen evolution. *Adv. Mater.* 25:5807–13.

Xu, J., H. Wu, L. Lu, et al. 2014. Integrated photo-supercapacitor based on bi-polar TiO_2 nanotube arrays with selective one-side plasma-assisted hydrogenation. *Adv. Funct. Mater.* 24:1840–46.

Yan, D., W. Wang, X. Luo, et al. 2018. $NiCo_2O_4$ with oxygen vacancies as better performance electrode material for supercapacitor. *Chem. Eng. J.* 334:864–72.

Yu, Z., L. Tetard, L. Zhai, et al. 2015. Supercapacitor electrode materials: Nanostructures from 0 to 3 dimensions. *Energy Environ. Sci* 8:702–30.

Zhai, T., S. Xie, M. Yu, et al. 2014. Oxygen vacancies enhancing capacitive properties of MnO_2 nanorods for wearable asymmetric supercapacitors. *Nano Energy* 8:255–63.

Zhang, C., L. Li, C.-C. Tuan, et al. 2018. A high-performance TiO_2 nanotube supercapacitor by tuning heating rate during H_2 thermal annealing. *J. Mater. Sci.: Mater. Electron.* 29:15130–37.

Zhou, H., and Y. Zhang. 2014. Electrochemically self-doped TiO_2 nanotube arrays for supercapacitors. *J. Phys. Chem. C* 118:5626–36.

Zou, R., K. Xu, T. Wang, et al. 2013. Chain-like $NiCo_2O_4$ nanowires with different exposed reactive planes for high-performance supercapacitors. *J. Mater. Chem. A* 1:8560–66.

4 Vanadium-Based Compounds for Supercapacitors

Xuan Pan, Wenyue Li, and Zhaoyang Fan

CONTENTS

4.1 Introduction 59
4.2 Vanadium Pentoxide 60
4.3 Vanadium Dioxide 63
4.4 Vanadium Trioxide 66
4.5 Vanadium Oxides with Mixed Valence States 67
4.6 Vanadium Nitrides 68
4.7 Other Vanadium Compounds 69
4.8 Conclusions 71
References 71

4.1 INTRODUCTION

Vanadium is the 22nd most abundant element in the earth crust. With five valence electrons ($3d^3 4s^2$), it has four oxidation states (V^{5+}, V^{4+}, V^{3+}, V^{2+}) that offer the readily available redox couples for facile faradic reactions. Therefore, vanadium-based compounds have been investigated for many electrochemical applications. In this chapter, we will highlight studies of vanadium-based oxides, nitrides, and other compounds for pseudocapacitance-based supercapacitors.

In general, pseudocapacitive energy storage is mainly through a surface and sub-surface-based facile charge transfer process, and the available surface area is a critical parameter in determining the material utilization and thus achievable specific capacitance, especially at high charge-discharge rates. Therefore, electrode material design must guarantee a large specific surface area (SSA), which also contributes to the surface-area related electric double-layer capacitance. A second consideration on electrode design is related with the overall high resistivity of the active inorganic compounds, which requires blending of the active materials with a conductive matrix, typically based on carbon or its allotropes, to facilitate surface charge transferring and double-layer charging/discharging. These two requirements call for nanostructure design of the electrode for achieving large-capacitance and high-power capability. Here, we will cover electrode nanostructure engineering of vanadium-based compounds.

Among the multiple vanadium oxide phases, the chemically stable vanadium pentoxide (V_2O_5), with a layered orthorhombic crystal structure, is the most commonly studied vanadate for ion intercalation- or pseudocapacitance-based energy storage and will be discussed in Section 4.2. Monoclinic VO_2(B), one of the polymorphs of vanadium dioxide (VO_2), which is formed by corner-shared VO_6 octahedra with an open lattice structure, is also an attractive electrode material and will be covered in Section 4.3. Pseudocapacitors based on metallic conducting V_2O_3 will be highlighted in Section 4.4. Other vanadium oxides, including mixed-valence oxides such as layered V_3O_7 and V_6O_{13} will be summarized in Section 4.4. Followed are metallic vanadium nitride-based electrode studies in Section 4.5, and then Section 4.6 discusses sulfide and other compounds.

It deserves to emphasize the electrolyte selection at the very beginning because faradic reaction-based pseudocapacitive charge storage, in general, requires high concentration of small ions (H^+, Na^+, K^+, etc.) in the electrolyte and hence aqueous-based solution is considered as the best electrolyte thus far. Unfortunately, similar with many other transitional metal oxides, vanadates have the general issue of dissolution in aqueous electrolytes, resulting in capacitance degradation and short-cycle lifetime. The dissolution problem of vanadates can be appreciated in their Pourbaix diagram, which maps out the possible equilibrium phases of vanadium in an aqueous electrochemical system (Pourbaix 1974). Electrolyte selection and the pH value thus become very critical when studying vanadium-based compounds. In an aqueous electrolyte, the proton (H^+) has the highest mobility, and its smallest size, comparing to Li^+, K^+, and Na^+, allows it to chemisorb to a single oxide ion. However, vanadium-based oxides can be easily dissolved in strong acids. Using other alkali metal salts, such as neutral KCl and Na_2SO_4 based electrolytes, is more manageable, even though the K^+ or Na^+ ion is too large to chemisorb at a single surface oxide ion.

4.2 VANADIUM PENTOXIDE

Lee and Goodenough (1999) were the first to investigate V_2O_5 for pseudocapacitors in 1999. Amorphous hydrous vanadium oxide was prepared by quenching V_2O_5 fine powder heated at 950°C into a bath of deionized water. The material showed an ideal capacitance curve under cyclic voltammetry (CV) measurements, and a specific capacitance of ~350 Fg^{-1} was obtained using a KCl electrolyte. In the followed studies, a variety of V_2O_5 powders with different morphologies, including nanowires, nanobelts, nanospheres, nanoribbons, nanosheets and nanotubes (Wang, Zhang et al. 2015, Qu et al. 2009, Lin et al. 2015), were synthesized to investigate their pseudocapacitive charge storage performance. Different approaches have been employed to synthesize these nanoscale V_2O_5 materials, of which, hydrothermal method is more versatile. By simply changing hydrothermal synthesis durations, V_2O_5 with interesting morphologies such as nanosheets, nanorods and nanowires could be produced (Wang, Zhang et al. 2015), while by changing the type of solvent, other structures including nanoflowers, nanoballs, nanowires and nanorods were also obtained (Mu et al. 2015).

In addition to these powders, 3D network structures of V_2O_5 are also interesting, as shown in Figure 4.1. An ordered bi-continuous double-gyroid vanadium pentoxide network (Wei et al. 2012) was fabricated by electrodeposition (Potiron et al. 1999)

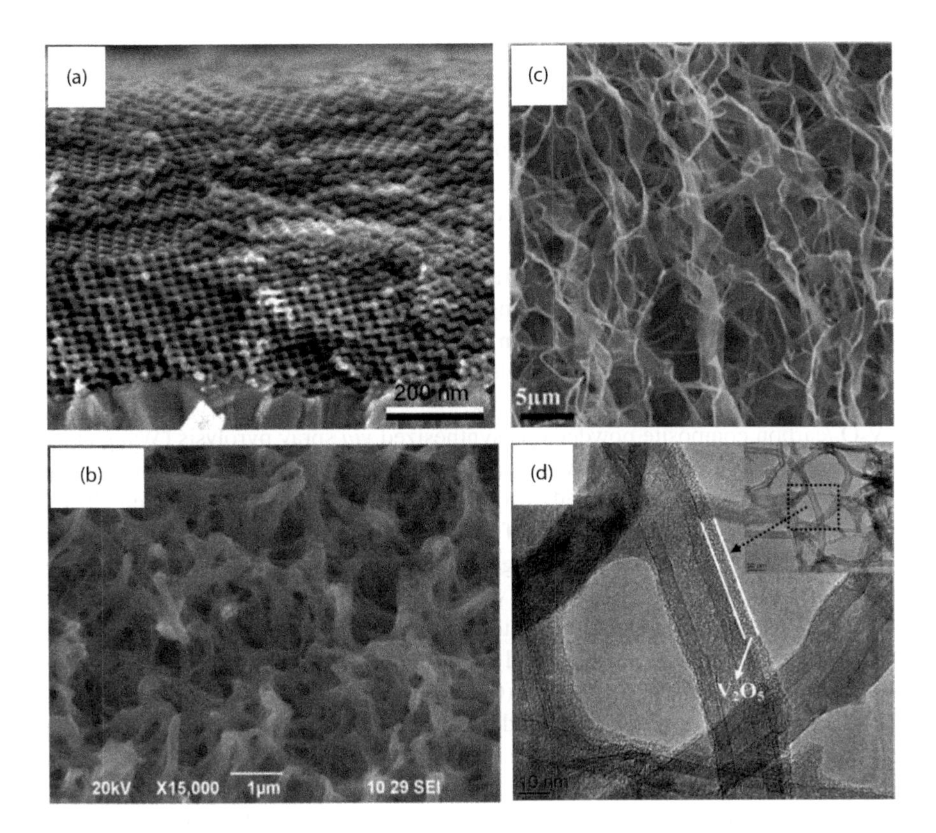

FIGURE 4.1 Examples of V_2O_5 based 3D nanostructures. (a) Cross-sectional SEM image of a mesoporous V_2O_5 double-gyroid film on a FTO substrate. (b) V_2O_5 nanoporous network. (c) 3D V_2O_5 architecture constructed from nanosheets. (d) TEM images show ultrathin V_2O_5 layer coated on the functionalized CNT surface. (a: Reprinted with permission from Wei, D. et al., *Nano Lett.*, 12, 1857–1862, 2012; b: Reprinted with permission from Saravanakumar, B. et al., *ACS Appl. Mat. Interfaces*, 4, 4484–4490, 2012; c: Reprinted with permission from Zhu, J. et al., *Nano Lett.*, 13, 5408–5413, 2013; d: Reprinted with permission from Sathiya, M. et al., *J. Am. Chem. Soc.*, 133, 16291–16299, 2011. Copyright 2011 American Chemical Society.)

into a self-assembled block copolymer template. V_2O_5 nanoporous network was synthesized *via* a simple capping-agent-assisted precipitation technique (Saravanakumar et al. 2012). V_2O_5 gel structure was obtained with V_2O_5 nanosheets *via* the combination of hydrothermal treatment and subsequent freeze-drying approach (Zhu et al. 2013). This flexible V_2O_5 gel structure with hierarchical pores possessed an SSA as large as 133 $m^2\ g^{-1}$. Using a Na_2SO_4 electrolyte, a capacitance of 451 Fg^{-1} was achieved with a capacitance retention more than 90% after 4000 cycles.

Ions from the electrolyte and electrons from the electrode, both must be transported to the same surface location of the active material for faradic reaction to occur. Electrodes with a large SSA can facilitate electrolytic ions migration to the active surface reaction spots, but the low electronic conductivity of V_2O_5 significantly retards electrons from reaching the same locations. To address the conductivity limitation of V_2O_5, composite electrode structures have been extensively investigated. Although physically mixing V_2O_5 nanomaterial with a conductive agent such as carbon black can enhance the electrode conductivity to a certain degree, more sophisticated nanocomposite powders or freestanding hybrid nanostructures have also been developed, where the second phase serves as a conductive network and provides structural integrity and stability. Such a well-designed nanocomposite powder or 3D hybrid nanostructure can offer several merits, such as hierarchical pores for easy access by an electrolyte, thus facilitating rapid diffusion of electrolytic ions within the electrode material; a short diffusion length for ions into the active material due to its nanoscale dimension; and a high electrical conductivity of the overall electrode through the embedded conductive network.

V_2O_5/carbon composite powder was synthesized *via* spray pyrolysis (Wang et al. 2009). Activated carbon powder was coated with V_2O_5 shell layer by mixing carbon powder with V_2O_5 sol (Kudo et al. 2002). Atomic layer deposition was employed for preparation of an ultrathin layer of V_2O_5 coating on the mesoporous activated carbon (Daubert et al. 2015). Interestingly, carbon was also coated on V_2O_5 as both a conductive shell and a protection layer, which gave a maximum specific capacitance of 417 Fg^{-1} with 100% capacitance retention even after 2000 continuous charge-discharge cycles (Balasubramanian and Purushothaman 2015).

One-dimensional carbon nanotubes (CNTs) and nanofibers (CNFs) are commonly used as the conductive support for oxide electrodes. Thin V_2O_5 film was coated on CNT surface using a wet chemical method (Shakir et al. 2013, Wu et al. 2015). Hydrous V_2O_5 was also coated on a freestanding CNT film via an electrodeposition process. With a small mass loading of 8.9 wt% of V_2O_5, a high specific Li-ion capacitance of 1230 Fg^{-1} and a high rate capability were obtained (Kim et al. 2006). Similarly, with controlled hydrolysis of vanadium alkoxide, a thin V_2O_5 layer with 4–5 nm thickness was deposited on CNTs, which demonstrated a large capacitive charge storage capability (Sathiya et al. 2011). An intertwined CNT and V_2O_5 nanowire composite matt was synthesized *via* a hydrothermal process using a solution of vanadium oxide and hydrophilic CNTs (Chen et al. 2011). A similar structure was also used in a flexible supercapacitor electrode (Perera et al. 2011). Freestanding electrospun carbon nanofibers-based (CNF) paper sheets have been commonly used as a conductive mesh for V_2O_5 loading. V_2O_5/CNF composites were prepared by an electrospinning method with V_2O_5 aggregates embedded in CNFs (Kim et al. 2012). Hierarchical porous V_2O_5 nanosheets were also grown on electrospun CNFs *via* a solvothermal method (Li, Peng et al. 2015) to fabricate flexible electrodes. Ultrathin V_2O_5 layer modified CNF papers were also successfully prepared by electrode position (Ghosh et al. 2011).

Two-dimensional graphene or reduced graphene oxide (rGO) has attracted much interest as a conductive facilitator for V_2O_5 based supercapacitor electrodes owing to its large surface area and high conductivity. One commonly employed method to synthesize V_2O_5/rGO nanocomposite is the solvothermal process with graphene oxide (GO) reduction to rGO and oxide deposition on rGO occurring simultaneously. In such composites, V_2O_5 with a variety of morphologies such as nanoparticles, nanorods (Li et al. 2013), nanobelts (Lee, Balasingam et al. 2015), nanowires

(Perera et al. 2013), nanofibers (Nagaraju et al. 2014), and nanosheets (Nagaraju et al. 2014), was intimately anchored on rGO sheets to achieve an enhanced electronic conductivity and an improved surface area, resulting in significantly promoted pseudocapacitive and electric double-layer capacitances as well as high-rate and high-power performances.

Considering the difference of activated carbon (AC), CNT, and graphene in their physical properties, it is not surprising that researches have been reported using more than one such carbon materials to form nanocomposites with V_2O_5 for their synergetic effects. For instance, with supercritical fluid adsorption followed by calcination, V_2O_5 thin layer was deposited on a CNT-AC structure (Do et al. 2014). Through layer by layer assembly technique, graphene sheets were alternatively inserted between 3 nm V_2O_5 coated CNT films (Shakir et al. 2014). Such a graphene spacer substantially enhanced the specific capacitance to an extraordinary value of ~2590 Fg^{-1}.

4.3 VANADIUM DIOXIDE

VO_2 has been widely studied for its metal-insulator transition (Zhao et al. 2011, 2012), related optical properties (Nazari et al. 2013, Karaoglan-Bebek et al. 2014), and its potential for electrochemical energy storage. In addition to the most stable monoclinic phase of VO_2(M1), VO_2 has a few other phases, of which VO_2(B) has attracted more attentions owning to its layered crystal structure and ionic diffusion channels. Similar to V_2O_5, designing VO_2-based electrode structures with a large SSA is also critical for high performance of pseudocapacitive charge storage. Different nanoscale morphologies, solid and porous spheres (Zhang et al. 2018), nanosheets (Rakhi et al. 2016), nanobelts (Wang et al. 2014), nanorods (Zheng et al. 2018), hexangular star fruit-like particles (Shao et al. 2012), and others have been synthesized for electrode applications. Enhancement of electrode conductivity is equally important, as demonstrated by our previous work (Pan et al. 2013). With VO_2 treated in H_2 for conductivity improvement, its capacitance can be increased by a factor of four. In particular, the supercapacitor demonstrated a specific capacitance of 300 Fg^{-1} and a specific energy of 17 Wh kg^{-1} at a specific current of 1 Ag^{-1}. It is interesting to note that hydrogenated VO_2 (R) can maintain its metallic phase at room temperature (Zhao et al. 2014), which might be exploited for high-rate supercapacitors.

Similar to the case for V_2O_5, hybridizing VO_2 with a variety of carbon allotropes to form composites, either in the form of powders or as freestanding structures, has also been extensively experimented to improve its capacity and cycling stability, because poor electrical conductivity and unfavorable structural stability are the common problems of transition metal oxides, including VO_2 (Deng et al. 2013, Zhang et al. 2016, Hosseini and Shahrokhian 2018b, Liang et al. 2013). As examples, 3D graphene/VO_2(B) nanobelts composite gel structure was synthesized in a one-step hydrothermal strategy (Xiao et al. 2015) with V_2O_5 and graphene oxide as precursors. In this structure, VO_2 nanobelts and graphene sheets formed interconnected porous microstructures. The composite gel exhibited a specific capacitance of 426 Fg^{-1} at a current density of 1 Ag^{-1}. In another study (Wang et al. 2014), the synergistic effect of VO_2 and graphene in the hybrid electrode improved the rate capability and the cycling stability. A similar structure was also applied for flexible supercapacitor (Lee, Wee, and Hong 2015). In a novel design, hydrogen molybdenum bronze (HMB) was electrochemically deposited on VO_2 nanoflakes that were grown in graphene foam (GF), thus forming a GF/VO_2/HMB electrode structure (Xia et al. 2015). Excellent rate performance was demonstrated with a capacitance of 485 Fg^{-1} at 2 Ag^{-1} and 306 Fg^{-1} at 32 Ag^{-1}. The HMB shell, due to its high electrical conductivity and ionic conductivity, may serve as a useful component in high-rate supercapacitors. With VO_2 particles anchored on freestanding vertical graphene network (Ren, Li et al. 2016), high-rate supercapacitors were demonstrated with a characteristic frequency of 15 Hz at −45° phase angle, and a relaxation time constant of 66.7 ms (Ren, Zhang, and Fan 2018). In particular, Chen et al. (2019) investigated VO_2/graphene@NiS_2 hybrid aerogel for all-solid-state asymmetric supercapacitors. As in Figure 4.2, the hybrid aerogel, consisting of VO_2 nanoparticles, graphene sheets, and NiS_2 nanoflakes, had a hierarchical mesoporous structure. It delivered a capacitance as large as 1280 Fg^{-1} at a current density of 1 Ag^{-1}. With the hybrid aerogel and a graphene aerogel as the positive and negative electrode, respectively,

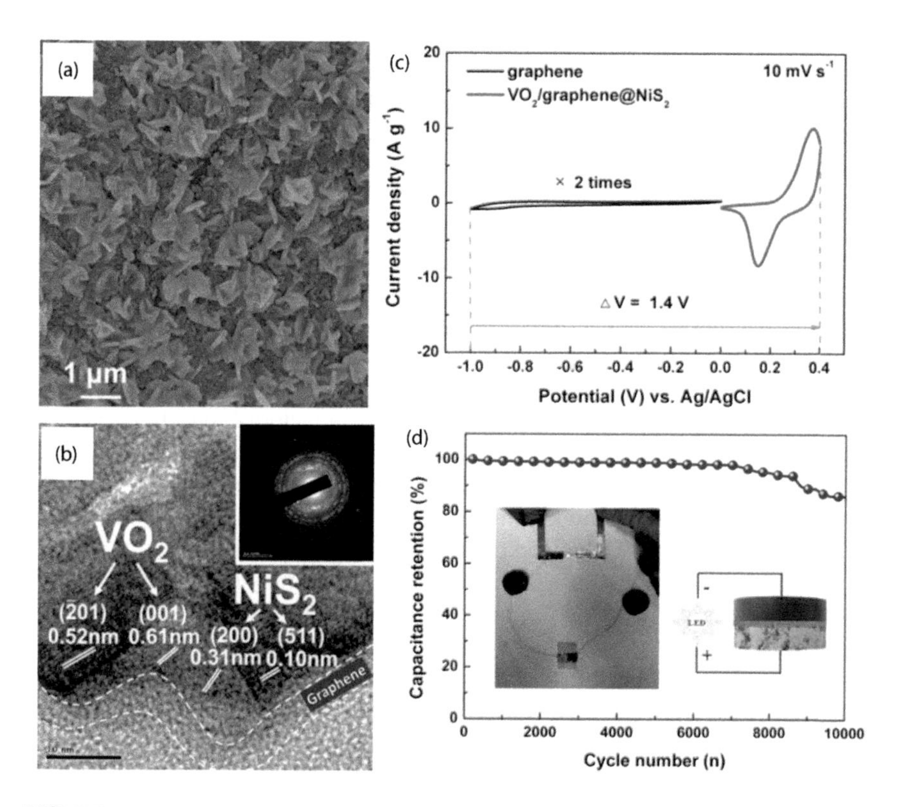

FIGURE 4.2 (a) SEM image and (b) TEM image of the VO_2/graphene@NiS_2 hybrid aerogel. Inset in (b) is the selected area electron diffraction pattern. (c) CV curves of VO_2/graphene@NiS_2 hybrid aerogel and graphene aerogel measured at 10 mV s^{-1}. (d) Cycling performance of the asymmetric supercapacitor tested at 10 mV s^{-1}. The inset shows a LED lighted up by the supercapacitor. (Reprinted with permission from Chen, H.-C. et al., *ACS Appl. Energy Mater.*, 2,459–467, 2019. Copyright 2019 American Chemical Society.)

and PVA/KOH gel as the electrolyte, solid-state asymmetric supercapacitors were fabricated, which provided an energy density of 60.2 Wh kg^{-1} at a power density of 350.0 W kg^{-1} with superior cycling stability.

Pseudocapacitive charge storage is generally considered as material surface or sub-surface related facile faradic process, without involving the ion intercalation into the "bulk" lattice. The latter is generally considered as the charge storage mechanism in batteries. However, intercalation-type pseudocapacitance has attracted considerable attentions in recent years (Lukatskaya, Dunn, and Gogotsi 2016). The argument is related to the intrinsic or extrinsic (nanoengineering related) ion intercalation kinetics, which is commonly reflected by the sweep rate (ν) dependence of cyclic voltammetry (CV) experiments. The current (i) in a battery electrode material is characterized by the classical semi-infinite diffusion, $i \sim \nu^{0.5}$, while it manifests by a linear dependence, $i \sim \nu$, in a supercapacitor. A number of metal oxides (Kim et al. 2017, Ren, Hoque et al. 2016) and others (Zhu et al. 2015) have been investigated recently as intercalation pseudocapacitor electrodes, one of which is VO_2(B).

VO_2(B) crystal consists of double-layered V_4O_{10} with tunnels formed along *b*-axis (Figure 4.3a). There are two types of tunnels—the large one is feasible for rapid Li^+ intercalation at 2.5 V vs. Li/Li^+, and the small one has extremely slow kinetics of Li^+ intercalation at 2.1 V vs. Li/Li^+. As a result, nanoengineering bulk material into atomically thin 2D sheets has demonstrated to be effective in lowering the intercalation barrier, and the charge storage in atomically thin sheets exhibits behaviors similar to the surface adsorption mechanism. In our study, VO_2(B) nanobelt forest (Ren et al. 2015) was synthesized on vertical graphene (Ren et al. 2014) substrate in a solvothermal process, with the belt thickness down to a few nanometers (Figure 4.3b). With a Li^+-ion based electrolyte, a stable discharge capacity of 178 mAh g^{-1} was demonstrated at a current density of 10 Ag^{-1}, while it remained at 100 mAh g^{-1} when the current density reached to 27 Ag^{-1}, corresponding to 300°C. The contributions from battery-like and pseudocapacitor-like charge storage mechanisms are plotted in Figure 4.3c. At high rates, the charge storage mainly came from the supercapacitor mechanisms, rather than that of the battery. In another study (Xia et al. 2018), VO_2(B) thickness was further reduced and it was demonstrated that there was no diffusion limitation to Li^+ in atomically thin VO_2(B) nanoribbons and a capacity of 140 mAh g^{-1} was achieved at a

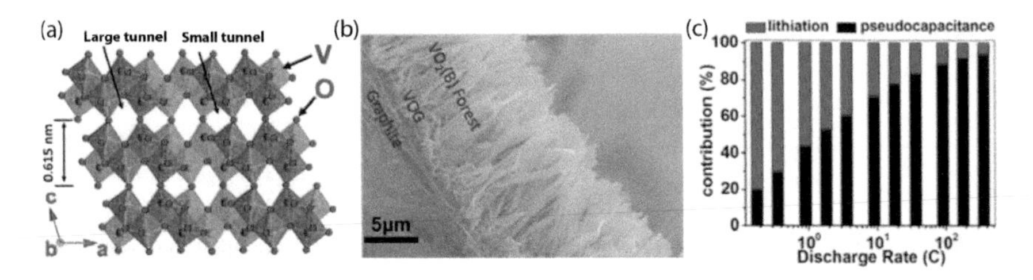

FIGURE 4.3 (a) The lattice structure of VO_2(B), (b) SEM image showing VO_2(B) nanobelt forest grown on vertical graphene substrate, and (c) contribution of non-diffusion-limited pseudo-capacitance and diffusion-limited intercalation to the total capacity at different discharge current densities. (Ren, G. et al., *J. Mater. Chem. A*, 3, 10787–10794, 2015. Reproduced by permission of The Royal Society of Chemistry.)

discharge rate of 100 C (36 s). Other similar observations were also reported (Chao et al. 2018, Wang et al. 2017). These observations suggested the potential of VO_2(B) for battery-like supercapacitors with much larger capacity.

4.4 VANADIUM TRIOXIDE

Of the many vanadium-based oxide phases, V_2O_3 exhibits quasi-metallic conductivity of around $10^3\ \Omega^{-1}\ cm^{-1}$, similar as that of RuO_2. This conductivity is much larger than that of VO_2 and V_2O_5 by several orders of magnitude. Therefore, V_2O_3 could be a better candidate for pseudocapacitive electrode material and has attracted attentions in recent years.

Using a solvothermal method, V_2O_3 flower-like nanostructure constructed by single crystalline nanoflakes was prepared (Liu et al. 2010). A capacitance of 218 Fg^{-1} at a current density of 0.05 Ag^{-1} was measured. The breakthrough of V_2O_3-based high-power ultrafast supercapacitor was reported from our study (Pan et al. 2014). As illustrated in Figure 4.4a, graphene sheets were used as the conductive framework,

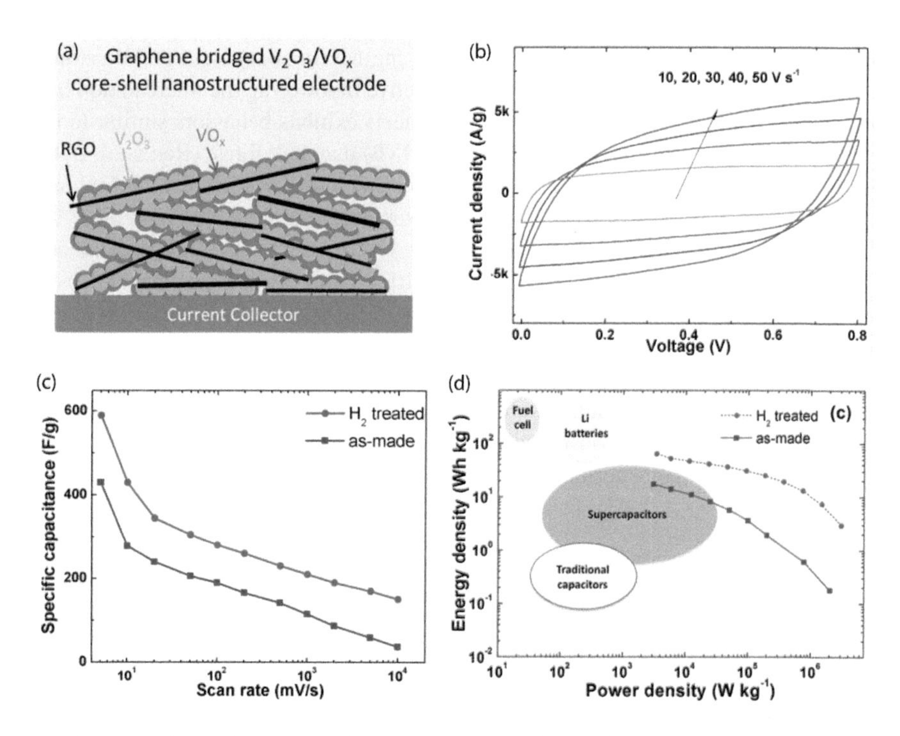

FIGURE 4.4 (a) Schematic of the graphene bridged V_2O_3/VO_x core-shell nanostructure. (b) Cyclic voltammograms of hydrogen treated composite over a voltage window between 0 and 0.8 V in 1 M Na_2SO_4 aqueous electrolyte at fast scan rates from 10 to 50 V s^{-1}. (c) The specific capacitances of the composite at different scan rates. (d) Ragone plot of the composite electrode performance. (From Pan. X. et al.: Fast supercapacitors based on graphene-bridged V_2O_3/VO_x core–shell nanostructure electrodes with a power density of 1 MW kg^{-1}. *Adv. Mater. Interfaces* 2014. 1.1400398. Copyright Wiley-VCH Verlag GmbH & Co. KGaA. Reproduced with permission.)

which is anchored with quasi-metallic conductive V_2O_3 as nano-cores. V^{3+} is easily oxidized into V^{4+} and V^{5+} after exposure to the air. So V_2O_3 nano-cores were further encapsulated with naturally formed VO_x that acted as electrochemical active oxides. This graphene bridged V_2O_3/VO_x core-shell nanostructure provides an electronically conductive highway, a large SSA, and suitable pore geometry. Together, such an electrode architecture, prepared via solvothermal synthesis and hydrogen thermal treatment, can provide both high specific power and high specific energy. Using 1 M Na_2SO_4 aqueous electrolyte, the fabricated cell exhibited a quasi-rectangular shape in CV measurements even at a scan rate as high as 50 V s^{-1} (Figure 4.4b). The electrode, based on the overall mass of the composite, provided a specific capacitance of 590 Fg^{-1} at a rate of 5 mV s^{-1} and remained at 150 Fg^{-1} when the rate increased to 10 V s^{-1} (Figure 4.4c). As shown by the Ragone plot (Figure 4.4d), our supercapacitor delivered an outstanding performance with a specific power more than 1 MW kg^{-1} at a specific energy of 10 Wh kg^{-1}, or a large energy density of 63 Wh kg^{-1} at a power density of 3.4 kW kg^{-1}. Such a performance was ascribed to its large electronic conductivity, the well-designed nanostructure, as well as a possibly facile redox mechanism.

Following this work, a similar study was conducted based on V_2O_3/VO_x core-shell structure, and the demonstrated cell exhibited ultrahigh capacitance and super-long cyclic durability of 100,000 cycles (Yu et al. 2015). V_2O_3 nanoflakes were also intertwined with nitrogen-doped rGO to form self-supported electrodes, which were used for flexible all-solid-state supercapacitors with LiCl in alkaline poly(vinyl alcohol) as a gel electrolyte (Hou et al. 2017). Core-shell composites (V_2O_3@C) with V_2O_3 nanorods encapsulated with thin carbon were fabricated. Such a V_2O_3@C nanorod structure showed a large specific surface area of 246 m^2g^{-1} (Hu, Liu, Zhang, Nie et al. 2018). V_2O_3@C nanosheet array was directly grown on a current collector through a hydrothermal method and then assembled into an ultrathin supercapacitor using a PVA-LiCl gel electrolyte. Such a cell showed a large voltage of 2.0 V owing to different redox reactions of vanadium ions in the anode and the cathode (Zhu et al. 2017). Hierarchical MoS_2-coated V_2O_3 composite nanosheet tubes were used as both the cathode and anode materials in supercapacitors (Peng et al. 2018). With a facile hydrothermal reaction followed by thermal annealing, V_2O_3 tubes with a diameter of ~1.5 μm were assembled by ultrathin nanosheets owing to thermal contract of the stacked sheets. The hollow tube backbone was decorated with vertical MoS_2 nanosheets. In this structure, V_2O_3 tubes prevented restacking of MoS_2 nanosheets and served as an electronic conductive scaffold. Such a composite electrode exhibited a wide potential window between −1.0 V and +1.0 V, and a large capacitance of 655 Fg^{-1}.

4.5 VANADIUM OXIDES WITH MIXED VALENCE STATES

Vanadium oxide can also crystallize with mixed valence states, manifested by the variety of Wadsley phases (V_nO_{2n+1}) formed by V^{5+} and V^{4+} states and the Magnéli phases (V_nO_{2n-1}) formed by V^{4+} and V^{3+} states (Bahlawane and Lenoble 2014). Of these large number of phases with mixed states, Wadsley phases have a layered crystal structure, permitting reversible ions intercalation. In particular, V_6O_{13} has a metallic conductivity and has attracted attentions for pseudocapacitive charge storage.

V_6O_{13} with sheet morphology was initially prepared by a thermal decomposition and quenching method with NH_4VO_3 as the precursor (Zeng et al. 2009). A limited specific capacitance was found due to a large thickness of the sheets (~200 nm). Hollow flowers-like V_6O_{13}, consisting of nanosheets, was further prepared *via* a facile sol-hydrothermal approach, which delivered a capacitance of 417 Fg^{-1} at a scan rate of 5 mV s^{-1} (Huang et al. 2011). Hierarchical oxide microspheres were synthesized, which contained 86.2 wt% V_6O_{13} with metallic conductivity and 13.8 wt% VO_2. A specific capacitance of 456 Fg^{-1} was measured at 0.6 Ag^{-1} (Li, Wei et al. 2015). Sulfur-doped V_6O_{13-x} nanowires with oxygen deficiencies, consisting of V^{5+}, V^{4+}, and V^{3+} states, were prepared as an anode electrode material for an asymmetric supercapacitor. Sulfur-doping improved the capacitive performance by reducing the charge-transfer resistance and increasing the Li ion diffusion coefficient. Significantly, it achieved a capacitance of 1353 Fg^{-1} (0.72 F cm^{-2}) at a current density of 1.9 Ag^{-1} (1 mA cm^{-2}) in 5 M LiCl electrolyte (Zhai et al. 2014).

Self-assembled nest-like V_3O_7 structure by porous V_3O_7 nanowires was fabricated, which was further coated with a layer of N-doped carbon (Zhao et al. 2018). The electrode delivered a capacitance of 660 Fg^{-1} at a current density of 0.5 Ag^{-1} and it remained at 188 Fg^{-1} when the current density reached 50 Ag^{-1}. 3D hierarchical porous $V_3O_7{\cdot}H_2O$ nanobelts/CNT/reduced graphene oxide composite was studied. The ternary composite exhibited a high specific capacitance of 685 Fg^{-1} at a current density of 0.5 Ag^{-1} and an excellent cycling stability with trivial capacitance loss after 10,000 cycles (Hu, Liu, Zhang, Chen et al. 2018).

Self-supported VO/VO_x/CNF electrode was also reported where vanadium monoxide (VO) coupled with amorphous VO_x was incorporated into CNFs by electrospinning and heat treatment (Tang et al. 2016). Ordered mesoporous carbon such as CMK-3 was also studied as another carbon based scaffold for oxides loading (Eftekhari and Fan 2017). Mixed-valence vanadium oxide (VOx)/CMK-3 composites were synthesized through a facile liquid-phase method followed by calcination, and used for supercapacitors (Hao et al. 2016).

4.6 VANADIUM NITRIDES

Except for a few special cases (e.g., V_2O_3), performances of pseudocapacitors based on transition metal oxides are limited by the low conductivity of oxides, poor rate capability and small specific power. Consequently, other transition metal compounds, such as nitrides and carbides have emerged as potential pseudocapacitive materials owning to their high metallic conductivities.

This is especially true for vanadium nitrides (VN), which has a conductivity of ~10^5 Ω^{-1} cm^{-1} and a very large specific capacitance. The extremely high conductivity of VN will be particularly useful for high-rate and high-power supercapacitors. However, it is emphasized that the pseudocapacitive charge storage mechanisms in nitrides are not very clear. Since the surface of transition metal nitrides can be easily oxidized in the air or in aqueous electrolytes, it has been suggested that the origin of charge storage, in fact, is still the redox reaction occurred in the surficial oxides, while the conductive nitride facilitates electron transportation (Choi, Blomgren, and Kumta 2006). In contrast, in another study (Bondarchuk et al. 2016) with pure oxygen-free vanadium nitride film that delivered a large capacitance of ~mF cm^{-2} at a rate of 1 V s^{-1}, redox reactions were found to play no or little roles. Instead, it was proposed that space charge accumulation in a subsurface layer contributed to the measured capacitance.

VN nanoparticles were first synthesized via ammonolysis reaction of VCl_4 in anhydrous chloroform, which demonstrated a specific capacitance of 1340 Fg^{-1} at a rate of 2 mV s^{-1} (Choi, Blomgren, and Kumta 2006). Since then, VN nano-powders, wires, rods, fibers, sheets, and other nanoscale morphologies have been reported for capacitive charge storage. Interestingly, mesocrystal nanosheets of VN were reported with a conductivity of 1.44×10^5 S m^{-1}, which delivered a superior volumetric capacitance of 1937 mF cm^{-3} (Bi et al. 2015).

Even with a high conductivity, VN may still need to form composite with carbon for ameliorating its poor electrochemical stability. VN nanoparticles in porous carbon nanospheres (Liu et al. 2016), VN nanoparticles in porous N-doped carbon (Fechler et al. 2014), freestanding mesoporous VN nanowires/CNT (Xiao et al. 2013), 3D porous VN nanowires–graphene (Wang, Lang et al. 2015), and many others have been investigated. VN/N-doped graphene composite provided a large specific energy of 81.73 Wh kg^{-1} and a high specific power of 28.82 kW kg^{-1} at 51.24 Wh kg^{-1}, and showed good stability with 98.6% retention after 10,000 cycles (Balamurugan et al. 2016). Interestingly, VN-TiN-based hybrid structures, such as TiN/VN core–shell composites (Dong et al. 2011, Pang et al. 2014) and vanadium titanium nitride/carbon nanofibers (Xu et al. 2015), were also reported as supercapacitor electrodes. With a flexible electrode structure based on 3D interweaved N-doped carbon encapsulated mesoporous VN nanowires and a KOH electrolyte, the supercapacitor exhibited 91.8% capacitance retention after 12,000 cycles (Gao et al. 2015).

4.7 OTHER VANADIUM COMPOUNDS

Mixed metal vanadates, like AlV_3O_9, ZnV_2O_4, $Zn_3V_2O_8$, $Ni_3V_2O_8$, $Co_3V_2O_8$, are another category of materials suitable for pseudocapacitive charge storage. As examples, amorphous AlV_3O_9 hierarchical microspheres were measured with a specific capacitance of 497 Fg^{-1} at 1 Ag^{-1}. They showed good stability with a retention capacity of 89% after 10,000 cycles (Yan et al. 2016). Urchin-shaped $Ni_3V_2O_8$ nanosphere electrode exhibited a specific capacity of 402.8 C g^{-1} at 1 Ag^{-1} (Kumar, Rai, and Sharma 2016). Core-shell nanostructure-based binder-free electrodes with $Co_3V_2O_8$-$Ni_3V_2O_8$ thin layers on carbon nanofibers were also reported (Hosseini and Shahrokhian 2018a).

Due to recent interests in layered transition metal dichalcogenides (TMDs), few-atomic-layer VS_2 sheets were investigated for on-chip micro-supercapacitor application by exploiting their high conductance, large surface area, and 2D permeable channels (Feng et al. 2011). A large area-specific capacitance of 4.76 mF cm^{-2} was measured. Hexagonal VS_2 anchored CNTs for flexible electrodes (Pandit, Karade, and Sankapal 2017), VS_2/rGO hybrid nanosheets (Pandit, Karade, and Sankapal 2017) and others have since been investigated.

With a layered crystal structure, vanadyl phosphate ($VOPO_4$) is suitable for ion intercalation. Due to enhanced ionicity of (V–O) bonds, V^{4+}/V^{5+} redox couple

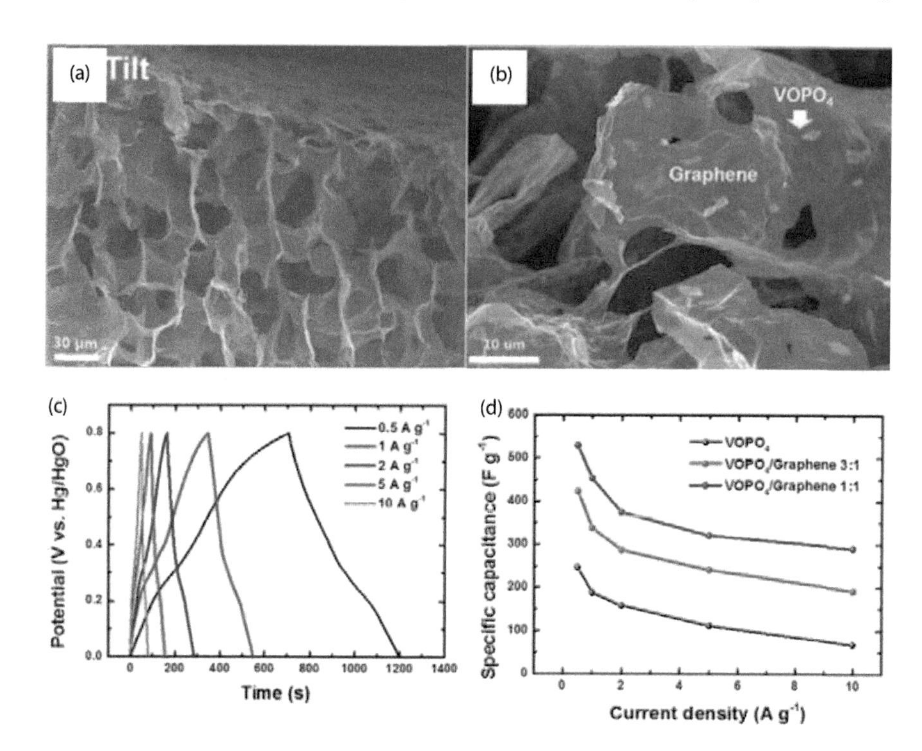

FIGURE 4.5 (a) SEM image (tilt view) of the vertically oriented porous 3D microstructures of the $VOPO_4$/RGO nanocomposite. (b) High-resolution SEM image of the porous $VOPO_4$/RGO nanocomposite showing distribution of $VOPO_4$ thin layers on RGO sheets. (c) Galvanostatic charge–discharge curves of 1:1 $VOPO_4$/RGO nanocomposite electrode at different current densities. (d) Specific capacitance as a function of current density for pristine $VOPO_4$, 1:1 VOPO4/RGO, and 3:1 $VOPO_4$/RGO composites. (Reproduced from Lee, K. et al., *Sci. Rep.*, 5, 13696, 2015 with a Creative Commons Attribution 4.0 International License.)

in $VOPO_4$ shows a higher potential than other vanadium oxides. In Lee's work (Figure 4.5), a facile ice-templated self-assembly process was adopted to fabricate a 3D porous nanocomposite of $VOPO_4$ nanosheets and graphene. This vertically oriented porous structure has a large surface area, an excellent electrical conductivity and facile ion diffusion paths. Such a structure delivered a capacitance of 527.9 Fg^{-1} at a current density of 0.5 Ag^{-1} with a superior cycling stability (Lee, Lee, et al. 2015). Amorphous $VOPO_4$/graphene composite was reported with an excellent rate capability (359 Fg^{-1} at 10 Ag^{-1}) (Chen et al. 2017). Mesoporous vanadium phosphate nanosheets were fabricated through a liquid crystal template method, and a capacity of 767 Fg^{-1} at 0.5 Ag^{-1} was reported. The charge storage mechanism was experimentally demonstrated to be based on the reversible two-step redox reactions between V(V) and V(III) in acidic medium (Mei et al. 2018).

4.8 CONCLUSIONS

Vanadium-based compounds have plenty of merits to be used as electrode materials for developing pseudocapacitors. Vanadium is an earth-abundant element and its low cost will facilitate fabrication of cost-effective pseudocapacitors for practical applications. The multiple oxidation states of vanadium offer the readily available redox couples for facile faradic reactions. The variety of vanadium oxides and other compound structures and phases, with different physical and chemical properties, provides many options in the search for a high-performance electrochemical active material. Like for other transition metal oxides, nanostructure engineering to control the nanoscale morphology of vanadium compounds for easy access by electrolyte and to achieve reasonable electronic conductivity is one of the key design considerations for better performances. Steady progresses are being made along this direction.

REFERENCES

Bahlawane, Naoufal, and Damien Lenoble. 2014. "Vanadium oxide compounds: Structure, properties, and growth from the gas phase." *Chemical Vapor Deposition* 20 (7-8-9):299–311.

Balamurugan, Jayaraman, Gopalsamy Karthikeyan, Tran Duy Thanh, Nam Hoon Kim, and Joong Hee Lee. 2016. "Facile synthesis of vanadium nitride/nitrogen-doped graphene composite as stable high performance anode materials for supercapacitors." *Journal of Power Sources* 308:149–157.

Balasubramanian, Sethuraman, and Kamatchi Kamaraj Purushothaman. 2015. "Carbon coated flowery V_2O_5 nanostructure as novel electrode material for high performance supercapacitors." *Electrochimica Acta* 186:285–291.

Bi, Wentuan, Zhenpeng Hu, Xiaogang Li, Changzheng Wu, Junchi Wu, Yubin Wu, and Yi Xie. 2015. "Metallic mesocrystal nanosheets of vanadium nitride for high-performance all-solid-state pseudocapacitors." *Nano Research* 8 (1):193–200.

Bondarchuk, Oleksandr, Alban Morel, Daniel Bélanger, Eider Goikolea, Thierry Brousse, and Roman Mysyk. 2016. "Thin films of pure vanadium nitride: Evidence for anomalous non-faradaic capacitance." *Journal of Power Sources* 324:439–446.

Chao, Dongliang, Chun-Han Lai, Pei Liang, Qiulong Wei, Yue-Sheng Wang, Changrong Zhu, Gang Deng, Vicky VT Doan-Nguyen, Jianyi Lin, and Liqiang Mai. 2018. "Sodium vanadium fluorophosphates (NVOPF) array cathode designed for high-rate full sodium ion storage device." *Advanced Energy Materials* 8 (16):1800058.

Chen, Hsieh-Chih, Yi-Cih Lin, Yan-Lin Chen, and Chi-Jen Chen. 2019. "Facile fabrication of three-dimensional hierarchical nanoarchitectures of VO_2/graphene@ NiS_2 hybrid aerogel for high-performance all-solid-state asymmetric supercapacitors with ultrahigh energy density." *ACS Applied Energy Materials* 2 (1):459–467.

Chen, Ningna, Jinhua Zhou, Qi Kang, Hongmei Ji, Guoyin Zhu, Yu Zhang, Shanyong Chen, Jing Chen, Xiaomiao Feng, and Wenhua Hou. 2017. "Amorphous vanadyl phosphate/graphene composites for high performance supercapacitor electrode." *Journal of Power Sources* 344:185–194.

Chen, Zheng, Veronica Augustyn, Jing Wen, Yuewei Zhang, Meiqing Shen, Bruce Dunn, and Yunfeng Lu. 2011. "High-performance supercapacitors based on intertwined CNT/V_2O_5 nanowire nanocomposites." *Advanced Materials* 23 (6):791–795.

Choi, Daiwon, George E Blomgren, and Prashant N Kumta. 2006. "Fast and reversible surface redox reaction in nanocrystalline vanadium nitride supercapacitors." *Advanced Materials* 18 (9):1178–1182.

Daubert, James S., Neal P. Lewis, Hannah N. Gotsch, J. Zachary Mundy, David N. Monroe, Elizabeth C. Dickey, Mark D. Losego, and Gregory N. Parsons. 2015. "Effect of meso- and micro-porosity in carbon electrodes on atomic layer deposition of pseudocapacitive V_2O_5 for high performance supercapacitors." *Chemistry of Materials* 27 (19):6524–6534.

Deng, Lingjuan, Gaini Zhang, Liping Kang, Zhibin Lei, Chunling Liu, and Zong-Huai Liu. 2013. "Graphene/VO_2 hybrid material for high performance electrochemical capacitor." *Electrochimica Acta* 112:448–457.

Do, Quyet H., Jesse Smithyman, Changchun Zeng, Chuck Zhang, Richard Liang, and Jim P Zheng. 2014. "Toward binder-free electrochemical capacitor electrodes of vanadium oxide-nanostructured carbon by supercritical fluid deposition: Precursor adsorption and conversion, and electrode performance." *Journal of Power Sources* 248:1241–1247.

Dong, Shanmu, Xiao Chen, Lin Gu, Xinhong Zhou, Haibo Wang, Zhihong Liu, Pengxian Han, Jianhua Yao, Li Wang, and Guanglei Cui. 2011. "TiN/VN composites with core/shell structure for supercapacitors." *Materials Research Bulletin* 46 (6):835–839.

Eftekhari, Ali, and Zhaoyang Fan. 2017. "Ordered mesoporous carbon and its applications for electrochemical energy storage and conversion." *Materials Chemistry Frontiers* 1 (6):1001–1027.

Fechler, Nina, Girum Ayalneh Tiruye, Rebeca Marcilla, and Markus Antonietti. 2014. "Vanadium nitride@ N-doped carbon nanocomposites: Tuning of pore structure and particle size through salt templating and its influence on supercapacitance in ionic liquid media." *RSC Advances* 4 (51):26981–26989.

Feng, Jun, Xu Sun, Changzheng Wu, Lele Peng, Chenwen Lin, Shuanglin Hu, Jinlong Yang, and Yi Xie. 2011. "Metallic few-layered VS_2 ultrathin nanosheets: High two-dimensional conductivity for in-plane supercapacitors." *Journal of the American Chemical Society* 133 (44):17832–17838.

Gao, Biao, Xingxing Li, Xiaolin Guo, Xuming Zhang, Xiang Peng, Lei Wang, Jijiang Fu, Paul K. Chu, and Kaifu Huo. 2015. "Nitrogen-doped carbon encapsulated mesoporous vanadium nitride nanowires as self-supported electrodes for flexible all-solid-state supercapacitors." *Advanced Materials Interfaces* 2 (13):1500211.

Ghosh, Arunabha, Eun Ju Ra, Meihua Jin, Hae-Kyung Jeong, Tae Hyung Kim, Chandan Biswas, and Young Hee Lee. 2011. "High pseudocapacitance from ultrathin V_2O_5 films electrodeposited on self-standing carbon-nanofiber paper." *Advanced Functional Materials* 21 (13):2541–2547.

Hao, Liang, Jie Wang, Laifa Shen, Jiajia Zhu, Bing Ding, and Xiaogang Zhang. 2016. "Synthesis and electrochemical performances of mixed-valence vanadium oxide/ordered mesoporous carbon composites for supercapacitors." *RSC Advances* 6 (30):25056–25061.

Hosseini, Hadi, and Saeed Shahrokhian. 2018a. "Advanced binder-free electrode based on core–shell nanostructures of mesoporous $Co_3V_2O_8$-$Ni_3V_2O_8$ thin layers@ porous carbon nanofibers for high-performance and flexible all-solid-state supercapacitors." *Chemical Engineering Journal* 341:10–26.

Hosseini, Hadi, and Saeed Shahrokhian. 2018b. "Vanadium dioxide-anchored porous carbon nanofibers as a Na^+ intercalation pseudocapacitance material for development of flexible and super light electrochemical energy storage systems." *Applied Materials Today* 10:72–85.

Hou, ZQ, ZY Wang, LX Yang, and ZG Yang. 2017. "Nitrogen-doped reduced graphene oxide intertwined with V_2O_3 nanoflakes as self-supported electrodes for flexible all-solid-state supercapacitors." *RSC Advances* 7 (41):25732–25739.

Hu, Tao, Yanyan Liu, Yifu Zhang, Meng Chen, Jiqi Zheng, Jie Tang, and Changgong Meng. 2018. "3D hierarchical porous V_3O_7 H_2O nanobelts/CNT/reduced graphene oxide integrated composite with synergistic effect for supercapacitors with high capacitance and long cycling life." *Journal of Colloid and Interface Science* 531:382–393.

Hu, Tao, Yanyan Liu, Yifu Zhang, Ying Nie, Jiqi Zheng, Qiushi Wang, Hanmei Jiang, and Changgong Meng. 2018. "Encapsulating V_2O_3 nanorods into carbon core-shell composites with porous structures and large specific surface area for high performance solid-state supercapacitors." *Microporous and Mesoporous Materials* 262:199–206.

Huang, Zhiyong, Hongmei Zeng, Li Xue, Xiangge Zhou, Yan Zhao, and Qiongyu Lai. 2011. "Synthesis of vanadium oxide, V_6O_{13} hollow-flowers materials and their application in electrochemical supercapacitors." *Journal of Alloys and Compounds* 509 (41):10080–10085.

Karaoglan-Bebek, Gulten, Md Nadim Ferdous Hoque, Mark Holtz, Zhaoyang Fan, and Ayrton A. Bernussi. 2014. "Continuous tuning of W-doped VO_2 optical properties for terahertz analog applications." *Applied Physics Letters* 105 (20):201902.

Kim, Bo-Hye, Chang Hyo Kim, Kap Seung Yang, Abdelaziz Rahy, and Duck J Yang. 2012. "Electrospun vanadium pentoxide/carbon nanofiber composites for supercapacitor electrodes." *Electrochimica Acta* 83:335–340.

Kim, Hyung-Seok, John B Cook, Hao Lin, Jesse S Ko, Sarah H Tolbert, Vidvuds Ozolins, and Bruce Dunn. 2017. "Oxygen vacancies enhance pseudocapacitive charge storage properties of MoO_3–x." *Nature Materials* 16 (4):454.

Kim, Il-Hwan, Jae-Hong Kim, Byung-Won Cho, and Kwang-Bum Kim. 2006. "Pseudocapacitive properties of electrochemically prepared vanadium oxide on carbon nanotube film substrate." *Journal of The Electrochemical Society* 153 (8):A1451–A1458.

Kudo, Togo, Yuji Ikeda, Takashi Watanabe, Mitsuhiro Hibino, Masaru Miyayama, Hiroyoshi Abe, and K Kajita. 2002. "Amorphous V_2O_5/carbon composites as electrochemical supercapacitor electrodes." *Solid State Ionics* 152:833–841.

Kumar, Rudra, Prabhakar Rai, and Ashutosh Sharma. 2016. "3D urchin-shaped Ni_3 (VO_4) 2 hollow nanospheres for high-performance asymmetric supercapacitor applications." *Journal of Materials Chemistry A* 4 (25):9822–9831.

Lee, Hee Y., and JB Goodenough. 1999. "Ideal supercapacitor behavior of amorphous V_2O_5 nH_2O in potassium chloride (KCl) aqueous solution." *Journal of Solid State Chemistry* 148 (1):81–84.

Lee, Kwang Hoon, Young-Woo Lee, Seung Woo Lee, Jeong Sook Ha, Sang-Soo Lee, and Jeong Gon Son. 2015. "Ice-templated self-assembly of $VOPO_4$–Graphene nanocomposites for vertically porous 3D supercapacitor electrodes." *Scientific Reports* 5:13696.

Lee, Minoh, Suresh Kannan Balasingam, Hu Young Jeong, Won G Hong, Byung Hoon Kim, and Yongseok Jun. 2015. "One-step hydrothermal synthesis of graphene decorated V_2O_5 nanobelts for enhanced electrochemical energy storage." *Scientific Reports* 5:8151.

Lee, Myungsup, Boon-Hong Wee, and Jong-Dal Hong. 2015. "High performance flexible supercapacitor electrodes composed of ultralarge graphene sheets and vanadium dioxide." *Advanced Energy Materials* 5 (7):1401890.

Li, Hong-Yi, Chuang Wei, Liang Wang, Qi-Sang Zuo, Xinlu Li, and Bing Xie. 2015. "Hierarchical vanadium oxide microspheres forming from hyperbranched nanoribbons as remarkably high performance electrode materials for supercapacitors." *Journal of Materials Chemistry A* 3 (45):22892–22901.

Li, Linlin, Shengjie Peng, Hao Bin Wu, Le Yu, Srinivasan Madhavi, and Xiong Wen Lou. 2015. "A flexible quasi-solid-state asymmetric electrochemical capacitor based on hierarchical porous V_2O_5 nanosheets on carbon nanofibers." *Advanced Energy Materials* 5 (17):1500753.

Li, Meili, Guoying Sun, Pingping Yin, Changping Ruan, and Kelong Ai. 2013. "Controlling the formation of rodlike V_2O_5 nanocrystals on reduced graphene oxide for high-performance supercapacitors." *ACS Applied Materials & Interfaces* 5 (21):11462–11470.

Liang, Liying, Haimei Liu, and Wensheng Yang. 2013. "Fabrication of VO_2 (B) hybrid with multiwalled carbon nanotubes to form a coaxial structure and its electrochemical capacitance performance." *Journal of Alloys and Compounds* 559:167–173.

Lin, Zongyuan, Xingbin Yan, Junwei Lang, Rutao Wang, and Ling-Bin Kong. 2015. "Adjusting electrode initial potential to obtain high-performance asymmetric supercapacitor based on porous vanadium pentoxide nanotubes and activated carbon nanorods." *Journal of Power Sources* 279:358–364.

Liu, Haimei, Yonggang Wang, Huiqiao Li, Wensheng Yang, and Haoshen Zhou. 2010. "Flowerlike vanadium sesquioxide: Solvothermal preparation and electrochemical properties." *ChemPhysChem* 11 (15):3273–3280.

Liu, Ying, Lingyang Liu, Lingbin Kong, Long Kang, and Fen Ran. 2016. "Supercapacitor electrode based on nano-vanadium nitride incorporated on porous carbon nanospheres derived from ionic amphiphilic block copolymers & vanadium-contained ion assembly systems." *Electrochimica Acta* 211:469–477.

Lukatskaya, Maria R., Bruce Dunn, and Yury Gogotsi. 2016. "Multidimensional materials and device architectures for future hybrid energy storage." *Nature Communications* 7:12647.

Mei, Peng, Yusuf Valentino Kaneti, Malay Pramanik, Toshiaki Takei, Ömer Dag, Yoshiyuki Sugahara, and Yusuke Yamauchi. 2018. "Two-dimensional mesoporous vanadium phosphate nanosheets through liquid crystal templating method toward supercapacitor application." *Nano Energy* 52:336–344.

Mu, Juyi, Jinxing Wang, Jinghua Hao, Pin Cao, Shuoqing Zhao, Wen Zeng, Bin Miao, and Sibo Xu. 2015. "Hydrothermal synthesis and electrochemical properties of V_2O_5 nanomaterials with different dimensions." *Ceramics International* 41 (10):12626–12632.

Nagaraju, Doddahalli H., Qingxiao Wang, Pierre Beaujuge, and Husam N. Alshareef. 2014. "Two-dimensional heterostructures of V_2O_5 and reduced graphene oxide as electrodes for high energy density asymmetric supercapacitors." *Journal of Materials Chemistry A* 2 (40):17146–17152.

Nazari, Mohammad, Yong Zhao, Vladimir V. Kuryatkov, Zhaoyang Fan, Ayrton A. Bernussi, and Mark Holtz. 2013. "Temperature dependence of the optical properties of VO_2 deposited on sapphire with different orientations." *Physical Review B* 87 (3):035142.

Pan, Xuan, Yong Zhao, Guofeng Ren, and Zhaoyang Fan. 2013. "Highly conductive VO_2 treated with hydrogen for supercapacitors." *Chemical Communications* 49 (38):3943–3945.

Pan, Xuan, Guofeng Ren, Md Nadim Ferdous Hoque, Stephen Bayne, Kai Zhu, and Zhaoyang Fan. 2014. "Fast supercapacitors based on graphene-bridged V_2O_3/VOx core–shell nanostructure electrodes with a power density of 1 MW kg^{-1}." *Advanced Materials Interfaces* 1 (9):1400398.

Pandit, Bidhan, Swapnil S. Karade, and Babasaheb R. Sankapal. 2017. "Hexagonal VS_2 anchored MWCNTs: First approach to design flexible solid-state symmetric supercapacitor device." *ACS Applied Materials & Interfaces* 9 (51):44880–44891.

Pang, Hongchang, Shu Jing Ee, Yongqiang Dong, Xiaochen Dong, and Peng Chen. 2014. "TiN@ VN nanowire arrays on 3D carbon for high-performance supercapacitors." *ChemElectroChem* 1 (6):1027–1030.

Peng, Huarong, Tianyu Liu, Yanhong Li, Xijun Wei, Xun Cui, Yunhuai Zhang, and Peng Xiao. 2018. "Hierarchical MoS_2-coated V_2O_3 composite nanosheet tubes as both the cathode and anode materials for pseudocapacitors." *Electrochimica Acta* 277:218–225.

Perera, Sanjaya D., Anjalee D. Liyanage, Nour Nijem, John P. Ferraris, Yves J. Chabal, and Kenneth J. Balkus Jr. 2013. "Vanadium oxide nanowire–graphene binder free nanocomposite paper electrodes for supercapacitors: A facile green approach." *Journal of Power Sources* 230:130–137.

Perera, Sanjaya D., Bijal Patel, Nour Nijem, Katy Roodenko, Oliver Seitz, John P. Ferraris, Yves J. Chabal, and Kenneth J. Balkus Jr. 2011. "Vanadium oxide nanowire–carbon nanotube binder-free flexible electrodes for supercapacitors." *Advanced Energy Materials* 1 (5):936–945.

Potiron, E., Anne Le Gal La Salle, A. Verbaere, Y. Piffard, and D. Guyomard. 1999. "Electrochemically synthesized vanadium oxides as lithium insertion hosts." *Electrochimica Acta* 45 (1–2):197–214.

Pourbaix, Marcel. 1974. *Atlas of Electrochemical Equilibria in Aqueous Solutions.* 2nd English edn. National Association of Corrosion: Houston, TX.

Qu, Q.T., Y. Shi, L.L. Li, Wan-Liang Guo, Y.P. Wu, H.P. Zhang, S.Y. Guan, and R. Holze. 2009. "V_2O_5 0.6 H_2O nanoribbons as cathode material for asymmetric supercapacitor in K_2SO_4 solution." *Electrochemistry Communications* 11 (6):1325–1328.

Rakhi, Raghavan Baby, Doddahalli H. Nagaraju, Pierre Beaujuge, and Husam N. Alshareef. 2016. "Supercapacitors based on two dimensional VO_2 nanosheet electrodes in organic gel electrolyte." *Electrochimica Acta* 220:601–608.

Ren, Guofeng, Md Nadim Ferdous Hoque, Jianwei Liu, Juliusz Warzywoda, and Zhaoyang Fan. 2016. "Perpendicular edge oriented graphene foam supporting orthogonal TiO_2 (B) nanosheets as freestanding electrode for lithium ion battery." *Nano Energy* 21:162–171.

Ren, Guofeng, Md Nadim Ferdous Hoque, Xuan Pan, Juliusz Warzywoda, and Zhaoyang Fan. 2015. "Vertically aligned VO_2 (B) nanobelt forest and its three-dimensional structure on oriented graphene for energy storage." *Journal of Materials Chemistry A* 3 (20):10787–10794.

Ren, Guofeng, Ruibo Zhang, and Zhaoyang Fan. 2018. "VO_2 nanoparticles on edge oriented graphene foam for high rate lithium ion batteries and supercapacitors." *Applied Surface Science* 441:466–473.

Ren, Guofeng, Shiqi Li, Zhao-Xia Fan, Md Nadim Ferdous Hoque, and Zhaoyang Fan. 2016. "Ultrahigh-rate supercapacitors with large capacitance based on edge oriented graphene coated carbonized cellulous paper as flexible freestanding electrodes." *Journal of Power Sources* 325:152–160.

Ren, Guofeng, Xuan Pan, Stephen Bayne, and Zhaoyang Fan. 2014. "Kilohertz ultrafast electrochemical supercapacitors based on perpendicularly-oriented graphene grown inside of nickel foam." *Carbon* 71:94–101.

Saravanakumar, Balakrishnan, Kamatchi Kamaraj Purushothaman, and Gopalan Muralidharan. 2012. "Interconnected V_2O_5 nanoporous network for high-performance supercapacitors." *ACS Applied Materials & Interfaces* 4 (9):4484–4490.

Sathiya Mariyappan, Annigere Prakash, K Ramesha, JM Tarascon, and Ashok Kumar Shukla. 2011. "V_2O_5-anchored carbon nanotubes for enhanced electrochemical energy storage." *Journal of the American Chemical Society* 133 (40):16291–16299.

Shakir, Imran, Joon Hyock Choi, Muhammad Shahid, Shaukat Ali Shahid, Usman Ali Rana, Mansoor Sarfraz, and Dae Joon Kang. 2013. "Ultra-thin and uniform coating of vanadium oxide on multiwall carbon nanotubes through solution based approach for high-performance electrochemical supercapacitors." *Electrochimica Acta* 111:400–404.

Shakir, Imran, Zahid Ali, Jihyun Bae, Jongjin Park, and Dae Joon Kang. 2014. "Layer by layer assembly of ultrathin V_2O_5 anchored MWCNTs and graphene on textile fabrics for fabrication of high energy density flexible supercapacitor electrodes." *Nanoscale* 6 (8):4125–4130.

Shao, Jie, Xinyong Li, Qunting Qu, and Honghe Zheng. 2012. "One-step hydrothermal synthesis of hexangular starfruit-like vanadium oxide for high power aqueous supercapacitors." *Journal of Power Sources* 219:253–257.

Tang, Kexin, Yuping Li, Yujiao Li, Hongbin Cao, Zisheng Zhang, Yi Zhang, and Jun Yang. 2016. "Self-reduced VO/VOx/carbon nanofiber composite as binder-free electrode for supercapacitors." *Electrochimica Acta* 209:709–718.

Wang, Bei, Konstantin Konstantinov, David Wexler, Hao Liu, and GuoXiu Wang. 2009. "Synthesis of nanosized vanadium pentoxide/carbon composites by spray pyrolysis for electrochemical capacitor application." *Electrochimica Acta* 54 (5):1420–1425.

Wang, Hong-En, Xu Zhao, Kaili Yin, Yu Li, Lihua Chen, Xiaoyu Yang, Wenjun Zhang, Bao-Lian Su, and Guozhong Cao. 2017. "Superior pseudocapacitive lithium-ion storage in porous vanadium oxides@ C heterostructure composite." *ACS Applied Materials & Interfaces* 9 (50):43665–43673.

Wang, Huanwen, Huan Yi, Xiao Chen, and Xuefeng Wang. 2014. "One-step strategy to three-dimensional graphene/VO_2 nanobelt composite hydrogels for high performance supercapacitors." *Journal of Materials Chemistry A* 2 (4):1165–1173.

Wang, Nannan, Yifu Zhang, Tao Hu, Yunfeng Zhao, and Changgong Meng. 2015. "Facile hydrothermal synthesis of ultrahigh-aspect-ratio V_2O_5 nanowires for high-performance supercapacitors." *Current Applied Physics* 15 (4):493–498.

Wang, Rutao, Junwei Lang, Peng Zhang, Zongyuan Lin, and Xingbin Yan. 2015. "Fast and large lithium storage in 3D porous VN nanowires–graphene composite as a superior anode toward high-performance hybrid supercapacitors." *Advanced Functional Materials* 25 (15):2270–2278.

Wei, Di, Maik R.J. Scherer, Chris Bower, Piers Andrew, Tapani Ryhänen, and Ullrich Steiner. 2012. "A nanostructured electrochromic supercapacitor." *Nano Letters* 12 (4):1857–1862.

Wu, Yingjie, Guohua Gao, Huiyu Yang, Wenchao Bi, Xing Liang, Yuerou Zhang, Guyu Zhang, and Guangming Wu. 2015. "Controlled synthesis of V_2O_5/MWCNT core/shell hybrid aerogels through a mixed growth and self-assembly methodology for supercapacitors with high capacitance and ultralong cycle life." *Journal of Materials Chemistry A* 3 (30):15692–15699.

Xia, Chuan, Zifeng Lin, Yungang Zhou, Chao Zhao, Hanfeng Liang, Patrick Rozier, Zhiguo Wang, and Husam N Alshareef. 2018. "Large Intercalation Pseudocapacitance in 2D VO_2 (B): Breaking through the Kinetic Barrier." *Advanced Materials* 30 (40):1803594.

Xia, Xinhui, Dongliang Chao, Chin Fan Ng, Jianyi Lin, Zhanxi Fan, Hua Zhang, Ze Xiang Shen, and Hong Jin Fan. 2015. "VO_2 nanoflake arrays for supercapacitor and Li-ion battery electrodes: Performance enhancement by hydrogen molybdenum bronze as an efficient shell material." *Materials Horizons* 2 (2):237–244.

Xiao, Xu, Xiang Peng, Huanyu Jin, Tianqi Li, Chengcheng Zhang, Biao Gao, Bin Hu, Kaifu Huo, and Jun Zhou. 2013. "Freestanding mesoporous VN/CNT hybrid electrodes for flexible all-solid-state supercapacitors." *Advanced Materials* 25 (36):5091–5097.

Xiao, Xuxian, Shun Li, Hua Wei, Dan Sun, Yuanzhan Wu, Guizhen Jin, Fen Wang, and Yingping Zou. 2015. "Synthesis and characterization of VO_2 (B)/graphene nanocomposite for supercapacitors." *Journal of Materials Science: Materials in Electronics* 26 (6):4226–4233.

Xu, Yunling, Jie Wang, Bing Ding, Laifa Shen, Hui Dou, and Xiaogang Zhang. 2015. "General strategy to fabricate ternary metal nitride/carbon nanofibers for supercapacitors." *ChemElectroChem* 2 (12):2020–2026.

Yan, Yan, Hao Xu, Wei Guo, Qingli Huang, Mingbo Zheng, Huan Pang, and Huaiguo Xue. 2016. "Facile synthesis of amorphous aluminum vanadate hierarchical microspheres for supercapacitors." *Inorganic Chemistry Frontiers* 3 (6):791–797.

Yu, Minghao, Yan Zeng, Yi Han, Xinyu Cheng, Wenxia Zhao, Chaolun Liang, Yexiang Tong, Haolin Tang, and Xihong Lu. 2015. "Valence-optimized vanadium oxide supercapacitor electrodes exhibit ultrahigh capacitance and super-long cyclic durability of 100 000 cycles." *Advanced Functional Materials* 25 (23):3534–3540.

Zeng, HM, Y Zhao, YJ Hao, QY Lai, JH Huang, and XY Ji. 2009. "Preparation and capacitive properties of sheet V_6O_{13} for electrochemical supercapacitor." *Journal of Alloys and Compounds* 477 (1–2):800–804.

Zhai, Teng, Xihong Lu, Yichuan Ling, Minghao Yu, Gongming Wang, Tianyu Liu, Chaolun Liang, Yexiang Tong, and Yat Li. 2014. "A new benchmark capacitance for supercapacitor anodes by mixed-valence sulfur-doped V_6O_{13}–x." *Advanced Materials* 26 (33):5869–5875.

Zhang, Yifu, Jiqi Zheng, Tao Hu, Fuping Tian, and Changgong Meng. 2016. "Synthesis and supercapacitor electrode of VO_2 (B)/C core–shell composites with a pseudocapacitance in aqueous solution." *Applied Surface Science* 371:189–195.

Zhang, Yifu, Xuyang Jing, Yan Cheng, Tao Hu, and Meng Changgong. 2018. "Controlled synthesis of 3D porous VO_2 (B) hierarchical spheres with different interiors for energy storage." *Inorganic Chemistry Frontiers* 5 (11):2798–2810.

Zhao, Danyang, Qiancheng Zhu, Dejian Chen, Xi Li, Ying Yu, and Xintang Huang. 2018. "Nest-like V_3O_7 self-assembled by porous nanowires as an anode supercapacitor material and its performance optimization through bonding with N-doped carbon." *Journal of Materials Chemistry A* 6 (34):16475–16484.

Zhao, Yong, Gulten Karaoglan-Bebek, Xuan Pan, Mark Holtz, Ayrton A Bernussi, and Zhaoyang Fan. 2014. "Hydrogen-doping stabilized metallic VO_2 (R) thin films and their application to suppress Fabry-Perot resonances in the terahertz regime." *Applied Physics Letters* 104 (24):241901.

Zhao, Yong, Ji Hao, Changhong Chen, and Zhaoyang Fan. 2011. "Electrically controlled metal–insulator transition process in VO_2 thin films." *Journal of Physics: Condensed Matter* 24 (3):035601.

Zhao, Yong, Joon Hwan Lee, Yanhan Zhu, M Nazari, Changhong Chen, Haiyan Wang, Ayrton Bernussi, Mark Holtz, and Zhaoyang Fan. 2012. "Structural, electrical, and terahertz transmission properties of VO_2 thin films grown on c-, r-, and m-plane sapphire substrates." *Journal of Applied Physics* 111 (5):053533.

Zheng, Jiqi, Yifu Zhang, Qiushi Wang, Hanmei Jiang, Yanyan Liu, Tianming Lv, and Changgong Meng. 2018. "Hydrothermal encapsulation of VO_2 (A) nanorods in amorphous carbon by carbonization of glucose for energy storage devices." *Dalton Transactions* 47 (2):452–464.

Zhu, Jixin, Liujun Cao, Yingsi Wu, Yongji Gong, Zheng Liu, Harry E Hoster, Yunhuai Zhang, Shengtao Zhang, Shubin Yang, and Qingyu Yan. 2013. "Building 3D structures of vanadium pentoxide nanosheets and application as electrodes in supercapacitors." *Nano Letters* 13 (11):5408–5413.

Zhu, Weihua, Ruizhi Li, Pan Xu, Yuanyuan Li, and Jinping Liu. 2017. "Vanadium trioxide@carbon nanosheet array-based ultrathin flexible symmetric hydrogel supercapacitors with 2 V voltage and high volumetric energy density." *Journal of Materials Chemistry A* 5 (42):22216–22223.

Zhu, Yue, Lele Peng, Dahong Chen, and Guihua Yu. 2015. "Intercalation pseudocapacitance in ultrathin $VOPO_4$ nanosheets: Toward high-rate alkali-ion-based electrochemical energy storage." *Nano Letters* 16 (1):742–747.

5 Future Prospects and Challenges of Inorganic-Based Supercapacitors

Vivian C. Akubude and Kevin N. Nwaigwe

CONTENTS

5.1 Concept of Supercapacitors 79
5.2 Materials for Supercapacitors 79
 5.2.1 Electrodes 80
 5.2.2 Separator 80
5.3 Organic versus Inorganic Based Supercapacitors 82
5.4 Future Prospects 83
5.5 Challenges 84
References 84

5.1 CONCEPT OF SUPERCAPACITORS

Supercapacitor is a multi-layered energy storage device that stores electrical energy in form of charges by adsorption of electrolyte ions onto the surface of electrode material. It is composed of two electrodes separated by an ion-porous medium in an electrolyte (El-Kady et al., 2012). It can also be referred to as electrochemical capacitor (EC capacitor), electrochemical double-layer capacitors (EDLCs) or ultra-capacitor. In charged state all the positive ions travel to the negative terminal and in discharged state all the ions are distributed randomly within the cell and vice versa (Kai and Matti, 2014)

5.2 MATERIALS FOR SUPERCAPACITORS

The properties/characteristics of a supercapacitors depends largely on the materials used in its building process especially the electrodes and the electrolytes as this determines the functionality in terms of thermal and electrical characteristics. These components work together to give the supercapacitor its overall performance. Therefore, careful selection of materials becomes imperative as this will also determine its application. Supercapacitors comprises of the following component parts.

5.2.1 Electrodes

Supercacitors are composed of two electrodes, which are usually made from porous and spongy material with high surface area. They are electrically connected to the conductive current/load collector. A good electrode must serve for the purpose of supercapacitor and should have good conductivity, high temperature stability, long-term chemical stability, high corrosion resistance, low cost and environmental friendly (Wikipedia, 2016). The electrode pores determine the capacitance, equivalent series resistance (ESR), specific power and the specific energy of the supercapacitor. High increase in capacitance, ESR, specific energy and decrease in specific power is achieved by smaller electrode pores. Supercapacitors are classified based on the electrode material used, as shown in Table 5.1.

5.2.2 Separator

This is a thin material used to disassociate the two electrodes to avoid short circuit by direct contact. They serve as electrolyte reservoirs and maintain ionic integrity. Separators used for building supercapacitors should be chemically inert, porous to the conducting ions and inexpensive. A very porous separator reduces the ESR and the electrolyte stability and conductivity is ensured by its inertness. If organic electrolytes are used, polymer (typically PP) or paper separators are applied. With aqueous electrolytes glass fiber separators as well as ceramic separators are possible (Adrian and Roland, 2000).

1. *Load/current collector:* This connects the electrodes to the capacitor terminals. This could be either sprayed onto the electrode, or metal foil may be used. A good collector should be able to distribute peak current (Wikipedia, 2016).
2. *Electrolytes:* These consist of solvents and chemicals that when dissolved separates into cations (–ve) and anions (+ve), which in turn creates electrical conductive between the two electrodes. The higher the ion content the better the conductivity of the electrolytes. Electrolyte used in supercapacitors are either organic (such as acetonitrile, propylene carbonate, tetrahydrofuran, diethyl carbonate, γ-butyrolactone and solutions with quaternary ammonium salts or alkyl ammonium salts) or inorganic (such as sulfuric acid, potassium hydroxide, quaternary phosphonium salts, sodium perchlorate, lithium perchlorate, lithium hexafluoride arsenate etc) in nature. Inorganic or aqueous electrolytes are the most basic type of electrolyte used in supercapacitors. The choice of electrolyte used will determine the operating voltage, temperature range, ESR and capacitance of the supercapacitor (Dong et al., 2015; Zhao et al., 2015). The choice between inorganic and organic electrolyte depends on the resistance, the capacitance, the manufacturability and the potential window size in which the system is electrochemically stable (Tanahashi et al., 1990).

TABLE 5.1
Classification of Supercapacitor Based on Electrode Type and Mechanism

Type	Description	Material Used	Benefits	Drawbacks	References
Electrochemical double-layer capacitors	The charge mechanism is non-faradaic and no redox reaction is involved.	Carbon and its derivatives (such as activated carbon, carbon nanotubes, carbon fiber, carbon gel, templated carbon).	Application of carbon and its derivatives will give a low electrical resistivity, high specific area and controllable pore size and distribution. Also, a higher electrochemical double-layer capacitance than pseudo-capacitance and relatively long cycle life.		Mazen and Drazen (2017), Manisha et al. (2013)
Pseudocapacitor	This mechanism involves a fast reversible faradaic reactions like electrosorption, redox reactions and intercalation processes at the electrode/electrolytic interface.	Ruthenium oxide, manganese oxide and electrically conducting.	High energy density.	Low-power density and poor cycling stability.	Li and Zhao (2009), Davies and Ping (2011), Manisha et al. (2013)
Hybrid capacitors	This involves combination of two different electrodes exhibiting different characteristics.	Carbon materials and conducting polymer, carbon material and methal oxide.	It exhibit both electrostatic and electrochemical capacitance. It exhibit high storage capacity and power/energy efficiency.	High fabrication cost for the electrode material.	Mazen and Drazen (2017), Manisha et al. (2013)

5.3 ORGANIC VERSUS INORGANIC BASED SUPERCAPACITORS

The use of organic electrolyte in supercapacitors ensures a good mobility of the ions, even at decreased temperatures (Novis and McCloskey, 2005). Its limitation is seen in the risk of explosion while working at extreme high temperatures of above 70°C.

For instance, organic electrolytes like tetraethylammonium tetrafluoroborate dissolved in acetonitrile or propylene carbonate, are presently used in commercial supercapacitors with activated carbon electrodes (Ruan et al., 2015; Huang et al., 2013; Jäckel et al., 2014). Organic solvent-based electrolytes tend to give wide potential and temperature windows (up to 3.5 V, from −50°C to 70°C). Also, though the cost of organic electrolytes is not low, but at least they are affordable for commercial purposes. Notwithstanding, they are usually highly flammable, which results to safety issues.

Also, research shows that a supercapacitor can be made using inorganic molten salt (alkaline metal nitrate salts) electrolytes (as shown in Table 5.2) that allows operation of the capacitor from temperatures greater than about 110°C (Kirk and Graydon, 2013). Therefore, using inorganic electrolyte helps to extend the positive temperature range, thus reducing the risk of explosion (Khomenko et al., 2006).

TABLE 5.2
Performance of Some Inorganic Electrolytes Used in Supercapacitors

Electrolyte	Potential Window (v)	Energy Density (Wh kg^{-1})	Power Density (kW kg^{-1})	Cycle Performance	References
Li_2SO_4	1.6	24	15	85%, 1000 cycles	Luo and Xia (2009)
Na_2SO_4	1.9	13.2	—	97%, 10,000 cycles	Qu et al. (2009)
Na_2SO_4	2	54.4	37.8	92%, 1000 cycles	Lu et al. (2016)
Na_2SO_4	1.7	34.8	21.0	84%, 4000 cycles	Liu et al. (2013)
H_2SO_4	1.8	26.5	17.8	83%, 3000 cycles	Yu et al. (2009)
KOH	1.5	43.75	—	88%, 5000 cycles	Kolathodi et al. (2015)
KOH	1.6	77.8	13.5	94.3%, 3000 cycles	Yan et al. (2012)
KOH	1.7	60.9	41.1	14.3%, 10,000 cycles	Peng et al. (2015)
KOH	1.8	50.3	—	100.9%, 6000 cycles	Li et al. (2016)
KOH	1.8	31	—	100%, 5500 cycles	Wang et al. (2015)
HQ/H_2SO_4	1	31.3	—	65%, 4000 cycles	Roldan et al. (2011)
KOH	1.8	21.1	3.59	87.42%, 10,000	Yu et al. (2014)
Na_2SO_4	1.8	11.3	9.1	85%, 4000 cycles	Xu et al. (2015)
NaOH	1.8	43.5	11.8	91.5%, 20,000 cycles	Zhou et al. (2013)
Na_2SO_4	1.9	19.5	0.13	97%, 10,000 cycles	Qu et al. (2009)
K_2SO_4	1.8	25.3	0.14	98%, 10,000 cycles	Qu et al. (2009)

(*Continued*)

TABLE 5.2 (*Continued*)
Performance of Some Inorganic Electrolytes Used in Supercapacitors

Electrolyte	Potential Window (v)	Energy Density (Wh kg^{-1})	Power Density (kW kg^{-1})	Cycle Performance	References
KOH	1.5	—	—	50%, 1000 cycles	Wang et al. (2008)
Na_2SO_4	2	18.2	10.1	92%, 2500 cycles	Gao et al. (2011)
Na_2SO_4	1.6	21.2	0.82	84.4%, 1000 cycles	Wu et al. (2014)
Na_2SO_4	—	30.2	14.5	83.4%, 5000 cycles	Gao et al. (2012)
KOH	1.6	31.2	396	82%, 3000 cycles	Xu et al. (2014)
$LiCLO_4$	1.8	86	25	85%, 2000 cycles	Hou et al. (2014)
KOH	1.6	22	19.5	80%, 4000 cycles	Wang et al. (2012a)
KOH	1.4	32	0.7	94%, 2000 cycles	Wang et al. (2012b)
NaOH	1.6	43.5	5.5	19.5%, 20,000 cycles	Zhou et al. (2013)
KOH	1.3	23.3	0.32	93%, 2500 cycles	Wang et al. (2012c)

The constraint of the inorganic-based supercapacitors is related to the nominal cell voltage level, which is lower when compared to organic-based supercapacitors. Also, the inorganic electrolytes are characterized by high value of conductivity than organic ones (Kötz and Carlen, 2000; Marie-Francoise et al., 2005). Research reviews that for the same activated carbon electrode an inorganic electrolyte achieves capacitance values of 160 F/g, while an organic electrolyte achieves only 100 F/g. The organic-based supercapacitor has high specific energy while the inorganic-based type has high specific power, and increased life cycle (Novis and McCloskey, 2005; Belyakov, 2008).

5.4 FUTURE PROSPECTS

Supercapacitors can be used either alone as a primary power source or as a supplementary one with rechargeable batteries for high-power applications, such as industrial mobile equipment and hybrid/electric vehicles (Dar et al., 2013). In automobile industry, supercapacitors act as support for batteries, power filtering, coupling and as buffer during acceleration and braking. This action helps to lower the operational cost and extend the battery life. It also protects the battery from the harmful effect of peak loads. Its application in regenerative braking helps to recover power in cars and electric mass transit vehicles that would otherwise lose braking energy as heat (Simon et al., 2014; Shukla et al., 2001). Cars with hybrid systems are also equipped with supercapacitors, which act as components of high power density. Researchers have developed hybrid gasolin-electric buses with available energy greater than 400 Wh. This technology offers at least 25% fuel saving and cutting the CO_2 emissions by up to 90% also, it is used as secondary power source in fuel cell vehicles for peak load leveling operations such as fuel starting, acceleration and braking, and this can improve the performance, efficiency and cleanliness in electric and hybrid vehicle technology (Shukla et al., 2001; Schöttle and Threin, 2000; Fuglevand, 2002; Raiser, 2006; Pearson, 2004). For rapid power delivery and recharging (i.e., high power density), supercapacitors have found applications in household appliances, electronic tools, mobile telephones, cameras, etc.

5.5 CHALLENGES

The constraints facing the use of inorganic electrolytes in supercapacitor design are highlighted below:

1. Low temperature range for several supercapacitor applications. This can be improved for supercapacitor operations by using some additives like ethylene glycol to decrease the lower temperature limit thereby increasing the working temperature (Roberts et al., 2013).
2. The performances of the supercapacitors are limited by the decomposition voltage of the electrolyte, which is 1.2 V for inorganic-based supercapacitors (Puscas et al., 2010). The electrochemical stable potential window of the electrolyte directly determines the supercapacitor operational cell voltage, which affects both the energy and power densities. Electrolytes with higher electrochemical stable potential window values allow increased cell voltage of the supercapacitor, which can significantly improve the energy density (Cheng et al., 2015).

REFERENCES

Adrian, S. and Roland, G. 2000. *Properties and Applications of Supercapacitors from the State-of-the-art to Future Trends. Proceeding PCIM*. Rossens, Switzerland.

Belyakov, A.I. 2008. *Asymetric Electrochemical Supercapacitors with Aqueous Electrolytes. ESSCAP*, Roma, Italy.

Cheng, Z., Yida, D., Wenbin, H., Jinli, Q., Lei, Z. and Jiujun, Z. 2015. A review of electrolyte materials and compositions for electrochemical supercapacitors. *Chem. Soc. Rev.* 44:7484–7539.

Dar, F.I., Moonooswamy, K.R. and Es-Souni, M. 2013. Morphology and property control of NiO nanostructures for supercapacitor applications. *Nanoscale Res. Lett.* 8:363.

Davies, A. and Ping, Y.A. 2011. Material advancements in supercapacitors: From activated carbon to carbon nanotube and grapheme. *Can. J. Chem. Eng.* 89 (6):1342.

Dong, L.B., Xu, C.J., Yang, Q., Fang, J., Li, Y and Kang, F. 2015. High-performance compressible supercapacitors based on functionally synergic multiscale carbon composite textiles. *J. Mater. Chem. A* 3:4729.

El-Kady, M.F., Strong, V., Dubin, S. and Kaner, R.B. 2012. Laser scribing of high-performance and flexible grapheme-based electrochemical capacitors. *Science* 335 (6074):1326–1330.

Fuglevand, W. 2002. Avista Laboratories. Fuel cell power system, method of distributing power, and method of operating a fuel cell power system. WO patent/2002/095851.

Gao, P.C., Lu, A.H. and Li, W.C. 2011. Dual functions of activated carbon in a positive electrode for MnO_2-based hybrid supercapacitor. *J. Power Sources* 196:4095–4101. doi:10.1016/j.jpowsour.2010.12.056.

Hou, Y., Chen, L., Liu, P., Kang, J., Fujita, T. and Chen, M. 2014. Nano-porous metal based flexible asymmetric pseudocapacitors. *J. Mater. Chem. A2* 10910–10916. doi:10.1039/c4ta00969j.

Huang, B., Sun, X., Zhang, X., Zhang, D. and Ma, Y. 2013. Organic electrolytes for activated carbon-based supercapacitors with flexible package. *Acta Phys. Chim. Sin.* 29:1998–2004.

Jäckel, N., Weingarth, D., Zeiger, M., Aslan, M., Grobelsek, I. and Presser, V. 2014. Comparison of carbon onions and carbon blacks as conductive additives for carbon supercapacitors in organic electrolytes. *J. Power Sources* 272:1122–1133. doi: 10.1016/j.jpowsour.2014.08.090.

Kai, V. and Matti, N. 2014. *Supercapacitors-Basics and Applications*. Skeletontech, Bautzen, Germany.

Khomenko, V., Raymundo-Pinero, E. and Béguin, F. 2006. Optimization of an asymmetric manganese oxide/activated carbon capacitor working at 2 V in aqueous medium. *Carbon* 153:183–190.

Kirk, D.W. and Graydon, J.W. 2013. A low temperature inorganic molten salt supercapacitor. *ECS Trans.* 53 (31):27–33.

Kolathodi, M.S., Palei, M. and Natarajan, T.S. 2015. Electrospun NiO nanofibers as cathode materials for high performance asymmetric supercapacitors. *J. Mater. Chem. A* 3 (14):7513–7522.

Kötz, R. and Carlen, M. 2000. Principles and applications of electrochemical capacitors. *Electrochim. Acta* 45:2483–2498.

Li, R., Lin, Z., Ba, X., Li, Y., Ding, R. and Liu, J. 2016. Integrated copper–nickel oxide mesoporous nanowire arrays for high energy density aqueous asymmetric supercapacitors. *Nanoscale Horiz.* 1:150–155.

Liu, X., Zhang, N., Ni, J. and Gao, L. 2013. Improved electrochemical performance of sol–gel method prepared $Na_4Mn_9O_{18}$ in aqueous hybrid Na-ion supercapacitor. *J. Solid State Electrochem.* 17 (7):1939–1944.

Lu, K., Song, B., Gao, X., Dai, H., Zhang, J. and Ma, H. 2016. High-energy cobalt hexacyanoferrate and carbon micro-spheres aqueous sodium-ion capacitors. *J. Power Sources* 303:347–353.

Luo, J.Y. and Xia, Y.Y. 2009. Electrochemical profile of an asymmetric supercapacitor using carbon-coated $LiTi_2(PO_4)_3$ and active carbon electrodes. *J. Power Sources* 186 (1):224–227.

Manisha, V., Tonya, P. and Li, J. 2013. Supercapacitors: Review of materials and fabrication methods. *J. Energ. Eng.* 139 (2):72–79.

Marie-Francoise, J.N., Gualous, H. Outbib, R. and Berthon, A. 2005. 42 V Power Net with supercapacitor and battery for automotive applications. *J. Power Sources* 143:275–283.

Mazen, Y. and Drazen, F. 2017. Performance of commercially available supercapacitors. *Energies* 10:1340. doi:10.3390/en10091340.

Novis, S.W. and McCloskey, L. 2005. Non-aqueous electrolytes for electrical storage devices, Patent, Philadelphia, PA.

Pearson, M. 2004. Ballard Power Systems. Power supply and ultracapacitor based battery simulator. US patent 2004/228055.

Peng, S., Li, L., Wu, H.B., Madhavi, S., Lou, X.W.D. 2015. Controlled growth of $NiMoO_4$ nanosheet and nanorod arrays on various conductive substrates as advanced electrodes for asymmetric supercapacitors. *Adv. Energy Mater.* 5 (2):1401172.

Puscas, A.M., Carp, M.C., Borza, P.N. and Coquery, G. 2010. Electric and thermal characterization of inorganic supercapacitors. *Bull. Transilvania* 3:52.

Qu, Q.T., Shi, Y., Tian, S., Chen, Y.H., Wu, Y.P. and Holze, R. 2009. A new cheap asymmetric aqueous supercapacitor: Activated carbon/$NaMnO_2$. *J. Power Sources* 194 (2):1222–1225.

Raiser, S. 2006. General Motors Corporation. Hybrid fuel cell system with battery capacitor energy storage system. WOpatent/2006/065364.

Roberts, A.J., De Namor, A.F.D. and Slade, R.C.T. 2013. Low temperature water based electrolytes for MnO_2/carbon supercapacitors. *Phys. Chem. Chem. Phys.* 15:3518–3526.

Roldan, S., Blanco, C., Granda, M., Menendez, R. and Santamaria, R. 2011. Towards a further generation of high-energy carbon-based capacitors by using redox-active electrolytes angewandte chemie. *Int. Ed.* 50 (7):1699–1701.

Ruan, D.B., Gu, G.S., Chen, Z.R., Yang, B. ISEE'Cap2015. *4th International Symposium on Enhanced Electrochem.* Capacitors, 2015 8–12 June, Montpellier, France.

Schottle, R. and Threin, G. 2000. Electrical power supply systems: Present and future. *VDI Berichte* 1547:449–476.

Shukla, A.K., Arico, A.S. and Antonucci, V. 2001. An appraisal of electric automobile power source. *Renew. Sust. Energ. Rev.* 5 (2):137–155.

Simon, P. and Gogotsi, Y. and Dunn, B. 2014. Where do batteries end and supercapacitors begin?. *Sci. Magaz.* 343 (6176):1210–1211.

Tanahashi, I., Yoshida, A. and Nishino, A. 1990. Comparison of the electrochemical properties of electric double-layer capacitors with an aqueous electrolyte and with a nonaqueous electrolyte. *Bull. Chem. Soc. Jpn.* 63:3611–3614.

Wang, X., Li, M., Chang, Z., Wang, Y., Chen, B., Zhang, L. and Wu, Y. 2015. Orientated Co_3O_4 nanocrystals on MWCNTs as superior battery-type positive electrode material for a hybrid capacitor. *J. Electrochem. Soc.* 162 (10):A1966–A1971.

Wang, X., Liu, W.S., Lu, X. and Lee, P.S. 2012c. Dodecyl sulfate-induced fast faradic process in nickel cobalt oxide–reduced graphite oxide composite material and its application for asymmetric supercapacitor device. *J. Mater. Chem.* 22:23114–23119. doi:10.1039/C2JM35307E.

Wang, X., Sumboja, A., Lin, M., Yan, J. and Lee, P.S. 2012a. Enhancing electrochemical reaction sites in nickel-cobalt layered double hydroxides on zinc tin oxide nanowires: A hybrid material for an asymmetric supercapacitor device. *Nanoscale* 4:7266–7272. doi:10.1039/c2nr31590d.

Wang, X., Wang, Y.Y., Zhao, C.M., Zhao, Y.X., Yan, B.Y. and Zheng, W.T. 2012b. Electrodeposited $Ni(OH)_2$ nanoflakes on graphite nanosheets prepared by plasma-enhanced chemical vapor deposition for supercapacitor electrode. *New J. Chem.* 36:1902–1906. doi:10.1039/C2NJ40308K.

Wikipedia. 2016. Supercapacitors. Retrieved from https://en.wikipedia.org/wiki/Supercapacitor (accessed February 16, 2016).

Xu, H., Hu, X., Yang, H., Sun, Y., Hu, C. and Huang, Y. 2015. Flexible asymmetric micro-supercapacitors based on Bi_2O_3 and MnO_2 nanoflowers: Larger areal mass promises higher energy density. *Adv. Energy Mat.* 5 (6):1401882.

Xu, K., Zou, R., Li, W., Liu, Q., Liu, X., An, L. and Hu, J. 2014. Design and synthesis of 3D interconnected mesoporous $NiCo_2O_4$@$CoxNi_{1-x}(OH)_2$ core-shell nanosheet arrays with large areal capacitance and high rate performance for supercapacitors. *J. Mater. Chem. A* 2:10090–10097. doi:10.1039/C4TA01489H.

Yan, J., Fan, Z., Sun, W., Ning, G., Wei, T., Zhang, Q., Zhang, R., Zhi, L. and Wei, F. 2012. Advanced asymmetric supercapacitors based on $Ni(OH)_2$/graphene and porous graphene electrodes with high energy density. *Adv. Funct. Mater.* 22 (12):2632–2641.

Yu, N., Gao, L., Zhao, S. and Wang, Z. 2009. Electrodeposited PbO_2 thin film as positive electrode in PbO_2/AC hybrid capacitor. *Electrochim. Acta* 54 (14):3835–3841.

Yu, X.Z., Lu, B.G. and Xu, Z. 2014. Super long-life supercapacitors based on the construction of nanohoneycomb-like strong coupled $CoMoO_4$-3D graphene hybrid electrode. *Adv. Mat.* 26 (7):1044–1051.

Zhao, C., Shu, K.W., Wang, C.Y., Gambhir, S. and Wallace, G.G. 2015. Reduced graphene oxide and polypyrrole/reduced graphene oxide composite coated stretchable fabric electrodes for supercapacitor application. *Electrochim. Acta* 172:12–19.

Zhou, C., Zhang, Y., Li, Y. and Liu, J. (2013). Construction of high-capacitance 3D CoO@ polypyrrole nanowire array electrode for aqueous asymmetric supercapacitor. *Nano Lett.* 13:2078–2085. doi:10.1021/nl400378j.

6 Tungsten Based Materials for Supercapacitors

Christelle Pau Ping Wong,
Chin Wei Lai, and Kian Mun Lee

CONTENTS

6.1 Introduction 89
6.2 Basic Principle of Supercapacitor 90
  6.2.1 Electrochemical Double-Layer Capacitors (EDLCs) 90
  6.2.2 Pseudocapacitor 91
  6.2.3 Performance Assessments 91
6.3 Tungsten Material for Supercapacitor 92
  6.3.1 Tungsten Trioxide (WO_3) 92
  6.3.2 Tungsten Disulfide (WS_2) 92
6.4 Hybrid-Supercapacitor 94
  6.4.1 Tungsten Trioxide-Carbon 94
  6.4.2 Tungsten Trioxide-Conducting Polymers 96
  6.4.3 Tungsten Disulfide-Carbon 96
  6.4.4 Tungsten Disulfide-Conducting Polymers 96
6.5 Conclusions 97
Acknowledgments 97
References 97

6.1 INTRODUCTION

The early work on capacitor technology was reported by Ewald Georg von Kleist in 1745, which inspired from the invention of Leyden jar. In 1876, Fitzgerald invented a wax-impregnated paper dielectric capacitor with foil electrodes and was used in radio receiver [1]. This conventional capacitor composed of two active materials that are separated by a dielectric. The dielectric (non-conducting) is interpolated between two electrodes, which can be air or paper with dielectric strength of 3×10^6 V m^{-1} and 16×10^6 V m^{-1}, respectively [2]. The overall capacitance and working potential of capacitor depend upon the dielectric. The discovery of conventional capacitor had inspired scientists and researchers, whom seek efficient energy storage devices.

After a few years, Charles Pollak patented a borax electrolyte aluminum electrolytic capacitor, which is then commercialized in late 1920s. These electrolytic capacitors were then considered as second generation capacitor, which show similar in the cell design to battery. Electrolytic capacitor, also known as polarized capacitor, is made up of two conductive electrodes (i.e., aluminum, tantalum and niobium) and a paper fully soaked in an electrolyte, acting as a separator between the electrodes [3]. The particular metal on electrode is coated with an insulating layer, acting as a dielectric in electrolytic capacitor. Because of the very thin dielectric oxide layer, a high volumetric capacitance can be achieved by combining sufficient dielectric strength. The destruction of electrolytic capacitor may happen if the polarity is reversed resulting in limited lifespan.

Taking this into account, electrochemical double-layer capacitors (EDLCs) are emerged as third generation of capacitor – until now. In 1853, the concept of electrical double-layer capacitance was first proposed by Hermann von Helmholtz. EDLCs stored charge at interface of electrode/electrolyte. The first EDLC was patented by General Electric Company in 1957 but did not commercialize their invention. In 1966, researchers from Standard Oil of Ohio (SOHIO) developed another EDLC and licensing their invention to Nippon Electric Corporation (NEC), which finally introduced to market in 1971, and was used as power backup in computer memory [2]. These capacitors use activated carbon (AC) as the active material for the anode and cathode in aqueous or organic electrolyte. This breakthrough has triggered the subsequent interests in energy storage devices research by scientists and researchers from all over the world on electrode materials as an important component in supercapacitor applications.

6.2 BASIC PRINCIPLE OF SUPERCAPACITOR

Supercapacitor is composed of three main basic components; they are electrode, electrolyte and separator. Electrode materials considered as the major factor to determine the performance of supercapacitor based on their storage mechanisms (EDLCs and pseudocapacitance). EDLCs store energy through the electrostatic charges accumulate at the electrode-electrolyte interface (non-faradaic). On the other hand, pseudocapacitors store energy by involving reversible oxidation/reduction reactions between electrolyte and electroactive materials (Faradaic) [4,5]. Table 6.1 summarizes electrode materials, storage mechanism, advantages and disadvantages of both capacitors.

6.2.1 Electrochemical Double-Layer Capacitors (EDLCs)

EDLCs, double-layer charge storage is a surface process and thus making it high reversible and longer the cycle life of supercapacitor. There is no electrochemical reaction involved between the interface of electrode and electrolyte. Several models were developed for the concept of electrical double layer, including Helmholtz model, Gouy-Chapman model and Stern model [7]. Helmholtz model is the earliest model of electrical double layer, which proposed that the charge at electrode surface is neutralized by forming a double layer of counter ions at a distance. Gouy and Chapman further modified the Helmholtz model, suggesting that the distribution of ion is determined by

TABLE 6.1
Different Types of Supercapacitor Devices

Supercapacitor	Electrode Material	Storage Mechanism	Advantage	Disadvantage
Electrochemical double-layer capacitors (EDLCs)	Carbonaceous: AC, CNT, graphene, rGO and carbon aerogel	Non-faradaic	Long cycle life	Low energy and power density
Pseudocapacitors	Transition metal oxides and conducting polymers	Faradaic	High specific capacitance, energy and power density	Short cycle life

Source: Sk, M.M. et al., *J. Power Sources*, 308, 121–140, 2016.

thermal motion, called diffusion layer. However, this model overestimated the electrical double-layer capacitance and thus Stern combined Helmholtz model and Gouy-Chapman model, suggesting two regions of ion distribution, which are inner Helmholtz plane (IHP) and outer Helmholtz plane (OHP). In Stern layer, specifically adsorbed ions accumulated on IHP while non-specifically adsorbed ions attached on OHP upon polarization [8]. EDLC normally utilizing material with high porosity and surface area, which are carbon-based materials such as activated carbon (AC), carbon nanotubes (CNTs), carbon aerogel and reduced graphene oxide (rGO). Among them, rGO is generally attractive as electrode materials since 2004 due to its incredible properties and structures that plays vital role in electrochemical performance.

6.2.2 Pseudocapacitor

Pseudocapacitor generates capacitance via three faradaic processes: (i) reversible adsorption; oxidation/reduction reaction of transition metal oxides; and (iii) electrochemical doping-undoping of conducting polymer [9]. Pseudocapacitor possess larger specific capacitance (10–100 times higher) and energy density as compared with EDLC. Nevertheless, redox reaction rate is usually slower than that of the non-faradaic process, which results in low power density and instability during cycling process. The most commonly known pseudocapacitance active materials are transition metal oxides such as RuO_2, ZnO, Fe_2O_3, MgO, WO_3, etc. and conducting polymers including PANI, PEDOT, PPy, etc.

6.2.3 Performance Assessments

Evaluating the capacitance value of supercapacitor is important to serve as a trademark and gives the information of where tested materials would meet the requirement as an electrode material. A series of tests and their equations are listed in Table 6.2 to calculate the capacitance, energy density and power density of supercapacitor [10].

TABLE 6.2
Performance Assessment of Supercapacitor Using Different Characterization Techniques

Performance Test	Equations	Remarks
Cyclic voltammetry (CV)	$C = \frac{\int idt}{dV \times SR}$	• Used in 3 electrode system • Observed the occurrence of redox peaks and repeating deviation
Galvanostatic charge/ discharge (GCD)	$C = \frac{I \times dt}{dV}$	• Best determines from slope of discharge curves
Energy density	$E = \frac{1}{2}CV^2$	• Describes how much the energy can be stored • Assessment of practical performance
Power density	$P_{max} = \frac{V^2}{4ESR} \text{ or } \frac{E}{t}$	• Describes how fast the energy can be delivered • Assessment of practical performance

6.3 TUNGSTEN MATERIAL FOR SUPERCAPACITOR

6.3.1 Tungsten Trioxide (WO_3)

Among the different transition metal oxides, tungsten trioxide (WO_3), an *n*-type semiconductor, is one of the most capable candidates for enhancement in electrochemical performance of supercapacitor applications because of its high intrinsic densities, high mechanical stability and good electrochemical redox characteristic [11]. It is a well-known fact that WO_3 has several distinct crystalline polymorphs, namely monoclinic, triclinic, tetragonal, orthorhombic, cubic and hexagonal phases. By comparing the crystal structure of WO_3 in Table 6.3, hexagonal WO_3 (*h*-WO_3) is the most notably material as electrode material for supercapacitor owing to its tunnels are effective for insertion of electrolyte ions and a high pseudocapacitance was obtained. *h*-WO_3 consists of WO_6 octahedra and aligned in six-membered ring by sharing corner oxygens to form hexagonal axis as illustrated in Figure 6.1 [12]. Stacking of such layers resulted in the formation of hexagonal tunnels.

Nonetheless, an obvious hindrance to the widespread use of WO_3 as an active electrode in capacity system is its low electrical conductivity and poor cycle stability. Considering this fact, incorporation of WO_3 with conductive carbon materials to fabricate composite as promising active electrode in supercapacitor is highly recommended.

6.3.2 Tungsten Disulfide (WS_2)

Tungsten disulfide (WS_2) has been widely investigated as electrode materials for supercapacitor. WS_2 has the advantages of large surface area and intrinsically layered texture, which favors rapid electron transportation. However, it suffers from

TABLE 6.3
Comparison of the Electrochemical Properties of WO_3 Supercapacitor and Its Different Crystal Structure

Sample	Synthesis Method	Morphology	Crystal Structure	Specific Capacitance (Current Density)	Cycling Stability (Cycles)	References
WO_3	Hydrothermal	Microspheres	Hexagonal	797.05 Fg^{-1} (0.5 Ag^{-1})	Retained 100.47% (1000 cycles)	[13]
h-WO_3	Hydrothermal	Aligned nanopillar bundles	Hexagonal	421.8 Fg^{-1} (0.5 Ag^{-1})	Retained almost 100% within 1000 cycles	[14]
WO_3	Hydrothermal	Nanofibers	Hexagonal	539.42 Fg^{-1} (2 Ag^{-1})	Retained 79.1% (6000 cycles)	[15]
h-$WO_3 \cdot nH_2O$	Hydrothermal	Nanorods	Hexagonal	496 Fg^{-1} at scan rate of 5 mV s^{-1}	–	[16]
WO_3-$WO_3 \cdot 0.5H_2O$	Microwave-assisted hydrothermal	Disordered nanorods	Hexagonal and cubic	293 Fg^{-1} at scan rate of 25 mV s^{-1}	Retained 72% (100 cycles)	[17]
m-WO_{3-x}	Template-assisted	Mesoporous	Cubic	199 Fg^{-1} (1 mA cm^{-2})	Remained almost 95% within 1200 cycles	[18]
WO_3	Microwave irradiation	Amorphous	Amorphous	231 Fcm^{-3} (1 Ag^{-1})	–	[19]
WO_3	Successive ionic layer adsorption and reaction (SILAR)	Irregular rods	Monoclinic	266 Fg^{-1} at scan rate of 10 mV s^{-1}	Remained 81% (1000 cycles)	[20]

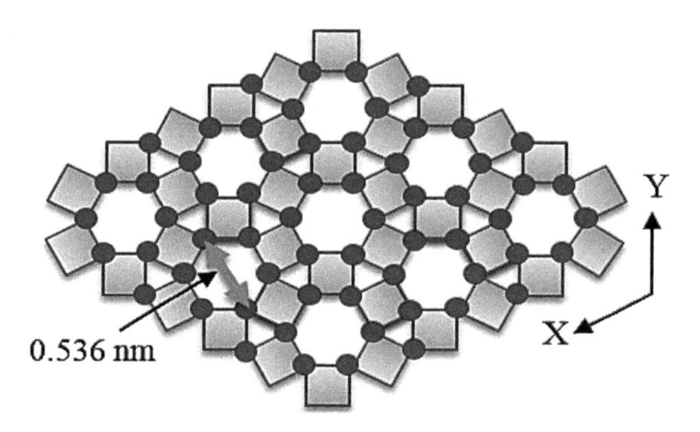

FIGURE 6.1 Crystal structure of h-WO_3. (From Liang, L. et al., *Sci. Rep.*, 3, 1–8, 2013.)

poor electronic conductivity. WS_2 has different phase depending on the location of S atoms where 2H phase is trigonal prismatic D_3h and 1T phase is octahedral Oh. Khalil et al. [21] reported that 1T phase with metallic properties which improves the electronic conductivity of WS_2. 1T phase WS_2 displayed 2813 μF cm^{-2} while 223.3 μF cm^{-2} for 2H phase of WS_2 at 0.5 A m^{-2}. Ghorai et al. [22] prepared WS_2 nanocrystals with the assisted of lithium-intercalation and sonication routes. The WS_2 quantum dots electrode showed high specific capacitance of 28 mF cm^{-2} at 0.1 mA m^{-2}. This was attributed to the large amounts of defect states occurs in WS_2.

6.4 HYBRID-SUPERCAPACITOR

Over the past few years, hybrid capacitor has been widely investigated because of the unique properties of high specific capacitance and excellent cycling performance. Generally, a hybrid capacitor comprised more than one active species in a single device, for example, incorporation of EDLC/EDLC, pseudo/pseudo or EDLC/pseudo in single electrode-symmetric hybrids. It is worth to mention that these hybrid capacitors able to overcome the major drawbacks of EDLC and pseudocapacitor. The operating voltage of hybrid capacitor can be larger than EDLC and pseudocapacitor leading to high energy and power density.

6.4.1 Tungsten Trioxide-Carbon

Reduced graphene oxide (rGO) has gained huge attention as an active electrode material owing to its good electrochemical stability, superior surface area (>2600 m^2g^{-1}), fast ions diffuse to its structure and good mechanical property [23]. Taking these facts into account, numerous literature reviews on rGO/WO_3 composites and their electrochemical performances are reported. A number of synthesis approaches such as hydrothermal, reflux and electrostatic self-assembly method, have been applied to prepare rGO/WO_3 composites for supercapacitor application.

Xing and her co-workers prepared WO_3 nanoparticles and dispersed it on rGO using hydrothermal technique, which demonstrating higher specific capacitance of 580 Fg^{-1} than pure WO_3 (255 Fg^{-1}) at 1 Ag^{-1} in 2 M KOH [24]. The improved capacitance of rGO/WO_3 composite ascribed to its high specific surface area and excellent electrical conductivity, which promotes ions diffusion and charge transfer kinetic. Ma et al. also synthesized rGO/WO_3 composite through hydrothermal method using sodium tungstate ($Na_2WO_4{\cdot}2H_2O$) as WO_3 precursor [25]. At 1 Ag^{-1}, the resulted $WO_3{\cdot}H_2O$ electrode exhibited a specific capacitance of 140 Fg^{-1} while composite demonstrated much larger specific capacitance of 244 Fg^{-1}. The improved electrochemical performance of composited is because of the synergistic effect of rGO and $WO_3{\cdot}H_2O$. In addition, Liu et al. [26] successfully grew WO_3 nanowires on graphene sheets using seed-mediated hydrothermal method and used as negative electrode. It delivered high-specific capacitance of 800 mF cm^{-2} at current density of 1 mA cm^{-2}. The great improvement in specific capacitance is because of the synergistic effects between graphene and WO_3 nanowires. Chu et al. [27] synthesized WO_3-RGO composite through electrostatic attraction between positive charged WO_3 and negative charged GO. The composites showed higher specific capacitance of 495 Fg^{-1} than pure WO_3 at 1 Ag^{-1} using 0.5 M H_2SO_4 electrolyte owing to the increasing of electron contact area of WO_3 in composite. The improved electrochemical performance is probably due to the increasing of electron contact area.

In addition, Chu et al. fabricated rGO/WO_3 composite via an electrostatic attraction between positive charge of surfactant modified-WO_3 and negative charge of GO [27]. The WO_3 was first prepared using hydrothermal method and $Na_2WO_4{\cdot}2H_2O$ used as precursor. The specific capacitance of rGO/WO_3 composite and pure WO_3 is 495 and 127 Fg^{-1} at 1 Ag^{-1} of current density, respectively. This is probably because of the synergistic effects of conductive rGO and WO_3. Furthermore, rGO/WO_3 also can be synthesized using reflux by adding glucose reduced rGO and WO_3 precursor ($Na_2WO_4{\cdot}2H_2O$), which exhibited specific capacitance of 140.8 Fg^{-1} at current density of 0.3 Ag^{-1} [28]. In contrast, the specific capacitance of WO_3-$WO_3{\cdot}H_2O$ mixture is 24.5 Fg^{-1}, which is almost six times lower than the composite. This results indicated that the addition of conductive rGO could enhance the electrolyte ion mobility in electrode. These reports suggest that the enhancement in specific capacitance of rGO/WO_3 composites owing to their new functionalities and properties as summarized below: (i) 2D graphene acts as support to facilitate the nucleation and growth of metal oxide with well-defined structures [29]; (ii) metal oxide anchored on rGO can avoid the restacking of rGO (vice versa), resulting in an increase of electroactive sites and long cycle life; and (iii) high conductive rGO also act as conductive framework to enhance the ion mobility.

Other than graphene, carbon aerogel (CA) possesses superior electrical conductivity and ordered mesoporous structure, which improved the mobility of electron into active materials. Liu et al. [30] incorporated WO_3 nanoparticles into carbon aerogel through solvent-immersion process and calcination route. This composite exhibited maximum specific capacitance of 140.5 Fg^{-1} while WO_3 nanoparticles showed only 25.8 Fg^{-1} at current density of 0.83 Ag^{-1}. The enhanced electrochemical performance of composite is because of the carbon aerogel served as template for WO_3 nanoparticles growing and prevent the agglomeration of WO_3 nanoparticles. Liu et al. [31]

prepared WO_3/CA composite via a one-pot route and exhibited specific capacitance of 609 Fg^{-1} at scan rate of 5 mV s^{-1}, which is 50% higher than pristine WO_3 nanowires. The incorporation of carbon aerogel showed capacity retention of 98% owing to the CA effectively preventing the agglomeration of WO_3 nanowires, resulting more contact surface for electron. Wang et al. [32] synthesized WO_3 nanoparticles and WO_3/CA composites yield specific capacitance of 54 and 700 Fg^{-1}, respectively at scan rate of 25 mV s^{-1}. The excellent specific capacitance of composites was attributed to relatively high electrical conductivity of CA leading to rapid transportation of electron into active material.

6.4.2 Tungsten Trioxide-Conducting Polymers

Conducting polymers including PANI, PPy, PEDOT and their derivatives are widely studied as electrode materials for supercapacitors owing to its high electrical conductivity, high chemical stability, easy preparation and high flexibility. Chemical bath deposition technique was used to synthesize PANI-WO_3 nanocom posite [33]. Both nanocomposite and structured metal oxides enhanced the surface area resulting in good specific capacitance (96 Fg^{-1}) at 5 mV s^{-1} in 0.5 M H_2SO_4 electrolyte. Zou et al. [34] reported that PANI-WO_3 composite prepared via electrodeposition techniques exhibited specific capacitance of 201 Fg^{-1} at current density of 1.28 mA cm^{-2}.

6.4.3 Tungsten Disulfide-Carbon

The advantages of graphene also can enhanced the electrochemical performance of WS_2. WS_2/active carbon fiber nanocomposite was successfully prepared through electrospinning and carbonization followed by hydrothermal methods [35]. The nanocomposite delivered a high capacitance of 600 Fg^{-1} at current density of 1 Ag^{-1} in 1M KOH electrolyte. This is due to thin WS_2 nanosheets and high conductivity of active carbon. Shang et al. [36] prepared WS_2 nanoplates on carbon fiber cloth (CFC) through solvothermal technique. WS_2/CSC displayed high specific capacitance of 399 Fg^{-1} at 1.0 Ag^{-1} and retained 99% of capacitance after 500 cycles due to the well disperse of WS_2 and conductive behaviors of CFC. Moreover, WS_2 also incorporated with carbon tubes via hydrothermal and calcination methods to form nanocomposite electrode [37]. The electrode exhibited high specific capacitance of 337 Fg^{-1} at 10 Ag^{-1} and good cyclability in 3 M KOH electrolyte. The porous structure of carbon tubes increase the contact surface for electron resulting in good electrochemical performances.

6.4.4 Tungsten Disulfide-Conducting Polymers

WS_2 also incorporated with conducting polymers to form electrode. WS_2/PEDOT:PSS film was prepared through vacuum filtration method [38]. This film displayed high capacitance of 86 mF cm^{-2} at scan rate of 40 mV s^{-1} in 1M H_2SO_4 electrolyte. WS_2/PEDOT:PSS film has the advantages of good flexibility, high conductivity, high electrochemical performance and good chemical stability.

6.5 CONCLUSIONS

In summary, tungsten-based carbon nanocomposites have attracted much concern to improve the performance of devices due to the advantages of each component and the synergistic effect between them tends to eliminate the demerits of each other. The conductive carbon acts as support to facilitate the nucleation and growth of tungsten-based material with well-defined structures and enhanced the electrical conductivity of composite. Carbon prevent the restacking of tungsten leads to an increase of active sites for ions contact and contributing high specific capacitance. Conducting polymers possess high electrical conductivity leading to rapid electron transportation. Tungsten disulfide has the advantages of layered structure and high electrochemical behaviors. Meanwhile, widening the operating voltage of the cell by using non-aqueous electrolyte (mixture of ionic liquid and organic electrolyte) is also an effective way to enhance its energy and power density. High cell voltage can reduce the number of cells that are used in series for high power systems as well as increase the reliability of the devices. The ionic liquid electrolytes can be modified by adding an organic electrolyte to reduce the viscosity of ionic liquids such as acetonitrile and propylene carbonate. Ionic liquids can lower the vapor pressure of organic electrolytes, and the high conductivity of acetonitrile is also able to enhance the conductivity of ionic liquids.

ACKNOWLEDGMENTS

This research work was financially supported by the Impact-Oriented Interdisciplinary Research Grant (No. IIRG018A-2019) and Global Collaborative Programme – SATU Joint Research Scheme (No. ST012-2019).

REFERENCES

1. M.M. Masoumi, S. Naderinezhad, One of two conductors in capacitor substitute by Mercury, *Bonfring International Journal of Power Systems and Integrated Circuits*, 4 (2014) 1.
2. P. Sharma, T. Bhatti, A review on electrochemical double-layer capacitors, *Energy Conversion and Management*, 51 (2010) 2901–2912.
3. M. Jayalakshmi, K. Balasubramanian, Simple capacitors to supercapacitors-an overview, *International Journal of Electrochemical Science*, 3 (2008) 1196–1217.
4. S. Faraji, F.N. Ani, Microwave-assisted synthesis of metal oxide/hydroxide composite electrodes for high power supercapacitors—A review, *Journal of Power Sources*, 263 (2014) 338–360.
5. J. Samdani, K. Samdani, N.H. Kim, J.H. Lee, A new protocol for the distribution of MnO_2 nanoparticles on rGO sheets and the resulting electrochemical performance, *Applied Surface Science*, 399 (2017) 95–105.
6. M.M. Sk, C.Y. Yue, K. Ghosh, R.K. Jena, Review on advances in porous nanostructured nickel oxides and their composite electrodes for high-performance supercapacitors, *Journal of Power Sources*, 308 (2016) 121–140.
7. L.L. Zhang, X. Zhao, Carbon-based materials as supercapacitor electrodes, *Chemical Society Reviews*, 38 (2009) 2520–2531.
8. E. Frackowiak, Q. Abbas, F. Béguin, Carbon/carbon supercapacitors, *Journal of Energy Chemistry*, 22 (2013) 226–240.

9. A. Burke, Ultracapacitors: Why, how, and where is the technology, *Journal of Power Sources*, 91 (2000) 37–50.
10. S. Faraji, F.N. Ani, The development supercapacitor from activated carbon by electroless plating—A review, *Renewable and Sustainable Energy Reviews*, 42 (2015) 823–834.
11. L. Liang, J. Zhang, Y. Zhou, J. Xie, X. Zhang, M. Guan, B. Pan, Y. Xie, High-performance flexible electrochromic device based on facile semiconductor-to-metal transition realized by $WO_3{\cdot}2H_2O$ ultrathin nanosheets, *Scientific Reports*, 3 (2013) 1–8.
12. W. Sun, M.T. Yeung, A.T. Lech, C.-W. Lin, C. Lee, T. Li, X. Duan, J. Zhou, R.B. Kaner, High surface area tunnels in hexagonal WO_3, *Nano Letters*, 15 (2015) 4834–4838.
13. J. Xu, T. Ding, J. Wang, J. Zhang, S. Wang, C. Chen, Y. Fang, Z. Wu, K. Huo, J. Dai, Tungsten oxide nanofibers self-assembled mesoscopic microspheres as high-performance electrodes for supercapacitor, *Electrochimica Acta*, 174 (2015) 728–734.
14. M. Zhu, W. Meng, Y. Huang, Y. Huang, C. Zhi, Proton-insertion-enhanced pseudocapacitance based on the assembly structure of tungsten oxide, *ACS Applied Materials & Interfaces*, 6 (2014) 18901–18910.
15. S. Yao, X. Zheng, X. Zhang, H. Xiao, F. Qu, X. Wu, Facile synthesis of flexible WO_3 nanofibers as supercapacitor electrodes, *Materials Letters*, 186 (2017) 94–97.
16. Z. Chen, Y. Peng, F. Liu, Z. Le, J. Zhu, G. Shen, D. Zhang, M. Wen, S. Xiao, C.-P. Liu, Hierarchical nanostructured WO_3 with biomimetic proton channels and mixed ionic-electronic conductivity for electrochemical energy storage, *Nano Letters*, 15 (2015) 6802–6808.
17. K.-H. Chang, C.-C. Hu, C.-M. Huang, Y.-L. Liu, C.-I. Chang, Microwave-assisted hydrothermal synthesis of crystalline WO_3–$WO_3{\cdot}0.5H_2O$ mixtures for pseudocapacitors of the asymmetric type, *Journal of Power Sources*, 196 (2011) 2387–2392.
18. S. Yoon, E. Kang, J.K. Kim, C.W. Lee, J. Lee, Development of high-performance supercapacitor electrodes using novel ordered mesoporous tungsten oxide materials with high electrical conductivity, *Chemical Communications*, 47 (2011) 1021–1023.
19. C.-C. Huang, W. Xing, S.-P. Zhuo, Capacitive performances of amorphous tungsten oxide prepared by microwave irradiation, *Scripta Materialia*, 61 (2009) 985–987.
20. N.M. Shinde, A.D. Jagadale, V.S. Kumbhar, T.R. Rana, J. Kim, C.D. Lokhande, Wet chemical synthesis of WO_3 thin films for supercapacitor application, *Korean Journal of Chemical Engineering*, 32 (2015) 974–979.
21. A. Khalil, Q. Liu, Q. He, T. Xiang, D. Liu, C. Wang, Q. Fang, L. Song, Metallic 1T-WS_2 nanoribbons as highly conductive electrodes for supercapacitors, *RSC Advances*, 6 (2016) 48788–48791.
22. A. Ghorai, A. Midya, S.K. Ray, Superior charge storage performance of WS_2 quantum dots in a flexible solid state supercapacitor, *New Journal of Chemistry*, 42 (2018) 3609–3613.
23. Z. Bo, Z. Wen, H. Kim, G. Lu, K. Yu, J. Chen, One-step fabrication and capacitive behavior of electrochemical double layer capacitor electrodes using vertically-oriented graphene directly grown on metal, *Carbon*, 50 (2012) 4379–4387.
24. L.-L. Xing, K.-J. Huang, L.-X. Fang, Preparation of layered graphene and tungsten oxide hybrids for enhanced performance supercapacitors, *Dalton Transactions*, 45 (2016) 17439–17446.
25. L. Ma, X. Zhou, L. Xu, X. Xu, L. Zhang, C. Ye, J. Luo, W. Chen, Hydrothermal preparation and supercapacitive performance of flower-like $WO_3{\cdot}H_2O$/reduced graphene oxide composite, *Colloids and Surfaces A: Physicochemical and Engineering Aspects*, 481 (2015) 609–615.
26. B. Liu, Y. Wang, H.-W. Jiang, B.-X. Zou, WO_3 nanowires on graphene sheets as negative electrode for supercapacitors, *Journal of Nanomaterials*, 2017 (2017) 1–9.

27. J. Chu, D. Lu, X. Wang, X. Wang, S. Xiong, WO_3 nanoflower coated with graphene nanosheet: Synergetic energy storage composite electrode for supercapacitor application, *Journal of Alloys and Compounds*, 702 (2017) 568–572.
28. Y. Cai, Y. Wang, S. Deng, G. Chen, Q. Li, B. Han, R. Han, Y. Wang, Graphene nanosheets-tungsten oxides composite for supercapacitor electrode, *Ceramics International*, 40 (2014) 4109–4116.
29. S. Iijima, Helical microtubules of graphitic carbon, *Nature*, 354 (1991) 56.
30. X. Liu, G. Sheng, M. Zhong, X. Zhou, Dispersed and size-selected WO_3 nanoparticles in carbon aerogel for supercapacitor applications, *Materials & Design*, 141 (2018) 220–229.
31. X. Liu, G. Sheng, M. Zhong, X. Zhou, Hybrid nanowires and nanoparticles of WO_3 in a carbon aerogel for supercapacitor applications, *Nanoscale*, 10 (2018) 4209–4217.
32. Y.-H. Wang, C.-C. Wang, W.-Y. Cheng, S.-Y. Lu, Dispersing WO_3 in carbon aerogel makes an outstanding supercapacitor electrode material, *Carbon*, 69 (2014) 287–293.
33. I.C. Amaechi, A.C. Nwanya, A.B. Ekwealor, P.U. Asogwa, R.U. Osuji, M. Maaza, F.I. Ezema, Electronic thermal conductivity, thermoelectric properties and supercapacitive behaviour of conjugated polymer nanocomposite (polyaniline-WO_3) thin film, *The European Physical Journal Applied Physics*, 69 (2015) 30901.
34. B. Zou, S. Gong, Y. Wang, X. Liu, Tungsten oxide and polyaniline composite fabricated by surfactant-templated electrodeposition and its use in supercapacitors, *Journal of Nanomaterials*, 2014 (2014) 3.
35. X. Qiu, L. Wang, L.-Z. Fan, Immobilization of tungsten disulfide nanosheets on active carbon fibers as electrode materials for high performance quasi-solid-state asymmetric supercapacitors, *Journal of Materials Chemistry A*, 6 (2018) 7835–7841.
36. X. Shang, J.-Q. Chi, S.-S. Lu, J.-X. Gou, B. Dong, X. Li, Y.-R. Liu, K.-L. Yan, Y.-M. Chai, C.-G. Liu, Carbon fiber cloth supported interwoven WS_2 nanosplates with highly enhanced performances for supercapacitors, *Applied Surface Science*, 392 (2017) 708–714.
37. B. Hu, X. Qin, A.M. Asiri, K.A. Alamry, A.O. Al-Youbi, X. Sun, WS_2 nanoparticles–encapsulated amorphous carbon tubes: A novel electrode material for supercapacitors with a high rate capability, *Electrochemistry Communications*, 28 (2013) 75–78.
38. A. Liang, D. Li, W. Zhou, Y. Wu, G. Ye, J. Wu, Y. Chang, R. Wang, J. Xu, G. Nie, Robust flexible WS_2/PEDOT:PSS film for use in high-performance miniature supercapacitors, *Journal of Electroanalytical Chemistry*, 824 (2018) 136–146.

7 Microwave-Assisted Inorganic Materials for Supercapacitors

M. Ramesh and Arivumani Ravanan

CONTENTS

7.1 Introduction ... 102
7.2 Literature Review ... 103
7.2.1 Role of Microwave in Energy Storage Devices ... 103
7.2.2 Microwaves Generation and Synthesis Techniques ... 103
7.2.3 Interaction of Microwaves with Materials ... 104
7.2.4 Methods for Synthesis of Inorganic Materials ... 106
7.2.4.1 Solid State Microwave Synthesis ... 106
7.2.4.2 Single Mode Solid-State Microwave Assisted Synthesis ... 107
7.2.4.3 Microwave Assisted Hydro-Thermal/Solvothermal Synthesis ... 108
7.2.4.4 Microwave Synthesis Combined with Sol-Gel or Combustion ... 109
7.2.5 Microwave-Assisted Inorganic Materials for Supercapacitors ... 109
7.2.5.1 3D Flower-Like $NiMnO_3$ Nano-balls ... 110
7.2.5.2 3D Flower-On-Sheet Nanostructure of NiCo LDHs ... 110
7.2.5.3 Stannous Ferrite Micro-cubes ... 111
7.2.5.4 Reflux Rapid Synthesis of MnO_2 Nanostructures ... 111
7.2.5.5 L-Arginine Capped α-$Ni(OH)_2$ Microstructures ... 111
7.2.5.6 Binder-Free Synthesis of 3D Ni-Co-Mn Oxide Nano-flakes at Ni Foam ... 112
7.2.5.7 Rapid Microwave-Assisted Synthesis of Graphene Nano-sheet/Co_3O_4 Composite ... 112
7.2.5.8 High-Rate FeS_2 Nano-particles Anchored on Graphene ... 113
7.2.5.9 Copper Tungstate Nano-powder ... 113
7.2.5.10 3D Hierarchical MnO_2 Microspheres ... 113
7.2.5.11 Hybrid Polymer Materials and Composites ... 113
7.3 Conclusions ... 114
References ... 114

7.1 INTRODUCTION

A huge attempt is needed for exploring and investigating the supercapacitor electrode materials such as metal oxides/hydroxides, ceramics, polymers, composites, mixed metal oxides/hydroxides. The microwave radiation is also known as volumetric heating and this technique is extended in the laboratory to industries and the field of inorganic and materials chemistry. Sutton [1] studied and described the microwave heating and its effect on ceramics materials in the 1970. The radiation of microwaves is ranged between 0.3 and 300 GHz in the low order region of the electromagnetic spectrum. The concept of electric dipoles in a material is lead to the understanding of microwaves. Dipolar molecules in materials do rotate, and the resistance to the movement produces a considerable amount of heat [2]. However, there are some inquiries and questions to uncover the basic concepts of the kinetic attributes and reaction mechanisms. Hence, this technique is often used as a trial and error method [3]. In microwave-assisted nanomaterials for supercapacitors, it is necessary to know about conventional heating, how far it varies from microwave heating principles, and about interaction between microwaves and solid matter. The demonstration of microwave-assisted synthesis in an efficient way and the different methods of performing the microwave-assisted synthesis, in particular, solid state, single-mode, microwave synthesis are combined with sol-gel and hydro-thermal routes. The wide range of various microwave-assisted synthesis methods revealed many scopes for the fabrication of inorganic nanoparticles and structures. This opens up the feasibility to enhance and fine-tune the morphology, chemical and physical characteristics of nano-materials.

Supercapacitors are known as direct electrical storage devices in which the electric power is stored without any energy conversion. Hence the quick supplying of the electric power is ensured. Moreover, a higher order of power and current for 1×10^6 charge/discharge cycles is very feasible by supercapacitors. Owing to the variation in the capability of stored energy and dynamic behavior of other electrical energy sources, introducing of supercapacitors compatible. For instance, the electric energy is stored in the Helmholtz layer, also known as an electrochemical double layer. It is composed of the interface of the solid electrolyte. This electrolyte comprises positive and negative ionic charges. The solid electrode surface accumulates these ionic charges; therefore, the losses at the electrode surface due to the electronic charges are compensated. The ion size and electrolyte concentration determine the double layer thickness. Microwave heating is fundamentally different from the conventional one in which thermal energy is delivered to the surface of the material by radiant and/or convection heating that is transferred to the bulk of material via conduction. In contrast, microwave energy is delivered directly to the material through molecular interaction with the electromagnetic field. In microwave heating, the thermal energy is obtained from electromagnetic energy. Only the energy conversion is happening in the microwave heating, and there is no phenomenon of transferring of heat. This chapter mainly discussed microwave heating and the electrode materials of supercapacitors.

7.2 LITERATURE REVIEW

7.2.1 Role of Microwave in Energy Storage Devices

The microwave-assisted method is a modern technique and is quickly developing the field in the research sector, especially in the synthesis of inorganic materials. It renders the efficient way to regulate the distribution the particle size, reduced the time for synthesizing and macroscopic morphology in the synthesis process [4]. The attempt to synthesizing the metal oxide, ceramic, composite, and polymer electrode materials for supercapacitors through microwave-assisted techniques has provided the expected result [5–8]. The metal oxide nanostructured materials are considered superior materials for attaining the higher specific capacitance, and they display pseudo-capacitance performance. The report about the microwave-assisted techniques are synthesis and examining the super capacitance performance of metal oxide, ceramic, polymer and hydroxide electrode materials. These materials will play a key role are energy storage techniques, in particular supercapacitor technology due to greater exposure in specific capacitance values.

Microwave heating is the transfer of electromagnetic energy to thermal energy and is an energy conversion rather than heat transfer; therefore, the microwaves penetrate the materials for supplying the energy. The heat is produced all around the surface of the material, which is also known as volumetric heating. It leads toward the rapid and uniform heating of materials. This causes the thermal gradient of material processed by microwave-assisted is the reverse of the conventional method of heating. When the materials with different dielectric properties are subjected to microwave heating, they would combine with the major loss tangent materials because the phenomenon of energy transfers is occurring on the molecular scale. Hence, heating of selective materials is a natural one when assisting the microwaves [9]. Dipole rotation and ionic conduction are the two mechanisms involved in the transferring of microwave energy into the materials [10]. The molecules of the materials have dipole moments due to the electrical field that may be permanent or induced, which leads to the dipole rotation. This is also referred to as spatial alignment. Movement of the dissolved ions is termed as ionic conduction, which has occurred in the applied electromagnetic field.

7.2.2 Microwaves Generation and Synthesis Techniques

In conventional heating, conduction and convection are the modes of transferring heat energy to sources that generate thermal gradients, whereas microwave heating interactivity occurs with the electromagnetic field in the molecular level for the transferring of heat energy. One of the statement says that during the synthesis of material, the transferring of energy is a critical one [11]. The penetration depth of the radiation may vary based on materials, size, magnetic characteristics, frequency of the microwave, density of the material, power and dielectric properties. When relating to the conventional synthesis, there is a huge reduction in the processing time by the microwaves. About 10–10000 times of chemical reaction is possible in microwave-assisted tasks when relating to the conventional methods [12]. This rapid heating and

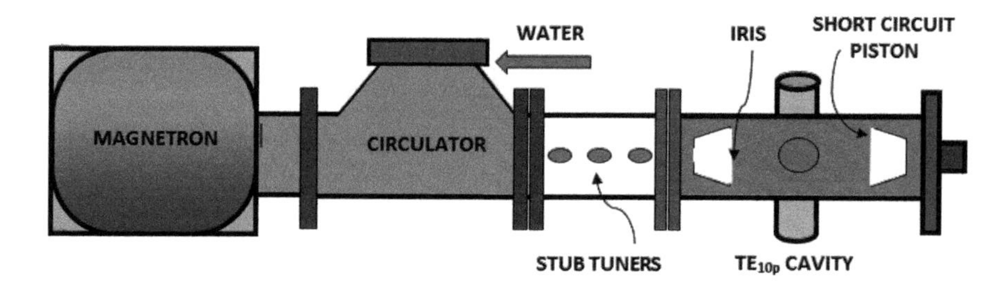

FIGURE 7.1 Schematic diagram of single-mode microwave apparatus.

non-homogeneous temperature outline becomes the reason for the generation of hot spots; hence, it may influence the system to damage [13,14]. At the same time, selective heating is possible in microwaves. The ability to combine in the magnetic field or/an electric field of the microwaves and based on the properties of the material, heating of the specific regions can be easily positioned (Figure 7.1).

Microwaves of 2.45 GHz at room temperature are the very often applied frequency, in which a few of these materials are unable to couple with each other. Hence, along with the precursor mixture, a radiation susceptor can be associated. The susceptor may be activated carbon, graphite or silicon carbide. An appreciable increment in temperature is ensured when the susceptor firmly paired with the radiation [15,16]. Nevertheless, many polar materials such as H_2O, NH_3, SO_2, C_2H_6O do combine with electro-magnetic waves firmly enough to achieve high temperatures; additives are not needed, and it takes a few minutes for the process to complete. There is also one more term from the researchers after studying so many anomalies which are participated with the heating process of micro-waves called non-thermal effects [17]. Anomalous effects between the interfaces of the particles and the electric field are the core topics of the present discussions. Owing to the second order effects, definitely has done its part in the plasma production and enhances the diffusion of the solid state [18,19].

7.2.3 Interaction of Microwaves with Materials

In relation with chemical conversions, the relative merits of microwave dielectric heating with conventional heating are as follows: (i) Microwave energy is initiated remotely into the reactor; hence the direct contact is avoided between the chemicals and energy source. (ii) There is no similar reaction of contaminated materials and chemicals found in the regularly used microwave frequencies for dielectric heating. Hence, microwave heating is very definitive and selective. (iii) These selective interactions are indicated that it is a perfect method under enhanced pressure situation, in order to accelerate the reactions chemically. (iv) In order to lead to much higher heating and to initiate the microwave energy into chemical reactions, at least one component is necessary with the capability to combine firmly with microwaves. (v) Finally, it is instant, volumetric and rapid heating without heat diffusion effects or walls.

Microwave generator, microwave cavity, mode stirrer, waveguide, turntable, and circulator are the six major elements in the microwave instruments needed to

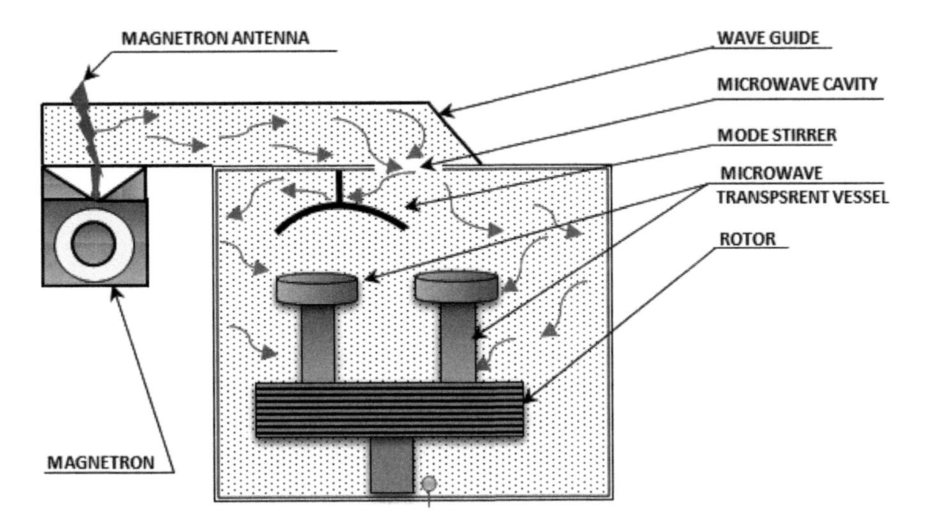

FIGURE 7.2 Schematic diagram of the basic elements of microwaves equipment for the synthesis of inorganic materials.

carry out the synthesis processes which is presented in Figure 7.2 [20]. The microwave generator is also known as magnetron, which is mounted to produce the fixed frequency microwaves in which the vacuum tubes and thermionic cathodes act as electron sources. Microwaves are generally guided toward the target from the microwave generator with the support of waveguides. Targets are located in the microwave cavities. The energy is directly forced into the microwave cavity. The mode stirrer in several directions governs the homogeneous dispensation of the received energy. The factor of dissipation and the sample size determine the percentage of absorbed incoming energy. In general, the regular microwave ovens are working at 2.45 GHz, and the output power obtained is less than 1 KW. Based on the reaction of materials with microwaves, those materials can be classified into three types [21].

Metals are used as microwave reflectors, which are supported in the preparation of waveguides. Low-loss materials are used as microwave transmitters. Those are transparent to microwaves such as fused quartz and Teflon. These materials are utilized as containers in order to achieve the chemical reactions and synthesis process in such short electro-magnetic waves. The high-loss materials are known for microwave absorbers that comprise the highly significant material for the microwave synthesis. Those extract the energy from the microwave field and achieve the hotness very quickly. During the synthesis of material assisted by microwaves, absorptive materials are adopted for the precursor solutions. Reflective materials are engaged for the fabrication of major components of the microwave system. Research-oriented microwave ovens are designed for monitoring purposes only. Pressure and temperature are the major parameters to be monitored when the specimen is under the heating process in the microwave. Probes are the only drawback in the monitoring system, which would not make trouble the microwave energy.

7.2.4 Methods for Synthesis of Inorganic Materials

Different methods and techniques are used in practices. Solid-state microwave, microwave-assisted hydro-thermal synthesis, single mode solid state microwave synthesis, and microwave synthesis combined with sol-gel or combustion are the different sub-types of microwave synthesis. Figure 7.3 indicates the methods of synthesis of materials for the supercapacitor.

7.2.4.1 Solid State Microwave Synthesis

Solid state microwave synthesis is also termed as microwave irradiation of solid precursors. Mixing the precursors is the first step in the synthesis procedure. The next step includes packing the mixed precursors into a pellet, followed by depositing the same in a suitable crucible and finally keeping the same in the microwave oven. These crucibles are usually made up of silicon carbide or porcelain or aluminum oxide materials. This oven is made with the multimode option in which the electromagnetic field is disorderly dispersed in the oven. The microwaves from the multiple modes reacted together with loads of cavity, owing to the reflections from the cavity walls. The density of the field is relatively lower in such big cavities. Attaining the heating rates in shorter intervals is also must. Hence, the power of microwave to be applied is maintained at a higher range [15,22,23]. Producing enough temperature initially to begin the reaction; hence the radiation susceptor is required in a few cases to carry on the process [24]. One of the drawbacks of these types of microwave ovens is lacking temperature control. Hence, the temperature of the reactions performed is not known to others.

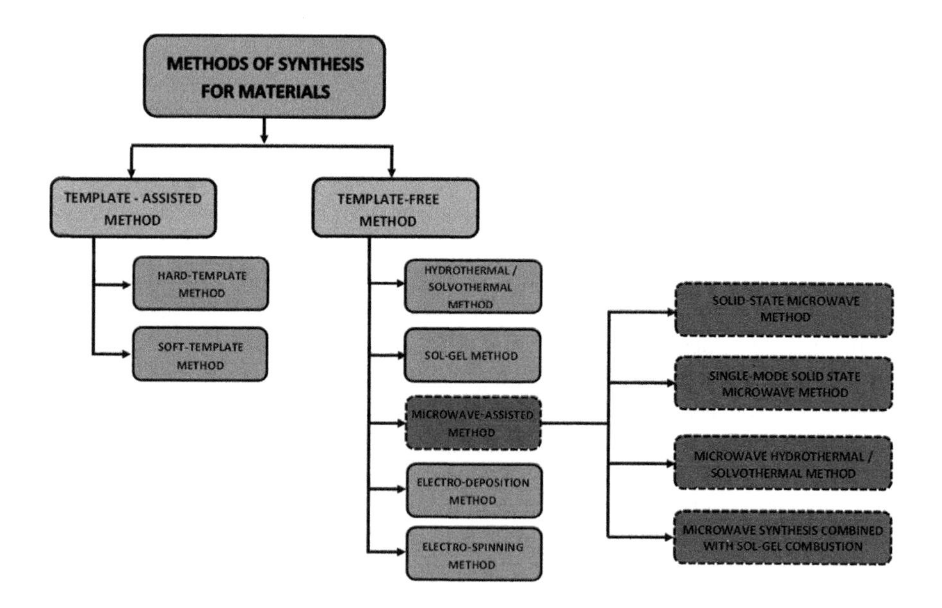

FIGURE 7.3 Block diagram of methods of synthesis for materials.

7.2.4.2 Single Mode Solid-State Microwave Assisted Synthesis

The single mode reactors are also called mono-mode reactors; moreover, this microwave apparatus is known as a second-generation instrument in which the microwave employed to heat up is polarized; across the cavity, it displays the peak and valley in the amplitude. The magnetic and electric fields are clearly determined and well balanced in their spatial alignment. In order to conduct either electrical heating or magnetic heating, proper position and orientation are required in the cavity for the material to be heated [25,26] (Figure 7.4).

As discussed earlier, standard 2.45 GHz frequency is generally maintained in the microwave generator. During the synthesis process, attention is needed for the location and position of the material. Hence, the distribution of the amplitude of magnetic elements as well as electric elements of the microwave can be effectively used in relation to the minima and maxima of electrical and magnetic modes. In order to work at various resonant modes, fine-tuning is demanded by adjusting in the cavity length and by operating on the short circuit piston and on the coupling iris. A pyrometer is supported to monitor the reaction temperature, which intercepts and indicates the measured thermal radiation.

In common, metal samples are known as conductive samples that can be heated in an effective manner in the magnetic field. At the same time, huge rates of heating are exhibited by ceramic materials such as alumina and zinc oxide in the electric field. Ceramics are well known for low conductivity and insulators. Heating of materials may be executed in both magnetic and electric fields when the mixed systems are used. It is nothing but a combination of metal and ceramic materials [15,27]. Sometimes, the heating mode can be chosen based on the constituent materials added in the mixed systems. Gupta et al. [16] reported that the influence

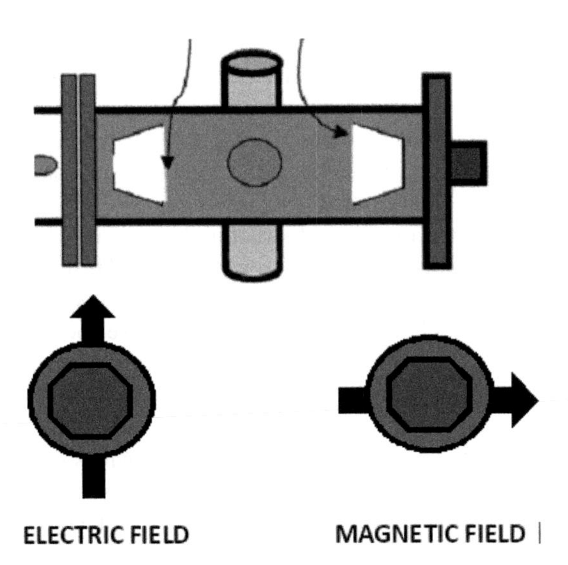

FIGURE 7.4 Schematic drawing of the magnetic and electric field orientation of the polarized wave.

of the electrical loss is much higher in the dielectric materials, for instance, pure oxide ceramics. These materials can be appreciably heated when kept in the electric field. In contrast, these materials exhibit almost nil or lower heating when kept in the magnetic field. Nevertheless, the magnetic loss would be an exception for particular metals and semiconductor materials; therefore, the magnetic field is used for heating of such kind of materials.

7.2.4.3 Microwave Assisted Hydro-Thermal/Solvothermal Synthesis

In general, MAHS (microwave-assisted hydro-thermal synthesis) is a homogeneous heating process. Direct transferring of heat from the water molecules to the precursor is happening in the hydro-thermal synthesis. Hence, the modified forms or new structured materials are made feasible [28]. The various reaction mechanisms mainly distinguish the microwave-assisted synthesizers of solid-state and hydrothermal processes.

The reactant molecules or/and ions react in thc solution is known as solvo-thermal or hydro-thermal reactions. In solid-state synthesis, diffusion of the raw materials is highly dependent on the reactions at their interfaces. The differences in the reaction mechanisms, even for the same reactants, would deliver the various microstructures of the materials [29]. Solvothermal synthesis is one of the famous synthesis processes in which non-aqueous solvents are used. Inorganic chemistry covers all the sectors by expanding the microwave-assisted solvothermal synthesis (MWSS) process [30]. The major demerit in the hydro-thermal synthesis method is a slow kinetic movement of solution irrespective of the given temperature. One researcher distinguished that hydro-thermal syntheses executed with conventional heating and hydro-thermal syntheses done by heating autoclaves with microwaves. Microwaves can be used for the solution heating because of the great coupling of the microwave with water molecules in the solution. Hence, the kinetics of the solution are achieved [29,31,32] (Figure 7.5).

Apart from time and power, the other parameters such as pressure, pH of the solution and temperature of reaction are involved in the microwave-assisted hydrothermal synthesis process. Hence, in order to regulate the temperature and pressure of water in the sub-critical region, advanced technology is needed in this synthesis process. Oxy-hydroxides, binary metallic oxides, ternary oxides, mesoporous materials and zeolites are more complicated structured materials and are some of the functional materials fabricated through microwave-assisted hydro-thermal synthesis process. It is a moderate pressure synthesis method and also known as the wet process in which the aqueous solution is heated in an autoclave about 100°C or more, at the same time the pressure is also increased simultaneously. Owing to this action, an enhancement in the dispersion of the system elements occurred. Moreover, it reacts very quickly.

If the temperature of the water is above 374°C is known as supercritical synthesis conditions, whereas if it is below 374°C is named as subcritical conditions of synthesis. During supercritical conditions, reaction process becomes hard to control because of the rapid increment of the pressure in an autoclave along with the temperature. Teflon is the material used very often for making the autoclaves where the case temperature is restricted to 250°C. By choosing, the suitable synthesis parameters

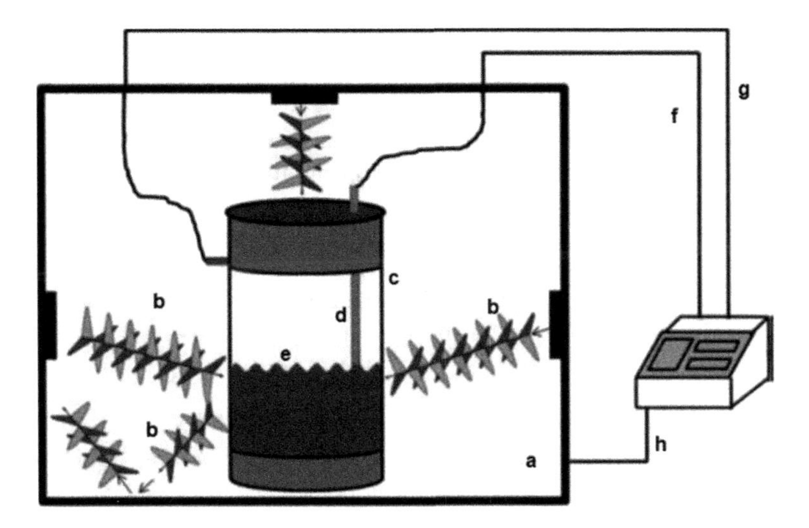

FIGURE 7.5 Schematic drawing of microwave-assisted hydrothermal synthesis process: (a) microwave oven, (b) microwave radiation, (c) reaction vessel, (d) temperature sensor, (e) reactants in solution, (f) temperature control line, (g) pressure control line, and (h) micro-wave power control line. (From Schmidt, R. et al., Microwaves: Microwave-assisted hydrothermal synthesis of nanoparticles, in *CRC Concise Encyclopedia of Nanotechnology*, 13, 2015.)

and proper tuning of the nucleation mechanisms based on the requirement lead to the controlling of particle size. Hence, this method is highly favorable for effective production of the nano-particles [34].

7.2.4.4 Microwave Synthesis Combined with Sol-Gel or Combustion

Synthesis of inorganic materials is usually done through wet-chemical techniques known as a gel process, especially for the metal oxides. A chemical solution initially performs as a precursor. This solution is evolved toward for the formation of polymeric network. The corresponding oxide is obtained through the thermal decomposition of respective gel. During the synthesis, the mixing system of oxide materials and solid fuel with proper proportion leads to the combustion reactions. An exothermic reaction takes place here in which the microwaves render the ignition to the system. The enormous amount of the gas produced with the rapid process together generates the nano-sized particles and porous sponge-shaped materials [34].

7.2.5 Microwave-Assisted Inorganic Materials for Supercapacitors

The use of microwave radiation in the synthesis of inorganic materials leads to energy and time saving. The involvement of microwave radiation in the synthesis process leads to produce the novel advanced materials such as thin films, porous ceramics, polymers and metal oxides. The rapid and volumetric heating of microwaves made the synthesis very quick and generated uniform materials.

7.2.5.1 3D Flower-Like $NiMnO_3$ Nano-balls

Hydro-thermal methods were used for preparing $NiMnO_3$ electrode at different deposition temperature with the assistance of microwaves. When the temperature was increased, the morphology of the electrode material ($NiMnO_3$) has transformed to nano-balls from nano-bulks. These $NiMnO_3$ nano-balls resemble as 3D flowers. Due to this structure of flower-like nano-balls, along with the increment in the surface area, electrolyte ion transfer ports are also adding more, which ended in the enrichment of electro-chemical properties. This experimental work also ensures that $NiMnO_3$ will play an important role in the electrode material of the supercapacitor material owing to its greater specific capacity and high-cycle stability [35] (Figure 7.6).

7.2.5.2 3D Flower-On-Sheet Nanostructure of NiCo LDHs

In the preparation of the electrodes for supercapacitors, two-dimensional nanostructures are suffered by the subsequent deterioration and aggregation of behavior in practical applications even though it has the superior energy storing capability. Therefore, it is advisable to have the assemblage of 2D materials into 3D nanostructures. In this relation, a proposal of using microwave-assisted synthesis of 2D structures with fine-adaptable 3D architectures in a manageable atmosphere has been registered. It is easily achieved by fine-tuned the proportion of water/ethylene glycol. Three dimensional hierarchical structures of nickel-cobalt double hydroxide with novel flower-on-sheet are gained at 40% of ethylene glycol, whereas 2D nano-sheets are achieved at 75% of ethylene glycol content and microspheres are at 0%. It is suggested that the nuclei are scattered and dispersed by the ethylene glycol molecules

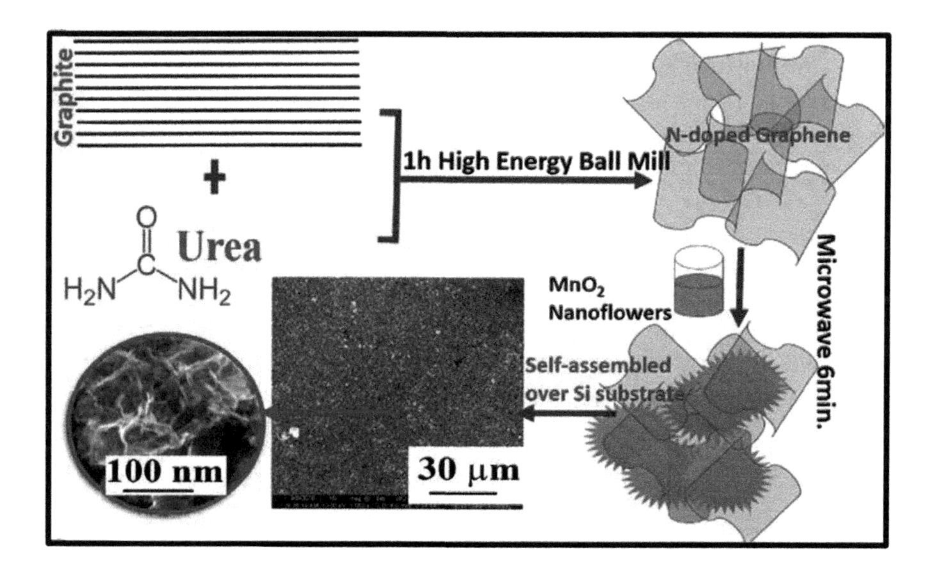

FIGURE 7.6 Experimental procedure for a typical microwave-assisted synthesis of N-doped Graphene-wrapped MnO_2 nanoflower. (From Duraia, E.S.M. et al., *The Min. Met. Mater. Soc.*, 47, 2, 2018.)

under the microwave irradiation which promote the initial development of 2D sheets, followed by nano-flowers on the sheets, which are produced due to the influence of the assemblage of hydro-phobicity. The flowers on the NiCo LDH sheet produce a superior stability and greater capacitance with the retention of 71% at 30 A/g. Due to this, the sheet has attributed the role of performing as buffer substrate, which leads to the more dynamic site of flower. Eventually, the solution indicates the controllable establishment of 2D extracted 3D nano-materials, which can be exerted for the energy storing elements to the forthcoming generation [37].

7.2.5.3 Stannous Ferrite Micro-cubes

It has been known well that the supercapacitors are the significant application of the electro-chemistry field. Bindu et al. [38] ensured the enrichment specific capacitance property for the supercapacitor applications by synthesizing the nano-rods of Fe_2O_3 and microtubes of $SnFe_2O_4$ through the simplified microwave-assisted technique. Microtubes of $SnFe_2O_4$ resulted in an improved specific capacitance than the nano-rods and ensured that $SnFe_2O_4$ was an efficient and cost-effective supercapacitor material.

7.2.5.4 Reflux Rapid Synthesis of MnO_2 Nanostructures

With the support of microwave-assisted reflux, α-MnO_2 urchin-like nanomaterials and γ-MnO_2 nanostructure have been synthesized, in which urchin-like nanomaterials were kept 5 min in acidic conditions, and γ-MnO_2 nanoparticles were kept under neutral conditions for about 5 min. The investigation of material character reported that γ-MnO_2 nanostructures exhibit small sized particles with the large pore volumes when comparing with the α-MnO_2 urchin-like nanomaterials. Galvanostatic charge-discharge techniques and cyclic voltammetry were used for the investigation of the electrochemical behavior of materials. Eventually, the γ-MnO_2 nanoparticles registered as a promising electrode material at a higher capacitance of 311 Fg^{-1} (at a current density of 0.2 Ag^{-1}) [39].

7.2.5.5 L-Arginine Capped α-$Ni(OH)_2$ Microstructures

In order to eliminate or reduce the troubles connected with the other synthetic methods, a new method known as green chemistry is recommended. The microwave is a path with an easy, simplistic, famous and quick one. William et al. [40] registered that under the existence of l-arginine through microwave irradiation technique 3D-connected α-$Ni(OH)_2$ sheets were fabricated, in which l-arginine acted as the surfactant. The fabricated structure keeps the retaining efficiency of about 87.3%, even after 10,000 cycles. It registered a specific capacity of about 549 C/g at 2 A/g. Furthermore, at a power density of 601 W/kg, another design of the hybrid supercapacitor setup, has the ability to yield the energy density about 57 Wh/kg, in which the α-$Ni(OH)_2$ is incorporated with the polyurethane foam –6 M KOH/activated carbon. Therefore, this hybrid material, arginine-supported α-Ni $(OH)_2$, would be aided as better energy storage for light electronic devices [40].

7.2.5.6 Binder-Free Synthesis of 3D Ni-Co-Mn Oxide Nano-flakes at Ni Foam

Good dimensional stability, lesser contact, greater dimensional stability, and highly stable structures are the best attributes of nickel foam. Moreover, while using this material as an electrode, which stops the structural collapse and enables for quick ion insertion. Lamiel et al. [41] have constructed a controlled pattern of Ni-Co-Mn oxide with NF through microwave irradiation in a productive and binder-free way. This procured structure contained hierarchical networked nano-flakes along with the voids, producing a firm and interconnected nano-flakes on the NF. The high specific capacitance of 2536 F/g at 6.49 A/g was the result of the consequences of synergistic integration of these three metal oxides. This superior structure of Ni-Co-Mn oxide nano-flakes enhanced the pseudo-capacitive behavior remarkably and ensuring this electrode material as a binder-free supercapacitor with good performance [41].

7.2.5.7 Rapid Microwave-Assisted Synthesis of Graphene Nano-sheet/Co_3O_4 Composite

Graphene nano-sheet (GNS)/Co_3O_4 composite electrode has been rapidly synthesized by an employed microwave-assisted method. Electro-chemical properties are described by cyclic voltammetry, galvanostatic charge/discharge, and electro-chemical impedance spectroscopy. The maximum specific capacitance of 243.2 Fg^{-1} has been obtained at a scan rate of 10 mV s^{-1} in 6 M KOH aqueous solution for GNS/Co_3O_4 composite. In addition, the composite exhibits eminent long cycle life along with ~95.6% specific capacitance retained after 2000 cycles [42] (Figure 7.7).

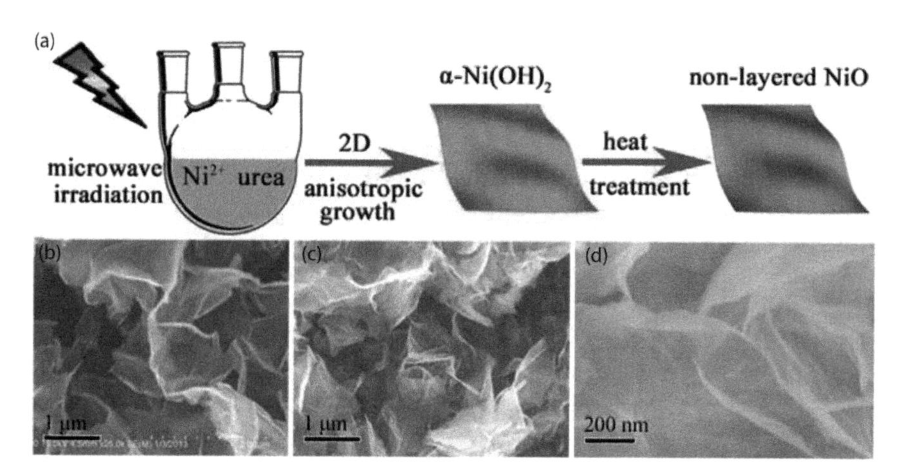

FIGURE 7.7 (a) Schematic diagram for synthesis of microwave-assisted typical ultrathin nickel hydroxide and oxide nanosheets, (b) Field emission SEM image of α-$Ni(OH)_2$, and (c) and (d) field emission SEM image of NiO nanosheets. (From Zhu, Y. et al., *Sci. Rep.*, 4, 5, 2014.)

7.2.5.8 High-Rate FeS_2 Nano-particles Anchored on Graphene

The expeditious and ease of application of the microwave-assisted hydro-thermal method has been used to synthesize FeS_2 deposited GNS. The attained FeS_2/GNS anode unveil a high specific capacity of 793 Cg^{-1} at 3 Ag^{-1} and an excellent rate capability of 82% at 30 Ag^{-1} in a greater potential range of −1.1 to 0V. The solid-state hybrid supercapacitor assembled with FeS_2/GNS as an anode and $Ni(OH)_2$ at Co_9S_8 core-shell as a cathode exhibits a high specific energy of 95.8 Wh kg^{-1} at an adequate power density of 949 W kg^{-1} and still retains 40 Wh kg^{-1} even at 15.8 k W kg^{-1} coupled with long cycling stability (86% retention after 5000 cycles), which is highly comparable with those of the recently documented iron oxide-based asymmetric supercapacitors. These verdicts contribute a highly propitious alternative anode material for supercapacitors with excellent energy density [44].

7.2.5.9 Copper Tungstate Nano-powder

Synthesis of copper tungstate nano-powder was developed through the microwave-assisted process. The XRD test resulted in the development of anorthic phase for calcined powders and orthorhombic phase for the as-prepared. Microwave-irradiated powder samples and Scherrer equations judged the average particle sizes. By using $CuWO_4$ nano-powder, nano-composite films on glassy carbon were prepared. This fabricated electrode material resulted in the specific capacitance of 77 F/g [45].

7.2.5.10 3D Hierarchical MnO_2 Microspheres

Microwave heating method was used to synthesize the ultra-thin nano-sheets and three-dimensional hierarchical MnO_2 microspheres, which had the specific surface area of 184.32 m^2/g. Active carbon and 3D hierarchical MnO_2 microspheres were used as anode and cathode materials, respectively. The electrolyte materials provided the improvement in the potential windows of the asymmetric device and individual electrode. It reported enhanced electrochemical behavior. At a power density of 1494 W/kg, the asymmetric device displayed an energy density of 105 Wh/kg with 80% retention efficiency after 6000 cycles. Hence, the prepared ionic electrolyte-based asymmetric supercapacitor device ensured the next level of energy storage applications [46].

7.2.5.11 Hybrid Polymer Materials and Composites

The synthesis of polymer-inorganic hybrid materials and composites under microwave irradiation satisfies the characteristics such as a reduction in processing time, uniform heating of materials, rapid curing of resins and effective crosslinking of composite materials. Furthermore, the advantages of microwave-assisted synthesis comprise smaller particle size, greater particle density and higher exfoliation degree, which significantly enhance the performance of the materials. The diminished size of the materials enhanced the electronic and optical devices [47]. Predominantly, these components are used as electrode materials for supercapacitor owing to display the superior specific capacitance, improves cyclic performance and increases the energy density and power density of the materials.

7.3 CONCLUSIONS

Changing the conventional techniques for the synthesis of inorganic materials toward an economic and eco-friendly environment would support the progress of sustainable chemistry. Also, the natural environment would be preserved. At present, various complicated inorganic materials are developed through novel microwave-assisted synthesis techniques, and they supports the solid-state chemistry field. Many inorganic materials are quickly synthesized due to the interaction between the microwave and materials such as oxide, hydroxide, ceramics and polymers. Microwave-assisted synthesis in a solid state is limited to only ternary components, especially perovskites whereas hydro-thermal synthesis, combustion/sol-gel methods lead to appreciable complex doped phases. Especially, the different morphologies and metastable phases are possible in the solvothermal synthesis technique.

Single-mode polarized microwave synthesis allows separating the magnetic and electric components and provides accurate control of the temperature together with much shorter reaction and processing times. The synthesis of a wide range of perovskite oxide materials was shown to be feasible by microwave techniques, where such materials include super-conducting, ferromagnetic, ferroelectric, dielectric and multiferroic perovskite systems. Chemical surface modifications are done to enhance the capacitive behavior and wettability of materials. For enhancing the specific capacitance of electrode components, metal oxides, perovskites, hydroxide particles, ceramics, and mixed metal oxides are accumulated on carbon. Various carbon materials with the appropriate design of pores, high surface area, and adaptable surface chemistry are anticipated. Those properties would provide superior electrode materials to improve the electrochemical double-layer capacitors with greater performance. Moreover, the high specific surface area can also be developed. Another significant role of these electrode materials to be used in supercapacitors will enrich the capacitance through the faradaic pseudo-capacitance effects. Furthermore, the metal used is important as an electrode in order to advance the capacitive response of carbon capacitors. This study indicates that the microwave-assisted synthesis approach can have a huge emphasis on refining and tuning the surface properties of various nanomaterials for energy storage applications.

REFERENCES

1. Sutton,W.H., Microwave processing of ceramic materials. *American Ceramic Society.* Bulletin, Westerville, 1989. **68**: p. 376.
2. Zhao, J., Yan, W., *Modern Inorganic Synthetic Chemistry.* Amsterdam, The Netherlands: Elsevier, 2011. **8**: p. 23.
3. Clark, D.E., Folz, D.C., Folgar, C.E., Mahmoud, M., *Microwave Solutions for Ceramic Engineers: The American Ceramics Society.* 2005, Westerville, OH: John Wiley & Sons. 494.
4. Park, S.-E., Chang, J.-S., Hwang, Y., Kim, D., Jhung, S., Hwang, J., Supramolecular interactions and morphology control in microwave synthesis of nanoporous materials. *Catalysis Surveys from Asia*, 2004. **8**(2): p. 20.
5. Wang, J., Niu, B., Du, G., Zeng, R., Chen, Z., Guo, Z., Dou, S., Microwave homogeneous synthesis of porous nanowire Co_3O_4 arrays with high capacity and rate capability for lithium ion batteries. *Materials Chemistry and Physics*, 2011. **126**(3): p. 8.

6. Vijayakumar, S., Ponnalagi, A.K., Nagamuthu, S., Muralidharan, G., Microwave assisted synthesis of Co_3O_4 nanoparticles for high-performance supercapacitors. *Electrochimica Acta*, 2013. **106**: p. 6.
7. Nagamuthu, S., Vijayakumar, S., Muralidharan, G., Synthesis of Mn_3O_4/Amorphous carbon nanoparticles as electrode material for high performance supercapacitor applications. *Energy & Fuels*, 2013. **27**(6): p. 8.
8. Ming, B., Li, J., Kang, F., Pang, G., Zhang, Y., Chen, L., Xu, J., Wang, X., Microwave–hydrothermal synthesis of birnessite-type MnO_2 nanospheres as supercapacitor electrode materials. *Journal of Power Sources*, 2012. **198**: p. 4.
9. Das, S., Mukhopadhyay, A.K., Datta, S., Basu, D., Prospects of microwave processing: An overview. *Bulletin of Materials Science*, 2008. **31**(7): p. 14.
10. Li, J.-M., Chang, K.-H., Wu, T.-H., Hu, C.-C., Microwave-assisted hydrothermal synthesis of vanadium oxides for Li-ion supercapacitors: The influences of Li-ion doping and crystallinity on the capacitive performances. *Journal of Power Sources*, 2013. **224**: p. 7.
11. Balaji, S.R.K., Mutharasu, D., Shanmugan, S., Subramanian, N.S., Ramanathan, K., Influence of Sm3+ ion in structural, morphological and electrochemical properties of $LiMn_2O_4$ synthesized by microwave calcination. *Ionics*, 2010. **16**: p. 351.
12. Parada, C., Morán, E., Microwave-assisted synthesis and magnetic study of nanosized Ni/NiO materials. *Chemistry of Materials*, 2006. **18**(11): p. 7.
13. Liu, S.-F., Abothu, I.R., Komarneni, S., Barium titanate ceramics prepared from conventional and microwave hydrothermal powders. *Materials Letters*, 1999. **38**(5): p. 7.
14. Komarneni, S., Katsuki, H., Nanophase materials by a novel microwave-hydrothermal process. *Pure Applied Chemistry*, 2002. **9**: p. 154.
15. Gupta, M., Leong, E.W.W., *Microwaves and Metals*. 2008, Singapore: John Wiley & Sons. 228.
16. Das, S., Mukhopadhyay, A.K., Datta, S., Basu, D., Prospects of microwave processing: An overview. *Bulletin of Materials Science*, 2009. **32**(1): p. 13.
17. Herrero, M.A., Kremsner, J.M., Kappe, C.O., Nonthermal microwave effects revisited: On the importance of internal temperature monitoring and agitation in microwave chemistry. *The Journal of Organic Chemistry*, 2008. **73**(12): p. 36.
18. Wroe, R., Rowley, A.T., Evidence for a non-thermal microwave effect in the sintering of partially stabilized zirconia. *Journal of Materials Science*, 1996. **31**(8): p. 8.
19. de la Hoz, A., Diaz-Ortiz, A., Moreno, A., Microwaves in organic synthesis: Thermal and non-thermal effects. *Chemical Society Reviews*, 2005. **34**: p. 15.
20. Michael P áMingos, D., Tilden lecture: Applications of microwave dielectric heating effects to synthetic problems in chemistry. *Chemical Society Reviews*, 1991. **20**: p. 47.
21. Rao, K.J., Bala, V., Ganguli, M., Ramakrishnan, P.A., Synthesis of inorganic solids using microwaves. *Chemistry of Materials*, 1999. **11**(4): p. 14.
22. Kingston, H.M., Haswell, S.J., (eds) *Microwave-Enhanced Chemistry: Fundamentals, Sample Preparation, and Applications*. Washington DC: American Chemical Society, 1997: p. 772.
23. Clark, D.E., Folz., D.C., West, J.K., Processing materials with microwave energy. *Materials Science and Engineering: A* 2000. **287**: p. 6.
24. Chandrasekaran, S., Basak, T., Ramanathan, S., Experimental and theoretical investigation on microwave melting of metals. *Journal of Materials Processing Technology*, 2011. **211**(3): p. 6.
25. Kitchen, H.J., Vallance, S., Kennedy, J.L., Tapia-Ruiz, N., Carassiti, L., Harrison, A., Whittaker, A.G., Drysdale, T.D., Kingman, S.W., Gregory, D.H., Modern microwave methods in solid-state inorganic materials chemistry: From fundamentals to manufacturing. *Chemical Reviews*, 2014. **114**(2): p. 206.

26. Katsuki, H., Furuta, S., Komarneni, S., Semi-continuous and fast synthesis of nanophase cubic $BaTiO_3$ using a single-mode home-built microwave reactor. *Materials Letters*, 2012. **83**: p. 3.
27. Cheng, J., Roy., R., Agrawal, D., Radically different effects on materials by separated microwave electric and magnetic fields. *Materials Research Innovations*, 2002. **5**: p. 8.
28. An, Z., Tang, W., Hawker, C.J., Stucky, G.D., One-Step microwave preparation of well-defined and functionalized polymeric nanoparticles. *Journal of American Chemical Society*, 2006. **128** (47): p. 2.
29. Feng, S., Hydrothermal and solvothermal syntheses. *Modern Inorganic Synthetic Chemistry*. 2011, Amsterdam, the Netherlands: Elsevier B.V. 31.
30. Gonzalez-Prieto, R., Herrero, S., Jiménez-Aparicio, R., Moran, R., Prado-Gonjal, J., Priego, J.L., Schmidt, R., Microwave-assisted solvothermal synthesis of inorganic compounds (molecular and non molecular). *Microwave Chemistry*. 2018.
31. Komarneni, S., Roy, R., Li, Q.H., Microwave-hydrothermal synthesis of ceramic powders. *Materials Research Bulletin*. 1993. **27**: p. 13 Berlin, Germany: Boston De Gruyter.
32. Komarneni, S., D'Arrigo., M.C., Leonelli, C., Pellacani, G.C., Katsuki, H., Microwave-hydrothermal synthesis of nanophase ferrites. *Journal of the American Ceramic Society*, 1998. **81**: p. 13. 45.
33. Schmidt, R., Prado Gonjal, J., Morán, E., Microwaves: Microwave assisted hydrothermal synthesis of nanoparticles. In. *CRC Concise Encyclopedia of Nanotechnology*, Boris Ildusovich Kharisov, Oxana Vasilievna Kharissova, Ubaldo Ortiz-Mendes, eds., 2015, Boca Raton, FL: CRC Press, p. 13.
34. Prado-Gonjal1, J., Schmidt, R., Moran, E., Microwave-assisted synthesis and characterization of perovskite oxides. In. *Perovskite: Crystallography, Chemistry and Catalytic Performance*, Zhang, J., and Li, H., eds. 2013, Hauppauge, NY: Novascience Publishers. p. 27.
35. Qiao, S., Huang, N., Sun, Y., Zhang, J., Zhang, Y., Gao, Z., Microwave-assisted synthesis of novel 3D flower-like $NiMnO_3$ nanoballs as electrode material for high-performance supercapacitor. *Journal of Alloys and Compounds*, 2019. **775**: p. 8.
36. Duraia, E.S.M., Fahami, A., Beall, G.W., Microwave-assisted synthesis of N-doped graphene-wrapped MnO_2 nanoflowers. *The Minerals, Metals & Materials Society*, 2018. **47** (12): p. 2.
37. Yanping, Z., Li, J., Yang, Y., Luo, B., Zhang, X., Fong, E., Chub, W., Huanga, K., Unique 3D flower-on-sheet nanostructure of NiCo LDHs: Controllable microwave-assisted synthesis and its application for advanced supercapacitors. *Journal of Alloys and Compounds*, 2019. **788**: p. 8.
38. Bindu, K., Sridharan, K., Ajith, K.M., Lim, H.N., Nagarajaa, H.S., Microwave assisted growth of stannous ferrite microcubes as electrodes for potentiometric non-enzymatic H_2O_2 sensor and supercapacitor applications. *Electrochimica Acta*, 2016. **217**: p. 11.
39. Zhang, X., Sun, X., Zhang, H., Zhang, D., Ma, Y., Microwave-assisted reflux rapid synthesis of MnO_2 nanostructures and their application in supercapacitors. *Electrochimica Acta*, 2013. **87**: p. 8.
40. William, J.J., Babu, I.M., Muralidharan, G., Microwave assisted fabrication of L-Arginine capped α-$Ni(OH)_2$ microstructures as an electrode material for high performance hybrid supercapacitors. *Materials Chemistry and Physics*, 2019. **224**: p. 12.
41. Lamiel, C., Nguyen, V.H., Kumar, D.R., Shima, J.J., Microwave-assisted binder-free synthesis of 3D Ni-Co-Mn oxide nanoflakes@Ni foam electrode for supercapacitor applications. *Chemical Engineering Journal*, 2017. **316**: p. 12.

42. Yan, Y., Wei, T., Qiao, W., Shao, B., Zhao, Q., Zhang, L., Fan, Z., Rapid microwave-assisted synthesis of graphene nanosheet/Co_3O_4 composite for supercapacitors. *Electrochimica Acta*, 2010. **55**(23): p. 6.
43. Zhu, Y., Cao, C., Tao, S., Chu, W., Wu, Z., Li, Y., Ultrathin nickel hydroxide and oxide nanosheets: Synthesis, characterizations and excellent supercapacitor performances. *Scientific Reports*, 2014. **4**: p. 5.
44. Sun, Z., Lin, H., Zhang, F., Yang, X., Jiang, H., Wang, Q., Qu, F., Rapid microwave-assisted synthesis of high-rate FeS_2 nanoparticles anchored on graphene for hybrid supercapacitors with ultrahigh energy density. *Journal of Chemistry A*, 2018. **6**(30): p. 11.
45. Kumar, R.D., Karuppuchamy, S., Microwave-assisted synthesis of copper tungstate nanopowder for supercapacitor applications. *Ceramics International*, 2014. **40**: p. 6.
46. Khalid, S., Cao, C., Naveed, M., Waqar, Y. 3D hierarchical MnO_2 microspheres: A prospective material for high performance supercapacitors and lithium-ion batteries. *Royal Society of Chemistry*, 2017. **1**: p. 10.
47. Bogdal, D., Bednarz, S., Matras-Postolek, K. Microwave-Assisted Synthesis of Hybrid Polymer Materials and Composites. In. *Microwave-Assisted Polymer Synthesis. Advances in Polymer Science*, Hoogenboom R., Schubert U., Wiesbrock F., eds, 2014, vol. 274., Cham, Switzerland: Springer, p. 241.

8 Tin-Based Materials for Supercapacitor

Muhammad Mudassir Hassan, Nawshad Muhmmad, Muhammad Sabir, and Abdur Rahim

CONTENTS

8.1 Introduction 119
8.2 Supercapacitor 120
8.3 Types of Supercapacitor 121
8.3.1 EDLCs 122
8.3.2 Pseudocapacitor 123
8.3.3 Hybrid Supercapacitor 123
8.4 Electrode Configuration 124
8.5 Electrode Materials for Supercapacitor 124
8.5.1 Carbon-Based Electrodes 124
8.5.2 Metal Oxides 125
8.5.3 Conductive Polymers 125
8.5.4 Composites Materials 126
8.5.5 Tin-Based Materials 126
8.6 Conclusion 129
References 129

8.1 INTRODUCTION

With the modern economic era and the industrial revolution, the non-renewable energy resources such as crude oil, coal, and diesel are leading the world towards an energy crisis and green-house effect. The total global energy demand for biofuels was estimated at 13.731 billion tonnes of oil equivalent in 2012 and is likely to reach 18.30 billion tonnes in 2035 (Singh et al. 2015). The removal and unequal placement of natural resources have already caused economic problems resulting in several issues such as generation of energy, storage, and industry-related problems. Therefore, the development of robust and powerful renewable energy resources is beneficial for mankind. Thus, solar energy, wind energy, and hydropower are the major sources of renewable energy. But, still, these resources are not reliable. The sun does not shine at night, winds are not controlled by us, and also the water resources are connected to the climate conditions (Luo et al. 2018). So, there is an extreme need to store energy for non-stop working of our electronic devices and electric vehicles. To date, an electrochemical method for energy storage is the most promising and extensively used technology among other energy storage devices. This is because of their exceptional benefits such as versatility, high efficiency, and flexibility. At the front of these energy storage systems, batteries and supercapacitors first come in mind. Both have a different storage mechanism, cycle life, and storage capability (Kim et al. 2015; Jiang and Kucernak 2002).

Supercapacitors also called electrochemical capacitors are one of the most favorable electrochemical energy storage devices due to their fast charging/discharging, long life cycle, safety, and high power. Due to these advantages supercapacitors have attracted a lot of attention in the recent years (Yu et al. 2013). In this chapter tin-based materials will be discussed for supercapacitor applications. As we all know tin is a lightweight material that can be very beneficial for automobiles and transportation. As compared to this, the lead acid batteries are very heavy, and they are difficult to move easily. Its use in energy storage material is due to its corrosion resistance and high conductivity, i.e., 8.7×10^6 S/m at 20°C. Another advantage of using tin is that it is an environmental friendly, nonpoisonous material. Composites of tin are widely used in supercapacitor electrodes and provide good energy density with high powers. Tin is a less expensive metal and therefore it also has wide applications other than supercapacitors. The main hindrance behind using tin is due to its low specific capacitance. More research is required to find out better tin-based composite materials for energy storage devices. Besides this tin has no major disadvantages, although some of the tin-organic compounds are found to be harmful for the aquatic life.

8.2 SUPERCAPACITOR

Before going to the brief description of supercapacitors, take a look at the background of supercapacitors. Lithium-ion batteries were first introduced by Whittingham, Scrosati, and Armand in 1990 under Sony labs (Saleem et al. 2016). These batteries were best in performance and have high energy density, but they lack in power density. Therefore, the world needed more suitable and efficient batteries.

In the 1960s, conventional capacitors were already in use, but supercapacitors were first introduced by H.I Becker of General Electric. Since 1980 many big manufacturers came on the stage and developed different supercapacitors with many improvements. Nowadays supercapacitors are widely utilized in portable electronics, electric vehicles, and power sources or backups devices (Libich et al. 2018).

A supercapacitor is basically a device that can store a charge via electric double layer or Helmholtz layer at the electrode-electrolyte interface. The energy storage capacity of a supercapacitor can be easily illustrated with a Ragone plot, as shown in Figure 8.1.

We can see clearly that supercapacitors provide high-power density more than batteries and very high energy density as compared to traditional capacitors (Ali et al. 2017). Thus, the supercapacitor is the only device that eliminates the gap between traditional capacitors and batteries.

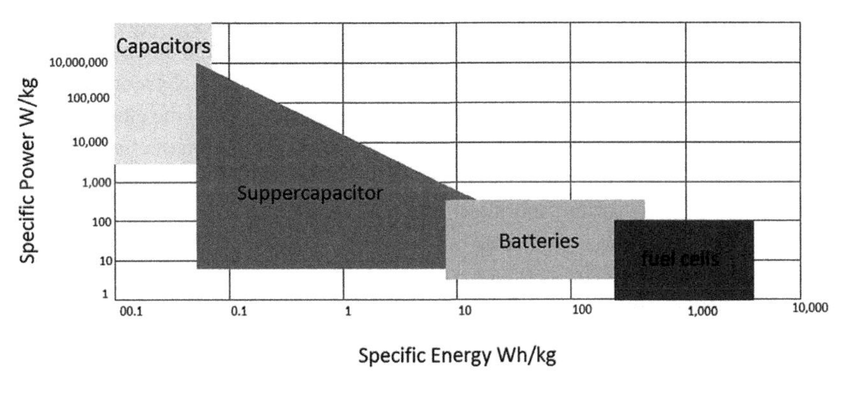

FIGURE 8.1 Ragone plot for energy and power density.

A supercapacitor basically consists of three parts:

1. Anode
2. Cathode
3. Electrolyte

8.3 TYPES OF SUPERCAPACITOR

Supercapacitors are mostly divided into three types, but when considering its storage mechanism, there are two main types. The third type of supercapacitor is basically a combination of first two types. These types are:

1. Electrochemical double-layer capacitors (EDLCs)
2. Pseudocapacitors
3. Hybrid supercapacitor

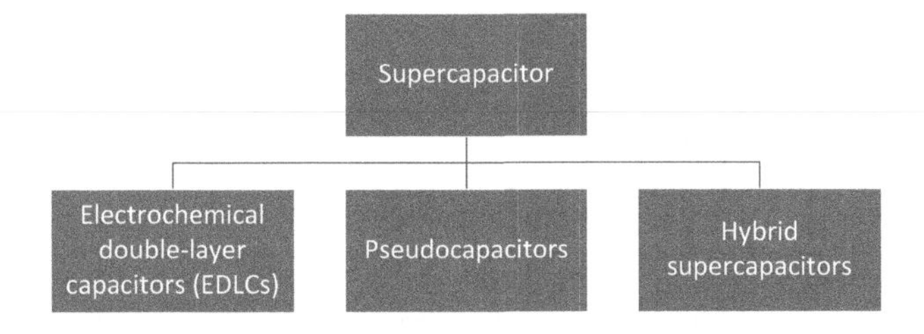

SCHEMATIC REPRESENTATION OF TYPES OF SUPERCAPACITORS

8.3.1 EDLCs

EDLCs are the most common types of supercapacitors. They have a high surface area and electrical conductivity. The charge storage mechanism is by the electrostatic accumulation of charges at the electrode and electrolyte interface. EDLCs store energy by the physical adsorption/desorption of charges from an electrolyte on the electrode. Thus the capacitance of EDLCs is barely based on the static accumulation of charges (Zhang and Zhao 2009). This is the same as Helmholtz described the double layer. The concept of Helmholtz was further modified by Gouy and Chapman. They described the double layer as a diffusive layer. Later on stern described this double layer effect and also named as stern layer. The stern layer was the combination of two layers as an inner layer and a diffusive layer and capacitance is due to the effect of both these layers. Specific capacitance of electric double layer is calculated by the following formula.

$$C = \varepsilon_r \varepsilon_0 A / d$$

where ε_r is the dielectric constant for the ectrolyte, ε_0 is dielectric constant in vacuum A is the surface area for electrode and d is the thickness of double layer.

The EDLCs are most commonly used in commercial markets. These supercapacitors use an electrolyte sandwiched between the electrodes instead of dielectric in the commonly used capacitors (Abbas et al. 2018) (Figure 8.2).

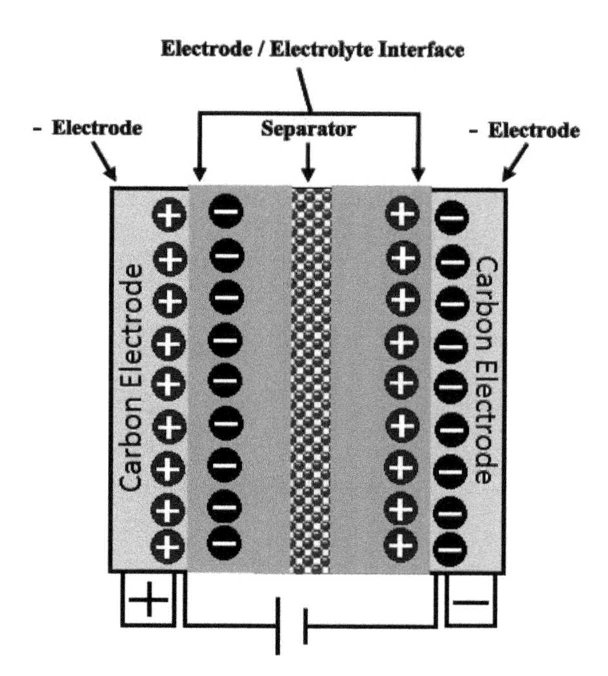

FIGURE 8.2 Storage mechanism of an electrochemical double-layer capacitor (EDLC). (From Abbas, Q. et al., Carbon/metal oxide composites as electrode materials for supercapacitors applications, 2018.)

8.3.2 Pseudocapacitor

The second type of supercapacitor is pseudocapacitor. It is based on reversible faradic reactions on the electrode materials to store energy. The redox reactions for charging/discharging takes place on the electrode electrolyte interface; therefore, the pseudocapacitors are somehow similar to batteries in storage mechanism, but they can be distinguished through the electroanalytical experiments. The most frequently used materials for pseudocapacitors are metal oxides, conducting polymers, and carbon-based materials. The pseudocapacitors show much larger capacitance values than double-layer capacitors. But they have low power density and cyclic stability due to the slow redox reactions occurring at the electrode. Remarkably, the response of pseudocapacitor is the same as EDLCs (Augustyn et al. 2014) (Figure 8.3).

8.3.3 Hybrid Supercapacitor

Hybrid supercapacitors include both faradic and double-layer mechanism; therefore, they show more energy density than EDLCs and longer cyclic stability. Commonly one electrode of the supercapacitor is of battery type and other is of supercapacitor material. Another type of hybrid supercapacitors is like one electrode is of EDLC and other of pseudocapacitor. This enhances the cell voltage, which is the most important parameter of supercapacitor. The most common setups of hybrid supercapacitors are AC//graphite, AC//$LiMn_2O_4$, and AC//$Ni(OH)_2$ (Zhang and Zhao 2012). This type of supercapacitor is getting more attention, and researchers are looking for its commercial preparation.

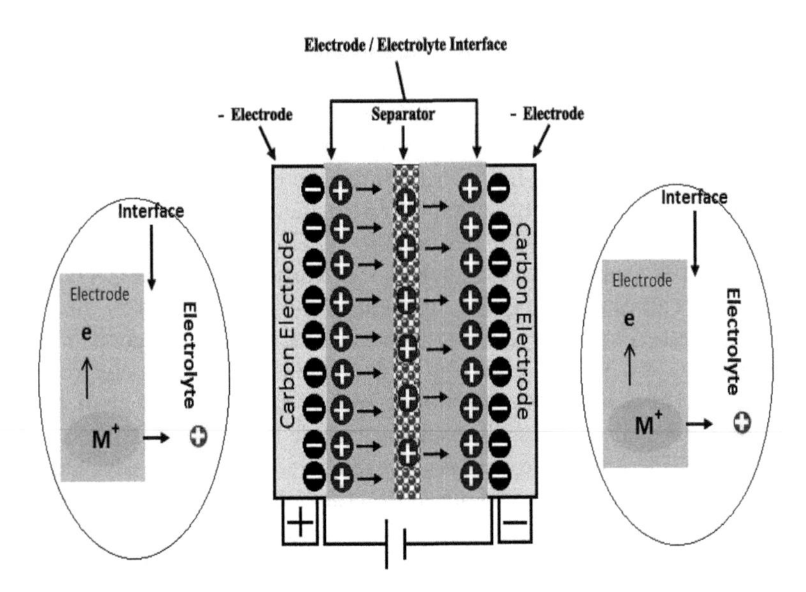

FIGURE 8.3 Storage mechanism of a pseudocapacitor. (From Abbas, Q. et al., Carbon/metal oxide composites as electrode materials for supercapacitors applications, 2018.)

8.4 ELECTRODE CONFIGURATION

Based on the electrode configuration, supercapacitors are divided into two categories:

1. Symmetric
2. Asymmetric

Generally, symmetric supercapacitors are made of the same electrode material while asymmetric supercapacitors are made of different electrode materials. Commonly a carbon electrode is used in asymmetric supercapacitor to achieve high capacitance and energy density (Ganesh et al. 2006). Generally, it is noted that asymmetric configuration provides high specific capacitance.

8.5 ELECTRODE MATERIALS FOR SUPERCAPACITOR

8.5.1 Carbon-Based Electrodes

Carbon-based materials, which include activated carbon, porous carbons, carbon nanotubes, and graphene, are of common use. They can be used both as positive and negative electrodes. The main advantages of carbon-based electrodes are that they have low density, high surface area, low cost, and stability. Activated carbon is mostly used in EDLCs, but it is also used in pseudocapacitors and hybrid supercapacitors for high energy density (Wang et al. 2013). Electrode materials like activated carbon have a lot of porosity and high specific surface area of more than 2000 m^2 per gram. The specific surface area of carbon materials increases with the capability of accumulation of charges, but only high surface area does not always increase the specific capacitance. So, porosity is the second major factor for increasing the specific capacitance. The pore size is importance and can be prepared in different sizes. For this purpose zeolites, metal-organic frameworks, silica, and some oxides are used as templates (Wang et al. 2015). Carbon nanotubes (CNTs) are hollow structures and have a diameter of nanoscale range. Depending on the number of tubes it is further divided into two types, i.e., single wall CNTs and multiwall CNTs. The specific surface area increases to about 2630 m^2 per gram. Capacitive behavior of carbon nanotubes CNTs is similar to activated carbon. The specific capacitance can be increased by the amount of carbon aerogel. Also, the cyclic performance of CNTs is very good. The electronic properties of CNTs are also very appreciable with excellent electron or ion transportation (Liu et al. 2008). Recently, graphene has attracted a great attention in the field of material sciences. Graphene is a single layer one atom thick carbon-based material. The unique properties of graphene include impressive electronic conductivity and high mechanical strength. Graphene can be prepared in many ways like CVD, PECVD, reduction of graphite oxides in solution and many more. Composites of graphene with other materials can also be obtained from active materials (Stankovich et al. 2006).

8.5.2 Metal Oxides

Metal oxides have been extensively studied for the storage capacity in energy storage devices like supercapacitors. The most commonly used metal oxide in a supercapacitor is RuO that has a specific capacitance of about 905 Fg^{-1} but due to its high cost, its commercial market is lowered and confined only to lab experiments. To overcome that problem noble metals were replaced by transition metal oxides such as iron oxide FeO, nickel oxide NiO, vanadium oxide V_2O_5, lead oxide PbO_2, cobalt oxide Co_3O_4, tin oxide SnO, and many others. The basic advantage of transition metal oxides using as an electrode is that they have a mesoporous structure, which makes it a promising candidate for supercapacitor electrodes. Metal oxides have been studied as both positive and negative electrodes and have shown excellent results. Their interaction between electrodes and electrolytes is very efficient. Metal oxides show fast ionic transportation at electrode-electrolyte interface (Raza et al. 2018). The major drawback coming from transition metal oxides is that they provide low specific capacitance when applied for practical applications. Therefore, more efficient and high specific transition metallic oxides need to be studied.

In recent years the mixed transition metallic oxides, another branch of metal oxide, came on the stage. They proved to be more cost efficient and stand out as good candidates for supercapacitor applications. Due to the effect of two distinct metals, they provide a large potential window and also more electroactive sites for high specific capacitance. Many mixed transition metallic oxides were reported, but the most fascinating among them is vanadium-based mixed metallic oxide. Others were also reported like metal cobaltites, metal molybdates; these have shown to have excellent specific capacitance (Low et al. 2019). Many methods were utilized for the synthesis of metallic oxides, but the most studied and facile method is the hydrothermal route. The synthesis of metal oxides with the required structure and morphology is still a task for researchers. Thus, there is a significant demand for the economical and environmental friendly materials that are easily synthesized for energy storage devices.

8.5.3 Conductive Polymers

Conducting polymers are basically organic-based polymers. They have the ability to conduct electricity through a bonded medium. Over the last two decades conducting polymers have been widely studied for supercapacitor applications. They have a high energy density, low cost, and show reversible redox reactions. In conducting polymers materials like polypyrrole (PPy), and polyaniline (PANI) have been extensively studied as supercapacitor electrodes. Conducting polymers have been developed and utilized in different nanostructured forms like nanorods, nanosheets, etc. These nanostructured polymers show excellent performance high surface area, porosity, and conductivity (Ryu et al. 2002).

Polyaniline PANI is found to be a good candidate for the supercapacitor electrodes due to higher electrical conductivity, excellent stability, and low synthesis cost.

Many nanocomposites were also tried with PANI for the energy storage electrodes, but the main challenge behind its commercial productivity is its fabrication into a cell. For the fabrication of electrodes, binders are required which are inert and thus distort the PANI nanostructure. Also, the cost becomes higher for the fabrication of cells. Thus a facile and single straightforward method is needed to fabricate PANI polymer nanostructures (Meng et al. 2017).

Polypyrrole (PPy) is another conducting polymer with unique features like fast charging-discharging, high thermal stability, high energy density, and conductivity. Polypyrrole is the most advancing conducting polymer and have received much interest among the researchers (Fan and Maier 2006).

8.5.4 Composites Materials

Carbon materials have been widely used as energy storage electrodes despite the fact they provide low specific capacitance. Metal oxides and conducting polymers have been broadly studied; they provide good specific capacitance and storage capacity, but they lag in cyclic performance during the charging-discharging. To overcome this problem composite materials were studied with higher energy and power densities. On concentrating this challenge Au/Ni core/shell nanocrystals were prepared by utilizing bimetallic heterostructures. The specific capacitance was reported to be 806.1 Fg^{-1} (Duan and Wang 2014). 3D printing is also being utilized to facilitate the process for composite structures on supercapacitor electrodes. Graphene composite materials were also reported in supercapacitor electrodes (Bai et al. 2011).

8.5.5 Tin-Based Materials

Tin is a chemical element with symbol Sn and atomic number 50. It is mostly acquired from the ore of cassiterite, which consists of stannic oxide SnO_2. It has two oxidation states—2^+ and 4^+. It is also an abundant element on earth with largest number of stable isotopes. At room temperature it has mainly two types of allotropes. Melting point of tin is 231.9°C, and it has very low toxicity; therefore, it is used in many alloys and other applications. Here, we discuss only its storage application in supercapacitors (Batzill and Diebold 2005).

Metal oxides show a lot of potential in supercapacitive performance. The specific capacitance shown by metal oxides is very high, but the challenge is to maintain long life cycle. Yang et al. prepared SnO on surface of reduced graphene oxide and its electrochemical behavior was tested. The SnO was used as anode while the cathode used was MoO_3 to make an asymmetric supercapacitor. The results showed the energy density of 33 Wh Kg^{-1} in aqueous electrolyte. It was due to high work function between the two electrodes. Moreover, the cyclic stability was reported to be 92.5% after 20,000 cycles (Yang et al. 2018).

Doping of metals increases the conductivity and offers better behavior than metals solely. Sn doped manganese oxide were prepared by hydrothermal method. Octahedral molecular sieve was prepared and checked for the electrochemical performance. Different concentrations of SnO_2 were doped, and its behavior was reported.

SEM images shown that with the increase in concentration of SnO_2 the length of nanorods increased. The supercapacitive performance was tested in 1 M Na_2SO_4 electrolyte. This showed 187 Fg^{-1} specific capacitance. It was also found that doping concentration greater than 10% showed the best performance (Sun et al. 2012).

The same tin-doped manganese oxide was prepared by coprecipitation method for electrode material for supercapacitor. Some 200–500 nm spherical particles were prepared with 100 nm length nanorods. Specific capacitance of 293 Fg^{-1} was reported by the authors. The increase in capacitance was due to increase in molar ratio of Mn to Sn which was taken 50:01. The cyclic performance first decreased but remained stable after 600 cycles (Pang et al. 2009).

Graphene (Gr)-based tin nanoparticles were prepared by thermal reduction of graphite oxide by adding tin powder to it. The tin nanoparticles were allocated on graphite sheet uniformly. Specific capacitance of Gr-based tin electrodes was reported to be 320 Fg^{-1} in 2 M KNO_3 aqueous electrolyte. The CV of the reported sample shows an electric double-layer behavior and fast charging in 25 s at 1 Ag^{-1} (Qin et al. 2012).

Mostly metal oxides show weak electrical conductivity behavior. To overcome that problem carbon-based nanostructured materials are hybridized to increase their conductivity. In the reported work CNT were used as template for the synthesis of SnO_2–Cu_2O nano composite. The specific capacitance of the material was reported to be 662 Fg^{-1}. To provide strong interaction, tin acted as an intermediate layer between carbon nanotubes and copper. The cyclic performance was also found to be excellent with 94% retention rate after 5000 cycles (Daneshvar et al. 2018).

Other than oxides, sulfides show outstanding properties due to its high theoretical capacitance, good carrier mobility, and semiconducting nature. It also forms a layered structure that has unit cells of hexagonal shape. Tin sulfides were widely reported for supercapacitor application. SnS_2 flower like nanostructures were synthesized by the solvothermal method. FE-SEM shows flower-like nanostructures with different sizes. The CV plots describe the pseudocapacitive behavior of the material, which explains that it has very high specific capacitance. The specific capacitance and power density were reported to be 524.5 Fg^{-1} and 12.3 W kg^{-1} respectively at the current density of 0.08 Ag^{-1} (Mishra et al. 2017).

Ni Co Mn doped SnS_2 graphene-based aerogel was prepared by solvothermal method. It was reported that 2D and 3D nanosheets structures were developed by this method. The electrochemical tests were conducted to study and calculate the specific capacitance of the composites. The specific capacitance was reported to be 523.51 Fg^{-1} at scan rate of 5 mV s^{-1}. The reported material also shows very good cyclic stability. The retention was 98.57% after 2000 cycles (Chu et al. 2018).

Changing the morphology of nanoparticles creates a big effect on the characteristics of a material. In this regard tin sulfides nanostructures with different morphologies were prepared and their electrochemical behavior was studied. For this sheet-like tin sulfide, flower-like tin sulfide, and ellipsoid-like tin sulfides were synthesized by hydrothermal and solvothermal method. All of them were studied at different scan rates and their behaviors were analyzed. The specific capacitances of sheet-like, ellipsoid-like, and flower-like nanostructures were 117.72, 390.38 and 431.82 Fg^{-1} respectively. Flower-like SnS_2 showed the best performance (Parveen et al. 2018).

Reduced graphene oxides rGO plays a very vital role in energy storage devices. It is very stable in organic solvents that makes it a strong candidates for electrodes in energy storage devices. Moreover it also provides a large surface area and is more conductive than graphene only. In this regard H Chauhan et al. synthesized tin sulfide with rGO a composite for supercapacitor electrode. The layered nanocomposite acts as a superior electrode. The SnS_2/rGO nanocomposite was synthesized by hydrothermal approach. The SEM showed the hexagonal tin sulfide nanoparticles with rGO sheets embedded on it. For electrochemical analysis symmetric configured cell was prepared with aqueous electrolyte. The specific capacitance reported was 488 F/g. The cyclic stability was found to be 95% after 1000 cycles (Chauhan et al. 2017).

A net-like structure was prepared of tin sulfide carbon composite. The tin sulfide nanoparticles were prepared by chemical precipitation. Then the composite was prepared at 650°C by carbonization method. The CV of the composite shows the pure electric double-layer effect. The net-like nanostructure shows the specific capacitance of 36.16 Fg^{-1}. It was reported that the carbon composite had shown excellent conductivity (Li et al. 2009).

Nanowires of tin oxide coated on MoO_3 were synthesized by hydrothermal method. The electrochemical properties were studied in 1 M Na_2SO_4 electrolyte solution. The specific capacitance of tin oxide/molybdenum oxide nanowires was reported to be 295 Fg^{-1}. The cyclic stability of the reported material was also very good with 97% retention after 1000 cycles (Shakir et al. 2012).

SnO_2 graphene nanocomposite were prepared through wet chemical method. Tetragonal structure of tin oxide nanostructures was formed on graphene with 50 nm size. The electrochemical performance was studied in 6 M KOH electrolyte solution. The reported specific capacitance was 818.6 Fg^{-1} (Velmurugan et al. 2016).

Two-dimensional tin oxide on graphene was synthesized by hydrothermal method. The tin oxide graphene composite provides more area than graphene solely. The more area means more porosity and specific area. Electrochemical measurements were reported to be 294 Fg^{-1} specific capacitance (Li et al. 2014).

SnO nanoporous films were fabricated anodically. The tin substrate was anodized in an aqueous electrolyte. Due to nanoporous structure the reported material provides high electrochemical properties. The specific capacitance of the reported film was 274 Fg^{-1}. The CV of the reported material shown more like a pseudocapacitive behavior (Shinde et al. 2013).

A ternary oxide of mixed metals was also studied. Ni Co Sn oxides thin film were prepared by Pechini technique. Xrd pattern shown in the literature was of thin metal sheets on titanium. SEM showed the cracks on the surface of the material, which increased the porosity. The electrochemical properties were analyzed in KOH solution as electrolyte. The typical redox behavior was noted in the CV diagrams, which shows the pseudocapacitive behavior of the thin film. The CV graph showed the same characteristic as of nickel oxide curves. The CV showed the specific capacitance of 328 Fg^{-1} and retention ratio of 86% after 600 cycles. The power density was calculated to be 345.7 W kg^{-1} (Ferreira et al. 2014).

Tin oxide thin films were studied by spraying different molarity of $SnCl_4$. These films were focused on only 200 planes. The CV of the prepared material was showing pseudocapacitive behavior. The specific capacitance was found out to be 150 Fg^{-1}. The authors claimed that change in molarity of solution produces more porous SnO_2 thin films (Yadav 2016).

SnO_2 graphene composite was prepared by solvothermal method. The prepared nanoparticle was measured by FESEM and its size was claimed to be only 10 nm. The specific capacitance of 363.5 Fg^{-1} was reported which was much higher than its rivalries (Lim et al. 2013).

Conducting polymers are of great interest in the storage devices. One of them PANI is the most promising material. It provides fast charging discharging and very high-power density, but it suffers from low cyclic stability. To address the problem a SnO_2 nanorods were prepared by seed aided hydrothermal method. These nanorods were synthesized on carbon fibers after that PANI layer was deposited on it through the electrode chemical deposition technique. These PANI/SnO_2 nanowires acted as directional nanorods during electrochemical reaction. A symmetric electrode cell was prepared to study the electrochemical behavior of the samples. The sulfuric acid PVA gel was used as an electrolyte. The specific capacitance of the reported composite was claimed to be 367.5 Fg^{-1} at current density of 0.5 Ag^{-1}. The retention ratio was 88.3% even after 2000 cycles which was a great achievement. The CV shows a rectangular shape, which is an obvious form of double layer effect (Xie and Zhu 2017).

8.6 CONCLUSION

In conclusion, various supercapacitors (EDLCs, pseudocapacitor, hybrid supercapacitor) have been used as energy storage devices, and there is a dire need for non-stop working of our electronic devices and electric vehicles. The different materials, like metal oxides, conductive polymers, composites, carbon-based materials, and tin-based materials are used in the fabrication of supercapacitors. Among these materials, tin-based composite materials and tin sulfides are promising materials for supercapacitor applications owing to its unique supercapacitive properties of tin.

REFERENCES

Abbas, Qaisar, Abdul G Olabi, Rizwan Raza, and Des Gibson. 2018. Carbon/metal oxide composites as electrode materials for supercapacitors applications. In: *Reference Module in Materials Science and Materials Engineering*, 2018, Oxford: Elsevier. pp. 1–10.

Ali, Faizan, Xiaohua Liu, Dayu Zhou et al. 2017. Silicon-doped hafnium oxide anti-ferroelectric thin films for energy storage. *Journal of Applied Physics* 122 (14):144105.

Augustyn, Veronica, Patrice Simon, and Bruce Dunn. 2014. Pseudocapacitive oxide materials for high-rate electrochemical energy storage. *Energy & Environmental Science* 7 (5):1597–1614.

Bai, Hua, Chun Li, and Gaoquan Shi. 2011. Functional composite materials based on chemically converted graphene. *Advanced Materials* 23 (9):1089–1115.

Batzill, Matthias, and Ulrike Diebold. 2005. The surface and materials science of tin oxide. *Progress in Surface Science* 79 (2–4):47–154.

Chauhan, Himani, Manoj K Singh, Praveen Kumar, Safir Ahmad Hashmi, and Sasanka Deka. 2017. Development of SnS_2/RGO nanosheet composite for cost-effective aqueous hybrid supercapacitors. *Nanotechnology* 28 (2):025401–025401.

Chu, Hang, Fangfang Zhang, Liyuan Pei, Zheng Cui, Jianfeng Shen, and Mingxin Ye. 2018. Ni, Co and Mn doped SnS_2-graphene aerogels for supercapacitors. *Journal of Alloys and Compounds* 767:583–591.

Daneshvar, Farhad, Atif Aziz, Amr M Abdelkader, Tan Zhang, Hung-Jue Sue, and Mark E Welland. 2018. Porous SnO_2–Cu x O nanocomposite thin film on carbon nanotubes as electrodes for high performance supercapacitors. *Nanotechnology* 30 (1):015401.

Duan, Sibin, and Rongming Wang. 2014. Au/Ni 12 P_5 core/shell nanocrystals from bimetallic heterostructures: In situ synthesis, evolution and supercapacitor properties. *NPG Asia Materials* 6 (9):e122.

FanFan, Li-Zhen, and Joachim Maier. 2006. High-performance polypyrrole electrode materials for redox supercapacitors. *Electrochemistry communications* 8 (6):937–940.

Ferreira, CS, RR Passos, and LA Pocrifka. 2014. Synthesis and properties of ternary mixture of nickel/cobalt/tin oxides for supercapacitors. *Journal of Power Sources* 271:104–107.

Ganesh, V, S Pitchumani, and V Lakshminarayanan. 2006. New symmetric and asymmetric supercapacitors based on high surface area porous nickel and activated carbon. *Journal of Power Sources* 158 (2):1523–1532.

Jiang, Junhua, and Anthony Kucernak. 2002. Electrochemical supercapacitor material based on manganese oxide: preparation and characterization. *Electrochimica Acta* 47 (15):2381–2386.

Kim, Brian Kihun, Serubbable Sy, Aiping Yu, and Jinjun Zhang. 2015. Electrochemical supercapacitors for energy storage and conversion. *Handbook of Clean Energy Systems*:1–25.

Li, Yang, Huaqing Xie, and Jiangping Tu. 2009. Nanostructured SnS/carbon composite for supercapacitor. *Materials Letters* 63 (21):1785–1787.

Li, Zijiong, Tongqin Chang, Gaoqian Yun, Jian Guo, and Baocheng Yang. 2014. 2D tin dioxide nanoplatelets decorated graphene with enhanced performance supercapacitor. *Journal of Alloys and Compounds* 586:353–359.

Libich, Jiří, Josef Máca, Jiří Vondrák, Ondřej Čech, and Marie Sedlaříková. 2018. Supercapacitors: Properties and applications. *Journal of Energy Storage* 17:224–227.

Lim, SP, NM Huang, and HN Lim. 2013. Solvothermal synthesis of SnO_2/graphene nanocomposites for supercapacitor application. *Ceramics International* 39 (6): 6647–6655.

Liu, CG, M Liu, F Li, and HM Cheng. 2008. Frequency response characteristic of single-walled carbon nanotubes as supercapacitor electrode material. *Applied Physics Letters* 92 (14):143108.

Low, Wei Hau, Poi Sim Khiew, Siew Shee Lim, Chiu Wee Siong, and Ejikeme Raphael Ezeigwe. 2019. Recent development of mixed transition metal oxide and graphene/mixed transition metal oxide based hybrid nanostructures for advanced supercapacitors. *Journal of Alloys and Compounds* 775:1324–1356.

Luo, Yangxi, Wenjing Hong, Zongyuan Xiao, and Hua Bai. 2018. A high-performance electrochemical supercapacitor based on a polyaniline/reduced graphene oxide electrode and a copper(ii) ion active electrolyte. *Physical Chemistry Chemical Physics* 20 (1):131–136.

Meng, Qiufeng, Kefeng Cai, Yuanxun Chen, and Lidong Chen. 2017. Research progress on conducting polymer based supercapacitor electrode materials. *Nano Energy* 36:268–285.

Mishra, Rajneesh Kumar, Geun Woo Baek, Kyuwon Kim, Hyuck-In Kwon, and Sung Hun Jin. 2017. One-step solvothermal synthesis of carnation flower-like SnS_2 as superior electrodes for supercapacitor applications. *Applied Surface Science* 425:923–931.

Pang, Xu, Zheng-Qing MA, and Lie Zuo. 2009. Sn Doped MnO_2 Electrode Material for Supercapacitors. *Acta Physico-Chimica Sinica* 25 (12):2433–2437.

Parveen, Nazish, Sajid Ali Ansari, Hatem R Alamri, Mohammad Omaish Ansari, Ziyauddin Khan, and Moo Hwan Cho. 2018. Facile synthesis of SnS_2 nanostructures with different morphologies for high-performance supercapacitor applications. *ACS Omega* 3 (2):1581–1588.

Qin, Z, ZJ Li, M Zhang, BC Yang, and RA Outlaw. 2012. Sn nanoparticles grown on graphene for enhanced electrochemical properties. *Journal of Power Sources* 217:303–308.

Raza, Waseem, Faizan Ali, Nadeem Raza et al. 2018. Recent advancements in supercapacitor technology. *Nano Energy* 52: 441–473.

Ryu, Kwang Sun, Kwang Man Kim, Nam-Gyu Park, Yong Joon Park, and Soon Ho Chang. 2002. Symmetric redox supercapacitor with conducting polyaniline electrodes. *Journal of Power Sources* 103 (2):305–309.

Saleem, Amin M, Vincent Desmaris, and Peter Enoksson. 2016. Performance enhancement of carbon nanomaterials for supercapacitors. *Journal of Nanomaterials* 2016, Article ID 1537269, 17 pages. https://doi.org/10.1155/2016/1537269.

Shakir, Imran, Muhammad Shahid, Muhammad Nadeem, and Dae Joon Kang. 2012. Tin oxide coating on molybdenum oxide nanowires for high performance supercapacitor devices. *Electrochimica Acta* 72:134–137.

Shinde, Dipak V, Deok Yeon Lee, Supriya A Patil et al. 2013. Anodically fabricated self-organized nanoporous tin oxide film as a supercapacitor electrode material. *RSC Advances* 3 (24):9431–9435.

Singh, Sonal, Shikha Jain, PS Venkateswaran et al. 2015. Hydrogen: A sustainable fuel for future of the transport sector. *Renewable and Sustainable Energy Reviews* 51:623–633.

Stankovich, Sasha, Dmitriy A Dikin, Geoffrey HB Dommett et al. 2006. Graphene-based composite materials. *Nature* 442 (7100):282.

Sun, Ming, Fei Ye, Bang Lan et al. 2012. One-step hydrothermal synthesis of Sn-doped OMS-2 and their electrochemical performance. *International Journal of Electrochemical Science* 7:9278–9289.

Velmurugan, V, U Srinivasarao, R Ramachandran, M Saranya, and Andrews Nirmala Grace. 2016. Synthesis of tin oxide/graphene (SnO_2/G) nanocomposite and its electrochemical properties for supercapacitor applications. *Materials Research Bulletin* 84:145–151.

Wang, Faxing, Shiying Xiao, Yuyang Hou, Chenglin Hu, Lili Liu, and Yuping Wu. 2013. Electrode materials for aqueous asymmetric supercapacitors. *RSC Advances* 3 (32):13059–13084.

Wang, Faxing, Zheng Chang, Minxia Li, and Yuping Wu. 2015. Nanocarbon-based materials for asymmetric supercapacitors. *Nanocarbons for Advanced Energy Storage* 1:379–415. https://doi.org/10.1002/9783527680054.ch14.

Xie, Yibing, and Feng Zhu. 2017. Electrochemical capacitance performance of polyaniline/tin oxide nanorod array for supercapacitor. *Journal of Solid State Electrochemistry* 21 (6):1675–1685.

Yadav, Abhijit A. 2016. Spray deposition of tin oxide thin films for supercapacitor applications: Effect of solution molarity. *Journal of Materials Science: Materials in Electronics* 27 (7):6985–6991.

Yang, Chongyang, Minqiang Sun, and Hongbin Lu. 2018. Asymmetric all-metal-oxide supercapacitor with superb cycle performance. *Chemistry–A European Journal* 24 (23):6169–6177.

Yu, Aiping, Victor Chabot, and Jiujun Zhang. 2013. *Electrochemical Supercapacitors for Energy Storage and Delivery: Fundamentals and Applications*. Boca Raton, FL: CRC press.

Zhang, Jintao, and XS Zhao. 2012. On the configuration of supercapacitors for maximizing electrochemical performance. *ChemSusChem* 5 (5):818–841.

Zhang, Li Li, and XS Zhao. 2009. Carbon-based materials as supercapacitor electrodes. *Chemical Society Reviews* 38 (9):2520–2531.

9 Inorganic Materials-Based Next-Generation Supercapacitors

Muhammad Aamir, Arshad Farooq Butt, and Javeed Akhtar

CONTENTS

9.1 Introduction 134
9.2 Capacitor 134
9.3 Classification of Supercapacitors 135
9.3.1 Electrochemical Double-Layer Supercapacitors 135
9.3.2 Pseudo Supercapacitors 136
9.3.3 Hybrid Supercapacitors 136
9.4 Mechanism of Capacitor Charging 136
9.5 Supercapacitors 137
9.6 Differences between Batteries and Supercapacitors 137
9.7 Electrode Materials 138
9.8 Metal Oxides/Hydroxide 138
9.9 Emerging Next-Generation Inorganic Materials for Supercapacitance 140
9.9.1 Metal-Organic Frameworks (MOFs) 140
9.10 MXenes 140
9.11 Metal Nitride (MNs) 141
9.12 Black Phosphorous (BP) 142
9.13 Perovskites 143
9.14 Fabrication of Perovskite Supercapacitors 143
9.15 Perovskite Oxide-Supercapacitance Performance 144
9.16 Halide Perovskite-Supercapacitance Performance 145
9.17 Conclusion 146
References 147

9.1 INTRODUCTION

In the current situation, the upsurge in global warming, depletion in fossil fuels and mankind reliance on energy-based appliances have driven the scientists and engineers to look for renewable energy alternatives. In this contest, the solar energy conversion to electrical energy has gained enormous attention for energy production. Concurrently, storage of electrical energy is highly in demand to overcome growing energy consumption. Thus, batteries, fuel cells and capacitors are gaining substantial consideration from researchers. Ever since discovery of "quantum size effect" in nanostructured materials, numerous efforts were made to fabricate energy storage devices using them. This has resulted not only miniaturization of such devices but also many folds enhancement in overall efficiency of these devices. In actual, batteries are postdated invention compared to the use of capacitors, which are simple devices consists of a glass jar coated with silver foil (Conway 2013). In this chapter, we describe some new exciting inorganic materials that have potential to build next-generation energy storage devices. We have highlighted first metal oxide/hydroxide materials used as electrode materials followed by emerging materials including, MOF, Mxene, metal nitride, black phosphorus and oxide/metal halide hybrid perovskites.

9.2 CAPACITOR

Generally speaking, a capacitor is a device that is capable of storing electrical energy. These are also called electrochemical capacitors, supercapacitors, ultracapacitors, or hybrid capacitors. Among all these terms, ultra-capacitor and supercapacitor refer to the devices in which charge storage takes place in the parallel placed electrical double layer (EDL). Moreover, the ultra-capacitors have high surface area of anode and cathode. In simplest form, it consists of two parallel plates of some conducting materials like metal, separated by an insulating layer named as dielectric. The separators are present between anode and cathode and act as physical barriers to avoid electrical shortage. The separators can be gelled electrolytes or porous materials including ceramic, mica, or waxed paper. These separators should be permeable to ions, but should remain inert during the electrochemical process. There is a large variety of metal plates in shape or sizes that are used to construct a capacitor, and it depends on capacitor ultimate use as well. In the case, when direct current (DC) is applied, the respective plates store charge in the form of voltage. After full charging, the current stops to flow through the circuit. However, when current is applied through alternate current (AC) circuit, after full charging, the current continues to flow in the circuit in either direction. In this case, actually there is no net current flow through the capacitor. The quantity of charge that a capacitor stores is known as capacitance, and it depends on following factors (Winter and Brodd 2004):

- Surface area of two plates; larger the surface area, the higher will be capacitance
- Separating distance between two plates. The lesser this distance, the higher will be capacitance and vice versa
- Permittivity of dielectric materials; the higher it has value, the greater will be capacitance.

9.3 CLASSIFICATION OF SUPERCAPACITORS

There are three broader classes of supercapacitors (Figure 9.1), including the following.

9.3.1 Electrochemical Double-Layer Supercapacitors

The immersion of electronically conducting electrodes into an ion conducting electrolyte solution results in the spontaneous charges assembling along the electrode surface and into the solution near the electrode surface, which is one of the common type, wherein charge is stored on respective plates using non-faradic method. Upon applying voltage, charge starts to accumulate at the surface of plates. Simultaneously ions in electrolytes also begins to move, thus establishing a double layer, sandwiched the charge on the electrode surface and the charges on the electrolyte. In this type of electrode, plates are made of activated carbon. The double layer forms and relax directly at the electrode surface and formations at constant time that is approximately 10^{-8} s. Therefore, EDL response is rapid with respect to the potential changes (Raza et al. 2018, Winter and Brodd 2004).

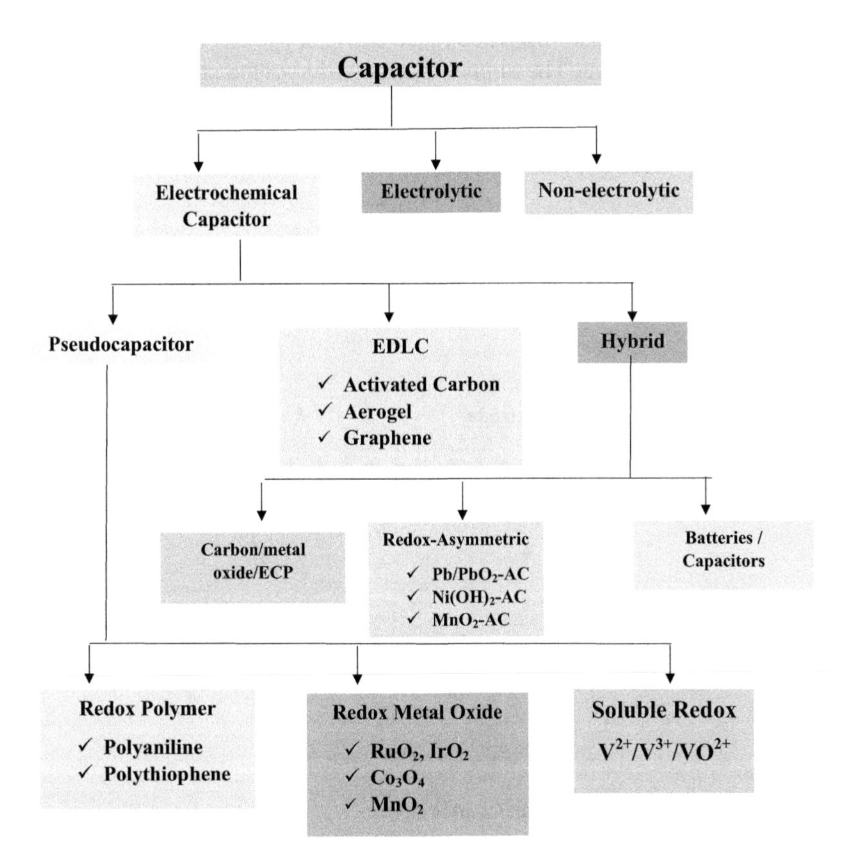

FIGURE 9.1 Schematic showing the classification of capacitor.

9.3.2 Pseudo Supercapacitors

In this type of supercapacitor, the charge accumulation mechanism is redox type (faradaic charge). A set of redox reactions, namely oxidation, causes reduction to take place on or near the surface of electrodes when ions move in electrolytes towards opposite direction. For example, T-Nb_2O_5, MoS_2 and $MoSe_2$ based supercapacitors fall in this case (Yan et al. 2017).

9.3.3 Hybrid Supercapacitors

The hybrid supercapacitor is a combination of both EDLC and pseudocapacitor, thus delivering more power and energy. In this case, one part of hybrid supercapacitor operates like EDLC while other follow pseudo-supercapacitors, thus making it more efficient (Winter and Brodd 2004).

9.4 MECHANISM OF CAPACITOR CHARGING

Two parallel electrodes sandwiched the dielectric materials are the constituents of the capacitor (Figure 9.2). The energy stored in these devices is in the form of electrostatic field. The potential difference found between two electrodes is responsible for charging of the device. The applied potential derive the charges (positive and negative) to travel towards the surface of corresponding oppositely charged electrodes. Capacitance is the measure of ratio of electric charge to the potential difference between two electrodes and is represented as:

$$C = \frac{Q}{V} \tag{9.1}$$

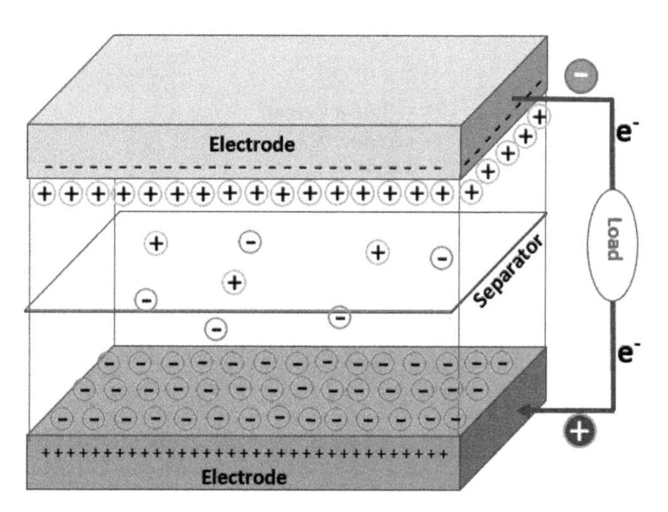

FIGURE 9.2 Schematic of supercapacitance device.

9.5 SUPERCAPACITORS

These are also renowned as electrochemical capacitors and or ultra-capacitors and are capable to store more charge (100 times more) than conventional capacitor. The demand of electricity storage for transport vehicles, portable electronic devices, industrial equipment and drones and for military applications etc has urged the scientists to search for next-generation supercapacitors. Capacitors have improved charge stability and superior cycle stabilities (Winter and Brodd 2004, Zhang and Zhao 2009).

Generally, capacitors take only few seconds for charging and discharging. At present, on the basis of specific dielectric use and physical state, various type of capacitors are available. Each class has specific characteristic and application, including small capacitors for electronics to large power capacitors that have high energy storage. The supercapacitor has a higher energy density compared to fuel cells and batteries. Currently, the specific energy density of commercially available supercapacitors is (<10 Wh Kg^{-1}) around 3–15 times less than Li-ion batteries (150 Wh Kg^{-1}). Therefore, increasing interest has been observed by scientist for research in supercapacitors to achieve the specific energy densities equal or larger than batteries.

9.6 DIFFERENCES BETWEEN BATTERIES AND SUPERCAPACITORS

Supercapacitors, batteries and fuel cells together make the energy conversion and storage system. However, technically, energy storage and conversion mechanisms are different, but they have some electrochemical similarities as well. The key features for the above-mentioned three electrochemical devices includes:

- The electrical energy is produced by redox reaction taking place at anode and cathode in fuel cell and battery. Furthermore, battery is a closed system and anode-cathodes act as active mass. On the contrary, a fuel cell is an open system that transfers charge through electrodes and required active masses fed from outside. In the capacitors (supercapacitors), redox reaction may not be involved in energy storage process, therefore, cathode and anode terms are not appropriate for these systems.
- In batteries, the charge rate is a kinetically controlled phenomena and reversible process.
- Typical cycle life of supercapacitors is about 30,000 h to 1 million h unlike batteries where ~500 h is maximum (Qu et al. 2016).
- The discharge times for supercapacitors is very low 1–10 s a/in 10–60 mins.
- Supercapacitors can efficiently work in a temperature range of –40°C to 100°C, while batteries –20°C to 60°C (Kirk and Graydon 2013).

9.7 ELECTRODE MATERIALS

Wide range of materials from pure metals to metal oxides, carbon nanotube to graphene and composites with amalgamation of polymers or additives have been used to construct supercapacitors. Following are key features for materials used as potential electrodes for supercapacitors.

- Low toxicity and environmentally friendly
- Cost-effective and should be based on earth's abundant materials
- Should have very low resistivity and light weight

9.8 METAL OXIDES/HYDROXIDE

Different materials can be used for the fabrication of supercapacitors, depending on the type applications and capacitance range required. There are three types of electrode materials based on their use for EDCL, pseudocapacitors and hybrid capacitor. At present, numerous electrode material have been available for supercapacitors. Among these materials, carbon is most widely used in various forms (Obreja 2008).

Apart from carbon-based materials, inorganic materials have gained enormous attention for supercapacitance applications. Among inorganic materials, metal oxide has shown high conductivity, therefore, can be used as electrode materials in supercapacitance application (Wang et al. 2018b). Due to this, these metal oxides-based supercapacitors have shown large specific capacitance and durability. Ruthenium oxide was the first generation electrode material for supercapacitors and it follows faradic charge transport reaction (Augustyn, Simon, and Dunn 2014). The maximum theoretical capacitance that can be estimated for RuO_2 and $RuO_2{\cdot}0.5\ H_2O$ was 1450 and 1360 Fg^{-1}, respectively (Dmowski et al. 2002). The charge storage performance of the ruthenium oxide electrodes was based on the proton insertion and extraction in the RuO_2 structure. As expected, the $RuO_2{\cdot}0.5\ H_2O$ material has shown enhanced proton conductivity that results in higher supercapacitance performance (850 Fg^{-1}) compared to anhydrous RuO_2 (125 Fg^{-1}). The surface area of the material plays an important role to determine the efficiency of material. The increased charge storage of such materials is due to more active sites available for redox reaction. The RuO_2 with mesoporous structure have shown enhanced capacitance value of 202 Fg^{-1} compared to RuO_2 with low surface area (146 Fg^{-1}) (Kuratani, Kiyobayashi, and Kuriyama 2009). Likewise, hierarchical porous RuO_2 materials allow more electrolyte to the active area of electrode, thus, rate capability of only 8.6% decrease was observed with an increase in current density ranging from 0.5 to 50 Ag^{-1} (Kuratani, Kiyobayashi, and Kuriyama 2009, Hu et al. 2006). The porosity of materials allow more penetration to the electrolyte and thus reduces proton diffusion length that enhanced the material's power density. Although, ruthenium oxides have provided a promising supercapacitance, cost and low abundance impede its commercial application.

In comparison to RuO_2 transition metal oxides are earth abundant, cheap and less toxic materials. Among transition metal oxides, manganese oxide (MnO_2) based electrodes are inexpensive and environmentally safe. MnO_2 also exhibit high theoretical capacitance of 1370 Fg^{-1} (Shi et al. 2014). MnO_2 have tunnel structured crystal lattice, thus provided a better cation intercalation, that ultimately could help to produces high capacitance (Brousse et al. 2006, Xia et al. 2014). MnO_2 nanotubes have shown highest specific capacitance around 461 Fg^{-1} at 5 mV s^{-1} (Zhu et al. 2012). Even after wide exploration, MnO_2 materials have shown low capacitance performance compared to RuO_2 material due to poor conductivity (Yan et al. 2014).

NiO and $Ni(OH)_2$ are another type of cheap and thermo-chemically stable electrode materials that have shown high theoretical capacitance of 2082–2584 Fg^{-1} (Kim, Thiyagarajan, and Jang 2014, Wang et al. 2014, Kate, Khalate, and Deokate 2018). NiO nanosheets have shown specific capacitance around 585 Fg^{-1} at 5 Ag^{-1} due to high surface area (Cao et al. 2011). However, NiO nanobelts synthesized by hydrothermal method have shown specific capacitance around 1126 Fg^{-1} at 2 Ag^{-1} and also show capacitance retention (95%) up to 2000 cycles (Wang et al. 2012). Additionally, micro-sphered α-$Ni(OH)_2$ have shown specific capacitance around 1789 Fg^{-1} at the 0.5 Ag^{-1} (Du et al. 2013). Cobalt oxide/hydroxide were also been tested for capacitance applications due to their high theoretical capacitance (3560–3700 Fg^{-1}), cyclic stability and corrosion resistance (Shi et al. 2014, Xia et al. 2014). The cobalt oxide-based hierarchical porous channels have large surface area, therefore, exhibit higher charge storage performance with high cyclic stability. Moreover, the nanoparticles of cobalt oxide have shown higher specific capacitance of around 2735 Fg^{-1} at 2 Ag^{-1}, highest capacitance performance reported for Co_3O_4 materials (Jagadale et al. 2012, Jiang et al. 2010), whereas, the $Co(OH)_2$ electrodeposited on Ni form has shown specific capacitance of 2018 Fg^{-1} at 2 Ag^{-1} (Xia et al. 2011). The specific capacitance was further enhanced to 2646 Fg^{-1} at 8 Ag^{-1} for the electrodeposition of $Co(OH)_2$ on the mesoporous substrates. At current density of 48 Ag^{-1}, this materials showed the specific capacitance of 2274 Fg^{-1} (Zhou et al. 2009). Despite high supercapacitance performance, the working potential for $Co(OH)_2$ was 0.5–0.6 V. The high working potential limits the practical application of cobalt oxide/hydroxide materials.

In addition to aforementioned inorganic materials for supercapacitance, Fe_2O_3, F_3O_4, V_2O_5, Bi_2O_3, SnO_2, and TiO_2 based electrode materials has also been studied for capacitance applications. However, metal oxide exhibits various individual limitations and lower capacitance performance (Ciszewski et al. 2015, Lim, Huang, and Lim 2013, Mitchell et al. 2017, Wang et al. 2006, Xie et al. 2011, Xiong et al. 2012). To overcome these issues, spinel metal oxides were explored for capacitance applications. These spinel-based metal oxides have two or more metal ions that can participate in redox reactions involved in capacitance mechanism. In this contest, various spinel metal oxides including $MnCo_2O_4$, $CoMn_2O_4$, $MnNi_2O_4$, $ZnCo_2O_4$, $NiCo_2O_4$, and $NiMn_2O_4$ have been explored for supercapacitance applications (An et al. 2014, Bhoyate Kahol, and Gupta 2019, Chen et al. 2014, Huang et al. 2013, Hui et al. 2016, Kuang et al. 2017, Liu et al. 2013, Mitchell et al. 2015, Wang et al. 2018a).

The Ni or Co-based spinel composites of manganese oxide have shown improved performance compared to individual manganese oxide. For example, $MnCo_2O_4$ nanowires have shown specific capacitance of around 2108 Fg^{-1}, whereas for $CoMn_2O_4$ nanowires, the capacitance of 1342 Fg^{-1} at 1 Ag^{-1} (An et al. 2014, Mitchell et al. 2015). On the other hand, $NiCo_2O_4$ based flower-like nanostructures have shown specific capacitance of 1191 Fg^{-1} at 1 Ag^{-1} (Mitchell et al. 2015).

9.9 EMERGING NEXT-GENERATION INORGANIC MATERIALS FOR SUPERCAPACITANCE

9.9.1 Metal-Organic Frameworks (MOFs)

Apart from the material intrinsic properties, the supercapacitance depends upon the porosity and surface area of the material. Among new inorganic materials for supercapacitance applications, MOFs have gained considerable interest due to structural diversity, large specific surface area, tunable porosity and wide range of active sites. MOFs consists of polyvalent organic carboxylates and metal units, which combined to form 3D structures having well-defined pore size distributions (Salunkhe, Kaneti, and Yamauchi 2017, Zhao, Li et al. 2015b, 2016). MOFs have a large surface area and high porosity that increases the contact area between electrode and electrolyte and fast transportation of electrons and electrolyte ions. These properties ultimately enhances the specific capacitance of the MOFs, however, various problems are associated with this materials; for example, at high charge-discharge rates, low electrical conductivity was observed, low cyclic stability, and diffusion distance for electrolytes in the materials (Salunkhe, Kaneti, and Yamauchi 2017).

The Co-based MOF films of thickness of around 5 μm have shown pseudocapacitance of around 206.76 Fg^{-1} at 0.6 Ag^{-1} in LiOH electrolytes. These Co-MOF exhibited an excellent cyclic stability by retaining 98.5% even after 1000 cycles. This high cyclic stability was due to good crystalline properties and their stability in electrolytes (Lee et al. 2012). The length of linker in MOF plays an important role to tune the specific surface area and pore size. The Co-MOF with linkers of variable length have been synthesized and explored for capacitance applications. The Co-MOF with longest linker showed an improvement in the specific surface area and enhancement in pore size and maximum capacitance of 179.2 Fg^{-1} at 10 mV s^{-1} scan rate. The retained its capacitance performance around 94% after 1000 cycle (Lee et al. 2013).

The Ni-based MOFs have also gained an interest due to their layered structure results in the excellent capacitance performance of these materials. Ni-MOF-24 has shown capacitance 1127 Fg^{-1} in KOH electrolyte at 0.5 Ag^{-1} current density. The cyclic stability shown by this material was found to be 90% after 3000 cycles.

In addition, the solid state asymmetric supercapacitor was also fabricate by using Ni-MOF as a positive electrode, which has shown capacitance of 230 mF cm^{-2} at 1 mA cm^{-2} current density and the 92.8% capacitance retention after 5000 cycles (Yan et al. 2016). The morphology of the MOF has also played an important role to improve the capacitance performance of these materials by enhancing their specific surface area (Zhao et al. 2018). The rod-like Ni-MOF has shown capacitance of about 1698 Fg^{-1} at 1.0 A^{-1} due to high surface area, good pore size distribution and efficient rapid charge transportation (Xu et al. 2016).

9.10 MXenes

MXenes are the layered transition metal nitrides, carbides or carbonitride, which provide conductivity and hydrophilicity with excellent mechanical properties (Chaudhari et al. 2017, Naguib et al. 2011). Various polar organic and metal ions can be easily intercalated the MXenes and play an important role in the energy storage of material. Among various MXenes, the 2D layered $Ti_3C_2T_x$ was extensively explored for supercapacitance application. The $Ti_3C_2T_x$ based flexible paper electrodes in acidic electrolytes have shown capacitance of 892 F cm^{-3} with enhanced cyclic stability (Fu et al. 2017). The capacitance results of $Ti_3C_2T_x$ was found superior than carbon-based double-layer capacitors and also found better than recently reported activated graphene-based electrodes (350 F cm^{-3}) (Ghidiu et al. 2014, Lukatskaya et al. 2013, Tao et al. 2013). $Ti_3C_2T_x$ has shown high charge and discharge rate. These MXenes have shown remarkable cyclability for 10,000 cycles (Ghidiu et al. 2014).

The cyclic voltammetry of MXenes does not produce redox peaks, and it looks like carbon-based double-layer capacitors. The surface area is not sufficient to explain the performance of the capacitor. There are various factors that control the volumetric capacitance of MXenes.

One factor is the density of the electrode, which is the conversion factor for gravimetric to volumetric performance. The MXene electrode densities is in the range of 3–4 g cm^{-3}, however, the MXene composites with polymer binders have lower densities (1–2.5 g cm^{-1}). Second factor is the surface chemistry of MXene that can affect the volumetric capacitance of this material (Anasori, Lukatskaya, and Gogotsi 2017). The fluorine and oxygen containing functional groups can increase the capacitance performance of the MXenes. The HF treated $Ti_3C_2T_x$ modified with N2H4, (Mashtalir et al. 2016) KOH, (Dall'Agnese et al. 2014) and DMSO (Lukatskaya et al. 2013, Xie et al. 2014) solutions have shown improved capacitance depending on the electrolytes. The $Ti_3C_2T_x$ synthesized by using HCl-LiF mixture have more oxygen functional groups, which results in 900 F cm^{-3} capacitance (Hope et al. 2016). Later, $Ti_3C_2T_x$ @ MnO_2 nanocomposites have posed a good supercapacitance were also fabricated to possess good cyclic stability (Rakhi et al. 2016).

On the other hand, the carbon nanotubes, or graphene, were integrated into the MXenes to synthesize the MXene-hybrid materials. These materials have shown improved performance due to the better ion movement in aqueous and organic electrolytes (Mashtalir et al. 2015, Zhao et al. 2015a). Likewise, the intercalation of MXene with polymers having polar functional groups has an advantage that such polymer prevent MXene from restacking and thus improve their mechanical properties without effecting the electrochemical performance of the material.

9.11 METAL NITRIDE (MNs)

Metal nitrides have shown good electrochemical properties, and high thermal stability. In this context, vanadium nitride has been the most extensively explored material for supercapacitance application. The high specific supercapacitance of about 1340 Fg^{-1} with 1.6×10^6 Sm^{-1} electrical conductance was reported for this material (Choi, Blomgren, and Kumta 2006). Electrochemical performance of the vanadium nitride depends upon the heat treatment of material during synthesis. For example, at elevated temperature of about 1000°C, the crystallite size of the material was increased, which results in decease of specific supercapacitance. The origin of supercapacitance performance of the metal nitride materials is based on the surface oxide or oxynitride groups and double-layer charging. VN nanoparticles undergo aggregation that results in poor electrochemical performance and ineffective contact; to overcome this issue, the 1D VN nanofibers were fabricated by an electrospinning method (Xu, Wang et al. 2015b). These nanofibers have shown high specific capacitance of about 291.5 Fg^{-1} at 0.5 Ag^{-1}. Thermally stable layered vanadium bronze was synthesized and has shown specific capacitance of about 1937 mF cm^{-3} (Bi et al. 2015). However, limited rate capability is associated with these metal nitrides. The rate capability can be increased by fabricating the VN/CNT composite, which has more ability to retain the specific capacitance compared to pure VN materials (Ghimbeu et al. 2011).

Although TiN are the porous layered materials and has high electrical conductance compared to VN, they show relatively low capacity. The TiN has pore sized varied in the range of 3–10 nm. Likewise, Fe_2N nanoparticles were also deposited on the graphene sheets and used for supercapacitance application. The Fe_2N/graphene composite has shown good flexibility, high rate capability and high capacitance retention over a 20,000 cycles (Zhou et al. 2011). On the other hand, TiN and VN composite fibers have been synthesized and have shown high specific capacitance of around 247.5 Fg^{-1} at 2 m Vs^{-1} and improved cyclic stability (Zhou et al. 2011). The key challenge associated with the metal nitride supercapacitors is their cyclic life. In this context, carbon coating is helpful o prolog their cyclic stability. For example, TiN@C have exhibited the supercapacitance of around 124.5 Fg^{-1} at 5 Ag^{-1} compared to pure TiN. The carbon coating has made this material retain its capacitance for about 15 000 cycles (Lu et al. 2012). The MNs materials have shown supercapacitance of 37 and 75 Fg^{-1} at 30 mAg^{-1} (Balogun et al. 2015, Das et al. 2015, Zhu et al. 2015).

9.12 BLACK PHOSPHOROUS (BP)

Recently, black phosphorous emerged as the potential material for energy storage applications. The black phosphorous in bulk state has week van der Waals interactions and is converted into thin sheets quite easily (Lin et al. 2017). The phosphorene has shown direct band gap of around 0.3–2.2 eV on the basis of number of layers. Furthermore, interlayer spacing in black phosphorous is found to be 5.3°A, which is more than graphite and comparable with MoS_2 phase (Ren et al. 2017). Owing of these properties, the phosphorene obtained by liquid exfoliation were tested a capacitor electrode materials, which have shown capacitance value of 13.75 F cm^{-3} with excellent cyclic stability of 30,000 cycles (Hao et al. 2016). Later, the black phosphorous/carbon nanotube composite electrodes were also fabricated. The carbon nanotubes have increased the electrolytic shuttling and limit the possibility of restacking of phosphorene sheets. These electrodes have shown enhanced capacitance value of 41.1 F cm^{-3} at 0.005 Vs^{-1} and remarkable cyclic stability of 91.5% (Yang et al. 2017). Likewise, BP@red phosphorous were also used for capacitance values and have shown supercapacitance of about 60.1 Fg^{-1} and retained high cyclic stability even after 2000 cycles (Chen et al. 2017). The improvement in the fabrication protocols for the black phosphorous/phosphorene electrode could make this material a hot choice for the wearable devices.

9.13 PEROVSKITES

Perovskite term is used for those materials, which represents the structure of prototype $CaTiO_3$ mineral and a general formula of ABX_3, where A is monovalent cation, B is divalent cation, and X is anion. In perovskite structures, the BX_3^- units form the 3D framework, and A cation occupies the interstices present in the octahedron (Akhtar, Aamir, and Sher 2018). Perovskite materials are classified into two major classes—one is oxides and the other is halides perovskites. Metal oxide perovskite have been used as ferroelectric, dielectric, piezoelectric, electrostrictive, magnetoresistant and multiferroic materials.

In metal halide perovskite, the B sites are divalent metal cations, A are monovalent metal/organic cations and X are halide ions. Metal halide perovskites are further divided into all inorganic metal halide perovskites, where A is Cs^+/Rb^+ cations, and hybrid perovskites, where A is small $CH_3NH_3^+$ cations (Akhtar, Aamir, and Sher 2018).

Recently, metal halide perovskite materials have gained enormous attention as an efficient light harvesting materials for photovoltaic applications due to tunable direct band gap, superior charge carrier mobility, elongated diffusion length and easy processability (Aamir, Adhikari et al. 2018, Aamir, Khan, Sher, Malik et al. 2017, Aamir, Khan et al. 2018, Aamir, Sher et al. 2017). Besides photovoltaic applications, these materials have shown applications in light emitting diodes, (Zhang et al. 2015) sensing, (Aamir, Sher, Malik, Akhtar et al. 2016, Aamir, Sher, Malik, Revaprasadu et al. 2016, Aamir, Khan, Sher, Bhosale et al. 2017) and photocatalysis (Aamir, Shah et al. 2017) etc. Recently, perovskite materials have gained attention as new electrode materials for energy storage systems (supercapacitors). Perovskite oxide has a wide range of properties such as tunable compositions, shapes and functionalities, which make them suitable for supercapacitor electrode. The metal halide perovskites have similar properties with high ionic conductivity; therefore, perovskite based electrode material for supercapacitor have been emerged as next-generation inorganic materials for energy storage applications.

9.14 FABRICATION OF PEROVSKITE SUPERCAPACITORS

The pioneered work in the utilization of halide perovskite for supercapacitance was based on the designing of solar cell integrated supercapacitance application. In this contest, the $CH_3NH_3PbI_{3-x}Cl_x$ perovskite was to fabricate mesoporous solar cell. Onto the surface of solar cell, $MoO_3/Au/MoO_3$ electrodes were evaporated with active area of 0.06 cm^2 (Zhou, Ren et al. 2016). Another device was fabricated by using $CH_3NH_3PbI_3$ material. The mesoporous heterojunction solar cell with hole-transport carbon layer and also as counter electrode was fabricated. The carbon layer was coated by doctoral blading method. The gel electrolyte layer was deposited onto the solar cell using 3.0 g conc. H_3PO_4 and 3.0 g PVA in deionized water. Finally, another carbon layer was also deposited using doctoral blading method (Liang et al. 2018).

Later, a supercapacitor was built using lead-free $(CH_3NH_3)_3Bi_2I_9$ material as electrode. The $(CH_3NH_3)_3Bi_2I_9$ perovskite, activated charcoal and PTFE binder were dispersed ultrasonicated in ethanol with weighted ratio of 85:10:5. The obtained dispersion was drop casted onto conductive carbon cloth and annealed at 60°C for 12 h under vacuum. The electrolyte solution consists of 30 mg/mL of methylammonium iodide in butanol was sandwiched to make complete device (Pious et al. 2017).

$CsPbI_3$ was also used as electrode material for supercapacitor. The device was fabricated by dispersing $CsPbI_3$, activated charcoal and PVDF (85:10:5 corresponding weight ratios) in ethanol. The electrode was deposited on two steel plates by spin coating and dried at 60°C for 12 hours. The filter paper was used as separator having 20 mg/mL cesium iodide in butanol electrolyte (Maji et al. 2018).

9.15 PEROVSKITE OXIDE-SUPERCAPACITANCE PERFORMANCE

Among various inorganic materials for supercapacitor application, perovskite oxides have gained enormous attention owing to high thermal stability, low cost, reversible redox reactions and high intrinsic capacity. The pseudocapacitance response of the inorganic materials could be further enhanced by reducing the crystalline dimensions to nanoscale. In this contest, specific capacitance, cyclic stability and rate capability are the major properties that are measured to estimate the pseudocapacitance response of the inorganic materials.

The first $LaMnO_{3\pm x}$ ($x = 0.09 \pm 0.02$) nanostructures-based pseudocapacitor was decorated by depositing thin film of $LaMnO_{3\pm x}$ with a loading of 30 wt% on the mesoporous carbon (Mefford et al. 2014). The resultant pseudocapacitor thin films of oxygen rich $LaMnO_{3.09}$ material or reduced oxygen $LaMnO_{2.91}$ material have showed near to ideal pseudocapacitive response appeared at $E_{1/2} = 0.3$ V. The oxygen rich $LaMnO_{3.09}$ material has shown increased specific capacitance compared to the material with reduced oxygen content, whereas, at significantly slow scan rate, the specific capacitance of $LaMnO_{3.09}$ material was found around 486.7 Fg^{-1} and 609.8 Fg^{-1} for $LaMnO_{2.91}$ material. The increased capacitance in these materials was due to increased oxygen vacancies as charge storage site (Mefford et al. 2014). $(La_{0.75}Sr_{0.25})_{0.95}MnO_{3-x}$ perovskite have shown high reduction catalytic activity, thermos-chemical stability with electrical conductance of about 44.9 Scm^{-1}. Therefore, the $(La_{0.75}Sr_{0.25})_{0.95}$ MnO_{3-x}–MnO_2 composite was synthesized by hydrothermal method (Lv et al. 2016). This material has shown capacitance of 108 Fg^{-1} and net capacitance of around 56 Fg^{-1} at 2 mV s^{-1}. To determine the conductance, the scan-rate plays a very important role. The current increases with increasing scan rate for this material, however, the specific capacitance was also decreases and CV curve devices from ideal rectangular shape. The low porosity of this perovskite oxide composite leads to the reduction of effective contact area with an electrolyte (Lv et al. 2016). The improvement in supercapacitance properties can be achieved by increasing surface area and improved charge transportation. In this contest, the high aspect ratio perovskite nanostructure was used. $La_xSr_{1-x}CoO_{3-\sigma}$ nanofibers have shown enhanced specific capacitance (Cao et al. 2015).

Particularly, this perovskite material with x = 0.7 have shown longer charging time and highest average capacitance. The specific capacitance shown by these nanofibers was found to be 747.75 Fg^{-1}. The Sr content in this perovskite oxide was found to have a role in the capacitance of this material. The increase in Sr content up to certain concentration increases the specific capacitance (Cao et al. 2015). Recently, the $LaNiO_3$ hollow spherical nanoparticles formed by solvothermal method were tested for capacitance and found that this materials has shown enhanced electrochemical performance of 422 Fg^{-1} at the current density of 1.0 Ag^{-1}. The electrochemical capacitance performance of this material was further increased to 501.9 Fg^{-1}. The $LaNiO_3$ nanomaterials were annealed for different times and they tested for electrochemical supercapacitance. The charging and discharging time was almost the same for the samples, which indicates the reversibility of redox reaction. Furthermore, the hollow structures have shown lowest internal resistance and charge transfer resistance (Li et al. 2017).

9.16 HALIDE PEROVSKITE-SUPERCAPACITANCE PERFORMANCE

The metal halide perovskite materials are thermally and chemically instable, which limit their application for supercapacitance. However, the scientist have explored their application in solar power conversion and storage devices. In this contest, the first parallel combined device based on $CH_3NH_3PbI_3$ perovskite photovoltaics to produce electric power generation and supercapacitor-a-2032-type cell (Xu, Li et al. 2015). Under irradiation, the capacitor voltage was increased to 0.3 V; however, the continuous irradiation leads to the slow increase in capacitor voltage to 0.710 V. The storage time of this combine system depends upon the active surface of perovskite solar cell (Xu, Li et al. 2015). Later, co-anode and co-cathode photochromic supercapacitors were reported (Zhou et al. 2014). The perovskite solar cell was vertically integrated with transparent electrode such as MoO_3/Au/MoO_3 and electrochromic supercapacitor. The co-anode photochromic supercapacitors was charged within 60 s and shown charge up to 0.61 V, whereas, co-cathode photochromic supercapacitors were charged up to 0.68 V in 85 s. Furthermore, energy density for co-cathode photochromic supercapacitors was found to be 24.5 mWh/m^2, power density was 377.0 mW/m^2 and specific areal capacitance was found to be 430.7 F/m^2. This application of metal halide perovskite-based photochromic supercapacitors have shown potential for multifunctional smart window applications (Zhou et al. 2014).

The metal halide perovskites such as $MAPbI_3$ have cornered shared inorganic network of PbI_6 octahedron, whereas the cuboctahedral interstices were occupied by organic cation in hybrid metal halide perovskites or inorganic monovalent cation in all inorganic metal halide perovskites. These inorganic networks can undergo deformation, results in structural transformations and phase transitions (Zhou et al. 2014). Apart from metal halide perovskite application in power generation-storage combine devices, the thin film electrochemical double-layer capacitor were also fabricated and test for $MAPbI_3$. The $MAPbI_3$ have sown high ionic conductivity compared to ionic conductivity, which make this materials suitable for capacitance applications (Yang et al. 2015). The electrochemical double-layer capacitor of

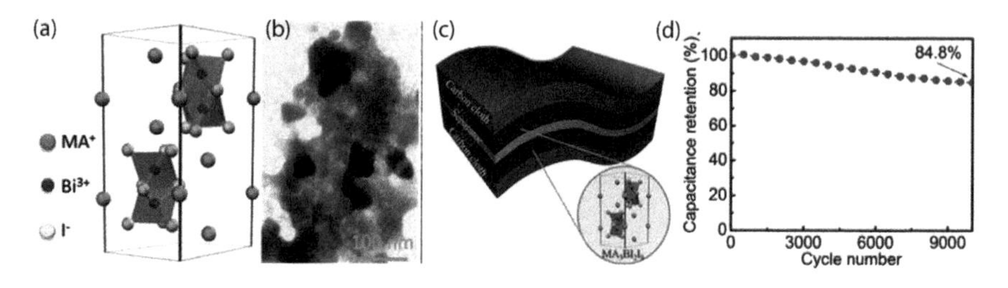

FIGURE 9.3 Represents the (a) $(CH_3NH_3)_3Bi_2I_9$ perovskite unit cell, (b) TEM image of the $(CH_3NH_3)_3Bi_2I_9$ film, (c) $(CH_3NH_3)_3Bi_2I_9$ based double-layer capacitor device architecture and (d) is the capacitance retention *vs* number of cycles. (Reprinted with permission from Pious, J.K. et al., *ACS Omega*, 2, 5798–5802, 2017. Copyright 2019 American Chemical Society.)

$MAPbI_3$ perovskite were fabricate, which shows capacitance of about 5.89 μF/cm^2 (Zhou, Li et al. 2016). The capacitance of these perovskite was related to the ion transportation in the perovskite layers. In this thin-film double-layer capacitors fabricated using $MAPbI_3$ perovskite, the thickness of film is very important, although thicker films produces more ions, however, the transportation of ions to reach the surface of perovskite layer is very limited (Yang et al. 2015). Lead-free zero-dimensional $(CH_3NH_3)_3Bi_2I_9$ has also shown energy storage potentials. The $(CH_3NH_3)_3Bi_2I_9$ perovskite is environmental friendly and thermally stable. It contains isolated $Bi_2I_9^{3-}$ units shaped by the face sharing of BiI_6^{3-} octahedrons, which are surrounded by methylammonium ions (Figure 9.3). The isolation of these $Bi_2I_9^{3-}$ units has high surface area and therefore, results in enhanced charge transportation and ultimately gave capacitance of around 5.5 mF/cm^2, 1000 times higher than the $MAPbI_3$ perovskite based capacitor of same architecture (Figure 9.3) (Pious et al. 2017). Recently, all-inorganic perovskite $CsPbI_3$ perovskite materials have been employed for electrochemical double-layer capacitance applications (Maji et al. 2018). This materials has shown maximum areal capacitance of 7.23 mF/cm^2 at 2 mV/s. The capacitance value of this material was decreased with increase in scan rate. The decrease in capacitance value is associated with ionic movement from electrolyte to electrode. At slow scan rate, the Cs^+ and I^- ions get enough time to get transported, therefore, maximum capacitance values were observed. This materials has shown 65.55% capacitance over 1000 charging and discharging cycles (Maji et al. 2018).

9.17 CONCLUSION

Supercapacitors have become the promising devices for the storing of electrical energy. In this contest, various types of materials have been employed ranging from organic to inorganic for electrodes preparation. Inorganic materials are very interesting due to their enhanced stability and high ionic conductivity. Among various inorganic materials, perovskites are the emerging next-generation inorganic materials for supercapacitors. These materials have high ionic conductivity. Metal halide perovskite are new candidate in this field. There are two type of metal halide perovskite based supercapacitors have been reported so far. One are the perovskite solar cell integrated supercapacitors, which are MoO_3 based or carbon based capacitor devices. Very recently, metal halide-based supercapacitors were also reported, which have shown promising capacitance performance. Therefore, it is expecting for metal halide perovskite as new inorganic electrode materials.

REFERENCES

Aamir, Muhammad, Malik Dilshad Khan, Muhammad Sher, Mohammad Azad Malik, Javeed Akhtar, and Neerish Revaprasadu. 2017. "Synthesis of hybrid to inorganic quasi 2D-layered perovskite nanoparticles." *Chemistry Select* 2 (20):5595–5599. doi:10.1002/slct.201700518.

Aamir, Muhammad, Malik Dilshad Khan, Muhammad Sher, Neerish Revaprasadu, Mohammad Azad Malik, and Javeed Akhtar. 2018. "Broadband emission in a new lead free all-inorganic 3D CsZnCl2I perovskite." *New Journal of Chemistry* 42 (21):17181–17184. doi:10.1039/c8nj04404j.

Aamir, Muhammad, Malik Dilshad Khan, Muhammad Sher, Sheshanath V Bhosale, Mohammad Azad Malik, Javeed Akhtar, and Neerish Revaprasadu. 2017. "A facile route to cesium lead bromoiodide perovskite microcrystals and their potential application as sensors for nitrophenol explosives." *European Journal of Inorganic Chemistry* 2017 (31):3755–3760.

Aamir, Muhammad, Muhammad Sher, Malik Dilshad Khan, Mohammad Azad Malik, Javeed Akhtar, and Neerish Revaprasadu. 2017. "Controlled synthesis of all inorganic CsPbBr2I perovskite by non-template and aerosol assisted chemical vapour deposition." *Materials Letters* 190:244–247.

Aamir, Muhammad, Muhammad Sher, Mohammad Azad Malik, Neerish Revaprasadu, and Javeed Akhtar. 2016. "A facile approach for selective and sensitive detection of aqueous contamination in DMF by using perovskite material." *Materials Letters* 183:135–138.

Aamir, Muhammad, Muhammad Sher, Muhammad Azad Malik, Javeed Akhtar, and Neerish Revaprasadu. 2016. "A chemodosimetric approach for the selective detection of $Pb2^+$ ions using a cesium based perovskite." *New Journal of Chemistry* 40 (11):9719–9724.

Aamir, Muhammad, Tham Adhikari, Muhammad Sher, Neerish Revaprasadu, Waqas Khalid, Javeed Akhtar, and Jean-Michel Nunzi. 2018. "Fabrication of planar heterojunction CsPbBr2I perovskite solar cells using ZnO as an electron transport layer and improved solar energy conversion efficiency." *New Journal of Chemistry* 42 (17):14104–14110.

Aamir, Muhammad, Zawar Hussain Shah, Muhammad Sher, Azhar Iqbal, Neerish Revaprasadu, Mohammad Azad Malik, and Javeed Akhtar. 2017. "Enhanced photocatalytic activity of water stable hydroxyl ammonium lead halide perovskites." *Materials Science in Semiconductor Processing* 63:6–11. doi:10.1016/j.mssp.2017.01.001.

Akhtar, Javeed, Muhammad Aamir, and Muhammad Sher. 2018. "Chapter 2—Organometal lead halide perovskite A2—Thomas, Sabu." In *Perovskite Photovoltaics*, edited by Aparna Thankappan, 25–42. Academic Press, Elsevier, UK.

An, Cuihua, Yijing Wang, Yanan Huang, Yanan Xu, Changchang Xu, Lifang Jiao, and Huatang Yuan. 2014. "Novel three-dimensional $NiCo_2O_4$ hierarchitectures: Solvothermal synthesis and electrochemical properties." *CrystEngComm* 16 (3):385–392.

Anasori, Babak, Maria R Lukatskaya, and Yury Gogotsi. 2017. "2D metal carbides and nitrides (MXenes) for energy storage." *Nature Reviews Materials* 2 (2):16098.

Augustyn, Veronica, Patrice Simon, and Bruce Dunn. 2014. "Pseudocapacitive oxide materials for high-rate electrochemical energy storage." *Energy & Environmental Science* 7 (5):1597–1614.

Balogun, Muhammad-Sadeeq, Weitao Qiu, Wang Wang, Pingping Fang, Xihong Lu, and Yexiang Tong. 2015. "Recent advances in metal nitrides as high-performance electrode materials for energy storage devices." *Journal of Materials Chemistry A* 3 (4):1364–1387. doi:10.1039/C4TA05565A.

Bhoyate, Sanket, Pawan K Kahol, and Ram K Gupta. 2019. "Nanostructured materials for supercapacitor applications." *Nanoscience* 5:1–29. The Royal Society of Chemistry.

Bi, Wentuan, Zhenpeng Hu, Xiaogang Li, Changzheng Wu, Junchi Wu, Yubin Wu, and Yi Xie. 2015. "Metallic mesocrystal nanosheets of vanadium nitride for high-performance all-solid-state pseudocapacitors." *Nano Research* 8 (1):193–200.

Brousse, Thierry, Mathieu Toupin, Romain Dugas, Laurence Athouël, Olivier Crosnier, and Daniel Bélanger. 2006. "Crystalline MnO_2 as possible alternatives to amorphous compounds in electrochemical supercapacitors." *Journal of The Electrochemical Society* 153 (12):A2171–A2180.

Cao, Chang-Yan, Wei Guo, Zhi-Min Cui, Wei-Guo Song, and Wei Cai. 2011. "Microwave-assisted gas/liquid interfacial synthesis of flowerlike NiO hollow nanosphere precursors and their application as supercapacitor electrodes." *Journal of Materials Chemistry* 21 (9):3204–3209.

Cao, Yi, Baoping Lin, Ying Sun, Hong Yang, and Xueqin Zhang. 2015. "Symmetric/asymmetric supercapacitor based on the perovskite-type lanthanum cobaltate nanofibers with Sr-substitution." *Electrochimica Acta* 178:398–406.

Chaudhari, Nitin K., Hanuel Jin, Byeongyoon Kim, Du San Baek, Sang Hoon Joo, and Kwangyeol Lee. 2017. "MXene: An emerging two-dimensional material for future energy conversion and storage applications." *Journal of Materials Chemistry A* 5 (47):24564–24579. doi:10.1039/C7TA09094C.

Chen, Hao, Linfeng Hu, Min Chen, Yan Yan, and Limin Wu. 2014. "Nickel–cobalt layered double hydroxide nanosheets for high-performance supercapacitor electrode materials." *Advanced Functional Materials* 24 (7):934–942.

Chen, Xinhang, Guanghua Xu, Xiaohui Ren, Zhongjun Li, Xiang Qi, Kai Huang, Han Zhang, Zongyu Huang, and Jianxin Zhong. 2017. "A black/red phosphorus hybrid as an electrode material for high-performance Li-ion batteries and supercapacitors." *Journal of Materials Chemistry A* 5 (14):6581–6588. doi:10.1039/C7TA00455A.

Choi, Daiwon, George E Blomgren, and Prashant N Kumta. 2006. "Fast and reversible surface redox reaction in nanocrystalline vanadium nitride supercapacitors." *Advanced Materials* 18 (9):1178–1182.

Ciszewski, Mateusz, Andrzej Mianowski, Piotr Szatkowski, Ginter Nawrat, and Jakub Adamek. 2015. "Reduced graphene oxide–bismuth oxide composite as electrode material for supercapacitors." *Ionics* 21 (2):557–563.

Conway, Brian E. 2013. *Electrochemical Supercapacitors: Scientific Fundamentals and Technological Applications*. Springer Science & Business Media, New York.

Dall'Agnese, Yohan, Maria R Lukatskaya, Kevin M Cook, Pierre-Louis Taberna, Yury Gogotsi, and Patrice Simon. 2014. "High capacitance of surface-modified 2D titanium carbide in acidic electrolyte." *Electrochemistry Communications* 48:118–122.

Das, Bijoy, M. Behm, G. Lindbergh, M. V. Reddy, and B. V. R. Chowdari. 2015. "High performance metal nitrides, MN (M=Cr, Co) nanoparticles for non-aqueous hybrid supercapacitors." *Advanced Powder Technology* 26 (3):783–788. doi:10.1016/j.apt.2015.02.001.

Dmowski, Wojtek, Takeshi Egami, Karen E Swider-Lyons, Corey T Love, and Debra R Rolison. 2002. "Local atomic structure and conduction mechanism of nanocrystalline hydrous RuO_2 from X-ray scattering." *The Journal of Physical Chemistry B* 106 (49):12677–12683.

Du, Hongmei, Lifang Jiao, Kangzhe Cao, Yijing Wang, and Huatang Yuan. 2013. "Polyol-mediated synthesis of mesoporous α-Ni (OH)2 with enhanced supercapacitance." *ACS Applied Materials & Interfaces* 5 (14):6643–6648.

Fu, Qishan, Jing Wen, Na Zhang, Lili Wu, Mingyi Zhang, Shuangyan Lin, Hong Gao, and Xitian Zhang. 2017. "Free-standing Ti_3C_2Tx electrode with ultrahigh volumetric capacitance." *RSC Advances* 7 (20):11998–12005. doi:10.1039/C7RA00126F.

Ghidiu, Michael, Maria R Lukatskaya, Meng-Qiang Zhao, Yury Gogotsi, and Michel W Barsoum. 2014. "Conductive two-dimensional titanium carbide 'clay' with high volumetric capacitance." *Nature* 516 (7529):78.

Ghimbeu, Camelia Matei, Encarnacion Raymundo-Piñero, Philippe Fioux, François Béguin, and Cathie Vix-Guterl. 2011. "Vanadium nitride/carbon nanotube nanocomposites as electrodes for supercapacitors." *Journal of Materials Chemistry* 21 (35):13268–13275.

Hao, Chunxue, Bingchao Yang, Fusheng Wen, Jianyong Xiang, Lei Li, Wenhong Wang, Zhongming Zeng, Bo Xu, Zhisheng Zhao, Zhongyuan Liu, and Yongjun Tian. 2016. "Flexible all-solid-state supercapacitors based on liquid-exfoliated black-phosphorus nanoflakes." *Advanced Materials* 28 (16):3194–3201. doi:10.1002/adma.201505730.

Hope, Michael A., Alexander C. Forse, Kent J. Griffith, Maria R Lukatskaya, Michael Ghidiu, Yury Gogotsi, and Clare P. Grey. 2016. "NMR reveals the surface functionalisation of Ti_3C_2 MXene." *Physical Chemistry Chemical Physics* 18 (7):5099–5102. doi: 10.1039/C6CP00330C.

Hu, Chi-Chang, Kuo-Hsin Chang, Ming-Champ Lin, and Yung-Tai Wu. 2006. "Design and tailoring of the nanotubular arrayed architecture of hydrous RuO_2 for next generation supercapacitors." *Nano Letters* 6 (12):2690–2695.

Huang, Liang, Dongchang Chen, Yong Ding, Shi Feng, Zhong Lin Wang, and Meilin Liu. 2013. "Nickel–cobalt hydroxide nanosheets coated on $NiCo_2O_4$ nanowires grown on carbon fiber paper for high-performance pseudocapacitors." *Nano Letters* 13 (7):3135–3139.

Hui, Kwun Nam, Kwan San Hui, Zikang Tang, VV Jadhav, and Qi Xun Xia. 2016. "Hierarchical chestnut-like $MnCo_2O_4$ nanoneedles grown on nickel foam as binder-free electrode for high energy density asymmetric supercapacitors." *Journal of Power Sources* 330:195–203.

Jagadale, Ajay D., VS. Jamadade, SN. Pusawale, and CD. Lokhande. 2012. "Effect of scan rate on the morphology of potentiodynamically deposited β-Co (OH)2 and corresponding supercapacitive performance." *Electrochimica Acta* 78:92–97.

Jiang, Jian, Jinping Liu, Ruimin Ding, Jianhui Zhu, Yuanyuan Li, Anzheng Hu, Xin Li, and Xintang Huang. 2010. "Large-scale uniform α-Co (OH)2 long nanowire arrays grown on graphite as pseudocapacitor electrodes." *ACS Applied Materials & Interfaces* 3 (1):99–103.

Kate, Ranjit S, Suraj A Khalate, and Ramesh J Deokate. 2018. "Overview of nanostructured metal oxides and pure nickel oxide (NiO) electrodes for supercapacitors: A review." *Journal of Alloys and Compounds* 734:89–111.

Kim, Sun-I, Pradheep Thiyagarajan, and Ji-Hyun Jang. 2014. "Great improvement in pseudocapacitor properties of nickel hydroxide via simple gold deposition." *Nanoscale* 6 (20):11646–11652.

Kirk, Donald W, and John W Graydon. 2013. "A low temperature inorganic molten salt supercapacitor." *ECS Transactions* 53 (31):27–33.

Kuang, Liping, Fengzhen Ji, Xuexue Pan, Dongliang Wang, Xinman Chen, Dan Jiang, Yong Zhang, and Baofu Ding. 2017. "Mesoporous $MnCo_2O_4$. 5 nanoneedle arrays electrode for high-performance asymmetric supercapacitor application." *Chemical Engineering Journal* 315:491–499.

Kuratani, Kentaro, Tetsu Kiyobayashi, and Nobuhiro Kuriyama. 2009. "Influence of the mesoporous structure on capacitance of the RuO_2 electrode." *Journal of Power Sources* 189 (2):1284–1291.

Lee, Deok Yeon, Seog Joon Yoon, Nabeen K. Shrestha, Soo-Hyoung Lee, Heejoon Ahn, and Sung-Hwan Han. 2012. "Unusual energy storage and charge retention in Co-based metal–organic-frameworks." *Microporous and Mesoporous Materials* 153:163–165. doi:10.1016/j.micromeso.2011.12.040.

Lee, Deok Yoon, Dipak V. Shinde, Eun-Kyung Kim, Wonjoo Lee, In-Whan Oh, Nabeen K. Shrestha, Joong Kee Lee, and Sung-Hwan Han. 2013. "Supercapacitive property of metal–organic-frameworks with different pore dimensions and morphology." *Microporous and Mesoporous Materials* 171:53–57. doi:10.1016/j.micromeso.2012.12.039.

Li, Zijiong, Weiyang Zhang, Haiyan Wang, and Baocheng Yang. 2017. "Two-dimensional perovskite $LaNiO_3$ nanosheets with hierarchical porous structure for high-rate capacitive energy storage." *Electrochimica Acta* 258:561–570.

Liang, Jia, Guoyin Zhu, Zhipeng Lu, Peiyang Zhao, Caixing Wang, Yue Ma, Zhaoran Xu, Yanrong Wang, Yi Hu, and Lianbo Ma. 2018. "Integrated perovskite solar capacitors with high energy conversion efficiency and fast photo-charging rate." *Journal of Materials Chemistry A* 6 (5):2047–2052.

Lim, S.P., N.M. Huang, and H.N. Lim. 2013. "Solvothermal synthesis of SnO_2/graphene nanocomposites for supercapacitor application." *Ceramics International* 39 (6):6647–6655.

Lin, Shenghuang, Yingsan Chui, Yanyong Li, and Shu Ping Lau. 2017. "Liquid-phase exfoliation of black phosphorus and its applications." *FlatChem* 2:15–37. doi:10.1016/j.flatc.2017.03.001.

Ling, Zheng, Chang E Ren, Meng-Qiang Zhao, Jian Yang, James M Giammarco, Jieshan Qiu, Michel W Barsoum, and Yury Gogotsi. 2014. "Flexible and conductive MXene films and nanocomposites with high capacitance." *Proceedings of the National Academy of Sciences* 111 (47):16676–16681.

Liu, X.Y., Y.Q. Zhang, X.H. Xia, S.J. Shi, Y. Lu, X. Li Wang, C.D. Gu, and J.P. Tu. 2013. "Self-assembled porous $NiCo_2O_4$ hetero-structure array for electrochemical capacitor." *Journal of Power Sources* 239:157–163.

Lu, Xihong, Gongming Wang, Teng Zhai, Minghao Yu, Shilei Xie, Yichuan Ling, Chaolun Liang, Yexiang Tong, and Yat Li. 2012. "Stabilized TiN nanowire arrays for high-performance and flexible supercapacitors." *Nano Letters* 12 (10):5376–5381.

Lukatskaya, Maria R, Olha Mashtalir, Chang E Ren, Yohan Dall'Agnese, Patrick Rozier, Pierre Louis Taberna, Michael Naguib, Patrice Simon, Michel W Barsoum, and Yury Gogotsi. 2013. "Cation intercalation and high volumetric capacitance of two-dimensional titanium carbide." *Science* 341 (6153):1502–1505.

Lv, Jingbo, Yaohui Zhang, Zhe Lv, Xiqiang Huang, Zhihong Wang, Xingbao Zhu, and Bo Wei. 2016. "Strontium doped lanthanum manganite/manganese dioxide composite electrode for supercapacitor with enhanced rate capability." *Electrochimica Acta* 222:1585–1591. doi:10.1016/j.electacta.2016.11.144.

Maji, Prasenjit, Apurba Ray, Priyabrata Sadhukhan, Atanu Roy, and Sachindranath Das. 2018. "Fabrication of symmetric supercapacitor using cesium lead iodide ($CsPbI_3$) microwire." *Materials Letters* 227:268–271.

Mashtalir, O, Maria R Lukatskaya, Alexander I Kolesnikov, E Raymundo-Pinero, Michael Naguib, MW Barsoum, and Y Gogotsi. 2016. "The effect of hydrazine intercalation on the structure and capacitance of 2D titanium carbide (MXene)." *Nanoscale* 8 (17):9128–9133.

Mashtalir, Olha, Maria R Lukatskaya, Meng-Qiang Zhao, Michel W Barsoum, and Yury Gogotsi. 2015. "Amine-assisted delamination of Nb_2C MXene for Li-ion energy storage devices." *Advanced Materials* 27 (23):3501–3506.

Mefford, J Tyler, William G Hardin, Sheng Dai, Keith P Johnston, and Keith J Stevenson. 2014. "Anion charge storage through oxygen intercalation in $LaMnO_3$ perovskite pseudocapacitor electrodes." *Nature Materials* 13 (7):726.

Mitchell, Elias, Ashley Jimenez, Ram K Gupta, Bipin Kumar Gupta, Karthik Ramasamy, Mohammad Shahabuddin, and Sanjay R Mishra. 2015. "Ultrathin porous hierarchically textured $NiCo_2O_4$–graphene oxide flexible nanosheets for high-performance supercapacitors." *New Journal of Chemistry* 39 (3):2181–2187.

Mitchell, James B, William C Lo, Arda Genc, James LeBeau, and Veronica Augustyn. 2017. "Transition from battery to pseudocapacitor behavior via structural water in tungsten oxide." *Chemistry of Materials* 29 (9):3928–3937.

Naguib, Michael, Murat Kurtoglu, Volker Presser, Jun Lu, Junjie Niu, Min Heon, Lars Hultman, Yury Gogotsi, and Michel W. Barsoum. 2011. "Two-dimensional nanocrystals produced by exfoliation of Ti_3AlC_2." *Advanced Materials* 23 (37):4248–4253. doi:10.1002/adma.201102306.

Obreja, Vasile VN. 2008. "On the performance of supercapacitors with electrodes based on carbon nanotubes and carbon activated material—A review." *Physica E: Low-dimensional Systems and Nanostructures* 40 (7):2596–2605.

Pious, Johnpaul K, ML Lekshmi, Chinnadurai Muthu, RB Rakhi, and Vijayakumar C Nair. 2017. "Zero-dimensional methylammonium bismuth iodide-based lead-free perovskite capacitor." *ACS Omega* 2 (9):5798–5802.

Qu, Gan, Shuangfeng Jia, Hai Wang, Fan Cao, Lei Li, Chen Qing, Daming Sun, Bixiao Wang, Yiwen Tang, and Jianbo Wang. 2016. "Asymmetric supercapacitor based on porous N-doped carbon derived from pomelo peel and NiO arrays." *ACS Applied Materials & Interfaces* 8 (32):20822–20830.

Rakhi, Raghavan Baby, Bilal Ahmed, Dalaver Anjum, and Husam N. Alshareef. 2016. "Direct chemical synthesis of MnO_2 nanowhiskers on transition-metal carbide surfaces for supercapacitor applications." *ACS Applied Materials & Interfaces* 8 (29):18806–18814. doi:10.1021/acsami.6b04481.

Raza, Waseem, Faizan Ali, Nadeem Raza, Yiwei Luo, Eilhann E Kwon, Jianhua Yang, Sandeep Kumar, Andleeb Mehmood, and Ki-Hyun Kim. 2018. "Recent advancements in supercapacitor technology." *Nano Energy* 52:441–473.

Ren, Xinlin, Peichao Lian, Delong Xie, Ying Yang, Yi Mei, Xiangrun Huang, Zirui Wang, and Xiting Yin. 2017. "Properties, preparation and application of black phosphorus/phosphorene for energy storage: A review." *Journal of Materials Science* 52 (17):10364–10386. doi:10.1007/s10853-017-1194-3.

Salunkhe, Rahul R., Yusuf V. Kaneti, and Yusuke Yamauchi. 2017. "Metal–organic framework-derived nanoporous metal oxides toward supercapacitor applications: Progress and prospects." *ACS Nano* 11 (6):5293–5308. doi:10.1021/acsnano.7b02796.

Shi, Fan, Lu Li, Xiu-li Wang, Chang-dong Gu, and Jiang-ping Tu. 2014. "Metal oxide/hydroxide-based materials for supercapacitors." *Rsc Advances* 4 (79):41910–41921.

Tao, Ying, Xiaoying Xie, Wei Lv, Dai-Ming Tang, Debin Kong, Zhenghong Huang, Hirotomo Nishihara, Takafumi Ishii, Baohua Li, and Dmitri Golberg. 2013. "Towards ultrahigh volumetric capacitance: Graphene derived highly dense but porous carbons for supercapacitors." *Scientific Reports* 3:2975.

Wang, Bao, Jun Song Chen, Zhiyu Wang, Srinivasan Madhavi, and Xiong Wen Lou. 2012. "Green synthesis of NiO nanobelts with exceptional pseudo-capacitive properties." *Advanced Energy Materials* 2 (10):1188–1192.

Wang, Chundong, Junling Xu, Muk-Fung Yuen, Jie Zhang, Yangyang Li, Xianfeng Chen, and Wenjun Zhang. 2014. "Hierarchical composite electrodes of nickel oxide nanoflake 3D graphene for high-performance pseudocapacitors." *Advanced Functional Materials* 24 (40):6372–6380.

Wang, Liang, Xinyan Jiao, Peng Liu, Yu Ouyang, Xifeng Xia, Wu Lei, and Qingli Hao. 2018. "Self-template synthesis of yolk-shelled $NiCo_2O_4$ spheres for enhanced hybrid supercapacitors." *Applied Surface Science* 427:174–181.

Wang, Shih-Yu, Kuo-Chuan Ho, Shin-Liang Kuo, and Nae-Lih Wu. 2006. "Investigation on capacitance mechanisms of Fe_3O_4 electrochemical capacitors." *Journal of The Electrochemical Society* 153 (1):A75–A80.

Wang, Yanfang, Chunyang Li, Weibin Zhou, Yusong Zhu, Lijun Fu, Yuping Wu, and Teunis van Ree. 2018. "Metal oxide-based superconductors in AC power transportation and transformation." In *Metal Oxides in Energy Technologies*, edited by Yuping Wu, 361–390. Elsevier, UK.

Winter, Martin, and Ralph J Brodd. 2004. What are batteries, fuel cells, and supercapacitors?: ACS Publications.

Xia, XH, JP Tu, YQ Zhang, YJ Mai, XL Wang, CD Gu, and XB Zhao. 2011. "Three-dimensional porous nano-Ni/Co $(OH)_2$ nanoflake composite film: A pseudocapacitive material with superior performance." *The Journal of Physical Chemistry C* 115 (45):22662–22668.

Xia, Xinhui, Yongqi Zhang, Dongliang Chao, Cao Guan, Yijun Zhang, Lu Li, Xiang Ge, Ignacio Mínguez Bacho, Jiangping Tu, and Hong Jin Fan. 2014. "Solution synthesis of metal oxides for electrochemical energy storage applications." *Nanoscale* 6 (10):5008–5048.

Xie, Keyu, Jie Li, Yanqing Lai, Wei Lu, Zhi'an Zhang, Yexiang Liu, Limin Zhou, and Haitao Huang. 2011. "Highly ordered iron oxide nanotube arrays as electrodes for electrochemical energy storage." *Electrochemistry Communications* 13 (6):657–660.

Xie, Yu, Yohan Dall'Agnese, Michael Naguib, Yury Gogotsi, Michel W Barsoum, Houlong L Zhuang, and Paul RC Kent. 2014. "Prediction and characterization of MXene nanosheet anodes for non-lithium-ion batteries." *ACS Nano* 8 (9):9606–9615.

Xiong, Qin-qin, Jiang-ping Tu, Yi Lu, Jiao Chen, Ying-xia Yu, Xiu-li Wang, and Chang-dong Gu. 2012. "Three-dimensional porous nano-Ni/Fe_3O_4 composite film: Enhanced electrochemical performance for lithium-ion batteries." *Journal of Materials Chemistry* 22 (35):18639–18645.

Xu, Juan, Chao Yang, Yufei Xue, Cheng Wang, Jianyu Cao, and Zhidong Chen. 2016. "Facile synthesis of novel metal-organic nickel hydroxide nanorods for high performance supercapacitor." *Electrochimica Acta* 211:595–602. doi:10.1016/j.electacta.2016.06.090.

Xu, Xiaobao, Shaohui Li, Hua Zhang, Yan Shen, Shaik M Zakeeruddin, Michael Graetzel, Yi-Bing Cheng, and Mingkui Wang. 2015. "A power pack based on organometallic perovskite solar cell and supercapacitor." *ACS Nano* 9 (2):1782–1787.

Xu, Yunling, Jie Wang, Laifa Shen, Hui Dou, and Xiaogang Zhang. 2015. "One-dimensional vanadium nitride nanofibers fabricated by electrospinning for supercapacitors." *Electrochimica Acta* 173:680–686.

Yan, Jun, Qian Wang, Tong Wei, and Zhuangjun Fan. 2014b. "Recent advances in design and fabrication of electrochemical supercapacitors with high energy densities." *Advanced Energy Materials* 4 (4):1300816.

Yan, Yan, Peng Gu, Shasha Zheng, Mingbo Zheng, Huan Pang, and Huaiguo Xue. 2016. "Facile synthesis of an accordion-like Ni-MOF superstructure for high-performance flexible supercapacitors." *Journal of Materials Chemistry A* 4 (48):19078–19085. doi:10.1039/C6TA08331E.

Yan, Yan, Tianyi Wang, Xinran Li, Huan Pang, and Huaiguo Xue. 2017. "Noble metal-based materials in high-performance supercapacitors." *Inorganic Chemistry Frontiers* 4 (1):33–51.

Yang, Bingchao, Chunxue Hao, Fusheng Wen, Bochong Wang, Congpu Mu, Jianyong Xiang, Lei Li, Bo Xu, Zhisheng Zhao, Zhongyuan Liu, and Yongjun Tian. 2017. "Flexible black-phosphorus nanoflake/carbon nanotube composite paper for high-performance all-solid-state supercapacitors." *ACS Applied Materials & Interfaces* 9 (51):44478–44484. doi:10.1021/acsami.7b13572.

Yang, Tae-Youl, Giuliano Gregori, Norman Pellet, Michael Grätzel, and Joachim Maier. 2015. "The significance of ion conduction in a hybrid organic–inorganic lead-iodide-based perovskite photosensitizer." *Angewandte Chemie International Edition* 54 (27):7905–7910.

Zhang, Huimin, Hong Lin, Chunjun Liang, Hong Liu, Jingjing Liang, Yong Zhao, Wenguan Zhang, Mengjie Sun, Weikang Xiao, and Han Li. 2015. "Organic–inorganic perovskite light-emitting electrochemical cells with a large capacitance." *Advanced Functional Materials* 25 (46):7226–7232.

Zhang, Li Li, and X.S. Zhao. 2009. "Carbon-based materials as supercapacitor electrodes." *Chemical Society Reviews* 38 (9):2520–2531.

Zhao, Meng-Qiang, Chang E Ren, Zheng Ling, Maria R Lukatskaya, Chuanfang Zhang, Katherine L Van Aken, Michel W Barsoum, and Yury Gogotsi. 2015. "Flexible MXene/carbon nanotube composite paper with high volumetric capacitance." *Advanced Materials* 27 (2):339–345.

Zhao, Yang, Xifei Li, Bo Yan, Dejun Li, Stephen Lawes, and Xueliang Sun. 2015. "Significant impact of 2D graphene nanosheets on large volume change tin-based anodes in lithium-ion batteries: A review." *Journal of Power Sources* 274:869–884. doi:10.1016/j.jpowsour.2014.10.008.

Zhao, Yang, Zhongxin Song, Xia Li, Qian Sun, Niancai Cheng, Stephen Lawes, and Xueliang Sun. 2016. "Metal organic frameworks for energy storage and conversion." *Energy Storage Materials* 2:35–62. doi:10.1016/j.ensm.2015.11.005.

Zhao, Yujie, Jinzhang Liu, Michael Horn, Nunzio Motta, Mingjun Hu, and Yan Li. 2018. "Recent advancements in metal organic framework based electrodes for supercapacitors." *Science China Materials* 61 (2):159–184.

Zhou, Feichi, Zhiwei Ren, Yuda Zhao, Xinpeng Shen, Aiwu Wang, Yang Yang Li, Charles Surya, and Yang Chai. 2016. "Perovskite photovoltachromic supercapacitor with all-transparent electrodes." *ACS Nano* 10 (6):5900–5908.

Zhou, Huanping, Qi Chen, Gang Li, Song Luo, Tze-bing Song, Hsin-Sheng Duan, Ziruo Hong, Jingbi You, Yongsheng Liu, and Yang Yang. 2014. "Interface engineering of highly efficient perovskite solar cells." *Science* 345 (6196):542–546.

Zhou, Shuang, Linkai Li, Hui Yu, Jizhang Chen, Ching-Ping Wong, and Ni Zhao. 2016a. "Thin film electrochemical capacitors based on organolead triiodide perovskite." *Advanced Electronic Materials* 2 (7):1600114.

Zhou, Wen-Jia, Mao-Wen Xu, Dan-Dan Zhao, Cai-Ling Xu, and Hu-Lin Li. 2009. "Electrodeposition and characterization of ordered mesoporous cobalt hydroxide films on different substrates for supercapacitors." *Microporous and Mesoporous Materials* 117 (1–2):55–60.

Zhou, Xinhong, Chaoqun Shang, Lin Gu, Shanmu Dong, Xiao Chen, Pengxian Han, Lanfeng Li, Jianhua Yao, Zhihong Liu, and Hongxia Xu. 2011. "Mesoporous coaxial titanium nitride-vanadium nitride fibers of core–shell structures for high-performance supercapacitors." *ACS Applied Materials & Interfaces* 3 (8):3058–3063.

Zhu, Changrong, Peihua Yang, Dongliang Chao, Xingli Wang, Xiao Zhang, Shi Chen, Beng Kang Tay, Hui Huang, Hua Zhang, Wenjie Mai, and Hong Jin Fan. 2015. "All metal nitrides solid-state asymmetric supercapacitors." *Advanced Materials* 27 (31):4566–4571. doi:10.1002/adma.201501838.

Zhu, Jixin, Wenhui Shi, Ni Xiao, Xianhong Rui, Huiteng Tan, Xuehong Lu, Huey Hoon Hng, Jan Ma, and Qingyu Yan. 2012. "Oxidation-etching preparation of MnO_2 tubular nanostructures for high-performance supercapacitors." *ACS Applied Materials & Interfaces* 4 (5):2769–2774.

10 Synthesis Approaches of Inorganic Materials

Bilal Akram, Arshad Farooq Butt, and Javeed Akhtar

CONTENTS

10.1 Hydrothermal Synthesis of Inorganic Materials ... 156
10.2 Classification of Hydrothermal Strategies ... 156
  10.2.1 Types of Synthesis on the Basis of Interior Reaction Conditions ... 156
    10.2.1.1 Template-Free Hydrothermal Synthesis ... 157
    10.2.1.2 Template-Mediated Hydrothermal Synthesis ... 159
  10.2.2 Hydrothermal Growth Strategies of Nanostructures Based on Exterior Reaction Environment Adjustment ... 169
  10.2.3 Microwave Assisted (MA) Route ... 170
  10.2.4 Magnetic Field Assisted (MFA) Route ... 171
10.3 Solvothermal Synthesis ... 173
10.4 Interface-Mediated Growth of Inorganic Nanostructures ... 173
  10.4.1 Instrumentation for Hydrothermal/Solvothermal Synthesis ... 175
10.5 Sol Gel Synthesis ... 176
10.6 Ultrasonic Nebulizer Assisted Chemical Vapors Deposition (UNA-CVD) ... 178
  10.6.1 Role of Precursors in UNA-CVD ... 178
  10.6.2 Role of Temperature in UNA-CVD ... 179
  10.6.3 Role of Solvent in UNA-CVD ... 179
  10.6.4 Different Variants of UNA-CVD ... 179
  10.6.5 Advantages of UNA-CVD ... 180
  10.6.6 Disadvantages of UNA-CVD ... 180
10.7 Colloidal Synthesis of Inorganic Materials ... 180
  10.7.1 Some Key Concepts in Colloidal Synthesis ... 180
    10.7.1.1 Nucleation Process ... 180
    10.7.1.2 Ostwald Repining ... 181
    10.7.1.3 Role of Capping Agents in Colloidal Synthesis ... 181
    10.7.1.4 Role of Precursor in Colloidal Synthesis ... 181
    10.7.1.5 Single Molecular Precursors ... 181
10.8 Two Steps Colloidal Synthesis of Inorganic Materials ... 182
  10.8.1 Advantages of Colloidal Synthesis ... 182
  10.8.2 Disadvantages ... 182
10.9 Conclusions ... 182
References ... 183

10.1 HYDROTHERMAL SYNTHESIS OF INORGANIC MATERIALS

Hydrothermal synthesis can be defined as synthesis of materials through chemical transformations and solubility affinities of chemical reagents in a sealed reactor containing water based solution over room temperature and pressure. This synthetic route involves chemical changes accompanied in solution-based growth, preparations, and assembly of certain materials. Notably, a lot of materials having special structures and properties may be obtained by following this strategy, which cannot be synthesized otherwise. Such synthetic strategies being convenient and mild, in some cases, may act as a substitute of solid-state reactions.

The primary characteristic of such solution-based hydrothermal strategies is their non equilibrium and non ideal conditions. High temperature and pressure states can activate the solvents of the system that may be aqueous or non aqueous. Hydrothermal-based strategies have become more and more significant approaches to obtain most inorganic nano, meso or microscale materials (Shi, Song et al. 2013). The easy operability and tunable behavior of these synthetic strategies is one more characteristic that bridges the physical properties and synthetic chemistry of as synthesized materials. As compared to other synthetic routes and techniques, these techniques have their predominance. This route can be adopted to obtain a diverse range of technologically significant crystalline or amorphous materials. The resulting materials have their own superior and specific chemistry along with distinct physical features. By using this technique, one may synthesize new materials with distinct electronic states, thermodynamically unstable structure, and ordered crystallographic orientations. Moreover, this technique is also useful to get thermodynamically equilibrium defected perfect single crystals, tunable size and structure, and in situ ion-incorporated materials.

10.2 CLASSIFICATION OF HYDROTHERMAL STRATEGIES

Precise control over different synthetic conditions during hydrothermal synthesis is a key to the accomplishment of the fabrication of structurally diverse inorganic materials. The controllable conditions may either be the variable states inside the reaction system or the external reaction environment such as input energy source. Hydrothermal/solvothermal synthesis has been categorized into two different types on the basis of above-mentioned conditions. On the basis of internal reaction conditions, the synthetic approach can either be template-free or template-assisted hydrothermal route. Based on the external reaction environment, i.e., mode of input energy the synthesis may either be microwave-assisted or magnetic field-assisted in addition to conventional heating temperature from an oven.

Herein, we discuss the typical hydrothermal synthesis of nanomaterials based on different conditions in order to provide a brief insight into the formation mechanism involved in the synthesis of inorganic nanostructures.

10.2.1 Types of Synthesis on the Basis of Interior Reaction Conditions

There are two main types of the strategies based on the internal reaction conditions: template-free hydrothermal synthesis and template assisted hydrothermal synthesis.

10.2.1.1 Template-Free Hydrothermal Synthesis

Among various hydrothermal approaches, the most commonly employed is the template-free synthetic routes, because of having advantages of convenient manipulation and product purity. These strategies primarily use the chemical interactions of reaction components during synthetic process without incorporation of templates to recognize the synthetic control of the inorganic nanomaterials. In the absence of any template the synthesis can proceed through any of the following pathways.

10.2.1.1.1 Recrystallization of Metastable Precursors (RMP) Route

Metastable precursors such as amorphous colloids or nanoparticles may be produced swiftly through a co-precipitation reaction in an aqueous solution by the use of vigorous agitation or highly energetic ultrasonication. High temperature and pressure during hydrothermal synthesis can transform these metastable precursors into different well-crystalline inorganic materials with a defined structure and geometry through a recrystallization process. This is the fundamental idea of RMP route.

10.2.1.1.2 Reshaping Bulk Materials (RBM) Route

In this route, mostly the commercial products are used as the precursor to get the target materials of desired morphology. Mostly, uniform nanostructures based on metal oxide or hydroxide can be obtained through this strategy. The precursors used in this strategy are usually commercial oxide products which usually own an irregular shape and large size. The hydrothermal reshaping bulk materials route was used to obtain TiO_2 nanotubes from commercially available rutile titania powder as the starting material that was first reported by Kasuga and co-workers in 1998 (Kasuga, Hiramatsu et al. 1998). Their pioneering work was being followed by many groups and results in the synthesis of titania nanotubes, nanosheets, nanowires, and nanobelts through same strategy.

10.2.1.1.3 Indirect-Supply Reaction Source (ISRS) Route

In the indirect-supply reaction source route, there is only a clear reaction solution having different dissoluble raw materials in the reaction container (autoclave) prior to the hydrothermal treatment. At definite reaction conditions, the controllable nucleation of the desired material starts through the chemical reaction. The chemical reaction occurs between ISRS and other ionic species presented in the medium that eventually results in the formation of final product that may be any inorganic nanostructured material. The ISRS may either be the ionic species in situ produced from the inorganic precursors through redox reaction or dissolved oxygen in water as solvent. In contrast to other routes, this is a homogeneous reaction-based route, which leads to the production of uniform shaped, high purity nanostructures with narrow size distribution.

10.2.1.1.4 Decomposition of Single-Source Precursor (DSSP) Route

In the decomposition of single-source precursor route, inorganic materials of varied structures can be grown and obtained through the decomposition of a single-source precursor under hydrothermal conditions. These precursors are distinct species having

TABLE 10.1
Basic States of the Four Template-Free Hydrothermal Routes

	RMP Route	RBM Route	ISRS Rout	DSSP Route
Pretreatment before hydrothermal reaction	Coprecipitation reactions	No use	Dissolution	Dissolution
State of precursors	Colloids or small-size nanocrystals	Bulk materials	Two or more dissoluble reactants	Single reaction source
State of reaction solution	Dispersed colloids or precipitation in solution	Bulk materials and aqueous solution	Clear aqueous solution	Mixtures of reaction source with water
Basic formation process of products during hydrothermal reaction	Nucleation and growth	Dissolution, nucleation and growth	Supply of reaction source, nucleation and growth	Decomposition of reaction source, nucleation and growth
Manipulation in whole process	Relatively complex	Relatively simple	Middle	Middle

those elemental components which are needed in the target inorganic nanostructured material. This strategy is comparatively cleaner and simpler as it may eradicate the need of multiple precursors. The use of multiple precursors is often expensive and leads to toxic byproducts hence this strategy is preferable as compared with the other three template-free hydrothermal routes. This route has also the advantage of avoiding the concern of complex reaction dynamics that can lead the target material impure.

A brief summary of the basic states of above-mentioned strategies is given in Table 10.1.

The template free hydrothermal strategy is often used for the formation of various inorganic materials. As a representative example of the materials synthesized via template-free hydrothermal method presented here are titania nanostructures of defined morphologies. The hydrothermal formation of various titania nanostructures from Degussa P25 as precursor at different alkali concentration and temperature was reported by Morgan and co-workers, which is displayed in Figure 10.1.

10.2.1.1.5 Growth Mechanism Involved in Template-Free Hydrothermal Synthesis

The template-free hydrothermal growth of inorganic materials can either be kinetically controlled or thermodynamically controlled and follows Ostwald ripening mechanism. Generally the morphology of target material is determined by the Ostwald ripening, as it is supposed that the primarily produced nuclei are free from defects (Penn and Banfield 1999). The chemistry behind this process

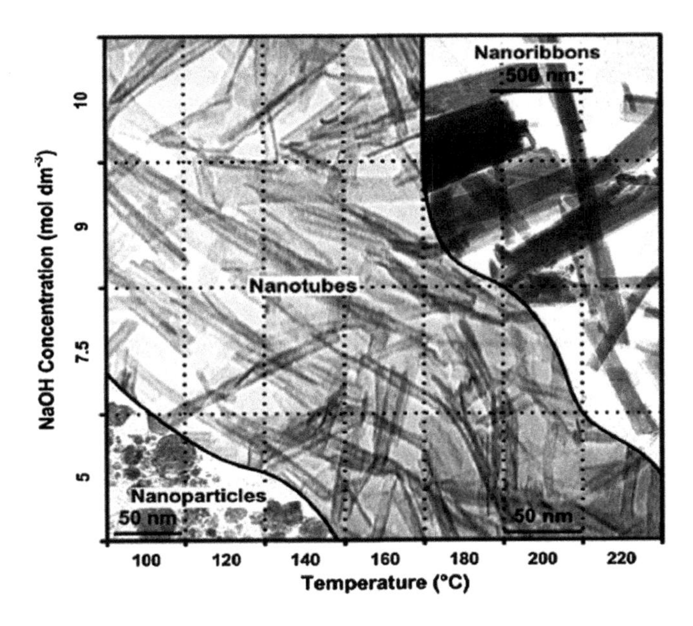

FIGURE 10.1 Morphological phase diagram of formed TiO_2 nanostructures after the hydrothermal treatment of Degussa P25 for 20 h. (Reproduced with permission from Morgan, D. L., et al. *Chem. Mater.*, 20, 3800–3802, 2008. Copyright 2008 American Chemical Society.)

is that smaller particles sacrificed via gradual dissolution in the mother liquor to give rise the larger one. This process might result in the evolution of specific morphology. Normally, a nanocrystal has different planes with distinct surface energy, hence nanocrystals having high surface energy planes tend to capture other monomers from the solution in order to minimize the surface energy which eventually leads to morphology evolution of resulting materials. The affinity changes based on the intrinsic crystal habit, ultimately generating inorganic materials having anisotropic morphologies, like nanodisks, nanowires, nanopolyhedrons, nanorods, and so on.

10.2.1.2 Template-Mediated Hydrothermal Synthesis

In comparison to above mentioned template-free synthesis, more uniformly distributed, controlled morphology having inorganic materials can be obtained through template assistance that may introduced in the reaction system during synthesis.

Templates used for the synthesis of inorganic materials may be generally categorized as soft template and hard template.

10.2.1.2.1 Soft-Templated Hydrothermal Synthesis Routes

Soft templates may include surfactants, biomolecules, ionic liquids, organic acids, or alcohols. The soft-template assisted hydrothermal process may involve the use of any of the above-mentioned agent hence can be categorized into five different strategies the detailed description of which is given below. A brief overview of the soft-template assisted routes is given in Figure 10.2.

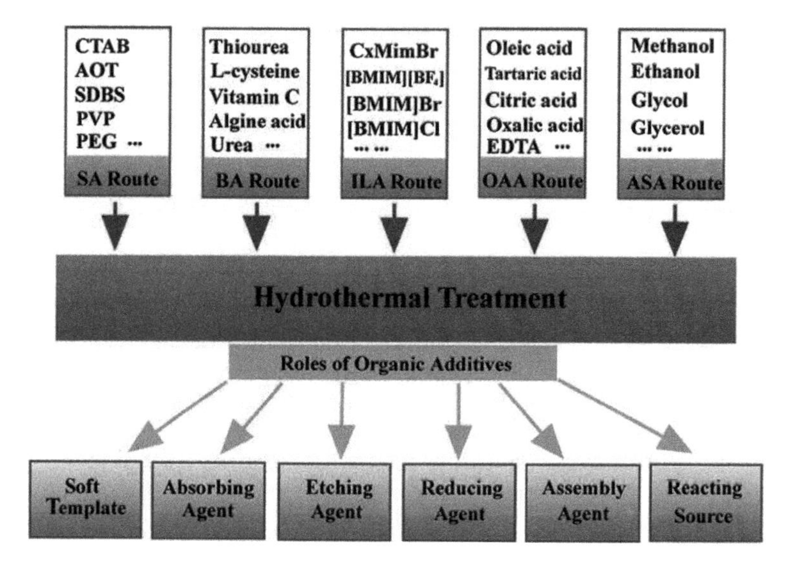

FIGURE 10.2 Five basic routes of the soft templates assisted hydrothermal synthetic strategies, namely SA route, BA route, ILA route, OAA route, and ASA route. (From Shi, W. et al., *Chem. Soc. Rev.*, 42, 5714–5743, 2013. Reproduced by permission of The Royal Society of Chemistry.)

10.2.1.2.2 Surfactants Assisted (SA)

Surfactants are generally amphiphilic species having organic nature. As a result, a surfactant has both oil soluble and water-soluble parts. In the SA hydrothermal approach, these molecules may have a direct consequence on the growing conditions of final target materials and attain the controlled manufacturing of such nanostructured materials. The introduced surfactants may act as soft template, etching and adsorbing agent.

Surfactant molecules being soft templates thoroughly investigated in the preparation of variety of void inorganic nanostructures (Chen, Sun et al. 2002, Cao, Hu et al. 2003a,b, Wang, Chen et al. 2004, Liu, Liu et al. 2005, Zhang, Sun et al. 2006, Song, Zhao et al. 2007, Li and He 2008). Li et al. developed a SA strategy for the preparation of VOx nanotubes. They had used cetyltrimethylammonium bromide (CTAB) as soft surfactant that helps in the aggregation of VO_3 to grow lamellar sheet like structures (Chen, Sun et al. 2002). These intermediate sheets get loosen at the ends and undergo self-rolling to form nanotubes under hydrothermal conditions, as shown in Figure 10.3. Moreover, Song et al. has reported the use of sodium dodecyl sulfate (SDS) as anionic soft surfactant during hydrothermal treatment to produce sea urchin-like MnO_2 nanostructures (Liu, Liu et al. 2005).

10.2.1.2.3 Biomolecules Assisted (BA)

A chemical compound found in living organisms is called a biomolecule. Based on the size of biomolecule it may either be a macromolecule or micromolecule. Macromolecules may include proteins, lipids, polysaccharides, and nucleic acids

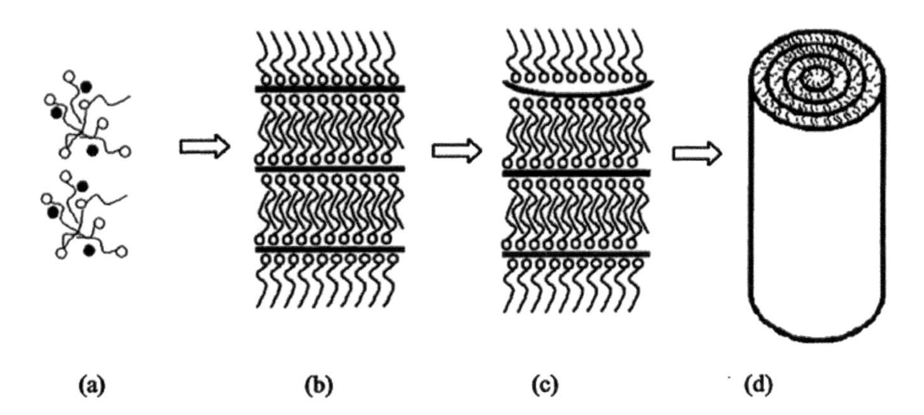

FIGURE 10.3 Schematic presentation of the rolling mechanism for the formation of vanadium oxide nanotubes: (a) the mixture of NH_4VO_3 and the template molecules, (b) layered structures formed through the hydrothermal treatment, (c) the beginning stage of the rolling process, and (d) the formed nanotubes. (Reproduced with permission from Chen, X. et al., *Inorg. Chem.*, 41, 4524–4530, 2002. Copyright 2002 American Chemical Society.)

whereas vitamins, glycolipids, hormones, phospholipids, etc. are categorized as micromolecules. The hydrothermal preparation assisted by such molecules has become a new focus in the synthesis of different unique materials. Biomolecules possess novel structural features that enables them to self-assemble other targeted functional materials. Due to their intrinsic properties, biomolecules can act as a source of an anion, self assembly and reducing agent during BA hydrothermal route as evidenced by many reports. As a consequence of these effects the formation of final materials is determined. These biomolecules may be gradually disintegrated to liberate various anions that further interact with metal ions to produce the N-doped oxide, sulfide, or pure oxide through the hydrothermal process.

BA hydrothermal route leads to the formation of a variety of inorganic materials having a definite shape and structure. In the upcoming lines we will briefly discuss the representative examples of the materials formed via BA route. L-cysteine, a biomolecule of amino acid origin has attained significant attention because of having simple hydrosulfide-group in the structure. It also contains three main functionalities, namely amino, carboxyl, and thiol, that possess a powerful metal cation coordinating affinity. Burford etal. have proven that cysteine can interact with Bi^{3+} ions to produce a complex. Due to presence of amino and carboxyl functional groups cysteine ligands leads to hydrogen bonding. The resulting hydrogen bonding bridge the individual cystein capped metal complexes. Different self assembled 3D metal sulfides nanostructures through hydrothermal synthesis mediated by cystein has been reported to have H-bonding in their assembled structure (such as In_2S_3 (Chen, Zhang et al. 2008), PbS (Zuo, Yan et al. 2008), CuS (Li, Xie et al. 2007), CoS (Bao, Li et al. 2008), ZnS (Tong, Zhu et al. 2007), MoS_2 (Chen, Li et al. 2012), Fe_3S_4 (Cao, Hu et al. 2009), Bi_2S_3 (Zhang, Ye et al. 2006), etc.). Moreover, the self-assembly states of CuI-cystien complex is affected by the concentration of cystein molecules

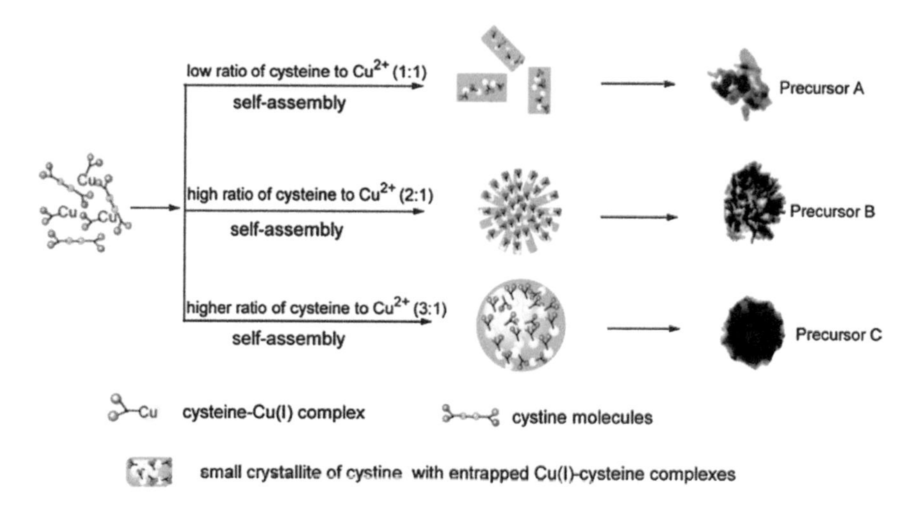

FIGURE 10.4 Schematic diagram of the formation of three typical precursors with different degrees of self-assembly for the synthesis of CuS nanostructures. (Reproduced with permission from Li, B. et al., *J. Phys. Chem. C*, 111, 12181–12187, 2007. Copyright 2007 American Chemical Society.)

in the solution as indicated by Xie et al. Change in concentration leads to the formation of different self-assembled products like dispersive flakes, solid microspheres, and flake assembled spheres. This is due to the strengthening of hydrogen bonding with increase in concentration (Figure 10.4) (Li, Xie et al. 2007). There are many other biomolecules like thioacetamide, glutathione (GSH), and thiourea, etc. which also played the same role as cystein during the BA preparation of complex sulfides (Lu, Gao et al. 2004, Qin, Fang et al. 2005, Fan and Guo 2008, Zuo, Yan et al. 2008).

10.2.1.2.4 Ionic Liquids Assisted (ILA)

A distinctive salt whose melting point lies in the range of +100°C to −100°C or even lower is called an IL that mostly exist in the liquid state with broad liquidus range (Wasserscheid and Keim 2000). As compared to conventional organic solvents, IL provides several advantages being a solvent during formation of different materials mostly having inorganic nature including low toxicity, good thermal stability, negligible vapor pressures, broad electrochemical potential windows, high ionic conductivity, and high synthetic flexibility.

In recent times, ILA approach has been used to obtain a variety of unique inorganic materials, such as oxides, chalcogenides and phosphates of various metals. Representative reports of the materials are hollow microspheres and nanorods of Fe_2O_3, nanoplates and peachstone-like complex structures of CuO, microspheres of In_2S_3, NiO, and MoS_2, self-assembled 3D nanoflowers of MoS, hierarchical dendrites of CdSe, $BiWO_6$ nests, nanoflakes of CuSe (Lou, Huang et al. 2011), BiOBr hollow microspheres (Cheng, Huang et al. 2011), 3D Bi_2S_3 nanoflowers (Jiang, Yu et al. 2005), AgBr microcrystals (Lou, Huang et al. 2011), REF3 (RE = La–Lu, Y) nano/microparticles (Lou, Huang et al. 2011), AgCl microspheres, $MnCO_3$ hollow

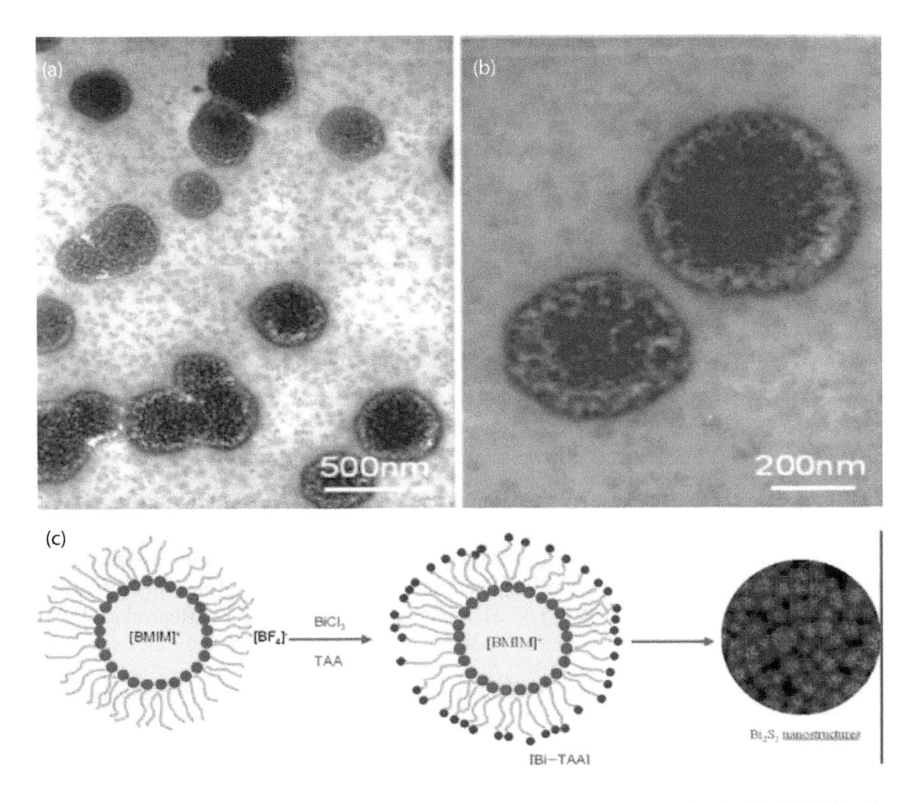

FIGURE 10.5 (a and b) TEM images of the vesicles formed in the IL [BMIM][BF4] solution (c) illustration of the formation mechanism of Bi_2S_3 flowers in IL solution. (Reproduced with permission from Jiang, J. et al., *Chem. Mater.* 17, 6094–6100, 2005. Copyright 2005 American Chemical Society.)

microspheres and $SmVO_4$ nanosheets. In these reports, they behave as soft templates in addition to their role as solvent or reactants. The soft template behavior of IL can be evidenced in a report by Yu et al. TEM images reveals that plenty of nanovesicles can be grown in an IL ([BMIM][BF_4]) diluted aqueous solution through hydrothermal synthetic strategy (Figure 10.5) (Jiang, Yu et al. 2005).

10.2.1.2.5 Organic Acids Assisted (OAA)

Organic acids are organic compounds having acidic characteristics. The most usual organic acids are carboxylic acids. These acids may generally interact with cationic part of the precursor salt such as metal ions in water to produce stable organic acid-metal complexes. Taking this unique tendency of organic acids into account, different nanomaterials, like ZnO self-assembled nanostructures, hollow spindles, nanocubes, and nanoplates of Fe_2O_3, nanospheres of La_2O_3 and In_2O_3, Mn_3O_4 nanooctahedrons, nanotubes of Sb_2S_3 and $NaSmF_4$, nanorods of $LnPO_4$, YVO_4:Eu^{3+}, Bi_2S_3, and Ag_2Se, $NaEu(MoO_4)_2$ microrugbies, InOOH hollow spheres, CeF_3 hollow nanocages/rings, nanoflowers of CaF_2 and Bi_2Te_3, Bi_2WO_6 nest-like nanostructures, and so on, have

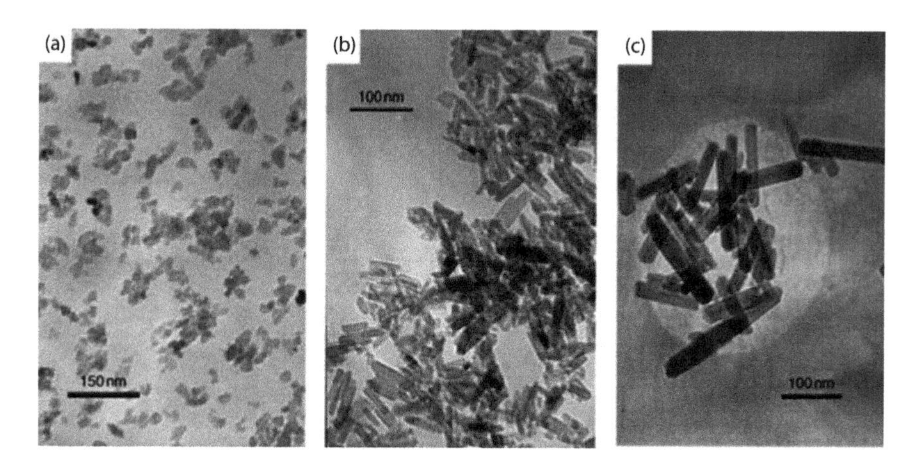

FIGURE 10.6 (a) TEM image of YPO_4 nanoparticles (0.1 g EDTA); (b) TEM image of YPO_4 nanoparticles (0.3 g EDTA); (c) TEM image of YPO_4 nanoparticles (0.5 g EDTA). (Yan, R., et al.: Crystal structures, anisotropic growth, and optical properties: Controlled synthesis of lanthanide orthophosphate one-dimensional nanomaterials. *Chemistry–A European Journal.* 2005. 11. 2183–2195. Copyright Wiley-VCH Verlag GmbH & Co. KGaA. Reproduced with permission.)

been obtained through OAA routes (Liang, Xu et al. 2004, Luo, Jia et al. 2005, Yan, Sun et al. 2005, Wang, Xu et al. 2009, Chen, Zhang et al. 2010, He, Wang et al. 2011). The organic acids used mainly include salicylic acid, oleic acid, ethylenediaminetetraacetic acid (EDTA), oxalic acid, malic acid, tartaric acid, and citric acid. It has been noticed that organic acids may act like a reducing agent, structure-directing agent, gas bubble release and assembly agent during the course of reaction.

For example, EDTA-assisted growth of different structures of YPO_4 are shown in Figure 10.6. It can be found that EDTA can restrain the anisotropic growth of different structures (Yan, Sun et al. 2005, Wang, Xu et al. 2009).

10.2.1.2.6 Alcoholic Solvent Assisted (ASA)

Organic compounds containing carbon bound hydroxyl moiety are alcohols. Polar nature of these alcohols arises as a result of hydroxyl groups. These groups also enable alcohols to do hydrogen bonding. Few short chain alcohols like ethanol, glycerol, and ethylene glycol (EG), etc., are utilized to mediate formation of nanostructured materials during hydrothermal treatment. They are acted as reducing and shape controlling agent through inhibiting the hydrolyzation. There are number of reports on the ethylene glycol-mediated shape-controlled hydrothermal synthesis of various nanostructures (Yang, Zhu et al. 2006, Li, Zhang et al. 2006, Yang, Li et al. 2008, Ashoka, Nagaraju et al. 2009, He, Wang et al. 2011).

In a typical example of this synthetic strategy consider the EG assisted formation of nanoflowers of Lu_2O_3. The introduced alcoholic solvent, i.e., EG protect the charges locating on the surface of Lu_2O_3 by restraining the adsorption of some ionic species, like sodium metal ion and acetate ion (Yang, Li et al. 2008). That is,

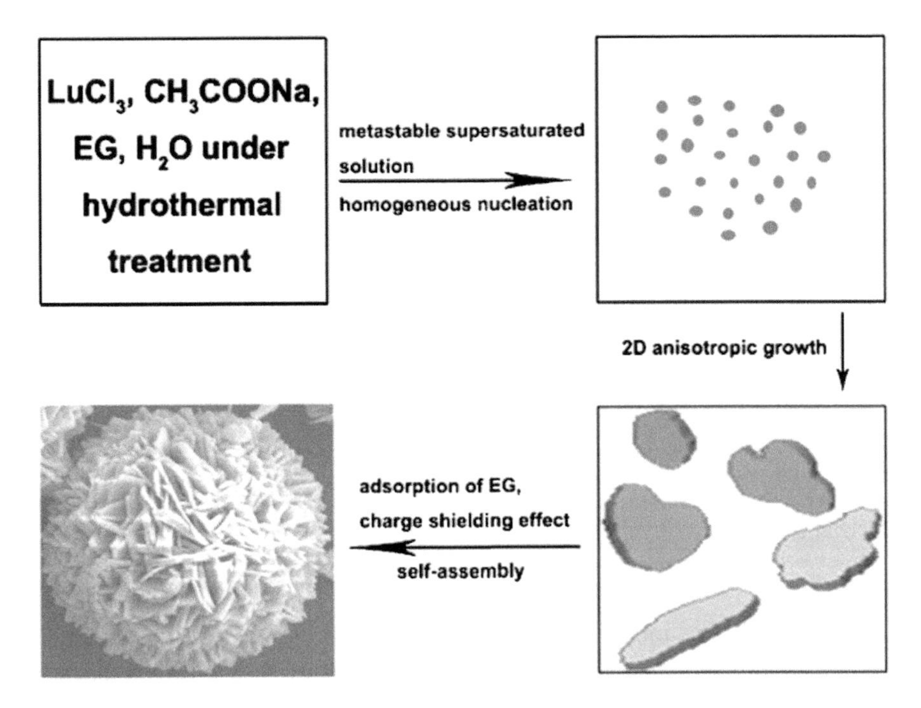

FIGURE 10.7 Illustration for the formation process of the flower-like Lu_2O_3 nanostructures. (Reproduced with permission from Yang, J. et al., *J. Phys. Chem. C*, 112, 12777–12785, 2008. Copyright 2008 American Chemical Society.)

the difference of charges between the edges and flat surfaces of a nanoflake are considerably decreased or protected. This imbalance in charges among different parts would results in a prime conjunction of edge to-surface in EG absence, therefore in the present EG assisted self-assembly, dual conjunctions, i.e., edge-to edge and edge-to-surfaces can be created in addition to few surface-to-surface conjunctions via the charge shielding consequence produced by ethylene glycol (Figure 10.7).

10.2.1.2.7 Hard-Templated Hydrothermal Synthetic Routes

In addition to the above discussed soft templates, a variety of hard templates can also be used to obtain well defined shapes having inorganic materials. Such templates have uniform size and pre-define shape that determine the size and shape of target material. So, the shape and size of target materials can be tuned by simply selecting an appropriate hard template of desired features. However, numerous problems are in practice while using such templates like materials and templates incompatibility, the selection or design of desired templates, and safe elimination of the templates. So, there is a need to understand the underlying mechanism intensely and surmount. On the basis of various approaches used to introduce and remove templates, the synthetic approaches can be categorized into following four routes the schematic of which is shown in Figure 10.8.

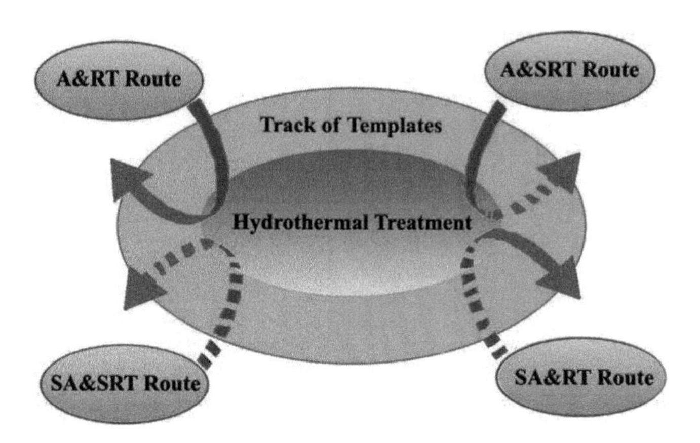

FIGURE 10.8 Basic routes of template-assisted hydrothermal synthetic strategies. (From Shi, W. et al., *Chem. Soc. Rev.*, 42, 5714–5743, 2013. Reproduced by permission of The Royal Society of Chemistry.)

10.2.1.2.8 Additive and Removed Template (A & RT) Route

In A & RT hydrothermal route, some well-known templates, such as carbon spheres, spherobacterium spheres, multiwalled carbon nanotubes (MWCNTS), and anodized aluminum oxide membranes (AAO) are being used. Before the hydrothermal treatments, such templates used to incorporate in the reaction solution which need to be removed artificially upon reaction completion through dissolution or calcination.

The representative example of the A & RT hydrothermal route is the formation of unique Fe_2O_3 cage-like hollow nanospheres demonstrated by Yu et al. that is displayed in Figure 10.9 (Yu, Yu et al. 2009).

10.2.1.2.9 Additive and Self-Removed Template (A & SRT) Route

A & SRT involves the use of templates which can react completely with other reactant during the hydrothermal treatment in order to provide them targeted structures. The templates used in this strategy have uniform size and shape. Typical templates may include Te nanowires, $Gd(OH)_3$ nanorods, CdO nanowires, and CdS nanorods, etc. Such templates are usually incorporated into the reaction medium prior to reaction initiation, and can automatically aloof upon reaction completion (Hu, Deng et al. 2001, Gu, Liu et al. 2008, Niu, Cao et al. 2008).

The advantages of this route are that it can avoid incompatibility problems between templates and targets and its removal from reaction system. As the involved templates directly take part in the synthesis, hence strategy may also recognize the production of nanomaterials having defined structure.

Single-crystalline $GdVO_4$:Eu and $GdVO_4$ nanorods have been synthesized via $Gd(OH)_3$ nanorods templated hydrothermal route as demonstrated by Chang et al. (Gu, Liu et al. 2008). Studies revealed that surface deposition mechanism is followed for production of nanorods. In this case, the deposition of VO_4 groups on the $Gd(OH)_3$ nanorods surface takes place which form $GdVO_4$ nanorods through

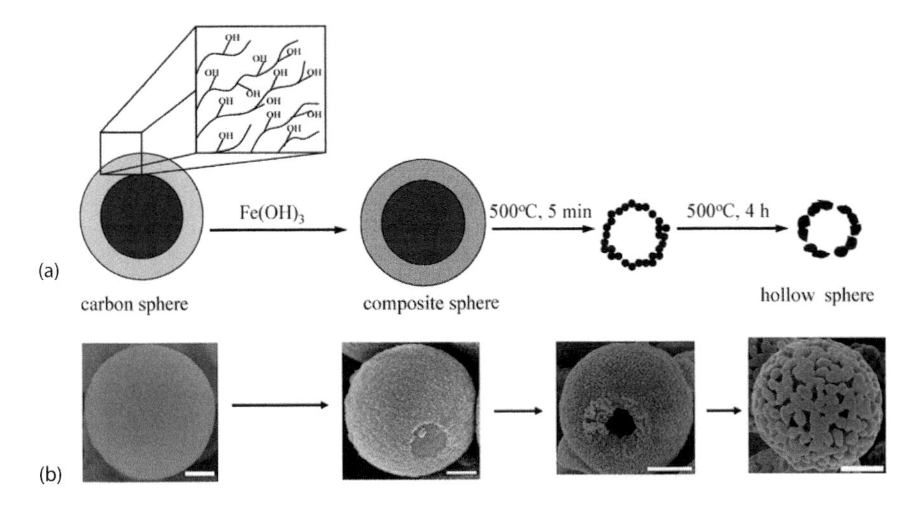

FIGURE 10.9 (a) Schematic illustration of the formation of cage-like Fe_2O_3 hollow spheres. (b) SEM images of carbon sphere, composite sphere, and Fe_2O_3 hollow sphere at 500°C for 5 min and 4 h. The scale bar is 200 nm. (Reproduced with permission from Yu, J. et al., *Cryst. Growth Des.*, 9, 1474–1480, 2009. Copyright 2009 American Chemical Society.)

progressive reaction between the two species. TEM and EDS analysis proves the core–shell nature of resulting materials (Figure 10.10). Similarly different structures of $La(OH)_3$ such as nanorods, hollow trapezohedrons and condensed nanospheres have been reported by Wan et al. They have employed lanthanum glycolate polyhedrons having 24 faces as starting material for controlled formation of above mentioned products under various hydrothermal hydrolysis rates (Niu, Cao et al. 2008).

10.2.1.2.10 Self-Additive and Self-Removed Template (SA & SRT) Route

In SA & SR template route there is in situ formation of well-defined shaped intermediate or uniformly dispersed gas bubbles released during reaction. These intermediates or gas bubbles induce the growth of target material through chemical reaction or heterogenous nucleation during hydrothermal synthesis. The limitation of this approach is to recognize and control products and templates balance.

During most of the chemical reactions there is production of gas bubbles like N_2, H_2 and O_2. These in situ produced gas bubbles behaves as a template in hydrothermal synthesis and assist the formation of different inorganic nanomaterials with hollow structures (Peng, Dong et al. 2003, Yang and Sasaki 2008, Li, Yang et al. 2008). For example, nanorod containing hierarchical hollow spheres of CoOOH has been synthesized by Sasaki and Yang through a one-step facile hydrothermal approach. They have claimed that during the hydrothermal treatment, O_2 gas bubbles are released which behave as templates (Yang and Sasaki 2008). In another report nanocrystal assembled ZnSe hollow spheres have been produced through hydrothermal treatment using N_2 gas bubbles as templates (Figure 10.11) (Peng, Dong et al. 2003). Moreover, Yang et al. too synthesized ZnO hollow microspheres composed

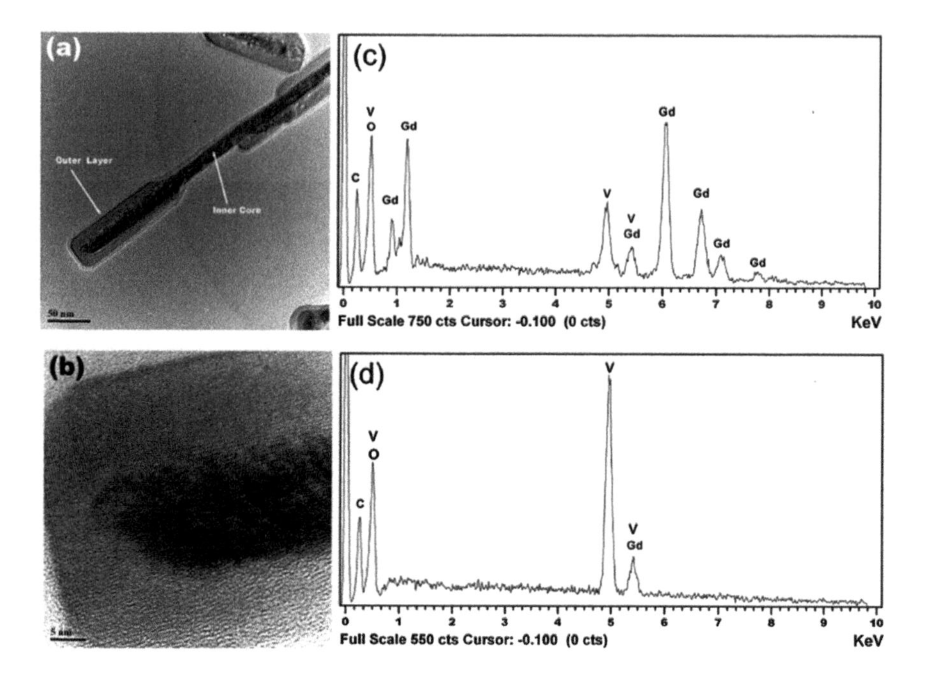

FIGURE 10.10 TEM micrograph and EDS results of the $Gd(OH)_3$–$GdVO_4$ core–shell nanorods prepared with hydrothermal reaction: (a) TEM micrograph, revealing nanorod with inner core and outer layer; (b) HRTEM for the tip of the nanorod, indicating poor crystallinity of the newly formed layer; (c) EDS of the inner core, showing the presence of Gd-rich in core and main; (d) EDS of the outer layer, implying that the outer layer is rich in V. (Reproduced with permission from Gu, M. et al., *Cryst. Growth Des.*, **8**, 1422–1425, 2008. Copyright 2008 American Chemical Society.)

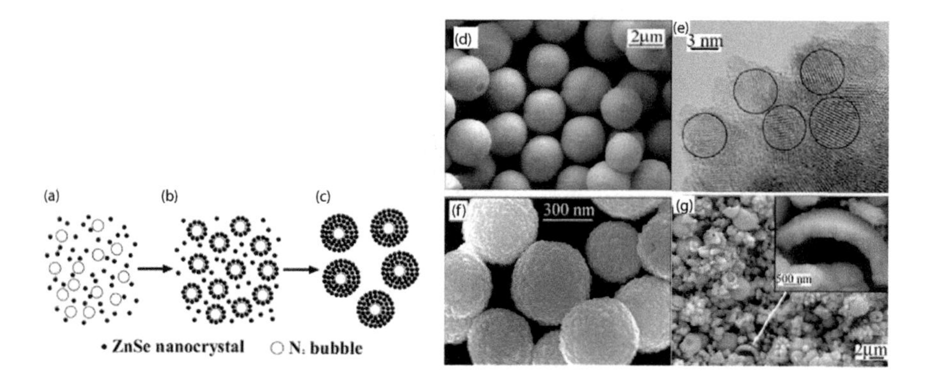

FIGURE 10.11 (a–c) Schematic representation of the formation mechanism of ZnSe microspheres (d,f) SEM image of ZnSe microspheres at different magnifications (e) HRTEM image of product after grinding, which indicates that the microspheres are the aggregation of small nanocrystals with the size of about 5–6 nm. (g) SEM image of the broken ZnSe microspheres. Inset shows an individual broken shell, which indicates that these microspheres are hollow inside.

of nanorods through H_2 gas as a template (Li, Yang et al. 2008). The advancement of such approaches provides an effective and unique route for growth of hollow nanostructured materials.

10.2.1.2.11 Self Additive and Removed Template (SA & RT route)

Thomas et al. has demonstrated a universal approach based on hydrothermal synthesis in 2006 to obtain a variety of oxide hollow spheres such as Fe_2O_3, CeO_2, NiO, CuO, Co_3O_4 and MgO (Titirici, Antonietti et al. 2006). In this route, an aqueous solution of different metal precursors and glucose was being subjected to hydrothermal treatment, resulting in situ formation of carbon spheres containing metal ions. Upon high temperature calcination, metal oxide hollow spheres can be attained with simultaneous removal of carbon spheres (Figure 10.12). Such SA & R template-mediated hydrothermal approach can generate most of the metal oxide based spherical hollow nanostructures. The shell thickness and surface area of resulting hollow spheres can be controlled by tuning the precursor and glucose ratios.

10.2.2 Hydrothermal Growth Strategies of Nanostructures Based on Exterior Reaction Environment Adjustment

In conventional hydrothermal synthetic approaches, the heating source is generally an oven which provides energy to the reaction system in the form of thermal conduction. However, there are few unavoidable issues, particularly in case of lower temperatures, such as slow rate and non-uniform conditions of reaction, and spiky heat gradients throughout the bulk solution. During the commercial-scale formation of nanomaterials, these problems, i.e., thermal gradients and inhomogeneous, non-uniform product, could be strongly exaggerated, that results in widened distributions of size. It is an utmost need to have some alternative source of

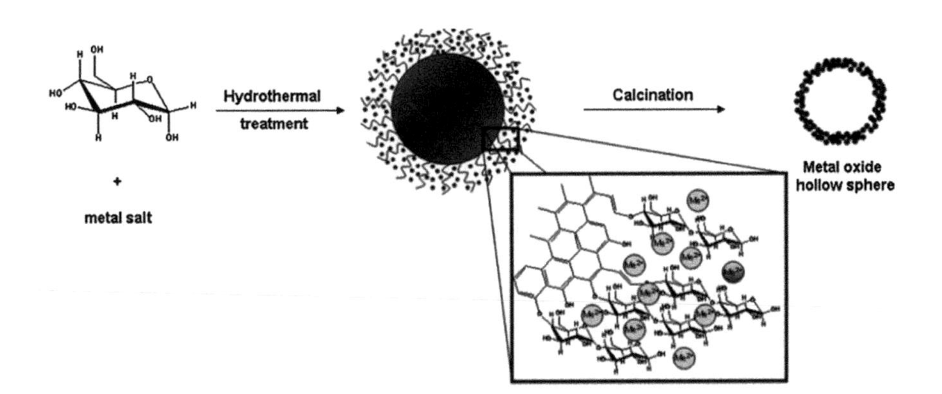

FIGURE 10.12 Schematic illustration of the synthesis of metal oxide hollow spheres from hydrothermally treated carbohydrate–metal salt mixtures. (Reproduced with permission from Titirici, M.-M. et al., *Chem. Mater.*, 18, 3808–3812, 2006. Copyright 2006 American Chemical Society.)

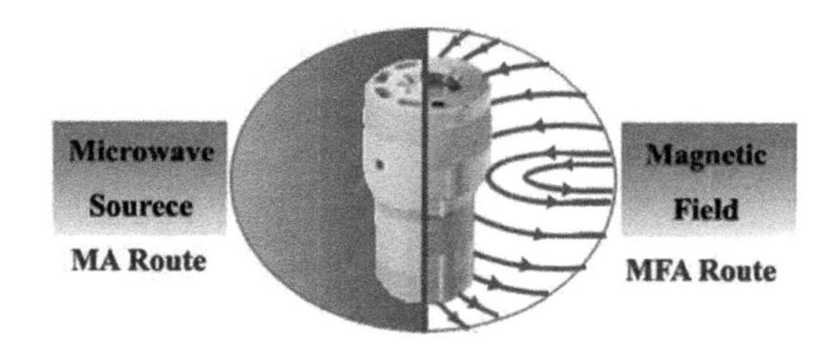

FIGURE 10.13 Hydrothermal growth routes of nanostructures based on exterior adjustment: MA route and MFA route. (From Shi, W. et al., *Chem. Soc. Rev.*, 42, 5714–5743, 2013. Reproduced by permission of The Royal Society of Chemistry.)

input energy in order to overcome these issues associated with conventional thermal induction. By taking these points into account, exterior reaction environment adjustment based following two hydrothermal approaches have been developed (Figure 10.13).

10.2.3 Microwave Assisted (MA) Route

Heating mode originated from microwave radiations is thought to be a uniform, fast and cost-effective way to commercially produce high-quality nanostructured inorganic materials. In this approach, these radiations are being employed as the source of heat instead of traditional heating from oven. Small-sized target materials, less reaction time and narrower particle size distribution are the obvious benefits of this strategy. The faster heating rate at initial stage provides additional benefits of energy savings. There is a digital programming of time and temperature in modern microwaves that further facilitate one to get an optimized reaction state more conveniently. This is much advantageous for lab-scale execution of several strategies offering formation of different high-quality nanostructured inorganic materials.

The MA hydrothermal approach is used to produce many inorganic materials with sub nanometer scale dimensions for example WO_3 nanowires with diameter of less than 10 nm has been successfully synthesized through the MA hydrothermal strategy. The as synthesized nanowires was much thinner in comparison with the one prepared via traditional hydrothermal route (Phuruangrat, Ham et al. 2010).

Different inorganic materials nanostructures, like simple metal oxides (ZnO, La_2O_3, TiO_2, Fe_2O_3, CuO, Fe_3O_4, SnO_2, CeO_2 & WO_3), composite oxides [$AgIn(WO_4)_2$, $BaTiO_3$, Bi_2WO_6, $BiFeO_3$, Bi_2MoO_6, Fe_2 $(MoO_4)_3$, $KNbO_3$, Zn_2SnO_4, $Cd_2Ge_2O_6$, $CaMoO_4$ & $MnWO_4$], simple metal hydroxides [$Ni(OH)_2$, GaOOH & $In(OH)_3$], phosphides (Ni_3P) and chalcogenide (FeSe) (Shi, Song et al. 2013) etc, have been prepared through this approach. Representative example of the materials synthesized via this route is sea urchin-like CuO microcrystals whose mechanistic pathway is given in Figure 10.14 (Volanti, Orlandi et al. 2010).

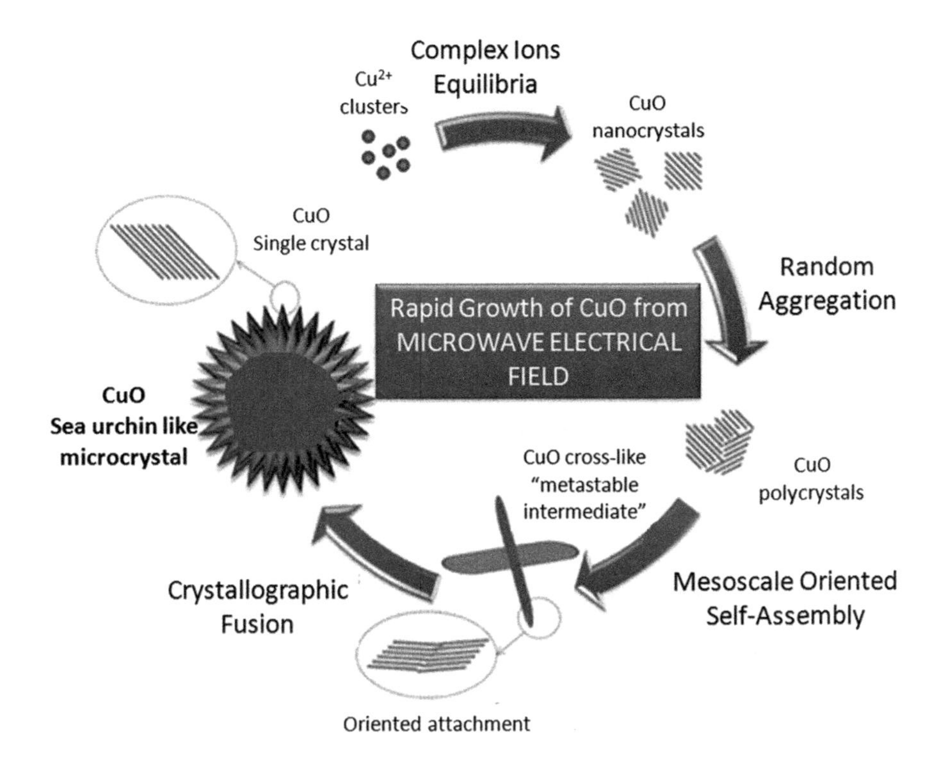

FIGURE 10.14 Schematic illustration of the crystal growth process of CuO sea urchin-like microcrystals via a mesoscale self-assembly under a microwave electrical field. (From Volanti, D.P. et al., *CrystEngComm.*, 12, 1696–1699, 2010. Copyright 2010. Reproduced with permission of The Royal Society of Chemistry.)

10.2.4 Magnetic Field Assisted (MFA) Route

MFA hydrothermal approaches have been currently employed to obtain variety of inorganic materials with distinct structure, morphology, and physical features, like Fe_3O_4 nanowires, $SrFe_{12}O_{19}$ nanowires, Fe_3S_4/FeS_2 microrods, Mn-doped ZnO nanocrystals, Co_3O_4 nanocubes/nanospheres, MnO_2 nanowires/urchins, and so on (Shi, Song et al. 2013). It has been established that the strength of externally provided magnetic field considerably affect the formation of inorganic nanostructured materials as indicated in report on MFA hydrothermal synthesis of Fe_3O_4 nanowires (Wang, Chen et al. 2004) (Figure 10.15). Single crystalline Fe_3O_4 nanowires were obtained under field strength of 0.25 T. Lessening the field strength to half result in poor yield of the product nanowires. Whereas in the absence of external magnetic field, no nanowires but nanoparticles of hexagonal and square geometry were being achieved. Detailed investigation about the effects of external magnetic field strength on the development of nanomaterials is still not become possible due to lack of techniques for in situ observation of product during the synthesis and it demands furthermore exploration. It is more interesting to know that the MFA hydrothermally

FIGURE 10.15 TEM images of the Fe_2O_3 samples obtained under (a) 0 T, (b) 0.15 T, (c) 0.25 T and (d) 0.35 T external magnetic fields. (From Wang, J. et al.: Magnetic-field-induced growth of single-crystalline Fe_3O_4 nanowires. *Advanced Materials.* 2004. 16. 137–140. Copyright 2004 Wiley-VCH Verlag GmbH & Co. KGaA. Reproduced with permission.)

synthesized nanostructured inorganic materials show different physical properties as compared to the nanostructures synthesized without use of magnetic field. One of such features is the saturated magnetization which is superior in case of materials synthesized via MFA route. The above-mentioned superiority was being evidenced by Chen et al. who reported different magnetic characteristics of same sized (15 nm) Co_3O_4 nanoparticles that were prepared in the absence of external magnetic field and in the presence of an external (0.2 T) magnetic field (Wang, Zeng et al. 2011). A clear difference in the properties of two materials was observed as the magnetization of MFA hydrothermally synthesized Co_3O_4 nanocrystals is 18 emu g^{-1}, which is much superior than those of the nanoparticles synthesized in the absence of magnetic field (2 emu g^{-1}).

10.3 SOLVOTHERMAL SYNTHESIS

Solvothermal synthesis involves the growth of materials through chemical reactions in a non-aqueous media at relatively high temperatures. This strategy is just like hydrothermal route, the only difference between two is that the precursor solution in solvothermal synthesis is usually non-aqueous. All the above discussion about hydrothermal equally applies on solvothermal synthesis.

10.4 INTERFACE-MEDIATED GROWTH OF INORGANIC NANOSTRUCTURES

By using solvothermal/hydrothermal route and selecting appropriate solvent system a variety of inorganic nanomaterials of varied structure can be achieved through interface mediated growth.

Two different phases having common boundaries give rise to interfaces. When a certain particle exerts a force on a region then a boundary is featured. Interfacial interactions enable the interfacial atoms to interact with each other to generate numerous valuable characteristics that become a cause of attention for scientist (Eisenthal 1993).

Interfaces are found to be the key factors that controlled the growth of novel low dimensional building blocks as highlighted by many researchers. The construction of the inorganic materials with sub-nanometer dimensions is a strategy which deals with the generation of interfaces among target nano solid and surrounding species. Scientists have devoted their great efforts to functionalize the region at interfaces for templated formation of different nanostructured materials. In a characteristic vapor–liquid–solid synthesis of nanowires, a few atomic thick gas–liquid and liquid–solid interface is generated in the reaction mixture during synthesis that assists the formation of nanowires. Monodispersed colloidal nanoparticles of any materials can be controllably grown by understanding and tuning the chemistry at interfaces. The colloidal nanocrystals mostly belong to composite class of materials having an inner core and a shell composed of nanocrystals and surfactant respectively. As a major route, the transporting features of atoms and ions across the nanocrystal and their boundaries were determined by the organic-inorganic interfaces hence affecting the development dynamic of inorganic nanocrystals. However, the engineering of interfaces is still the main challenge in this emerging field. An accurate interfacial control can result in the crystallization of nanostructured materials. Generally, careful generation of the few atomic thick interfaces among various matter states could eventually leads to a library of unique structures for example, (1) interfaces between gas and a liquid state may be used as the aggregation or nucleation centre for nanoparticles; (2) interfaces between two solids (solid–solid) can create different materials having core–shell or tape like structures; and (3) interfaces between an organic and inorganic layer might create various monodispersed colloidal nanocrystals or nanocomposite materials (Wang, Peng et al. 2007).

We will include some examples of different inorganic materials formed via interface-mediated growth following different interface interactions. The preparation of hollow spheres of ZnSe is assisted by an interface developed between gas and a liquid, systematically shown in Figure 10.11.

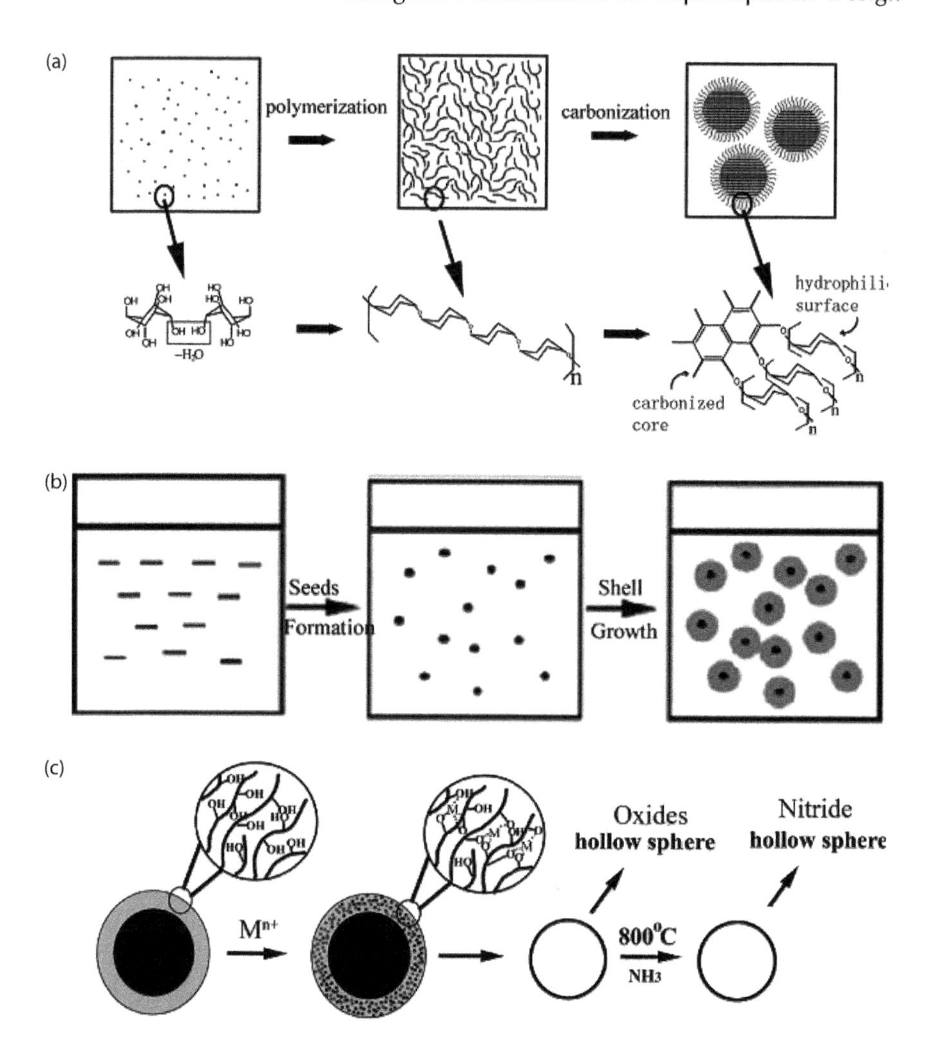

FIGURE 10.16 (a) Schematic growth model for carbonaceous spheres. (b) Schematic illustration of the formation of Ag@C core–shell structured nanospheres. (c) Schematic mechanism for the formation of GaN hollow spheres using carbon spheres as templates. (Reproduced with permission from Wang, X. et al., *Acc. Chem. Res.*, 40, 635–643, 2007. Copyright 2007 American Chemical Society.)

The second route involves the growth of a nanocrystal via formation of an interface between two solids. An ensemble of hollow spheres of metal oxides and core–shell structures of metal carbide can be obtained following this mechanism. Figure 10.16 shows the formation mechanism of hollow spheres.

The third pathway proceeds via the liquid–solid–solution phase transfer. A variety of ultrathin nanostructures can be obtained by this strategy. Figure 10.17 reveals the different nanostructures of various inorganic materials obtained via organic additives assisted solvothermal method and their growth accompanied via liquid–solid–solution phase assisted pathway (Wang, Zhuang et al. 2005).

FIGURE 10.17 TEM images of nanocrystals with nearly round shapes synthesized via the LSS strategy: (a) Ag (b) Au, (c) Rh, and (d) Ir. TEM images of nanocrystals with different shapes synthesized via the LSS strategy: (e) $NaYF_4$ nanoparticles, (f) $NaYF_4$ nanorods, (g) $LaVO_4$ nanocrystals with a square shape, and (h) YPO_4. 0.8 H_2O nanocrystals with a hexagonal shape. (Reproduced with permission from Wang, X. et al., *Acc. Chem. Res.*, 40, 635–643, 2007. Copyright 2007 American Chemical Society.)

The manipulation of a nanoobjects is still an issue that needs to be addressed in the field of nanoscience. In order to fully address this issue, scientists must need to devote their efforts to understand the chemistry of interfaces and their effect on structure determination.

10.4.1 Instrumentation for Hydrothermal/Solvothermal Synthesis

The basic equipment of solvothermal and hydrothermal synthetic experiments is a high-pressure container, commonly known as an autoclave. Advances in these syntheses depends mainly on the availability of equipment. Hydrothermal/Solvothermal experimentation need such facilities which can be availed routinely and consistently. An ideal autoclave should have following characteristics;

1. An excellent acid, alkali, and oxidant resistance
2. Mechanically strong enough to withhold extreme conditions of temperature and pressure
3. Structurally simple and convenient to use and operate
4. An appropriate size to attain a required heat gradient
5. Best sealing performance to maintain the desired pressure and temperature

The above requirements of an autoclave can be meet through internal lining of the container with an inert material. The most frequently used material for linings is Teflon.

Figure 10.18 displays the most frequently used Teflon-lined, stainless steel autoclave. Such autoclaves are generally used to conduct mild hydrothermal/solvothermal reactions at up to 270°C.

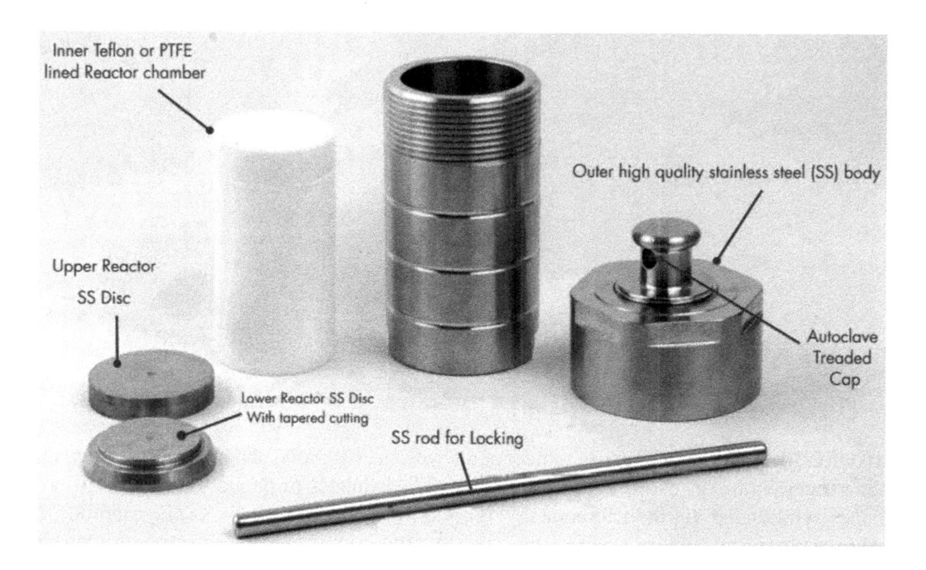

FIGURE 10.18 Generally used reactor (autoclave) during hydrothermal/solvothermal synthesis.

10.5 SOL GEL SYNTHESIS

The sol–gel routes to inorganic nanostructures relied on principle of polymerization of precursors of molecular nature like metal alkoxides. M. Ebelmen, a French chemist, carried out the first sol gel synthesis of silica in 1846. He noticed that hydrated silica is obtained by the moisture induced slow hydrolysis of silicic esters. Schott Glaswerke used this method to produce coatings over the glasses and brought this technique towards the industry in the half of twentieth century. However, its science developed in the late times. Since 1981, sol–gel chemistry get a new boom through publications of thousands of research papers and has provided new opportunities to the scientists working in this field (Ciriminna, Fidalgo et al. 2013).

Three different approaches can be used to obtain different materials via sol gel route:

1. Formation of gel from colloidal powders solution;
2. Hydrolysis of metal salt like alkoxide or nitrate followed by polycondensation. The obtained gel was finally dried at hypercritical temperature in order to get final target material;
3. Hydrolysis and polycondensation of metal precursors having alkoxide counterion followed by aging and drying at ambient temperature and pressure (Hench and West 1990).

Figure 10.19 shows brief overview of this synthetic route.

The under-discussion route has been very successful to get an insight about preparation of oxide based inorganic materials. These materials can be obtained through

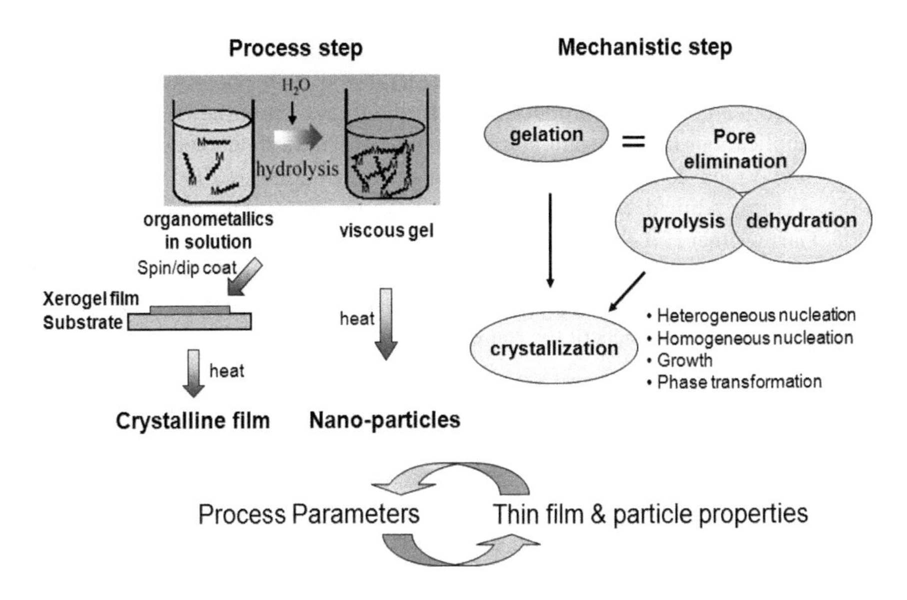

FIGURE 10.19 Sol gel synthesis; an overview.

simple hydrolysis and polycondensation of precursors via oxo bridges formation. This synthesis route has many advantages among which the most prominent are; (i) product purity and control over composition, (ii) control over texture, (iii) control over structure and homogeneity of the product. The sole adaptability of this synthetic approach is because of many tunable parameters, such as precursor`s nature and concentration, catalysis, solvent, water content, temperature, drying and aging environment. However, the concurrent control over composition, texture, and homogeneity in case of composite oxides cannot be achieved through straightforward ways and mostly needs complex, time-taking or costly operations. The primary issue encountered in such synthesis is the discrepancy in the rate of reaction. Certainly, in multicomponent systems the rate of hydrolysis and condensation of different raw materials directly affect the homogeneity of the system. Hence a homogeneous gel of the product can only be obtained when the rate of reaction of different starting materials is harmonized well with each other. This can be achieved by various strategies, like modifications of different reactive precursors. The modification can be made simply by interacting reactive precursor with acetic acid or acetylacetone to control the rate of hydrolysis. Another major issue is the crumple of the network formed by pores upon drying through evaporation because of capillary actions. This problem might be handled through occurrent hypercritical drying. One more strategy is to use templates to get organized porous materials. Multi-scale texturation can be obtained by a combined sol gel and aerosol processing (Debecker and Mutin 2012).

During last decades, numerous substitute to sol–gel strategies were anticipated, that involves the origination of the non-aqueous oxo bridges and the term "non-hydrolytic sol–gel" (NHSG) was introduced (Corriu et al. 1992a, 1992b, 1992c). The other oxygen sources may be either alkoxides, ethers, or alcohols.

The NHSG process is non-aqueous and its mechanism involves entirely diverse condensation process. These conditions considerably affect the homogeneity, texture, and surface features of the products. Current literature about the NHSG synthesis and its applications has been summarized in many reviews about oxide (Vioux 1997, Mutin and Vioux 2009), oxide nanoparticles (Jun, Choi et al. 2006, Niederberger 2007, Akram, Ahmad et al. 2018), and nanostructures (Pinna and Niederberger 2008, Garnweitner and Niederberger 2008).

The major benefit of sol gel synthesis is the capability to molecular level control of the composition and structure. Control of the texture, easy and one step preparation are also precious assets of the route.

10.6 ULTRASONIC NEBULIZER ASSISTED CHEMICAL VAPORS DEPOSITION (UNA-CVD)

Chemical vapors deposition (CVD) is widely used approach to produce pin-hole free inorganic materials as thin films for fabrication of transistors and other technological important functional devices (Jones and Hitchman 2009). Ultrasonic nebulizer assisted CVD is an improved modified CVD approach, in which aerosol is generated and used as medium to transport precursor/s in heated zone, where pyrolysis of precursor takes place. The use of unreactive environment by means of nitrogen gas, or argon for precursor decomposition offers superior way to produce diverse materials including metal-chalcogenides, metal oxides, zeolites, metal-organic frame work (MOF) and 2D materials with simple to complex geometries (Akhtar, Malik et al. 2010, Afzaal, Malik et al. 2010).

The use of ultrasonic nebulizers in UNA-CVD is an energy efficient, low-cost and greener method to produce "aerosol." The as- formed aerosols usually have low velocity and short life. In a typical approach, high frequency ultrasound waves are generated and passed through a precursor solution (precursor + organic solvent). These high frequency sound waves agitate "precursor solution" to yield liquid-gas interface that finally forms an aerosol of micro-sized droplets. We can call these micro-droplets as "microreactors" which when enter inside "hot zone" of furnace reactions take place. The reaction takes place either inside or on surface of droplets. Unlike sonochemical approach, ultrasound waves used in UNA-CVD do not initiate reactions but generate low velocity aerosol (Jones and Hitchman 2009).

10.6.1 Role of Precursors in UNA-CVD

Precursors choice is a key factor to grow thin films of inorganic materials by UNA-CVD. There is wide range of organometallic precursors, single molecular precursors and dual source is available. The choice of precursor is very much depending on solubility in organic volatile solvent. Toluene, acetonitrile, tetrahydrofuran (THF) are most popular choice where most of precursors are soluble. In some cases, a pair of solvents are used (Dunnill, Aiken et al. 2009, Marchand, Hassan et al. 2013, Ozkan, Crick et al. 2016).

Thermal stability of precursors is also key factor to be considered for synthesis of inorganic materials as thin films via UNA-CVD (Afzaal, Malik et al. 2010). Lead(II) xanthates and thiocarbamates precursors are most popular choice due to lower decomposition temperature (Malik and O'brien 2008, Boadi, Malik et al. 2012, Kevin, Malik et al. 2015).

10.6.2 Role of Temperature in UNA-CVD

Temperature influences two ways in UNA-CVD. It assists and provide required energy to decompose the precursor and second it controls the morphology of as-deposited thin films. It also affects texture and uniformity of thin films on the substrates. Thin films of PbS obtained at 350°C from precursors $[PbS_2CNRR')_2]$ (where R, R′ = Me, benzyl, heptyl, octadecyl, dioctyl, Hex, Et or nPr) had a dense granular microstructure whereas at 425°C, the as-deposited films consisted of non-uniform particles. Further increasing the pyrolysis temperature to 450°C resulted in changed the morphology to a mixture of acicular platelets and non-uniform particles (Akhtar, Malik et al. 2010).

10.6.3 Role of Solvent in UNA-CVD

The formation of "aerosol mist" and its evaporation in UNA-CVD set-up is also important parameter to prepare thin films of inorganic materials. The nature of solvents, volatility, boiling point, miscibility with precursor plays crucial role in execution of UNA-CVD as well as in controlling the crystallographic phases of inorganic deposited materials (Edusi, Sankar et al. 2012). For example, using ethanol, hexane, dichloromethane and isopropanol as solvent and titanium(IV) isopropoxide produced exclusively thin films of steel substrates having only anatase phase, while in methanol rutile phased TiO_2 obtained (Edusi, Sankar et al. 2012).

10.6.4 Different Variants of UNA-CVD

- *Electrostatic spray assisted vapor deposition (ESAVD)*
  In this case, vaporization of a liquid precursor is performed first and then charging of droplets is carried out by means of induction charging. These as-charged droplets are then transported towards a heated substrate with the help of an electric field. A much uniform and better-quality films of inorganic materials can be obtained.
- *Electrostatic aerosol assisted jet deposition (EAAJD)*
  The operation of EAAJD is identical to ESAVD, however a carrier gas (argon) is used to transport droplets in electric field and subsequently in hot zone where deposition of thin films takes place. This method also produces good quality thin films of inorganic materials.
- *Electric field assisted aerosol (EFAA) CVD*
  In this case, a potential difference between electrodes on a deposition substrate produces an electric field and remaining procedure is similar to ESAVD.

10.6.5 Advantages of UNA-CVD

- UNA-CVD is one of best cost-effective way to prepare desire thickness of thin films on wide range of substrates.
- Synthesis of inorganic materials takes place at higher temperature in inert atmosphere, so no post synthetic calcination or annealing of material is required.
- There are substantial less chances of impurities in resulting deposited inorganic materials as pyrolysis of precursor under inert conditions occurs at elevated temperature.
- Choice of wide range or precursors and also freedom to prepare multi-phased inorganic materials.
- A number of organic solvents (more volatile to less volatile) can be used.
- The shape of as-deposited inorganic materials can be tuned and controlled by judicial choicc of deposited temperature, solvent, concentration of precursor.
- There is choice to use functionalized substrates to prepare thin films.

10.6.6 Disadvantages of UNA-CVD

A dedicated CVD set-up is required to prepare thin films of inorganic materials, and solvents of high purity grade are essential for growing impurity free thin films.

10.7 COLLOIDAL SYNTHESIS OF INORGANIC MATERIALS

Among solution methods, colloidal synthesis (CS) is most popular technique to prepare inorganic materials in desire shape and size. The ease of simplicity in process, the use of wide range or precursors and shape directing/size controlling surface passivating agents are the key features that make this technique more attractive (Malik, Revaprasadu et al. 2012, Yuwen and Wang 2013, Taylor and Ramasamy 2017). In principle, CS utilizes chemical reduction or thermal decomposition of precursors to generate inorganic materials.

10.7.1 Some Key Concepts in Colloidal Synthesis

10.7.1.1 Nucleation Process

In CS, nucleation is first and crucial step, which involves coalescence of small ions/atoms/molecules to generate "nucleus" (Malik, Revaprasadu et al. 2012). This nucleus provides a site for other particles to agglomerate and form a crystal. The formation of nucleus can be explained by theory developed by Becker et al. using work done by Gibbs. Thus, this theory explains that "nucleation" is thermodynamic entity and follows the route to minimize the Gibbs free energy during the formation of the nucleus. However, this theory doesn't explain the formation of monodisperse inorganic materials during colloidal synthesis.

10.7.1.2 Ostwald Repining

Ostwald repining is an undesirable process in colloidal synthesis of inorganic materials. Accordingly, small crystals formed are exceptionally thermodynamically unstable due to high surface energy and large proportion of surface atoms, whose valency or coordination number is missing. These instantly formed particles and dissolve and redeposit into large particles. This is kinetically controlled and dynamic process happening in solution which ultimately results in decrease in number of small particles and corresponding large particles increase. The bottom line, average particle size increases. Thus, the effective strategy to produce monodisperse inorganic particles is use "sudden burst" (Taylor and Ramasamy 2017).

10.7.1.3 Role of Capping Agents in Colloidal Synthesis

Colloidal synthesis of inorganic materials is mostly performed in high boiling organic solvents like trioctylephosphine oxide (TOPO), hexadecylamine (HDA). These also acts as stabilizing capping agents for inorganic materials and passivate their surface. They also prevent the aggregation of small nanosized particles into clusters. Another advantage of using such solvents is ability to process inorganic materials at higher temperature (200°C–250°C). They also plays role in controlling size and shape of solution processed inorganic materials (Malik, Revaprasadu et al. 2012, Fu 2018).

10.7.1.4 Role of Precursor in Colloidal Synthesis

The choice of precursor plays a vital role in the synthesis of inorganic materials of desire shape and size in colloidal synthesis. In CS, formation of free nuclei and their growth in particles is crucial step which relies on the decomposition of precursor, how quickly and easily it decomposes and provides such monomers. Therefore, the nature of precursor is very important (Yuwen and Wang 2013).

10.7.1.5 Single Molecular Precursors

Single molecular precursors where metal is bonded with ligands are frequently used in colloidal synthesis. In colloidal synthesis, it undergoes decomposition following first order kinetic as follows.

SMP
Monomer

The formation of monomers will depend on the temperature of the reaction solution and the activation energy. The activation energy can be related to binding energy of the ligands with metal. If the binding energy of the ligand with metal is high, higher decomposition temperature is required and vice versa (Akhtar, Malik et al. 2010, Boadi, Malik et al. 2012). $-S_2COR$ (where R = alky chain, aromatic/aliphatic or cyclic) is an important ligand which have been used successfully to synthesize metal sulfides in via colloidal chemical route as low as room temperature (Kevin, Malik et al. 2015).

10.8 TWO STEPS COLLOIDAL SYNTHESIS OF INORGANIC MATERIALS

This method is widely used to prepare metal-chalcogenide inorganic materials. Metal salts (acetate, oxide, nitrates) are mixed and heated with high boiling organic solvents, trioctylphosphine, trioctylephosphine oxide (TOPO) or oleylamine or oleic acid in closed inert atmosphere. Once the desire temperature attained, chalogenide source (sulfur-TOP, Se-TOP, Te-TOP, TMS, TMSe) is added instantly and newly generated metal-chalogenide materials is centrifuged and washed (Hines and Scholes 2003, Abargues, Navarro et al. 2019). Scholes and co-workers used this route to prepare ultra-small size PbS nanoparticles with emission in 1200–1400 for mid-infrared applications (Hines and Scholes 2003). The shape of small PbS prepared in this method observed under TEM after injection appear angular and facet, whoever prolonged heating results in spherical shape of particles. On the other hand, particles obtained larger in size have symmetrical shape.

10.8.1 Advantages of Colloidal Synthesis

- This method provides best way to optimise kinetic and thermodynamic conditions to control shape and size of nanoparticles.
- In this method, surfactants can be employed to grow nanoparticles in desire shape.
- In colloidal synthesis, there is choice of wide range of precursors available (i.e; single molecular precursors to dual source).
- Surface modification of as-prepared inorganic materials can be easily performed in colloidal synthesis.

10.8.2 Disadvantages

Polydispersity is a major drawback of this approach, which can be overcome by suppressing Ostwald's ripening.

Pyrophoric and moisture sensitive precursors (metal-alkyls) need special protection and procedure to be used in colloidal synthesis.

10.9 CONCLUSIONS

The hydrothermal/solvothermal synthetic approaches to inorganic materials features possess process simplicity, mildness, and scalability. The techniques also hold potential of tuning crystal structure, chemical composition, size and morphology of the materials via controllable growth and nucleation by using established chemistry. Sol-gel is another versatile approach to prepare good quality inorganic oxide materials. The use of surfactants and mild temperature conditions make it attractive method to prepare technological important materials like TiO_2.

Ultrasonic nebulizer chemical vapors deposition (UNA-CVD) is a growing low-cost approach to prepare thin films of inorganic materials on a number of substrates.

Such thin films are vital components of functional devices (solar cells, supercapacitors). Lastly, colloidal synthesis is best solution-based approach to synthesize inorganic materials on any scale.

REFERENCES

Abargues, R., et al. (2019). "Enhancing the photocatalytic properties of PbS QD solids: The ligand exchange approach." *Nanoscale* **11**(4): 1978–1987.

Afzaal, M., et al. (2010). "Chemical routes to chalcogenide materials as thin films or particles with critical dimensions with the order of nanometres." *Journal of Materials Chemistry* **20**(20): 4031–4040.

Akhtar, J., et al. (2010). "Controlled synthesis of PbS nanoparticles and the deposition of thin films by aerosol-assisted chemical vapour deposition (AACVD)." *Journal of Materials Chemistry* **20**(29): 6116–6124.

Akram, B., et al. (2018). "Low-temperature solution-phase route to sub-10 nm titanium oxide nanocrystals having super-enhanced photoreactivity." *New Journal of Chemistry* **42**(13): 10947–10952.

Ashoka, S., et al. (2009). "Ethylene glycol assisted hydrothermal synthesis of flower like ZnO architectures." *Materials Letters* **63**(11): 873–876.

Bao, S.-J., et al. (2008). "Biomolecule-assisted synthesis of cobalt sulfide nanowires for application in supercapacitors." *Journal of Power Sources* **180**(1): 676–681.

Boadi, N. O., et al. (2012). "Single source molecular precursor routes to lead chalcogenides." *Dalton Transactions* **41**(35): 10497–10506.

Cao, F., et al. (2009). "3D Fe_3 S_4 flower-like microspheres: High-yield synthesis via a biomolecule-assisted solution approach, their electrical, magnetic and electrochemical hydrogen storage properties." *Dalton Transactions* **42**: 9246–9252.

Cao, M., et al. (2003a). "Selected-control synthesis of PbO_2 and Pb_3O_4 single-crystalline nanorods." *Journal of the American Chemical Society* **125**(17): 4982–4983.

Cao, M., et al. (2003b). "The first fluoride one-dimensional nanostructures: Microemulsion-mediated hydrothermal synthesis of BaF_2 whiskers." *Journal of the American Chemical Society* **125**(37): 11196–11197.

Chen, G.-Y., et al. (2010). "Synthesis and characterization of single-crystal Sb_2S_3 nanotubes via an EDTA-assisted hydrothermal route." *Materials Chemistry and Physics* **123**(1): 236–240.

Chen, L.-Y., et al. (2008). "Self-assembled porous 3D flowerlike β-In_2S_3 structures: Synthesis, characterization, and optical properties." *The Journal of Physical Chemistry C* **112**(11): 4117–4123.

Chen, X., et al. (2002). "Self-assembling vanadium oxide nanotubes by organic molecular templates." *Inorganic Chemistry* **41**(17): 4524–4530.

Chen, X., et al. (2012). "Biomolecule-assisted hydrothermal synthesis of molybdenum disulfide microspheres with nanorods." *Materials Letters* **66**(1): 22–24.

Cheng, H., et al. (2011). "One-pot miniemulsion-mediated route to BiOBr hollow microspheres with highly efficient photocatalytic activity." *Chemistry–A European Journal* **17**(29): 8039–8043.

Ciriminna, R., et al. (2013). "The sol–gel route to advanced silica-based materials and recent applications." *Chemical Reviews* **113**(8): 6592–6620.

Corriu, R. J., et al. (1992a). "Materials chemistry communications: Preparation of monolithic metal oxide gels by a non-hydrolytic sol–gel process." *Journal of Materials Chemistry* **2**(6): 673–674.

Corriu, R. J., et al. (1992b). "Preparation of monolithic gels from silicon halides by a non-hydrolytic sol-gel process." *Journal of Non-Crystalline Solids* **146**: 301–303.

Corriu, R., et al. (1992c). "Preparation of monolithic binary oxide gels by a nonhydrolytic sol-gel process." *Chemistry of Materials* **4**(5): 961–963.

Debecker, D. P. and P. H. Mutin (2012). "Non-hydrolytic sol–gel routes to heterogeneous catalysts." *Chemical Society Reviews* **41**(9): 3624–3650.

Dunnill, C. W., et al. (2009). "White light induced photocatalytic activity of sulfur-doped TiO_2 thin films and their potential for antibacterial application." *Journal of Materials Chemistry* **19**(46): 8747–8754.

Edusi, C., et al. (2012). "The effect of solvent on the phase of titanium dioxide deposited by aerosol-assisted CVD." *Chemical Vapor Deposition* **18**(4–6): 126–132.

Eisenthal, K. B. (1993). "Liquid interfaces." *Accounts of Chemical Research* **26**(12): 636–643.

Fan, L. and R. Guo (2008). "Fabrication of novel $CdIn_2S_4$ hollow spheres via a facile hydrothermal process." *The Journal of Physical Chemistry C* **112**(29): 10700–10706.

Fu, H. (2018). "Environmentally friendly and earth-abundant colloidal chalcogenide nanocrystals for photovoltaic applications." *Journal of Materials Chemistry C* **6**(3): 414–445.

Garnweitner, G. and M. Niederberger (2008). "Organic chemistry in inorganic nanomaterials synthesis." *Journal of Materials Chemistry* **18**(11): 1171–1182.

Gu, M., et al. (2008). "Preparation and photoluminescence of single-crystalline $GdVO_4$: Eu^{3+} nanorods by hydrothermal conversion of Gd $(OH)_3$ nanorods." *Crystal Growth and Design* **8**(4): 1422–1425.

He, D., et al. (2011). "Self-assembled 3D hierarchical clew-like Bi_2WO_6 microspheres: Synthesis, photo-induced charges transfer properties, and photocatalytic activities." *CrystEngComm* **13**(12): 4053–4059.

Hench, L. L. and J. K. West (1990). "The sol-gel process." *Chemical Reviews* **90**(1): 33–72.

Hines, M. A. and G. D. Scholes (2003). "Colloidal PbS nanocrystals with size-tunable near-infrared emission: Observation of post-synthesis self-narrowing of the particle size distribution." *Advanced Materials* **15**(21): 1844–1849.

Hu, J., et al. (2001). "Synthesis and characterization of $CdIn_2S_4$ nanorods by converting CdS nanorods via the hydrothermal route." *Inorganic Chemistry* **40**(13): 3130–3133.

Jiang, J., et al. (2005). "Morphogenesis and crystallization of Bi_2S_3 nanostructures by an ionic liquid-assisted templating route: Synthesis, formation mechanism, and properties." *Chemistry of Materials* **17**(24): 6094–6100.

Jones, A. C. and M. L. Hitchman (2009). "Overview of chemical vapour deposition." *Chemical Vapour Deposition: Precursors, Processes and Applications* 1–36.

Jun, Y. W., et al. (2006). "Shape control of semiconductor and metal oxide nanocrystals through nonhydrolytic colloidal routes." *Angewandte Chemie International Edition* **45**(21): 3414–3439.

Kasuga, T., et al. (1998). "Formation of titanium oxide nanotube." *Langmuir* **14**(12): 3160–3163.

Kevin, P., et al. (2015). "The controlled deposition of $Cu_2(Zn_yFe_{1-y})$ SnS_4, $Cu_2(Zn_yFe_{1-y})$ $SnSe_4$ and Cu_2 (Zn_yFe_{1-y}) Sn (S_xSe_{1-x}) 4 thin films by AACVD: Potential solar cell materials based on earth abundant elements." *Journal of Materials Chemistry C* **3**(22): 5733–5741.

Li, B. and J. He (2008). "Multiple effects of dodecanesulfonate in the crystal growth control and morphosynthesis of layered double hydroxides." *The Journal of Physical Chemistry C* **112**(29): 10909–10917.

Li, B., et al. (2007). "Controllable synthesis of CuS nanostructures from self-assembled precursors with biomolecule assistance." *The Journal of Physical Chemistry C* **111**(33): 12181–12187.

Li, L., et al. (2008). "Synthesis and photoluminescence of hollow microspheres constructed with ZnO nanorods by H_2 bubble templates." *Chemical Physics Letters* **455**(1–3): 93–97.

Li, S., et al. (2006). "Shape-control fabrication and characterization of the airplane-like FeO (OH) and Fe_2O_3 nanostructures." *Crystal Growth & Design* **6**(2): 351–353.

Liang, L., et al. (2004). "Hydrothermal synthesis of prismatic $NaHoF_4$ microtubes and $NaSmF_4$ nanotubes." *Inorganic Chemistry* **43**(5): 1594–1596.
Liu, Q., et al. (2005). "Nanometer-sized nickel hollow spheres." *Advanced Materials* **17**(16): 1995–1999.
Lou, Z., et al. (2011). "One-step synthesis of AgBr microcrystals with different morphologies by ILs-assisted hydrothermal method." *CrystEngComm* **13**(6): 1789–1793.
Lu, Q., et al. (2004). "Biomolecule-assisted synthesis of highly ordered snowflakelike structures of bismuth sulfide nanorods." *Journal of the American Chemical Society* **126**(1): 54–55.
Luo, F., et al. (2005). "Chelating ligand-mediated crystal growth of cerium orthovanadate." *Crystal Growth & Design* **5**(1): 137–142.
Malik, M. A. and P. O'brien (2008). "Basic chemistry of CVD and ALD precursors." *Chemical Vapour Deposition* 207–271.
Malik, M. A., et al. (2012). "Nanomaterials for solar energy." *Nanoscience* **1**:29–59.
Marchand, P., et al. (2013). "Aerosol-assisted delivery of precursors for chemical vapour deposition: Expanding the scope of CVD for materials fabrication." *Dalton Transactions* **42**(26): 9406–9422.
Morgan, D. L., et al. (2008). "Determination of a morphological phase diagram of titania/titanate nanostructures from alkaline hydrothermal treatment of Degussa P25." *Chemistry of Materials* **20**(12): 3800–3802.
Mutin, P. H. and A. Vioux (2009). "Nonhydrolytic processing of oxide-based materials: Simple routes to control homogeneity, morphology, and nanostructure." *Chemistry of Materials* **21**(4): 582–596.
Niederberger, M. (2007). "Nonaqueous sol–gel routes to metal oxide nanoparticles." *Accounts of Chemical Research* **40**(9): 793–800.
Niu, F., et al. (2008). "$La(OH)_3$ hollow nanostructures with trapezohedron morphologies using a new Kirkendall diffusion couple." *The Journal of Physical Chemistry C* **112**(46): 17988–17993.
Ozkan, E., et al. (2016). "Copper-based water repellent and antibacterial coatings by aerosol assisted chemical vapour deposition." *Chemical Science* **7**(8): 5126–5131.
Peng, Q., et al. (2003). "ZnSe semiconductor hollow microspheres." *Angewandte Chemie* **115**(26): 3135–3138.
Penn, R. L. and J. F. Banfield (1999). "Morphology development and crystal growth in nanocrystalline aggregates under hydrothermal conditions: Insights from titania." *Geochimica et Cosmochimica Acta* **63**(10): 1549–1557.
Phuruangrat, A., et al. (2010). "Synthesis of hexagonal WO_3 nanowires by microwave-assisted hydrothermal method and their electrocatalytic activities for hydrogen evolution reaction." *Journal of Materials Chemistry* **20**(9): 1683–1690.
Pinna, N. and M. Niederberger (2008). "Surfactant-free nonaqueous synthesis of metal oxide nanostructures." *Angewandte Chemie International Edition* **47**(29): 5292–5304.
Qin, A.-M., et al. (2005). "Formation of various morphologies of covellite copper sulfide submicron crystals by a hydrothermal method without surfactant." *Crystal Growth & Design* **5**(3): 855–860.
Shi, W., et al. (2013). "Hydrothermal synthetic strategies of inorganic semiconducting nanostructures." *Chemical Society Reviews* **42**(13): 5714–5743.
Song, X. C., et al. (2007). "Synthesis of MnO_2 nanostructures with sea urchin shapes by a sodium dodecyl sulfate-assisted hydrothermal process." *Crystal Growth & Design* **7**(1): 159–162.
Taylor, R. A. and K. Ramasamy (2017). "Colloidal quantum dots solar cells." *Nanoscience* **4**: 142–168.
Titirici, M.-M., et al. (2006). "A generalized synthesis of metal oxide hollow spheres using a hydrothermal approach." *Chemistry of Materials* **18**(16): 3808–3812.

Tong, H., et al. (2007). "Self-assembled ZnS nanostructured spheres: Controllable crystal phase and morphology." *The Journal of Physical Chemistry C* **111**(10): 3893–3900.

Vioux, A. (1997). "Nonhydrolytic sol–gel routes to oxides." *Chemistry of Materials* **9**(11): 2292–2299.

Volanti, D. P., et al. (2010). "Efficient microwave-assisted hydrothermal synthesis of CuO sea urchin-like architectures via a mesoscale self-assembly." *CrystEngComm* **12**(6): 1696–1699.

Wang, J., et al. (2004). "Magnetic-field-induced growth of single-crystalline Fe_3O_4 nanowires." *Advanced Materials* **16**(2): 137–140.

Wang, J., et al. (2009). "Na_2EDTA-assisted hydrothermal synthesis and luminescent properties of YVO_4: Eu^{3+} with different morphologies in a wide pH range." *Materials Science and Engineering: B* **156**(1–3): 42–47.

Wang, M., et al. (2011). "Controlled synthesis of Co_3O_4 nanocubes under external magnetic fields and their magnetic properties." *Dalton Transactions* **40**(3): 597–601.

Wang, X., et al. (2004). "Synthesis of β-FeOOH and α-Fe_2O_3 nanorods and electrochemical properties of β-FeOOH." *Journal of Materials Chemistry* **14**(5): 905–907.

Wang, X., et al. (2005). "A general strategy for nanocrystal synthesis." *Nature* **437**(7055): 121.

Wang, X., et al. (2007). "Interface-mediated growth of monodispersed nanostructures." *Accounts of Chemical Research* **40**(8): 635–643.

Wasserscheid, P. and W. Keim (2000). "Ionic liquids—New "solutions" for transition metal catalysis." *Angewandte Chemie International Edition* **39**(21): 3772–3789.

Yan, R., et al. (2005). "Crystal structures, anisotropic growth, and optical properties: Controlled synthesis of lanthanide orthophosphate one-dimensional nanomaterials." *Chemistry–A European Journal* **11**(7): 2183–2195.

Yang, J. and T. Sasaki (2008). "Synthesis of CoOOH hierarchically hollow spheres by nanorod self-assembly through bubble templating." *Chemistry of Materials* **20**(5): 2049–2056.

Yang, J., et al. (2008). "Self-assembled 3D flowerlike Lu_2O_3 and Lu_2O_3: Ln^3 + (Ln = Eu, Tb, Dy, Pr, Sm, Er, Ho, Tm) microarchitectures: Ethylene glycol-mediated hydrothermal synthesis and luminescent properties." *The Journal of Physical Chemistry C* **112**(33): 12777–12785.

Yang, L. X., et al. (2006). "A facile hydrothermal route to flower-like cobalt hydroxide and oxide." *European Journal of Inorganic Chemistry* **2006**(23): 4787–4792.

Yu, J., et al. (2009). "Hydrothermal synthesis and visible-light photocatalytic activity of novel cage-like ferric oxide hollow spheres." *Crystal Growth and Design* **9**(3): 1474–1480.

Yuwen, L. and L. Wang (2013). Nanoparticles and quantum dots. In: F. Devillanova, W.-W. Du Mont (eds), *Handbook of Chalcogen Chemistry: New Perspectives in Sulfur, Selenium and Tellurium Volume 2*, Cambridge: Royal Society of Chemistry, pp. 232–269.

Zhang, B., et al. (2006). "Biomolecule-assisted synthesis and electrochemical hydrogen storage of Bi_2S_3 flowerlike patterns with well-aligned nanorods." *The Journal of Physical Chemistry B* **110**(18): 8978–8985.

Zhang, D.-F., et al. (2006). "Size-controllable one-dimensional SnO_2 nanocrystals: Synthesis, growth mechanism, and gas sensing property." *Physical Chemistry Chemical Physics* **8**(42): 4874–4880.

Zuo, F., et al. (2008). "l-Cysteine-assisted synthesis of PbS nanocube-based pagoda-like hierarchical architectures." *The Journal of Physical Chemistry C* **112**(8): 2831–2835.

11 Metal-Organic Frameworks Derived Materials for Supercapacitors

E. Heydari-Soureshjani, Ali A. Ensafi, and Ahmad R. Taghipour-Jahromi

CONTENTS

11.1 Introduction 187
11.2 History of MOFs 189
11.3 Pristine MOFs 191
11.3.1 Pristine 3D-MOFs 191
11.3.2 Pristine 2D-MOFs 192
11.3.3 Pristine 0D and 1D-MOFs 193
11.4 MOF-Derived Materials 195
11.4.1 MOF-Derived Synthesis Strategies 195
11.4.2 MOF-Derived Transition Metal Oxide for Supercapacitors 197
11.4.3 MOF-Derived Transition Metal Sulfide for Supercapacitors 199
11.4.4 MOF-Derived Transition Metal Selenide for Supercapacitors 200
11.4.5 MOF-Derived Transition Metal Phosphide/Phosphate for Supercapacitors 201
11.4.6 MOF-Derived Carbon Composite for Supercapacitors 202
11.5 Combining MOFs and MOF-Derived with Conductive Materials for Supercapacitors 203
11.5.1 MOF-Derived Transition Metal Oxide Composites for Supercapacitors 203
11.5.2 MOF-Derived/Carbon Nanomaterials for Supercapacitors 204
11.5.3 MOF-Derived/Conductive Polymers for Supercapacitors 205
11.6 Summary 206
References 207

11.1 INTRODUCTION

Rapid development of human society and moving to modern life have been created a critical societal problem by more consumption of fossil fuel [1]. The rise in economic expansion around the world has led to boosting the productivity of energy-based appliances, which finally results in the high global energy consumption [1–3]. Eventually, the societal problem and environmental pollutions are evacuating the fossil fuels and changing the climate (global warming and melting glaciers) [4,5]. Besides, new demands have emerged in recent years such as demand for green energy sources, portable and wearable electronic devices, miniaturization and flexibilization of devices, hybrid and electric cars and smart grids [1,2,5–7]. In order to realize the sustainable development of human survival and society, the pursuit for renewable, environmentally-friendly energy sources and sustainable storage technologies are developing incessantly and tremendously [4,6,8]. So, the expansion in designing green and viable energy conversion in the fuel cells and storage in the secondary battery and supercapacitors (SCs) technologies became a hot research topic in the twenty-first century [9]. Secondary batteries and high-performance SCs are recognized as imperatively electrochemical energy storage devices. Presently, batteries and electrochemical capacitors (ECs) are being the most used. The comparison between the formal batteries and ECS show that: formal batteries have a higher energy density, but they have a low level of power at the discharge time; ECs have a higher power density than formal batteries, but their energy density are low. Therefore, the efforts for combination of higher energy and power density led to the usage of SCs. SCs superiority to batteries and ECs are the combination of energy density of batteries with a power density of ECs. Scheme 11.1 shows the comparison between SCs and other energy storage devices [10].

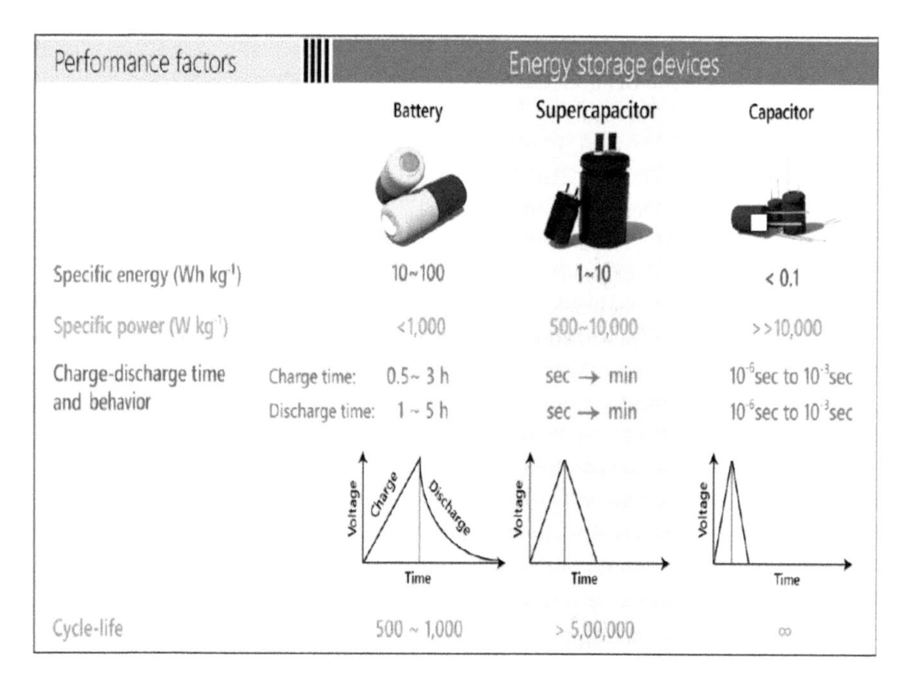

SCHEME 11.1 Comparison of Supercapacitors with other energy to storage devices in terms of different performance factors. Reprinted with permission from (Salunkhe, R.R. et al., *Acc Chem. Res.*, 49, 2796–2806, 2016.). Copyright (2019) American Chemical Society.

SCs offer high power density, fast energy yield, quick charge/discharge time, long cycle life time, low maintenance cost and operation in a wide range of temperature. SCs have capability to store energy during the accumulation of charge or reversible redox reactions. On that basis, SCs are arranged into the three categories rely on storage process: electrochemical double-layer capacitance (EDCL), pseudocapacitors (PCs) and their combination as a hybrid SCs [11]. It is clear that the efficient electrochemical performance of SCs greatly relies on electrode materials. Therefore, development of porous structure with high surface area and numerous electrochemical active sites have been attracted intense attention [12]. In the last two decades, the metal-organic frameworks (MOFs) are emerging in the electrochemical system for SCs. MOFs materials have indicated great advantages such as high surface area, controllable structures, and good thermal stability, high and tunable porosity [9,13,14]. The organic ligands in MOF structures (which include carbon groups and heteroatoms such as S, O, P and N) not only prevent the accumulation nanoparticles, but also form the uniform metal/cluster nanoparticles network [3,12]. Therefore, MOF structures attract immense interest for using as a template and/or precursor to prepare various carbon-based nanomaterials or transition metal-based functional nanomaterials (for instance: oxides, carbides, chalcogenides, sulfides, selenides, and phosphides) [15–17]. Development of derivative form of MOF structures shows the remarkable electrical performance as electrode materials for SCs application [3,18].

11.2 HISTORY OF MOFs

In the mid-1990s, Yaghi and colleagues used a term of MOFs [19,20]. MOFs are crystalline and porous coordination networks, which are created by connecting metal ions or their clusters (as an inorganic fraction) with linkers (as an organic fraction) [21].

The clusters are designed and used as secondary building units (SBUs). The linkers are organic compounds with union units such as carboxylate, phosphonate, sulfonate, and heterocyclic compounds [22]. According to the availability of large number of metal anions, clusters and organic linkers, thousands of MOFs have been synthesized with different structures, shapes, sizes and porosity (Scheme 11.2) [22,23]. Therefore, scientists' attention in academic and industry have been attracted to the MOFs for application in various fields like catalysis, separation, sensors, drag delivery and energy storage [24,25] (Scheme 11.3). At the same time, with the increasing use of MOFs in various fields, many efforts have been made to improve the MOFs properties, which lead to the formation of MOFs hybrids and derived. These modifications can be accursed before synthesis (via linker design), [22] during the synthesis and after the synthesis of MOFs [26].

This chapter provides a broad overview for synthesis and application of MOFs as electrode materials for SCs in three parts: (1) Pristine MOFs, (2) Converted MOFs, and (3) Combining MOFs with conductive materials.

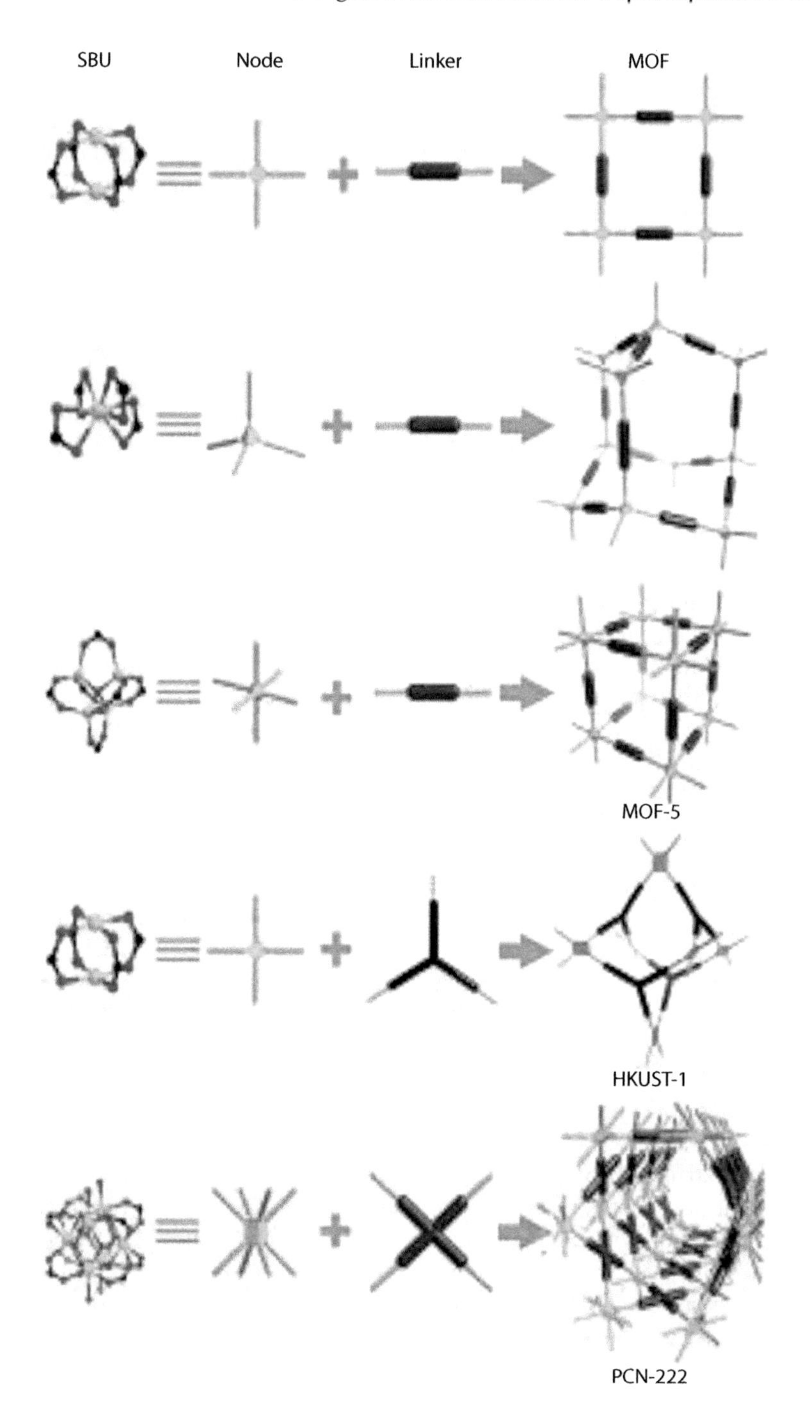

SCHEME 11.2 Graphical illustration of the construction of some representative coordination polymers/MOFs from SBUs and rigid linkers. (From Lu, W. et al., *Chem. Soc. Rev.*, 43, 5561–5593, 2014; Royal Society of Chemistry, reprinted with permission)

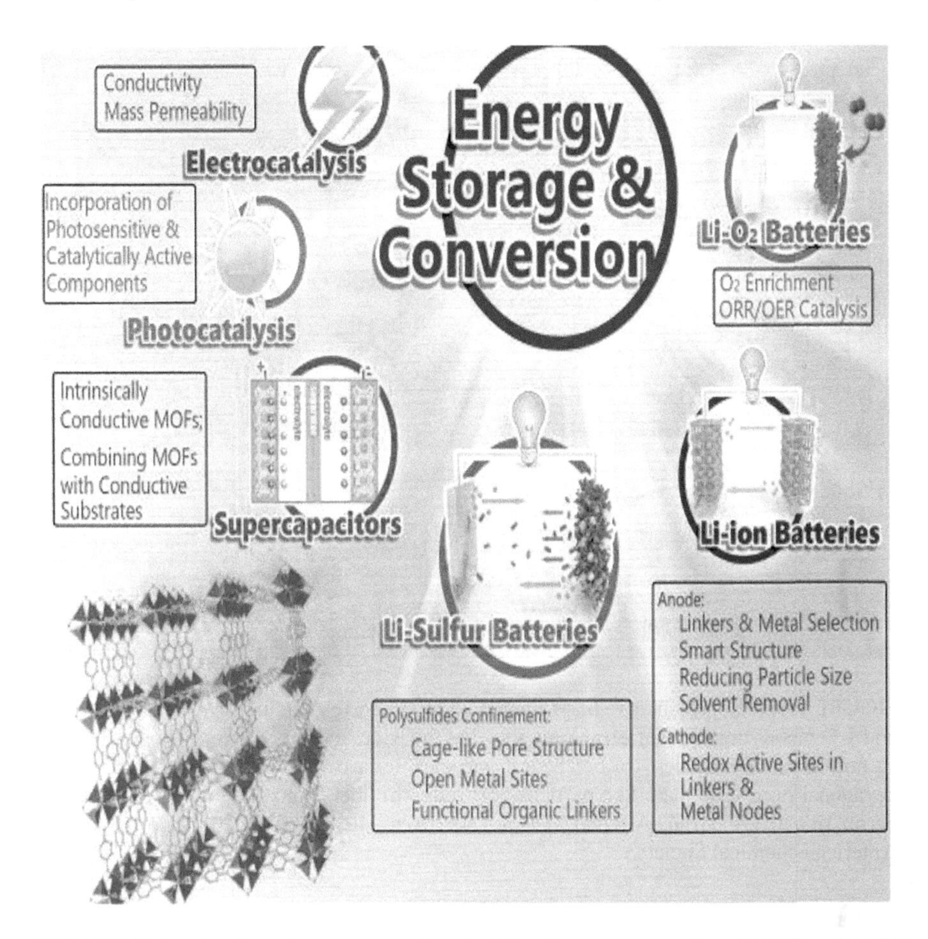

SCHEME 11.3 A schematic summary of typical development strategies of MOFs and MOF composites for energy storage and conversion applications. (From Liang, Z. et al., *Adv. Mater.*, 30, 1702891, 2018.)

11.3 PRISTINE MOFs

In general, pristine MOFs are divided into four categories include 3D (spherical, cubic, etc.), 2D (nanosheets), 1D (nanorods, nanotubes and nanofibers) and 0D (nanoparticles) (Figure 11.1) [27,28]. According to important of electrolyte-accessible surface areas, for achieving to excellent performance of power density and energy storage capability in SCs, MOFs (as a new class of porous materials with large surface areas and incorporated redox metal centers) have been attracted much attention electrode materials for PCs. Therefore, various pristine MOFs, with high surface area and tailored pore size were used as the base of PCs electrode materials.

11.3.1 Pristine 3D-MOFs

3D-MOFs are the first pristine MOFs, which were used as PCs electrode materials. Although 3D-MOFs have a large surface area and EDCL [25], they have some

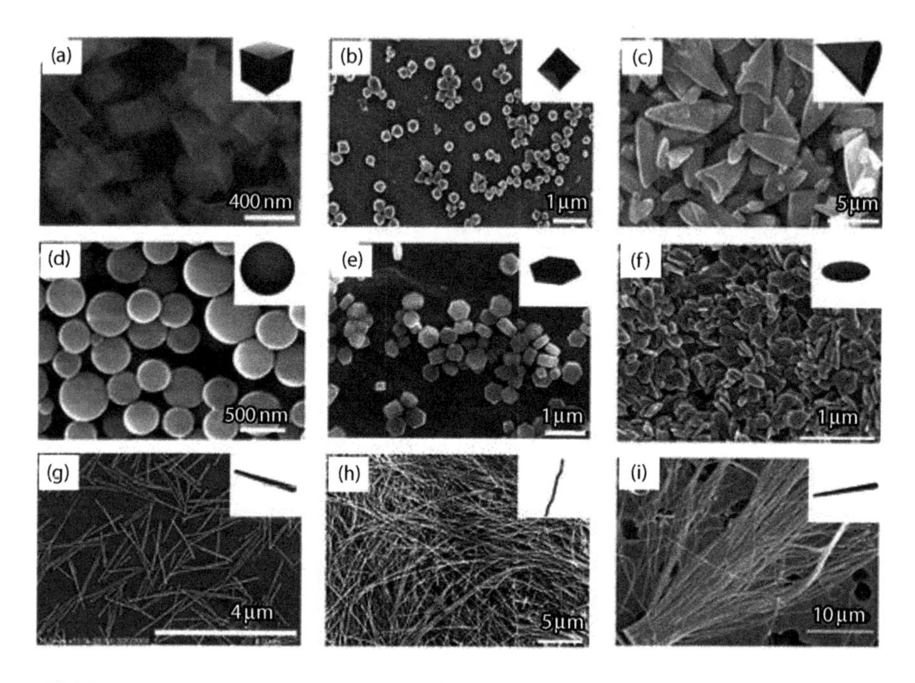

FIGURE 11.1 Representative SEM images and their corresponding schematic figures of 0- and 1-D metal–organic nanostructures showing the wide diversity of morphologies obtained so far: (a) cubic particles, (b) octahedral particles, (c) arrow-like particles, (d) spheres, (e) hexagonal lumps, (f) plate-like particles, (g) rods, (h) fibers, and (i) tubes. (Reprinted with permission from Carne, A. et al., *Chem. Soc. Rev.*, 40, 291–305, 2011. Copyright 2011 American chemical Society.)

limitations in practical applications as electrode materials for PCs. They have not high conductivity [29,30], stability and need activation treatment for expel solvents residual inside the pores or channels and do not show a good specific capacitance [29,31]. For the first time, Díaz et al. was used the Co-Zn MOF (Co8-MOF-5) as an electrode material of a SC in non-aqueous media (tetrabutylammonium hexafluorophosphate in acetonitrile). The obtained specific capacitance was very low due to low conductivity of this pristine 3D-MOF [32]. Concurrently, Lee et al. were investigated the Co-MOF-71 as an electrode material of a SC in different aqueous media (LiOH, KCl, LiCl and KOH). In different electrolyte, the maximum specific capacitance was obtained in LiOH (206.76 Fg^{-1} at 0.6 Ag^{-1}) [33]. Therefore, attempts to increase the surface area and availability of electroactive materials to boost PCs specific capacitance led to the production of low pristine dimensional MOFs including 2D, 1D and 0D.

11.3.2 Pristine 2D-MOFs

In recent year, some strategy has been used for the synthesis of 2D-MOFs introduced in two categories: "top-down" and "bottom-up" [28,31,34]. Top-down method is including mechanical, sonication, chemical and Li-intercalation exfoliation, whereas bottom-up method is including modulated, surfactant-assisted, interfacial

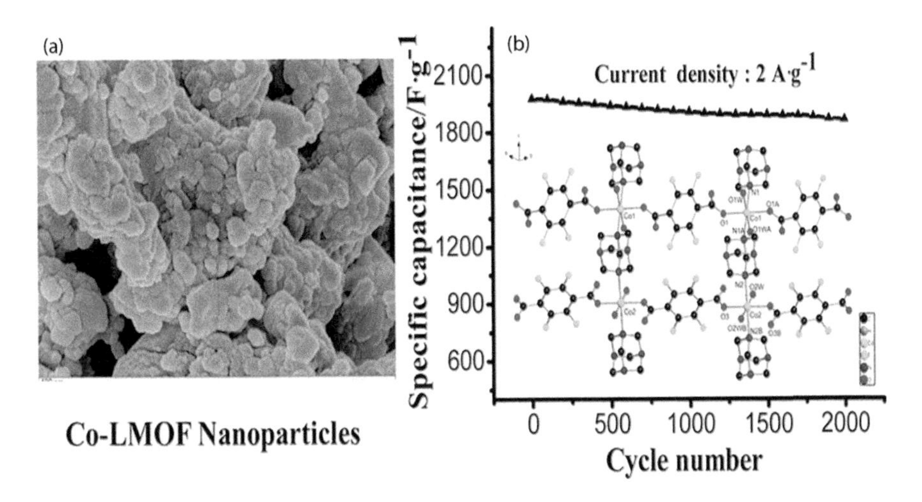

FIGURE 11.2 (a) Representative FE-SEM image and (b) cycle stability of Co-LMOF. Reprinted with permission from (Liu, X. et al., *ACS Appl. Mater Interfaces*, 8, 4585–4591, 2016. Copyright (2019) American Chemical Society.)

synthesis, sonication and three-layer synthesis methods [28,31,34]. However, all synthesis methods of 2D-MOFs include chemical/physical exfoliation of 3D-MOFs or directly synthesized in the presence of surfactant with a high molecular weight (like polyvinylpyrrolidone) [35] for prevention of growth in the third dimension [31]. By moving from 3D-MOFs to 2D-MOFs, the surface area and available active material of PCs electro-materials increase. As a result, the capacitance and power density of PCs increase, too. For instance, Liu et al. was synthesized the Co-based layered MOF (Co-LMOF) as a SC electrode material in 1.0 mol L KOH. According to this report (Figure 11.2), the electrode material display high specific capacitance and cycling durability due to the nanoscale size of Co-LMOF, proper space for diffusion of ions and increase the available active sites of electrode material [36].

11.3.3 Pristine 0D and 1D-MOFs

0D and 1D-MOFs can be synthesized by limiting of MOFs growth on other dimensional using of two strategies: (1) using emulsions or templates; and (2) increasing nucleation (by use of supercritical CO_2 as a solvent, [37] poor solvents, microwave radiation, ultrasounds or temperature) [27]. 0D and 1D-MOFs have more advantages than 2D-MOFs for use in PCs such as (a) They have more active sites and surface area, which reduce the path of mass/charge transfer and increase double layer capacitance, respectively; (b) They can be easily grown on thin films to form self-standing flexible energy storage devices [38,39]. For the first time, the coin-type symmetric SC (Scheme 11.4) was fabricated by Kang and Yaghi et al. [40]. The various organic ligands and metal ions were used to construct 23 different nanocrystals MOFs (nMOFs) with diverse structure backbones. Among them, the zirconium nMOF, $Zr_6O_4(OH)_4(BPYDC)_6$, display highest stack and areal capacitance of 0.64 and 5.09 mF cm^{-2} [40].

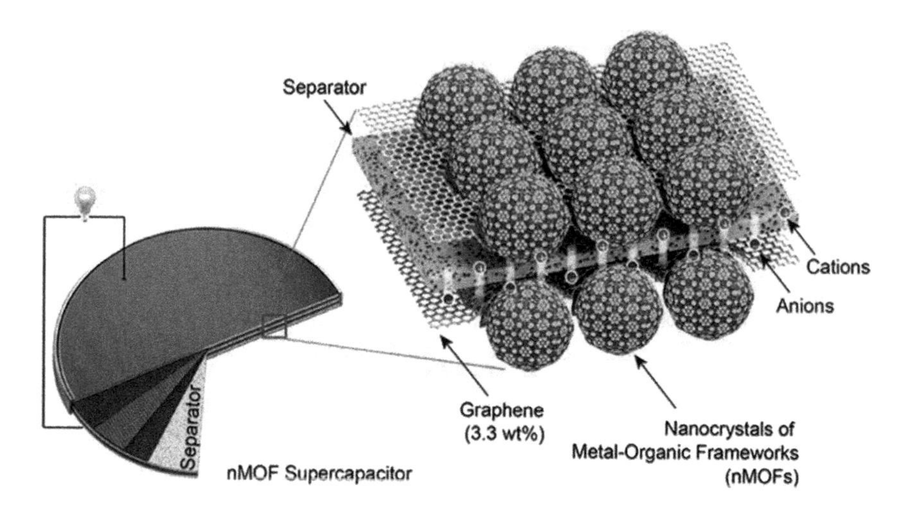

SCHEME 11.4 Construct for nMOF supercapacitors. Reprinted with permission (Choi, K.M. et al., *ACS Nano*, 8, 7451–7457, 2014. Copyright (2019) American Chemical Society.)

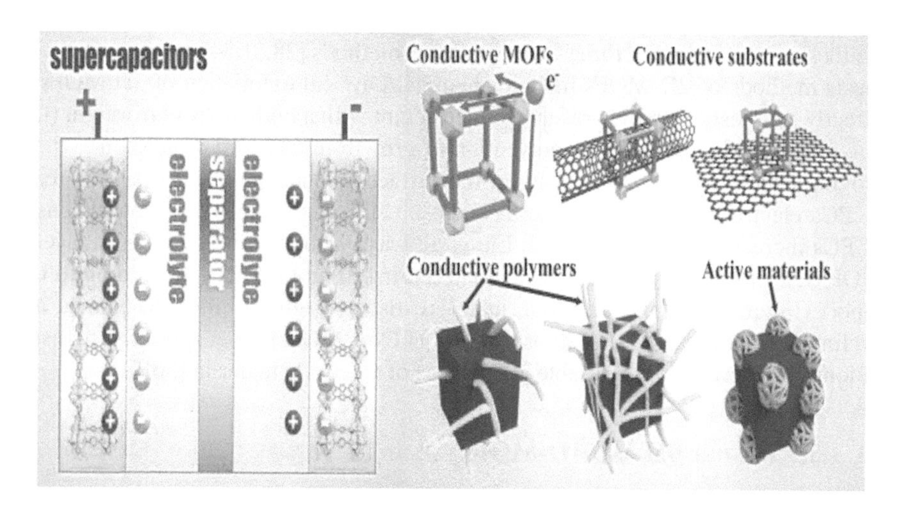

SCHEME 11.5 Schematic illustrations of a typical EDLC supercapacitor and development strategies of MOFs and MOF composites for supercapacitors. (From Liang, Z. et al., *Adv. Mater.*, 30, 1702891, 2018.)

As a result, by reducing the dimensions and size of MOFs as PCs electrode materials, although surface area and accessible activate site for redox reaction and ion/electron transfer increase, but low conductivity and stability of MOFs remind. Therefore, these problems hinder their further development of MOFs as electrode materials for PCs. In order to resolve these problems and use of MOFs as electrode materials for PCs, two feasible routes have been proposed: (1) converting MOFs to porous carbon, metal oxide, metal sulfide or other composite materials via post-treatment process; and (2) combining MOFs with conductive materials (Scheme 11.5).

11.4 MOF-DERIVED MATERIALS

In recent year, converting of MOFs to MOF-derived nanostructures has attracted much attention, due to insufficient electronic conductivity, thermal and chemical durability of most pristine MOFs. Moreover, MOF-derived nanostructures are playing significant roles in energy-related technologies such as SCs. For electrochemical energy production and storage application, MOF-derived nanostructures have several superiorities than pristine MOFs. These are: (1) carbon moiety produced by organic ligands, which improved the electrical conductivity through rapid electron transfer; (2) heteroatom (e.g., S, O, N, P) presented in the organic ligand parts, which can create a more active site and suitable substrate for decoration of various nanoparticles; (3) homogenous dispersion of metal atom or cluster in the carbon network due to in-situ synthesis of MOF structures; (4) proper morphology, size, porosity and structure can be produced by controlling the synthesis strategy; (5) many properties of pristine MOFs can be preserved in the MOF-derived materials (e.g., porosity and large surface area) [41]; (6) porous structures of MOF-derived can not only increase the effective contact of electrode/electrolyte and cause to reduce the diffusion lengths of electrolyte ions, but also improve the redox reaction by increasing the active edge sites [42]. In following, the MOF-derived synthesis strategies and different type and the effect of MOF-derived materials in SCs application were discussed.

11.4.1 MOF-Derived Synthesis Strategies

Synthesis of MOF-derived nanostructures includes two important steps, which are select appropriate MOF as a precursor or template and after that post-treatment of MOF structure [41]. It is worth noting that MOFs with different structures and morphology can be created by changing the synthesis conditions and methods (Scheme 11.6). In the following, the overall synthesis methods are discussed shortly.

To synthesize controllable MOF structures, some procedures have been explored such as solvothermal [13], hydrothermal [13], microwave [43], sonochemical [44], electrochemical [45] and mechanochemical [46] methods. The most straightforward and widely synthesis methods for MOF structures are solvothermal or hydrothermal, in which the reactions are occurring at high temperature or pressure in any solvent or water, respectively [47]. Due to applying high pressure in solvothermal process, the product has a higher yield and greater crystallinity [47,48]. Nevertheless, these synthesized methods typically require special equipment (e.g., autoclaves or sealed containers) and long reaction times (several hours or month) [20].

Microwave method for the first time was used in 2005 [47,49]. In the microwave-assisted method, a solution mixture in a proper solvent is transferred to a sealed container, put in a microwave device and heated for specific time and temperature. Oscillating electric field and permanent dipole moment of the molecules are coupled and lead to rapid liquid phase heating [50,51]. In the microwave-assisted method, large-scale synthesis of MOFs can carry out in a few minutes. Therefore, it is possible to decrease the time of the reaction. The crystal size of MOF structures can control by adjusting the temperature and concentration of the reagent in the reaction [52].

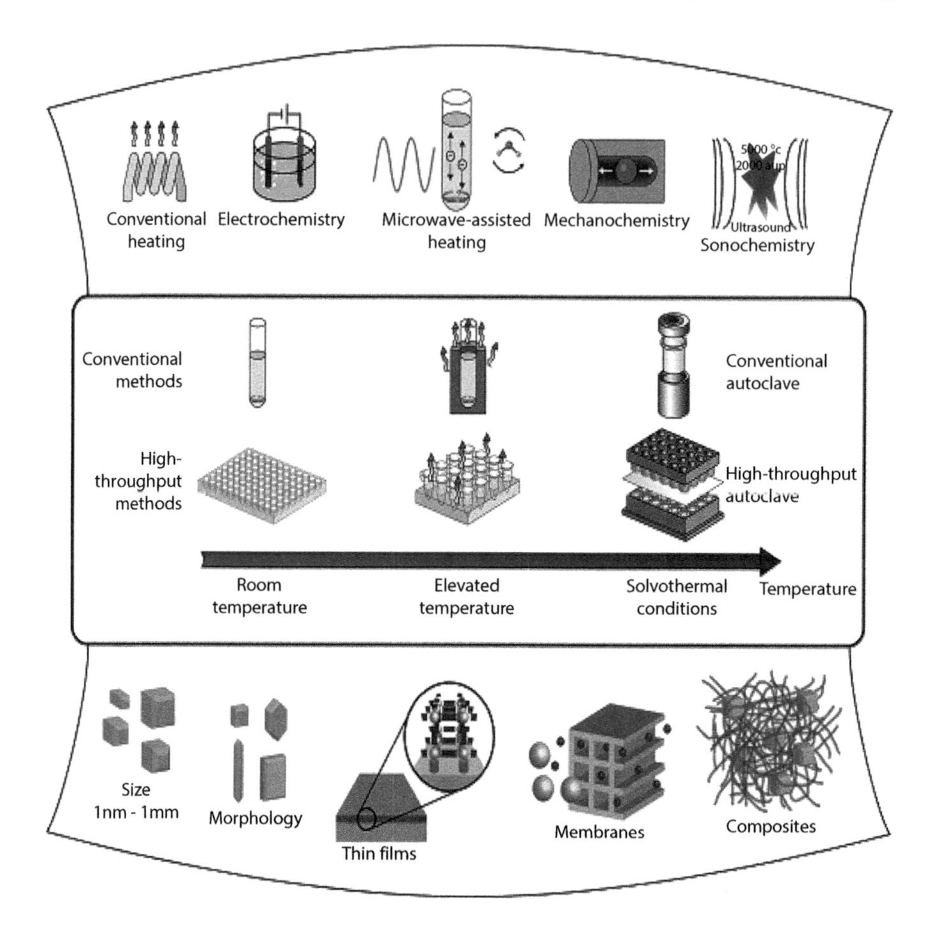

SCHEME 11.6 Overview of synthesis methods, possible reaction temperatures, and final reaction products in MOF synthesis. Reprinted with permission from (Stock, N. and Biswas, S., *Chem. Rev.*, 112, 933–969, 2011. Copyright (2019) American Chemical Society.)

In sonochemical method, a solution mixture is placed in the horn-type Pyrex container, which is equipped with an adjustable power sonicator bar without external cooling [13]. Growth and collapse of bubbles under the influence of sonication, denominated as acoustic cavitation, create very high positional temperatures (~5000 K) and pressures (~1000 bar) [53,54]. Finally, fine crystals are generated under extremely fast heating and cooling rates (>10^{10} K/s) [55]. In sonochemical method due to homogeneous and rapid nucleation, crystallization time and MOF particles size are decreased than usual solvothermal method [53,56].

Scholars at BASF first introduced electrochemical approach for the synthesis of MOF structures. In this method, metal ions are continuously introduced to the reaction medium (including dissolved linker and electrolyte) during the anodic dissociation via an electrochemical process. Directly deposition of metal on the cathode is prevented by using protic solvent and in this case, H_2 gas is released [13,47,55]. Electrochemical route has a various advantage such as elimination of nitrate, perchlorate, or chloride as a byproduct during large-scale production processes, feasibility to operate consecutive process and to procure a higher solids content compared to the normal batch reactions [13,47,55].

Mechanochemical synthesis of MOF structures was first reported in 2006 [57]. It involves force to mechanical breakage of intramolecular bonds and drives chemical transformation (e.g., by milling in ball mills) [13,47,55]. The interest in mechanochemical synthesis of MOF structures has several reasons. It has an environmentally friendly process so that the reaction takes place on the ambient temperature and solvent-free situation. Moreover, products have a small particle size, which obtains at the short reaction time. On some occasions, metal oxides can replace metal salts, which water produces as a by-product [13,47,55]. In addition, various synthesis strategies have been explored to construct MOF-derived materials such as porous carbons, metals, metal oxides, metal sulfides, metal phosphide/phosphate, a metal selenide and their multicomponent hybrids [58]. According to the reaction mechanism through conversion of MOFs to MOF-derived materials, the synthesis strategy can be classified into four categories: (1) self-pyrolysis in an inert atmosphere; (2) chemical reaction with gases or vapors; (3) chemical reaction with solutions; and (4) post chemical etching [59].

Self-pyrolysis of pristine MOFs under inert atmosphere (e.g., N_2 and Ar) usually cause metal/carbon-based composite through an in-situ carbonization procedure [60]. The final nanostructures typically display porous structure and maintain primary morphology of pristine MOFs. These nanostructures may consist of metal, metal oxide, or metal carbide materials in porous carbon frameworks [59,61].

In the attendance of reactive gas or vapor, solid-gas phase chemical reaction simply transforms the pristine MOF template to various MOF-derived nanostructure [59,62]. For example, hollow structured metal-oxides [59,63] and metal-based composite involving phosphides [59,64], sulfides [59,65] and selenides [59,66] can be easily produced via controlling pristine MOFs degradation at high temperatures and reaction between the pristine MOFs and corresponding gas/vapor, respectively.

Likewise, in the solid-liquid phase, a chemical reaction occurs with pristine MOFs and diverse reactant in a solution, which produced various composition with tunable structure [59,67]. For an instant, ion exchange can happen between the pristine MOF and ions in the solution, if the coordination band in the MOF structure between the metal ion and organic framework is proportionately week.

In addition to the above methods, post-treatment has been developed to modify the MOF-derived compositions or nanostructures. For instance, metal based component removes from self-pyrolysis pristine MOF via post chemical etching by acidic solution [59,68].

11.4.2 MOF-Derived Transition Metal Oxide for Supercapacitors

Electrodes based on transition metal oxide materials (such as RuO_2, MnO_2, TiO_2, Fe_3O_4, V_2O_5, Co_3O_4, NiO and ZnO) are working on the concept of pseudocapacitive mechanism. Due to the pseudocapacitive mechanism, these electrode materials exhibit a higher value of specific capacitance than EDCL. Despite that, these electrode materials do not have enough electrical conductivity, adequate specific surface area and pore size distribution. While by calcination of certain pristine MOFs, MOF derived transition metal oxides have been obtained. MOF-derived transition metal oxides have some specific superiority than other same materials such as high porosity, chemical/thermal durability and available internal surface areas. The ions can easily diffuse into the porous structure. Therefore, the electrochemical activity and charge storage capacities improve. Furthermore, by selecting the suitable pristine MOFs, the various MOF-derived transition metal oxides with favorable structure (such as 1D, 2D and 3D) have been constructed (Scheme 11.7), which affect the energy and power density of SCs [69].

Nevertheless, poor electrochemical conductivity, small potential window and unstable structural durability owing to severe accumulation, are limiting parameters to achieve high theoretical specific capacitance [17,70,71]. In order to solve these problems, carbon contained in the organic ligands can be maintained during the annealing process and/or metal oxide can be integrated with other metal oxide materials or conductive materials [17,70–72]. Integration of metal oxide with another one can be accrued during various methods such as immersion, grinding, stirring, and ultrasonic mixing. However, the uniform distribution phase of two metal oxide can be difficultly obtained via pyrolysis of pristine MOFs, when uncontrollable and physical immersion process was used for synthesis [17]. The uniform distribution of two metal oxide depends on the homogeneity of pristine MOFs [17]. For example, Guan et al. were prepared Co_3O_4 and $NiCo_2O_4$ from 2D Co-MOF as SCs electrode

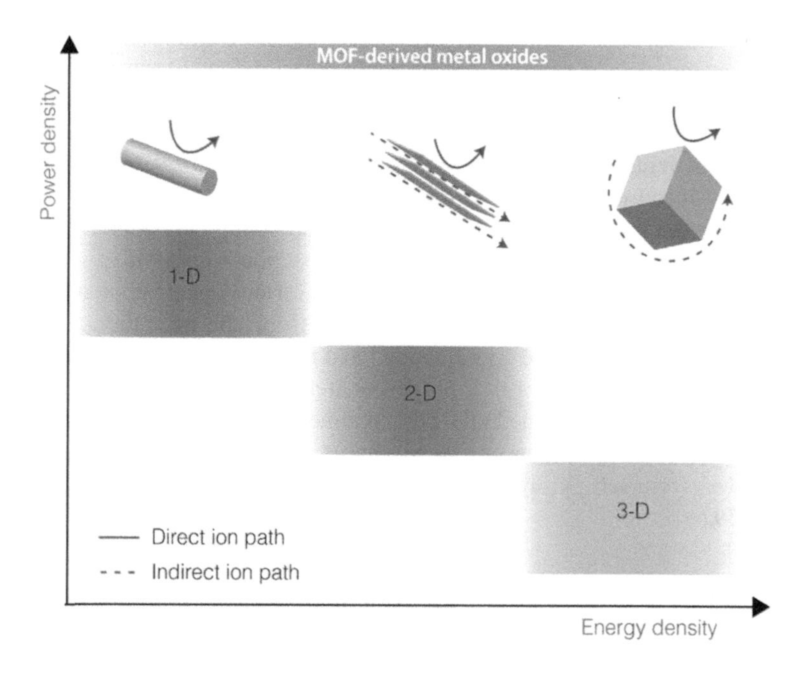

SCHEME 11.7 Schematic representation displaying the flexibility of MOF-derived metal oxide design for obtaining the desired performance in terms of energy and power densities. Reprinted with permission from (Salunkhe, R.R. et al., *ACS Nano*, 11, 5293–5308, 2017. Copyright (2019) American Chemical Society.)

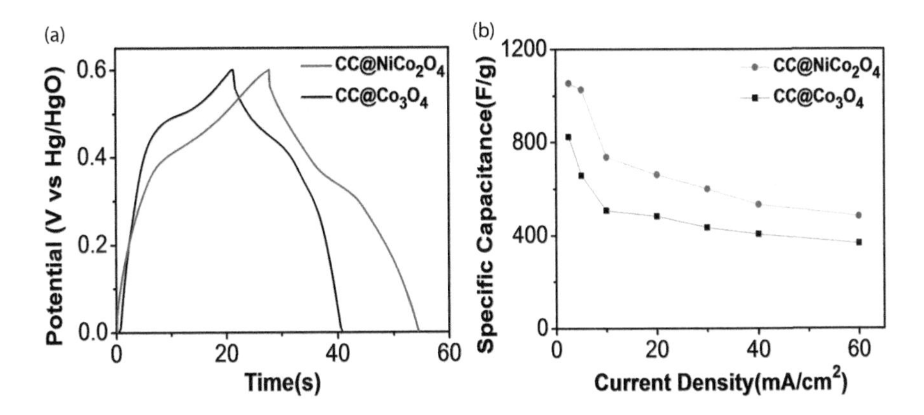

FIGURE 11.3 (a) Comparison of GCD curves of CC@$NiCo_2O_4$ and CC@Co_3O_4 at 10 mA cm^{-2} and (b) The plots of specific capacitances versus current densities for CC@$NiCo_2O_4$ and CC@Co_3O_4. (From Guan, C. et al., *Adv. Energy Mater*, 7, 1602391, 2017.)

materials. The results show that by integration of Ni to Co-MOF, the discharge time and specific capacitance improve at the same current density and potential window (Figure 11.3a and b) [73]. Xia et al. was synthesized 2D MOF nanoflake-assembled spherical microstructures, Ni/Co-MOF, as an electrode material of a supercapacitors. The Ni/Co-MOF nanoflake shown the higher specific capacitance of 530.4 Fg^{-1} than Ni-MOF and ZIF-67 at 0.5 Ag in 1.0 mol L^{-1} LiOH [74].

11.4.3 MOF-Derived Transition Metal Sulfide for Supercapacitors

Recently, MOF-templates sulfidation strategy has been explored to fabricate electrode materials of SCs [75]. These structures have been attracting much attention to energy storage and conversion owning to their crystal architecture, nanocrystalline morphology, multitude stoichiometries compositions and valence states. In addition, they have unique chemical/physical features, high mechanical/thermal stability and better electrochemical conductivity than their corresponding metal oxide counterparts do [75–78]. Their better redox behavior cause to receive high specific capacity/capacitance than electrode materials involving metal oxide and carbon [75–78]. Moving from traditional bulk to nano-architecture metal sulfide derived materials from MOF structures represent many advantages for SCs, including (1) boosts the contact area of electrolyte/electrode per unit mass, which leads to higher specific capacitances; (2) shorter diffusion distance for ions and electrons, which leads to higher power density; (3) good mechanical and structural features created by ion insertion/extraction, which leads to longer cycle life; and (4) appearance of new reactions that aren't feasible for the bulk electrode materials [75–78]. Compare to the MOF-derived monometal sulfides, mixed metal sulfides show a better redox reaction and electrochemical conductivity. Hence, expansion of alternative electrode materials base on MOF-derived metal sulfide are important [76,78,79]. For instance, Cao et al. were used MOF

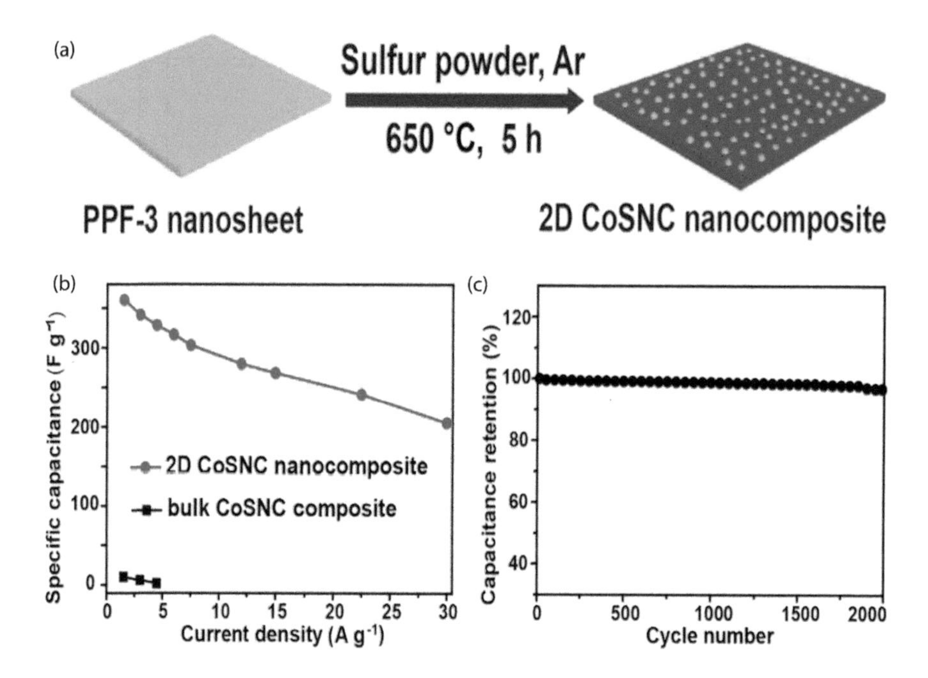

FIGURE 11.4 (a) Schematic illustration for the synthesis process of 2D CoSNC nanocomposites, (b) Specific capacitance of 2D CoSNC nanocomposite and bulk CoSNC composite as a function of current density, and (c) Cycling stability of 2D CoSNC nanocomposite electrode measured at 12.0 Ag^{-1} in 2.0 M KOH. Reprinted with permission from (Cao, F. et al., *J. Am. Chem. Soc.*, 138, 6924–6927, 2016. Copyright (2019) American Chemical Society.)

nanosheets as precursors to synthesize 2D $CoS_{1.097}$/nitrogen-carbon nanocomposite. The nanocomposite was obtained by simultaneous sulfidation and carbonization of MOF nanosheets (Figure 11.4a) and applied as an electrode material for a SC. The nanocomposite shows the high specific capacitance, rate capability and stability (Figure 11.4b and c) [35].

11.4.4 MOF-Derived Transition Metal Selenide for Supercapacitors

MOF-derived transition metal selenides have been developed as good positive electrode materials for SCs application, owing to special geometric architecture, various oxidation state and good electrochemical conductivity [80,81]. Thereby, introducing selenide atoms into the electrode materials enhance the active edge sites and electrochemical activity [82,83]. The selenide has a larger ionic radius and lower electronegativity than oxygen and sulfide [84,85], so it can create a more flexible structure and prevent decomposition during the charge/discharge process [86]. Chen et al. were prepared porous $CoSe_2$ from Co-MOF template with etching and selenization reaction. The bulk and porous form of $CoSe_2$ were examined as a SCs electrode material in 3.0 mol L^{-1} KOH. The porous $CoSe_2$ has a higher discharge time, specific capacitance and rate capability than bulk form (Figure 11.5a–c). Also, the porous

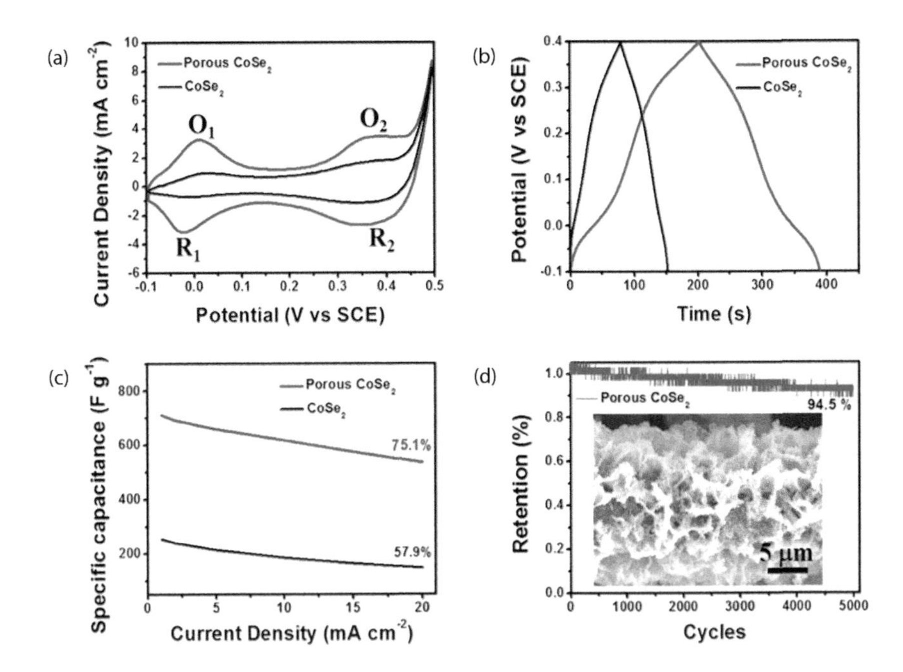

FIGURE 11.5 (a) Comparison of CV curves for porous $CoSe_2$ and $CoSe_2$ at scan rate of 5 mV s^{-1}, (b) GCD curves of as-prepared electrode at current densities of 1 mA cm^{-2}, (c) Specific capacitances of synthesized electrode materials at different current density and (d) Cycling performance of synthesized electrode materials for 5000 cycles at a current density of 5 mA cm^{-2} and corresponding SEM image. Reprinted with permission from (Chen, T. et al., *ACS Appl. Mater Interfaces*, 9, 35927–35935, 2017. Copyright (2019) American Chemical Society.)

structure shows the 94.5% retention in specific capacitance during 5000 cycles at current density of 5 mA cm^{-2} (Figure 11.5d) [82].

Moreover, transition metal selenide materials have some superiority such as higher initial coulomb efficiency than transition metal oxide and more stable cycling life than transition metal sulfide [83]. Nevertheless, transition metal selenide, like transition metal oxide or sulfide, has an intrinsic low electrical conductivity [87]. Therefore, development of hybrid materials, such as bimetallic selenide or composition with conductive materials, show better electrochemical activity than pristine MOF-derived metal selenide [88].

11.4.5 MOF-Derived Transition Metal Phosphide/Phosphate for Supercapacitors

MOF-derived transition metal phosphide/phosphate has been used in an energy storage device [89–91]. Phosphorus is a multivalent non-elemental and donor atom from nitrogen group, which is used in coordination chemistry [90]. Phosphorus can interact with more element of the periodic table and create various metal phosphides/phosphates [90]. These materials have high flexibility, acid-base stability in a pH range of 0–14, different composition with defined stoichiometry, various crystalline structure, abundance and low cost [90,92,93]. Also, phosphidation of MOF-derived materials decrease the electrical resistance, increase the electrocatalytic reactions and enhanced the faradic reaction across the electrode/electrolyte interface. These advantages are owing to the presence of P-P bond in the MOF-derived transition metal phosphide and the existence of a strong bond between the metal and P compared with nitrides and carbides [90,94]. Compare with MOF-derived transition metal oxide/hydroxide, these materials demonstrate more metalloid features owing to the electronic structures and the presence of multi-electron orbitals [95]. Therefore, MOF-derived transition metal phosphides/phosphates have high capacitance and pseudocapacitance features due to large internal surface area, channels and porosity for ion/charge conductivity, and presence of transition metal phosphide, respectively [90]. For the first time, Bendi et al. demonstrated a strategy to develop MOFs-templated hollow porous $Ni_xP_yO_z$ composite as a SC electrode material in 2.0 mol L^{-1} KOH. Figure 11.6a show the capacitance and rate capability of pristine Ni-MOF and MOF-derived $Ni_xP_yO_z$. According to the literature [89], the MOF-derived $Ni_xP_yO_z$ has higher capacitance and rate capability than pristine Ni-MOF. The Nyquist plots (Figure 11.6b) show the smaller semicircle diameter and steeper slope for MOF-derived $Ni_xP_yO_z$ than pristine Ni-MOF, which indicate high conductivity and charge transfer at the electrode/electrolyte interface [89].

11.4.6 MOF-Derived Carbon Composite for Supercapacitors

The intrinsic electrochemical activity of metal nanoparticles is limited due to instability, aggregation and poor reversibility [94]. So, carbon-based nanomaterials can be used as support to protect the nanoparticles. In this regard, MOF structures have been widely used to overcome these limitations [89,94]. Hence, the carbon layers

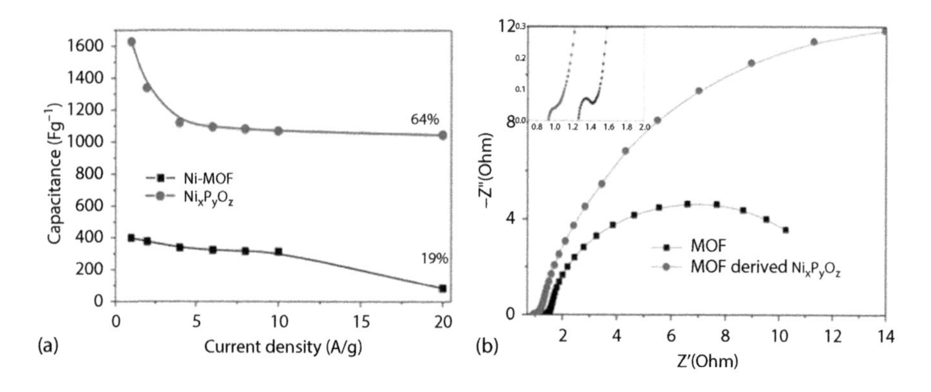

FIGURE 11.6 (a) Specific capacitance of MOF-derived $Ni_xP_yO_z$ electrode at different current densities and (b) The Nyquist plots of pristine MOF and MOF-derived $Ni_xP_yO_z$. (From Bendi, R. et al., *Adv. Energy Mater*, 6, 1501833, 2016.)

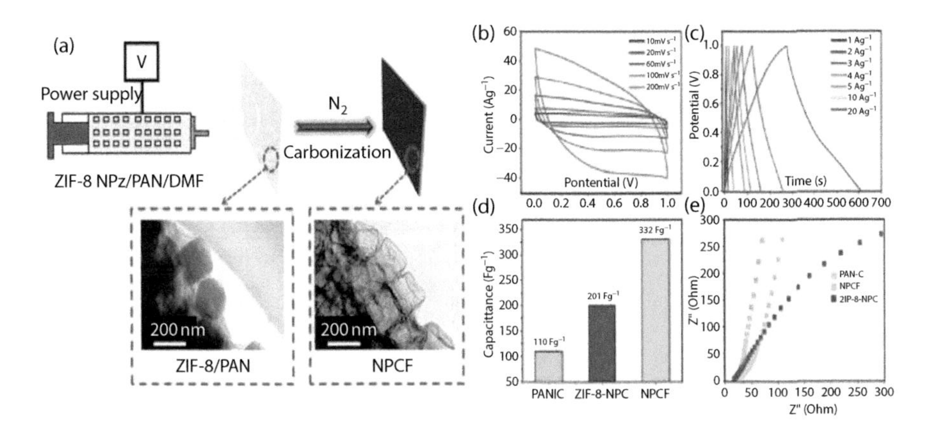

FIGURE 11.7 (a) Preparation process of nanoporous carbon fibers (NPCFs). (b) Cyclic voltammetry (CV) curves and (c) galvanostatic charge–discharge curves of NPCFs. (d) Capacitance and (e) Nyquist electrochemical impedance spectra of PAN-derived carbon (PAN-C), ZIF-8 derived nanoporous carbon (ZIF-8-NPC), and NPCFs. (From Wang, C. et al., *Chem. Commun.*, 53, 1751–1754, 2017; Royal Society of Chemistry, reprinted with permission)

from organic linkers must be maintained in the annealing process. As prepared, metal/porous carbon composites exhibited a greater specific capacitance, electrochemical conductivity and long cycling life due to the existence of porous carbon structures [14]. Porous carbon structure can obtain from different carbon sources and show various morphology and capacitance efficiency. First, the MOF structures are used as carbon sources and direct pyrolysis produced porous carbon structures [12]. Second, introducing of another organic material (e.g., glucose, ethylene glycol, glycerol, polyacrylonitrile (PAN) and furfuryl alcohol as carbon sources) to the porosity of MOF structures before the pyrolysis produce porous carbon structures [12,96]. Figure 11.7 shows a MOF-derived carbon-based electrode material for PCs application [96].

11.5 COMBINING MOFs AND MOF-DERIVED WITH CONDUCTIVE MATERIALS FOR SUPERCAPACITORS

MOF hybrids/composites have been successfully made with conductive/active species to improve the properties of MOFs as PCs electrode materials. These conductive/active species include: (1) metal oxides/hydroxides (nanoparticles/quantum dots/nanorods/nanoflowers); (2) conductive polymers; and (3) carbon nanomaterials (like graphene/carbon nanotubes (CNTs)) [29,97].

11.5.1 MOF-Derived Transition Metal Oxide Composites for Supercapacitors

Integration of converting MOFs or pristine MOFs with any bare of 0D, 1D, 2D and 3D metal oxides/hydroxides compounds show the synergistic effects on capacitance,

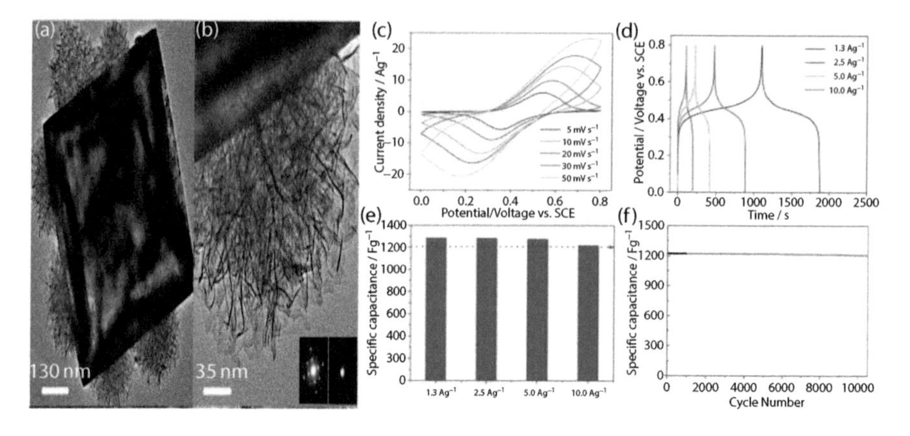

FIGURE 11.8 (a and b) TEM images of MnO_x–MHCF in different magnitude; (c) Cyclic voltammogram curves of MnO_x–MHCF electrode at different scan rates in the range of 5–50 mV s^{-1}; (d) The galvanostatic charge–discharge curves of MnO_x–MHCF electrode at current densities of 1.3–10.0 Ag^{-1} in 1.0 M Na_2SO_4 solution; (e) Specific capacitances of MnO_x MHCF nanocube electrodes derived from the discharging curves at the current density of 1.3–10.0 Ag^{-1} in 1.0 M Na_2SO_4 solution; (f) Cycling performance of the MnO_x–MHCF nanocube electrode measured at the current density of 10.0 Ag^{-1} in 1.0 M Na_2SO_4 solution for 10000 cycles. (From Zhang, Y. et al., *Adv. Mater*, 28, 5242–5248, 2016.)

stability and power density of PCs [3,69,98–100]. For example, Zhang et al. showed that by placing nanoflowers of MnO_2 on a manganese hexacyanoferrate hydrate (MHCF) MOF as a PCs electrode material, the capacitance could be increased up to three times as much than pure MHCF MOF (Figure 11.8) [98]. In addition to pure metal oxides/MOFs composites, hybrid of MOFs with binary metal oxide [69,101] and polyoxometals [17,102] were used for SCs applications because oxides/hydroxides support have multiple redox reactions, increase potential windows and electron transfer between participating cations [69,101].

Although a hybrid of MOFs with metal oxides, mixed metal oxides or hydroxides were shown a good electrochemical performance as PCs electrode materials, but the stability, energy and power densities of them are not very efficient [69]. Therefore, more effort is needed to improve these features and their practical application of MOF metal oxide composites as PCs electrode materials.

11.5.2 MOF-Derived/Carbon Nanomaterials for Supercapacitors

MOF-derived/carbon composites include MOFs, MOF-derived/CNTs (carbon nanotubes), MOF-derived/PC (porous carbon), MOF-derived/GO (graphite or graphene oxide), MOF-derived/CQDs (carbon quantum dots), MOF-derived/CF (carbon fiber) and MOF-derived/fullerene. Three different methods are available for synthesis of MOFs-derived/carbon nanomaterial hybrids, which were named: in-situ synthesis approaches, ex-situ synthesis approaches and other specific approaches [103]. Carbon nanomaterials especially CNTs and reduced graphene oxide (rGO) have a high surface area, conductivity and stability in different solutions. Carbon nanomaterials have been used as an electrode material for EDCLs and increase power density [104]. Therefore, the composites of MOFs and MOF-derived with carbon nanomaterials can theoretically improve various properties of MOFs-derived as PCs electrode materials [103]. For example, recently Xin et al. have synthesized hierarchical Zn-MOF-derived nitrogen-doped porous carbon/graphene composite with 90% retention of initial capacitance after 10,000 charge-discharge cycles (superb cycling stability) and superior rate capability [105].

According to the features mentioned for MOFs and MOFs-derived carbon nanomaterial hybrids and conductive polymers, it can be concluded that for improving the electrochemical properties and stability of MOFs and MOFs-derived as PCs electrode materials, MOFs-derived/nanocarbon materials and conductive polymer hybrids must be used with together [106]. Figure 11.9 display the combination of nanocarbon materials and conductive polymer hybrids with MOF-derived as a PCs electrode material.

11.5.3 MOF-Derived/Conductive Polymers for Supercapacitors

MOFs and MOF-derived, with high porosity and large specific surface area, have good potential in energy storage, but are limited by poor conductivity. One of the ways to reduce these shortcomings is to use MOFs with conductive polymers (CPs) such as polypyrrole (Ppy), polyaniline (PANI), polydopamine (PDA), poly o-aminophenol (POAP) and polyethylene dioxythiophene (PEDOT) [29,107–109]. CPs have good

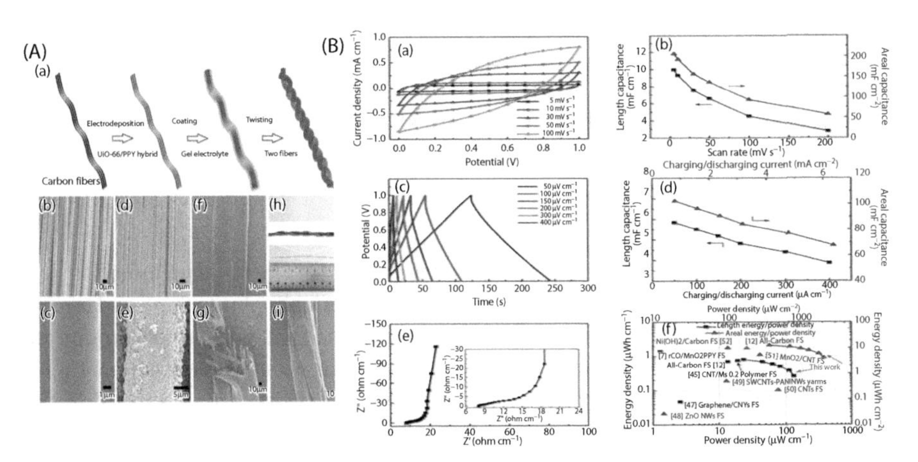

FIGURE 11.9 (A): (a) Illustration of the fabrication of UiO-66/PPY-based flexible fiber supercapacitor device. SEM images of (b, c) carbon fibers, (d, e) UiO-66/PPY-coated carbon fibers, and (f, g) surface and inside of gel electrolyte-coated fiber electrodes. (h) Digital photo and (i) SEM image of the fiber supercapacitor. (B): Electrochemical performance of the fiber supercapacitor device: (a) CV curves and (b) length and areal capacitances at different scan rates. (c) Galvanostatic charging/discharging curves and (d) length and areal capacitances at different current densities. (e) Electrochemical impedance spectroscopy (EIS). (f) Ragone plots with length and areal energy and power densities, compared with some fiber supercapacitors in references. Reprinted with permission from (Qi, K. et al., *ACS Appl. Mater Interfaces*, 10, 18021–18028, 2018. Copyright (2019) American Chemical Society.)

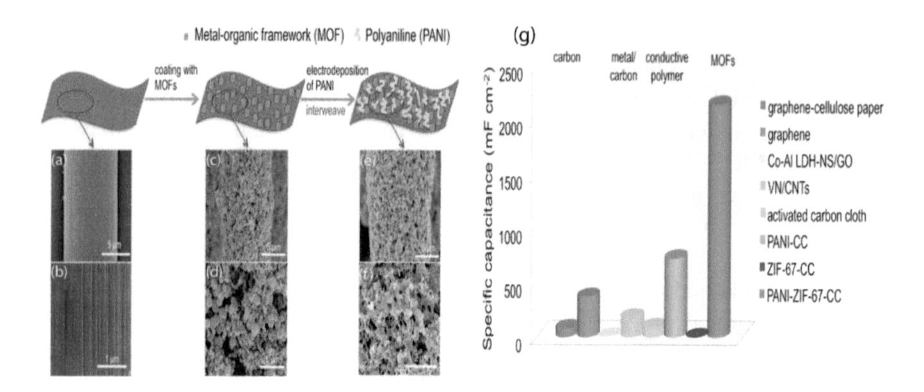

FIGURE 11.10 Schematic illustration of the two-step fabrication process of PANI-ZIF-67-CC electrode and SEM images of (a, b) the carbon cloth fibers, (c, d) after coating with ZIF-67, (e, f) after electropolymerization of aniline, and (g) Areal capacitances of the literature reported supercapacitors and that of PANI-ZIF-67-CC presented in this Communication. The PANI-CC and ZIF-67-CC are control experiments. Co–Al LDH-NS/ GO: Co–Al layered double hydroxide nanosheets (Co–Al LDH-NS) and graphene oxide (GO). VN/CNTs: VN/ carbon nanotubes. Reprinted with permission from (Wang, L. et al., *J. Am. Chem. Soc.*, 137, 4920–4923, 2015. Copyright (2019) American Chemical Society.)

capacitance, base of faradic mechanism, and stability [108]. Therefore, by the conjunction of CPs with pristine or derived MOFs, the electrochemical performance, capacitance, conductivity and durability of MOFs/CPs increase as PCs electrode materials [29,107–109]. Wang et al. were electropolymerized aniline on the Co-MOF crystal (ZIF-67) – carbon cloth (Figure 11.10). The electrode material shows the high areal capacitance of 2146 mF cm^{-2} at 10 mV s^{-1}, which is one of the highest capacitance among the MOF-based SCs [110].

11.6 SUMMARY

MOFs materials have indicated great advantages for SCs electrode materials such as high surface area, controllable structures, and good thermal stability, high and tunable porosity. The organic ligands in MOF structures (which include carbon groups and heteroatoms such as S, O, P and N) not only prevent the accumulation nanoparticles, but also form the uniform metal/cluster nanoparticles network. The pristine MOFs do not show a good stability and specific capacitance as PCs electrode materials because the linkers and metals dissolve during the reduction cycle and electric conductivity of linkers are low, respectively. Therefore, the MOF structures attract immense interest in using as a template and/or precursors to prepare various carbon-based nanomaterials or transition metal-based functional nanomaterials. Therefore, scientists decided to use MOFs derivations as PCs electrode materials to improve their electrochemical and sustainability characteristics. After production of MOFs derivations, efforts have been increased to improve the conductivity, stability, specific capacitance, power and energy density of MOFs (as PCs electrode materials) which lead to synthesis of MOFs derived/nanocarbon materials/conductive polymer hybrids.

REFERENCES

1. Mehtab T, Yasin G, Arif M, Shakeel M, Korai RM, Nadeem M, et al. Metal-organic frameworks for energy storage devices: Batteries and supercapacitors. *J Energy Storage* 2019;21:632–646.
2. Salunkhe RR, Tang J, Kamachi Y, Nakato T, Kim JH, Yamauchi Y. Asymmetric supercapacitors using 3D nanoporous carbon and cobalt oxide electrodes synthesized from a single metal–organic framework. *ACS Nano* 2015;9:6288–6296.
3. Deng X, Li J, Zhu S, He F, He C, Liu E, et al. Metal–organic frameworks-derived honeycomb-like Co_3O_4/three-dimensional graphene networks/Ni foam hybrid as a binder-free electrode for supercapacitors. *J Alloys Compd* 2017;693:16–24.
4. Du W, Bai Y-L, Xu J, Zhao H, Zhang L, Li X, et al. Advanced metal-organic frameworks (MOFs) and their derived electrode materials for supercapacitors. *J Power Sources* 2018;402:281–295.
5. Muzaffar A, Ahamed MB, Deshmukh K, Thirumalai J. A review on recent advances in hybrid supercapacitors: Design, fabrication and applications. *Renew Sustain Energy Rev* 2019;101:123–145.
6. Campagnol N, Romero-Vara R, Deleu W, Stappers L, Binnemans K, De Vos DE, et al. A hybrid supercapacitor based on porous carbon and the metal-organic framework MIL-100 (Fe). *ChemElectroChem* 2014;1:1182–1188.
7. Cao X, Zheng B, Shi W, Yang J, Fan Z, Luo Z, et al. Reduced graphene oxide-wrapped MoO_3 composites prepared by using metal–organic frameworks as precursor for all-solid-state flexible supercapacitors. *Adv Mater* 2015;27:4695–4701.
8. Wang L, Han Y, Feng X, Zhou J, Qi P, Wang B. Metal–organic frameworks for energy storage: Batteries and supercapacitors. *Coord Chem Rev* 2016;307:361–381.
9. Zhou S, Hao C, Wang J, Wang X, Gao H. Metal-organic framework templated synthesis of porous $NiCo_2O_4$/$ZnCo_2O_4$/Co_3O_4 hollow polyhedral nanocages and their enhanced pseudocapacitive properties. *Chem Eng J* 2018;351:74–84.
10. Salunkhe RR, Kaneti YV, Kim J, Kim JH, Yamauchi Y. Nanoarchitectures for metal–organic framework-derived nanoporous carbons toward supercapacitor applications. *Acc Chem Res* 2016;49:2796–2806.
11. Ensafi AA, Heydari-Soureshjani E, Rezaei B. Using $(t\text{-}Bu)_5[PW_{11}CoO_{39}]$ to fabricate a sponge graphene network for energy storage in seawater and acidic solutions. *Electrochim Acta* 2018;289:13–20.
12. Zhang J, Xue J, Li P, Huang S, Feng H, Luo H. Preparation of metal-organic framework-derived porous carbon and study of its supercapacitive performance. *Electrochim Acta* 2018;284:328–335.
13. Lee Y-R, Kim J, Ahn W-S. Synthesis of metal-organic frameworks: A mini review. *Korean J Chem Eng* 2013;30:1667–1680.
14. Zhao Y, Song Z, Li X, Sun Q, Cheng N, Lawes S, et al. Metal organic frameworks for energy storage and conversion. *Energy Storage Mater* 2016;2:35–62.
15. Zhou Y, Mao Z, Wang W, Yang Z, Liu X. In-situ fabrication of graphene oxide hybrid Ni-based metal–organic framework (Ni–MOFs@GO) with ultrahigh capacitance as electrochemical pseudocapacitor materials. *ACS Appl Mater Interfaces* 2016;8:28904–28916.
16. Zhang H, Nai J, Yu L, Lou XWD. Metal-organic-framework-based materials as platforms for renewable energy and environmental applications. *Joule* 2017;1:77–107.
17. Zhang Y, Lin B, Wang J, Han P, Xu T, Sun Y, et al. Polyoxometalates@ metal-organic frameworks derived porous MoO_3@CuO as electrodes for symmetric all-solid-state supercapacitor. *Electrochim Acta* 2016;191:795–804.

18. Du P, Dong Y, Liu C, Wei W, Liu D, Liu P. Fabrication of hierarchical porous nickel based metal-organic framework (Ni-MOF) constructed with nanosheets as novel pseudo-capacitive material for asymmetric supercapacitor. *J Colloid Interface Sci* 2018;518:57–68.
19. Yaghi OM, Li G, Li H. Selective binding and removal of guests in a microporous metal–organic framework. *Nature* 1995;378:703.
20. Yaghi OM, Li H. Hydrothermal synthesis of a metal-organic framework containing large rectangular channels. *J Am Chem Soc* 1995;117:10401–10402.
21. Zhou H-C, Long JR, Yaghi OM. Introduction to metal–organic frameworks. Chem Rev 2012;112(2):673–674.
22. Lu W, Wei Z, Gu Z-Y, Liu T-F, Park J, Park J, et al. Tuning the structure and function of metal–organic frameworks via linker design. *Chem Soc Rev* 2014;43:5561–5593.
23. Kaskel S. Editor, *The Chemistry of Metal-Organic Frameworks*, Synthesis, Characterization, and Applications 2 Volume Set. vol. 1. Hoboken, NJ: John Wiley & Sons; 2016.
24. Rezaei B, Taghipour Jahromi AR, Ensafi AA. Porous magnetic iron-manganese oxide nanocubes derived from metal organic framework deposited on reduced graphene oxide nanoflake as a bi-functional electrocatalyst for hydrogen evolution and oxygen reduction reaction. *Electrochim Acta* 2018;283:1359–1365.
25. Liang Z, Qu C, Guo W, Zou R, Xu Q. Pristine metal–organic frameworks and their composites for energy storage and conversion. *Adv Mater* 2018;30:1702891.
26. Farrusseng D, Canivet J, Quadrelli A. Design of functional metal–organic frameworks by post-synthetic modification. In: Farrusseng D, Editor, *Metal-Organic Frameworks: Applications from Catalysis to Gas Storage*, pp. 23–48. Wiley-VCH Verlag GmbH & Co. doi:10.1002/9783527635856.ch2.
27. Carne A, Carbonell C, Imaz I, Maspoch D. Nanoscale metal–organic materials. *Chem Soc Rev* 2011;40:291–305.
28. Zhao M, Huang Y, Peng Y, Huang Z, Ma Q, Zhang H. Two-dimensional metal–organic framework nanosheets: Synthesis and applications. *Chem Soc Rev* 2018;47:6267–6295.
29. Sundriyal S, Kaur H, Bhardwaj SK, Mishra S, Kim K-H, Deep A. Metal-organic frameworks and their composites as efficient electrodes for supercapacitor applications. *Coord Chem Rev* 2018;369:15–38.
30. Xu Y, Li Q, Xue H, Pang H. Metal-organic frameworks for direct electrochemical applications. *Coord Chem Rev* 2018;376:292–318.
31. Liu J, Yu H, Wang L, Deng Z, Nazir A, Haq F. Two-dimensional metal-organic frameworks nanosheets: Synthesis strategies and applications. *Inorganica Chim Acta* 2018;483:550–564.
32. Díaz R, Orcajo MG, Botas JA, Calleja G, Palma J. Co8-MOF-5 as electrode for supercapacitors. *Mater Lett* 2012;68:126–128.
33. Lee DY, Yoon SJ, Shrestha NK, Lee S-H, Ahn H, Han S-H. Unusual energy storage and charge retention in Co-based metal–organic-frameworks. *Microporous Mesoporous Mater* 2012;153:163–165.
34. Zhao M, Lu Q, Ma Q, Zhang H. Two-dimensional metal–organic framework nanosheets. *Small Methods* 2017;1:1600030.
35. Cao F, Zhao M, Yu Y, Chen B, Huang Y, Yang J, et al. Synthesis of two-dimensional $CoS_{1.097}$/nitrogen-doped carbon nanocomposites using metal–organic framework nanosheets as precursors for supercapacitor application. *J Am Chem Soc* 2016;138:6924–6927.
36. Liu X, Shi C, Zhai C, Cheng M, Liu Q, Wang G. Cobalt-based layered metal–organic framework as an ultrahigh capacity supercapacitor electrode material. *ACS Appl Mater Interfaces* 2016;8:4585–4591.
37. López-Periago A, Vallcorba O, Frontera C, Domingo C, Ayllón JA. Exploring a novel preparation method of 1D metal organic frameworks based on supercritical CO_2. *Dalt Trans* 2015;44:7548–7553.

38. Wang C, Kaneti YV, Bando Y, Lin J, Liu C, Li J, et al. Metal–organic framework-derived one-dimensional porous or hollow carbon-based nanofibers for energy storage and conversion. *Mater Horizons* 2018;5:394–407.
39. Wei Q, Xiong F, Tan S, Huang L, Lan EH, Dunn B, et al. Porous one-dimensional nanomaterials: Design, fabrication and applications in electrochemical energy storage. *Adv Mater* 2017;29:1602300.
40. Choi KM, Jeong HM, Park JH, Zhang Y-B, Kang JK, Yaghi OM. Supercapacitors of nanocrystalline metal–organic frameworks. *ACS Nano* 2014;8:7451–7457.
41. Liu J, Zhu D, Guo C, Vasileff A, Qiao S. Design strategies toward advanced MOF-derived electrocatalysts for energy-conversion reactions. *Adv Energy Mater* 2017;7:1700518.
42. Cao X, Tan C, Sindoro M, Zhang H. Hybrid micro-/nano-structures derived from metal–organic frameworks: Preparation and applications in energy storage and conversion. *Chem Soc Rev* 2017;46:2660–2677.
43. Ni Z, Masel RI. Rapid production of metal–organic frameworks via microwave-assisted solvothermal synthesis. *J Am Chem Soc* 2006;128:12394–12395.
44. Kim J, Yang S-T, Choi SB, Sim J, Kim J, Ahn W-S. Control of catenation in CuTATB-n metal–organic frameworks by sonochemical synthesis and its effect on CO_2 adsorption. *J Mater Chem* 2011;21:3070–3076.
45. Hartmann M, Kunz S, Himsl D, Tangermann O, Ernst S, Wagener A. Adsorptive separation of isobutene and isobutane on $Cu_3(BTC)_2$. *Langmuir* 2008;24:8634–8642.
46. Friščić T, Reid DG, Halasz I, Stein RS, Dinnebier RE, Duer MJ. Ion- and liquid-assisted grinding: Improved mechanochemical synthesis of metal–organic frameworks reveals salt inclusion and anion templating. *Angew Chemie* 2010;122:724–727.
47. Butova VV, Soldatov MA, Guda AA, Lomachenko KA, Lamberti C. Metal-organic frameworks: Structure, properties, methods of synthesis and characterization. *Russ Chem Rev* 2016;85:280.
48. Huang X, Lin Y, Zhang J, Chen X. Ligand-directed strategy for zeolite-type metal–organic frameworks: Zinc (II) imidazolates with unusual zeolitic topologies. *Angew Chemie Int Ed* 2006;45:1557–1559.
49. Jhung SH, Lee JH, Chang JS. Microwave synthesis of a nanoporous hybrid material, chromium trimesate. *Bull Korean Chem Soc* 2005;26:880–881.
50. Kerner R, Palchik O, Gedanken A. Sonochemical and microwave-assisted preparations of PbTe and PbSe. A comparative study. *Chem Mater* 2001;13:1413–1419.
51. Xu Y, Tian Z, Wang S, Hu Y, Wang L, Wang B, et al. Microwave-enhanced ionothermal synthesis of aluminophosphate molecular sieves. *Angew Chemie Int Ed* 2006;45:3965–3970.
52. Chalati T, Horcajada P, Gref R, Couvreur P, Serre C. Optimisation of the synthesis of MOF nanoparticles made of flexible porous iron fumarate MIL-88A. *J Mater Chem* 2011;21:2220–2227.
53. Gedanken A. Using sonochemistry for the fabrication of nanomaterials. *Ultrason Sonochem* 2004;11:47–55.
54. Suslick KS, Hammerton DA, Cline RE. Sonochemical hot spot. *J Am Chem Soc* 1986;108:5641–5642.
55. Stock N, Biswas S. Synthesis of metal-organic frameworks (MOFs): Routes to various MOF topologies, morphologies, and composites. *Chem Rev* 2011;112:933–969.
56. Suslick KS, Choe S-B, Cichowlas AA, Grinstaff MW. Sonochemical synthesis of amorphous iron. *Nature* 1991;353:414.
57. Pichon A, Lazuen-Garay A, James SL. Solvent-free synthesis of a microporous metal–organic framework. *CrystEngComm* 2006;8:211–214.
58. Wang H, Zhu Q-L, Zou R, Xu Q. Metal-organic frameworks for energy applications. *Chem* 2017;2:52–80.

59. Guan BY, Yu XY, Wu HB, Lou XW. Complex nanostructures from materials based on metal–organic frameworks for electrochemical energy storage and conversion. *Adv Mater* 2017;29:1703614.
60. Chen Y, Wang C, Wu Z, Xiong Y, Xu Q, Yu S, et al. From bimetallic metal-organic framework to porous carbon: High surface area and multicomponent active dopants for excellent electrocatalysis. *Adv Mater* 2015;27:5010–5016.
61. You B, Jiang N, Sheng M, Drisdell WS, Yano J, Sun Y. Bimetal–organic framework self-adjusted synthesis of support-free nonprecious electrocatalysts for efficient oxygen reduction. *ACS Catal* 2015;5:7068–7076.
62. Wu R, Qian X, Rui X, Liu H, Yadian B, Zhou K, et al. Zeolitic imidazolate framework 67-derived high symmetric porous Co_3O_4 hollow dodecahedra with highly enhanced lithium storage capability. *Small* 2014;10:1932–1938.
63. Guo W, Sun W, Wang Y. Multilayer CuO@NiO hollow spheres: Microwave-assisted metal–organic-framework derivation and highly reversible structure-matched stepwise lithium storage. *ACS Nano* 2015;9:11462–11471.
64. He P, Yu X, Lou XWD. Carbon-incorporated nickel–cobalt mixed metal phosphide nanoboxes with enhanced electrocatalytic activity for oxygen evolution. *Angew Chemie* 2017;129:3955–3958.
65. Wu R, Wang DP, Rui X, Liu B, Zhou K, Law AWK, et al. In-situ formation of hollow hybrids composed of cobalt sulfides embedded within porous carbon polyhedra/carbon nanotubes for high-performance lithium-ion batteries. *Adv Mater* 2015;27:3038–3044.
66. Hu H, Zhang J, Guan B, Lou XW. Unusual formation of CoSe@carbon nanoboxes, which have an inhomogeneous shell, for efficient lithium storage. *Angew Chemie Int Ed* 2016;55:9514–9518.
67. Hu H, Guan BY, Lou XWD. Construction of complex CoS hollow structures with enhanced electrochemical properties for hybrid supercapacitors. *Chem* 2016;1:102–113.
68. Zhang P, Sun F, Xiang Z, Shen Z, Yun J, Cao D. ZIF-derived in situ nitrogen-doped porous carbons as efficient metal-free electrocatalysts for oxygen reduction reaction. *Energy Environ Sci* 2014;7:442–450.
69. Salunkhe RR, Kaneti Y V, Yamauchi Y. Metal–organic framework-derived nanoporous metal oxides toward supercapacitor applications: Progress and prospects. *ACS Nano* 2017;11:5293–5308.
70. Gholivand MB, Heydari H, Abdolmaleki A, Hosseini H. Nanostructured CuO/PANI composite as supercapacitor electrode material. *Mater Sci Semicond Process* 2015;30:157–161.
71. Guo XL, Li G, Kuang M, Yu L, Zhang YX. Tailoring kirkendall effect of the KCu_7S_4 microwires towards CuO@MnO_2 core-shell nanostructures for supercapacitors. *Electrochim Acta* 2015;174:87–92.
72. Fang G, Zhou J, Liang C, Pan A, Zhang C, Tang Y, et al. MOFs nanosheets derived porous metal oxide-coated three-dimensional substrates for lithium-ion battery applications. *Nano Energy* 2016;26:57–65.
73. Guan C, Liu X, Ren W, Li X, Cheng C, Wang J. Rational design of metal-organic framework derived hollow $NiCo_2O_4$ arrays for flexible supercapacitor and electrocatalysis. *Adv Energy Mater* 2017;7:1602391.
74. Xia H, Zhang J, Yang Z, Guo S, Guo S, Xu Q. 2D MOF nanoflake-assembled spherical microstructures for enhanced supercapacitor and electrocatalysis performances. *Nano-Micro Lett* 2017;9:43.
75. Yu X, Yu L, Lou XW. Metal sulfide hollow nanostructures for electrochemical energy storage. *Adv Energy Mater* 2016;6:1501333.
76. Yuan H, Kong L, Li T, Zhang Q. A review of transition metal chalcogenide/graphene nanocomposites for energy storage and conversion. *Chinese Chem Lett* 2017;28:2180–2194.

77. Rui X, Tan H, Yan Q. Nanostructured metal sulfides for energy storage. *Nanoscale* 2014;6:9889–9924.
78. Yu XY, Lou XW. Mixed metal sulfides for electrochemical energy storage and conversion. *Adv Energy Mater* 2018;8:1701592.
79. Peng H, Ma G, Sun K, Mu J, Wang H, Lei Z. High-performance supercapacitor based on multi-structural CuS@polypyrrole composites prepared by in situ oxidative polymerization. *J Mater Chem A* 2014;2:3303–3307.
80. Kirubasankar B, Palanisamy P, Arunachalam S, Murugadoss V, Angaiah S. 2D $MoSe_2$-$Ni(OH)_2$ nanohybrid as an efficient electrode material with high rate capability for asymmetric supercapacitor applications. *Chem Eng J* 2019;355:881–890.
81. Yang T, Liu Y, Yang D, Deng B, Huang Z, Ling CD, et al. Bimetallic metal-organic frameworks derived Ni-Co-Se@C hierarchical bundle-like nanostructures with high-rate pseudocapacitive lithium ion storage. *Energy Storage Mater* 2019;17:374–384.
82. Chen T, Li S, Wen J, Gui P, Fang G. Metal–organic framework template derived porous $CoSe_2$ nanosheet arrays for energy conversion and storage. *ACS Appl Mater Interfaces* 2017;9:35927–35935.
83. Xu X, Liu J, Liu J, Ouyang L, Hu R, Wang H, et al. A general metal-organic framework (mof)-derived selenidation strategy for in situ carbon-encapsulated metal selenides as high-rate anodes for na-ion batteries. *Adv Funct Mater* 2018;28:1707573.
84. Kirubasankar B, Vijayan S, Angaiah S. Sonochemical synthesis of a 2D–2D MoSe 2/graphene nanohybrid electrode material for asymmetric supercapacitors. *Sustain Energy Fuels* 2019;3:467–477.
85. Kirubasankar B, Murugadoss V, Angaiah S. Hydrothermal assisted in situ growth of CoSe onto graphene nanosheets as a nanohybrid positive electrode for asymmetric supercapacitors. *RSC Adv* 2017;7:5853–5862.
86. Kirubasankar B, Murugadoss V, Lin J, Ding T, Dong M, Liu H, et al. In situ grown nickel selenide on graphene nanohybrid electrodes for high energy density asymmetric supercapacitors. *Nanoscale* 2018;10:20414–20425.
87. Liu X, Zhang J-Z, Huang K-J, Hao P. Net-like molybdenum selenide–acetylene black supported on Ni foam for high-performance supercapacitor electrodes and hydrogen evolution reaction. *Chem Eng J* 2016;302:437–445.
88. Wang X, Li F, Li W, Gao W, Tang Y, Li R. Hollow bimetallic cobalt-based selenide polyhedrons derived from metal–organic framework: An efficient bifunctional electrocatalyst for overall water splitting. *J Mater Chem A* 2017;5:17982–17989.
89. Bendi R, Kumar V, Bhavanasi V, Parida K, Lee PS. Metal organic framework-derived metal phosphates as electrode materials for supercapacitors. *Adv Energy Mater* 2016;6:1501833.
90. Li X, Elshahawy AM, Guan C, Wang J. Metal phosphides and phosphates-based electrodes for electrochemical supercapacitors. *Small* 2017;13:1701530.
91. Mei P, Kim J, Kumar NA, Pramanik M, Kobayashi N, Sugahara Y, et al. Phosphorus-based mesoporous materials for energy storage and conversion. *Joule* 2018;2:2289–2306.
92. Yan L, Jiang H, Wang Y, Li L, Gu X, Dai P, et al. One-step and scalable synthesis of Ni_2P nanocrystals encapsulated in N, P-codoped hierarchically porous carbon matrix using a bipyridine and phosphonate linked nickel metal–organic framework as highly efficient electrocatalysts for overall water splitting. *Electrochim Acta* 2019;297:755–766.
93. Jiao L, Zhou Y-X, Jiang H-L. Metal–organic framework-based CoP/reduced graphene oxide: High-performance bifunctional electrocatalyst for overall water splitting. *Chem Sci* 2016;7:1690–1695.
94. Veeramani V, Matsagar BM, Yamauchi Y, Badjah AY, Naushad M, Habila M, et al. Metal organic framework derived nickel phosphide/graphitic carbon hybrid for electrochemical hydrogen generation reaction. *J Taiwan Inst Chem Eng* 2019;96:634–638.

95. Elshahawy AM, Guan C, Li X, Zhang H, Hu Y, Wu H, et al. Sulfur-doped cobalt phosphide nanotube arrays for highly stable hybrid supercapacitor. *Nano Energy* 2017;39:162–171.
96. Wang C, Liu C, Li J, Sun X, Shen J, Han W, et al. Electrospun metal–organic framework derived hierarchical carbon nanofibers with high performance for supercapacitors. *Chem Commun* 2017;53:1751–1754.
97. Zhu Q-L, Xu Q. Metal–organic framework composites. *Chem Soc Rev* 2014;43:5468–5512.
98. Zhang Y, Cheng T, Wang Y, Lai W, Pang H, Huang W. A simple approach to boost capacitance: Flexible supercapacitors based on manganese oxides@MOFs via chemically induced in situ self-transformation. *Adv Mater* 2016;28:5242–5248.
99. Yao M, Zhao X, Jin L, Zhao F, Zhang J, Dong J, et al. High energy density asymmetric supercapacitors based on MOF-derived nanoporous carbon/manganese dioxide hybrids. *Chem Eng J* 2017;322:582–589.
100. Gao Y, Wu J, Zhang W, Tan Y, Zhao J, Tang B. The electrochemical performance of SnO_2 quantum dots@zeolitic imidazolate frameworks-8 (ZIF-8) composite material for supercapacitors. *Mater Lett* 2014;128:208–211.
101. Chen S, Xue M, Li Y, Pan Y, Zhu L, Zhang D, et al. Porous $ZnCo_2O_4$ nanoparticles derived from a new mixed-metal organic framework for supercapacitors. *Inorg Chem Front* 2015;2:177–183.
102. Tian J, Lin B, Sun Y, Zhang X, Yang H. Porous WO_3@CuO composites derived from polyoxometalates@metal organic frameworks for supercapacitor. *Mater Lett* 2017;206:91–94.
103. Liu X-W, Sun T-J, Hu J-L, Wang S-D. Composites of metal–organic frameworks and carbon-based materials: Preparations, functionalities and applications. *J Mater Chem A* 2016;4:3584–3616.
104. Sevilla M, Mokaya R. Energy storage applications of activated carbons: Supercapacitors and hydrogen storage. *Energy Environ Sci* 2014;7:1250–1280.
105. Xin L, Liu Q, Liu J, Chen R, Li R, Li Z, et al. Hierarchical metal-organic framework derived nitrogen-doped porous carbon/graphene composite for high performance supercapacitors. *Electrochim Acta* 2017;248:215–224.
106. He L, Liu J, Yang L, Song Y, Wang M, Peng D, et al. Copper metal–organic framework-derived CuO_x-coated three-dimensional reduced graphene oxide and polyaniline composite: Excellent candidate free-standing electrodes for high-performance supercapacitors. *Electrochim Acta* 2018;275:133–144.
107. Qi K, Hou R, Zaman S, Qiu Y, Xia BY, Duan H. Construction of metal–organic framework/conductive polymer hybrid for all-solid-state fabric supercapacitor. *ACS Appl Mater Interfaces* 2018;10:18021–18028.
108. Kim J, Lee J, You J, Park M-S, Al Hossain MS, Yamauchi Y, et al. Conductive polymers for next-generation energy storage systems: Recent progress and new functions. *Mater Horizons* 2016;3:517–535.
109. Naseri M, Fotouhi L, Ehsani A, Dehghanpour S. Facile electrosynthesis of nano flower like metal-organic framework and its nanocomposite with conjugated polymer as a novel and hybrid electrode material for highly capacitive pseudocapacitors. *J Colloid Interface Sci* 2016;484:314–319.
110. Wang L, Feng X, Ren L, Piao Q, Zhong J, Wang Y, et al. Flexible solid-state supercapacitor based on a metal–organic framework interwoven by electrochemically-deposited PANI. *J Am Chem Soc* 2015;137:4920–4923.

12 Surface Morphology Induced Inorganic Materials for Supercapacitors

K. Chandra Babu Naidu, D. Baba Basha, S. Ramesh, N. Suresh Kumar, Prasun Banerjee, and K. Srinivas

CONTENTS

12.1 Introduction 213
12.2 Principle and Theory 216
12.3 Inorganic Materials for Supercapacitors 217
  12.3.1 Perovskites 217
  12.3.2 Spinels-Based Composites 218
  12.3.3 Oxides/Hydroxides/Polymer Composites 223
12.4 Summary and Conclusions 231
Acknowledgments 232
References 232

12.1 INTRODUCTION

It was an established fact that the increasing energy demand attracted many scientists to develop various materials for supercapacitor (SC) applications. Indeed, this attention was evolved owing to the admirable properties of supercapacitors viz., high specific power, fast charge or discharge capacity, and long-run life cycles rather than the conventional energy storage devices [1]. However, basically, the materials for SCs can be of two types, such as inorganic and organic. In the literature, several organic materials were reported for SC applications [2–12]. On the other hand, there were ample inorganic materials in bulk as well as nanoform for potential SC applications. In view of this, there were two significant parameters that can contribute to serve as the efficient SC like physical adsorption and desorption of electrolyte ions [1,2]. Using these two parameters, one can estimate the capacity of SC in charging and discharging, respectively. That is, the higher the value of adsorption or desorption of

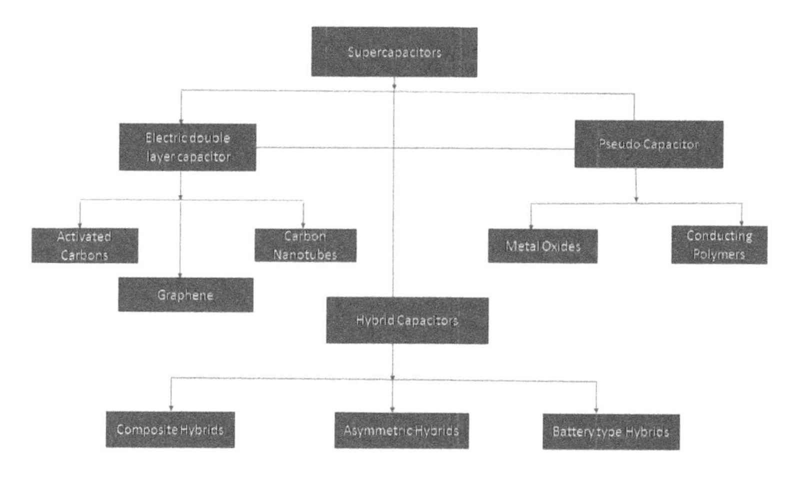

FIGURE 12.1 The classification of supercapacitors.

electrolyte ions, the greater the efficiency of that particular material in charging or discharging within SC. As far as the division of SCs is concerned, it was found that there were three kinds of SCs: (i) electrochemical double-layer capacitor (EDLC), (ii) pseudo capacitor (PC), and (iii) hybrid capacitor (HC) (as shown in Figure 12.1). Herein, the first device can store the energy by means of the accumulation of electrolyte ions at the nanomaterial interface. On the other hand, the latter device can preserve the energy owing to the faradaic redox reaction which can be happened at electrode or electrolyte interface [3]. In particular, it was an evidenced fact that the charge will be stored in EDLC devices depending on the electrostatic interaction, whereas the PC devices will keep the energy to store as a result of the transfer of electron at the electrode or electrolyte interface, which can in turn reveal the electron-transfer mechanism [1]. In case of HC, both electrostatic and electrochemical reactions take place. Furthermore, few intrinsic properties such as high specific surface area, high electrical conductivity, and high thermal conductivity can make the inorganic materials ideal for ultracapacitors (supercapacitors). The comparison among various parameters of normal battery, capacitors, and supercapacitor is shown in Table 12.1.

TABLE 12.1
Comparison of Battery, Fuel Cell, Capacitor, and Supercapacitor

Parameters	Capacitor	Supercapacitor	Battery	Fuel Cell
Charge time (sec)	10^{-6} to 10^{-3}	1 to 30	0.3 to 3	Depends on fuel supply [13]
Discharge time (sec)	10^{-6} to 10^{-3}	1 to 30	1 to 5	Depends on fuel supply [13]
Energy density (Wh/kg)	<0.1	1 to 10	20 to 10^2	>10^2 [13]
Power density (W/kg)	>10^4	10^3 to 2×10^3	50 to 2×10^2	5 to 150 [13]
Cycle life	>5×10^5	>10^5	5×10^2 to 2×10^3	Depends on fuel supply [13]
Efficiency	~1.0	0.90 to 0.95	0.70 to 0.85	0.6 to 0.7 [13]

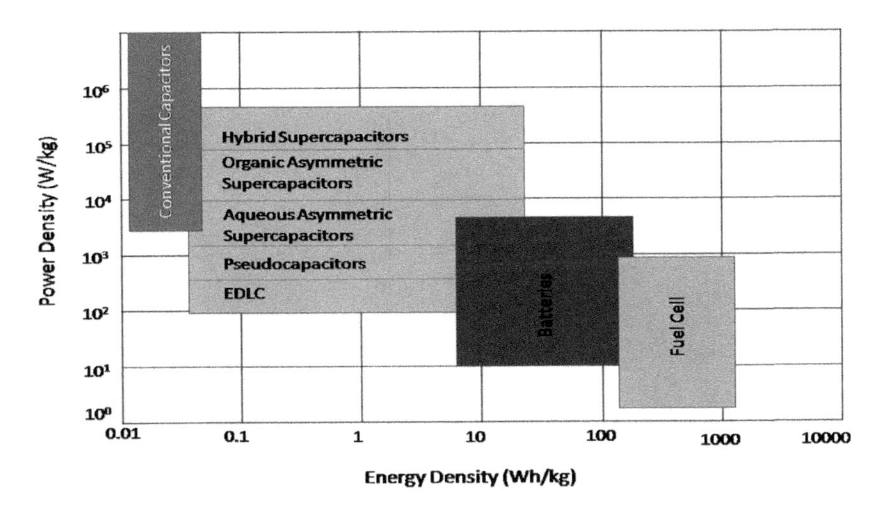

FIGURE 12.2 Ragone plot for various energy storage devices.

TABLE 12.2
Summarization and Comparison of Important Attributes of Supercapacitors

	Supercapacitor		
Attribute	**EDLC**	**Pseudo**	**Hybrid [14,15]**
Charge time (s)	1 to 10	1 to 10	100 [14]
Cell voltage (V)	2.7	2.3 to 2.8	3.6 [14]
Cycle life	10^6	10^5	5×10^5 [14,16]
Specific energy (Wh/kg)	3 to 5	10	180 [14]
Cost per kWh (USD)	~10^4	~10^5 [14]	—
Operating temperature (°C)	-40 to 65	-40 to 65	-40 to 65 [14]
Self discharge per month (%)	60	60 [14]	—
Type of electrolyte	Aprotic/protic	protic	Aprotic [14]

Figure 12.2 describes the pictorial representation of Ragone plot for different energy storage devices. Furthermore, the same comparison was made (shown in Table 12.2) for three kinds of supercapacitors in terms of distinct parameters related to SC applications. The photographs of EDLC SC were shown in Figure 12.3.

FIGURE 12.3 The photographs of EDLC SCs with their usage.

12.2 PRINCIPLE AND THEORY

In general, there are three main parts of the SCs, such as electrodes, electrolyte, and separator. The storage of energy usually depends on the development as well as separation of electric charges accumulated at the interface between electrode and electrolyte surfaces. Herein, the interface is referred as electric double layer interface. In addition, if the normal conductor is introduced into the electrolyte, the charges will be produced continuously. These charges can in turn develop an electric double layer at the interface of electrode and electrolyte (Figure 12.4). Thus, the electric double layer of few nanometers width can work like a physical capacitor with the charge carriers of electrode and electrolyte. Usually, the

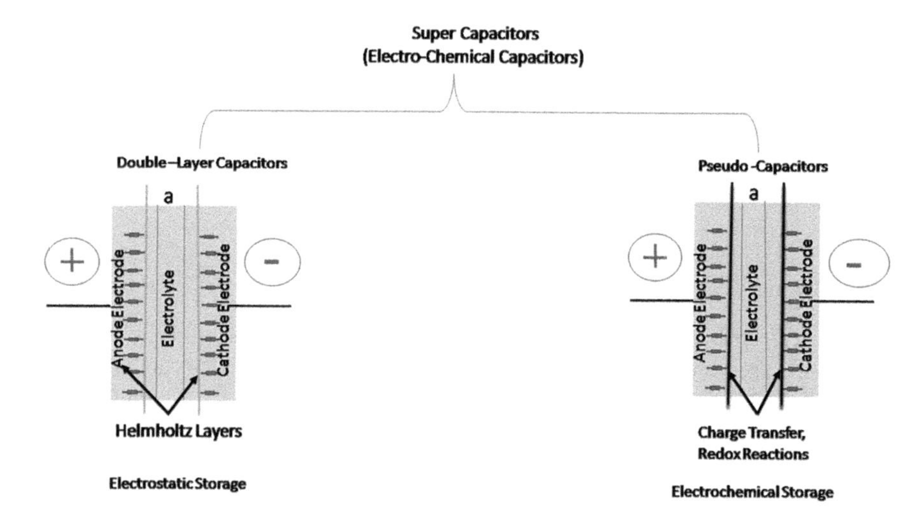

FIGURE 12.4 Internal parts of electrochemical SCs.

evolution and the thickness of electric double layer depend upon the electrode surface structure, type of electrolyte, and the voltage applied between the interface of electrode and electrolyte. Moreover, this electric double layer can be formed and relaxed during the same time interval. Therefore, one can expect a timely response of electric double layer to the input voltage introduced between electrode and electrolyte without taking place of chemical reaction. In addition, the electrode material must possess the large surface area, high stability, long cyclability, and high resistance to electrochemical redox reactions [17]. In order to avoid the short circuit, the separator is provided between the electrodes. Further, the energy density (E) as well as power density (P) of SC can be altered as a function of input voltage between electrode and electrolyte. The mathematical representation for evaluating the E and P parameters can be written as: $E = 0.5\ CV^2$, and $P = V^2/4R_s$, where C is associated to capacitance, V is related to the input voltage, and R_s is referred to series resistance [17].

12.3 INORGANIC MATERIALS FOR SUPERCAPACITORS

It was an established fact that the inorganic materials were classified into different types such as perovskites, spinels, various oxides, hydroxides, inorganic composites, polymer composites, etc. Herein, the perovskites were of ABO_3 structure, where A is a divalent element and B refers to tetravalent element [18]. In the same fashion, the spinel structured materials will have a general chemical formula of AB_2O_4; herein A is associated to divalent metal ion and B is a trivalent metallic element [19]. Moreover, the oxide/hydroxides were mixed to different inorganic materials in order to form the inorganic composites while the same materials can form the polymer composites after adding to polymer precursors. In this chapter, we thoroughly elucidated the efficiency of various parameters of different materials, which can be well suited for SC applications. In particular, the materials of distinct families were described at length using their morphological behavior, change of current density, concentration of the dopant/substituent concentration, synthesis method, etc.

12.3.1 PEROVSKITES

Nan et al. [20] described the synthesis, characterization, and electrochemical performance of various perovskite materials. It was also reported that the surface morphology, synthesis method, and other attributes influenced predominantly the electrochemical performance of these perovskites. For instance, the $La_{0.85}Sr_{0.15}MnO_3$, and $LaMnO_3$ perovskites revealed the flowers like surface morphology as reported in reference [21]. In addition, the high specific capacitance was obtained with increase of potential. As reported by Suresh et al. [22], it was also evidenced that the flower, fiber, and wire-like surface morphology can increase the electrical conductivity, and specific capacitance due to having the more number of grains, and less number of grain boundaries. Latter, these materials were used for even good electrodes in SCs. It was also investigated that if the elements like Ni, Co, and Mn were doped to $LaSrO_3$ perovskites, the huge value of specific capacity (1341 F/g at 0.5 A/g in 6 M KOH electrolyte) was achieved. Apart from this, the energy density was noted to be around

63.5 Wh/kg at power density 900 W/kg pertaining the 200% retention after 10^4 cycles at 20 A/g (current density) [20]. As a whole, it was confirmed that the perovskites can reveal high electrochemical performance as well upon doping/substituting the cations of good conductivity to perovskites. Furthermore, the $La_{0.85}Sr_{0.15}MnO_3$ indicated the current density versus potential curves of various concentrations of Sr-element into the $LaMnO_3$ (LSM) perovskite system [23]. In the same way, $LaMnO_3$ showed the Cole-Cole plots of distinct LSM contents. It was clear that the Z' versus Z'' plots [23] expressed high electrical conductivity through grains rather than the grain boundaries. This was understood from the intersecting position of the first semicircular arc at the Z'-axis as reported in the literature [24–28]. In Table 12.3, we listed different attributes of SCs from the perovskite materials including the compound/composite, synthesis approach, type of morphology (viz., powder, nanofiber, nanorod, nanosheet, nanospheres, nanoflowers, nanowires, nanoporous structure, etc.), specific surface area (m^2/g), specific capacitance (F/g), power density (Wh/kg), and cyclic stability. It was established from the Table 12.3 that the perovskites that exhibited the surface morphology containing the naofibers, naoflowers, nanowires, nanorods, and nanosheets revealed the high specific capacitance as well as high power density.

12.3.2 Spinels-Based Composites

Herein, the spinel materials along with their composites were discussed for electrochemical performance in SCs. In the literatures, specifically, the Co-based inorganic spinel ferrites, and iron oxide materials were studied at larger extent for SC applications. Even in the case of spinel-based composites, the surface morphology played a vital role. For example, recently few electrodes were developed using the NiO nanosheets anchored to $NiCo_2O_4$ carbon cloth (CC) via the hydrothermal method [58]. The obtained results indicated that the scanning electron microscopy (SEM) pictures revealed the 3D-helical CC/$NiCo_2O_4$-N@NiO as well as CC/$NiCo_2O_4$-S@NiO [58]. In addition, some defects (pores) were detected in the interior portions [58]. Therefore, one can predict that the electrolyte can be diffused easily and further it may offer high specific capacitance. As we expected, the similar kind of high values of specific capacitance, and high power densities were observed in Table 12.4. The numerical values of specific capacitance, and power densities were noticed to be 709.1 F/g at 2 mA/cm^2 & 6000 W/kg at 28 Wh/kg (helical CC/$NiCo_2O_4$-N@NiO composites), and 775.1 F/g at 2 mA/cm^2 & 5692 W/kg at 20.1 Wh/kg (helical CC/$NiCo_2O_4$-S@NiO composites) respectively with efficient retention after 10^4 cycles. In over all, it was understood that topography of prepared materials dominated in acquiring the efficient attributed for SC applications.

In the same fashion, the nanospheres as shown in reference [59] were formed in case of the SEM pictures of NiCo-precursor, $NiCo_2O_4$, and nitrogen and graphene-doped $NiCo_2O_4$ composites (prepared using the hydrothermal technique). Especially, the $NiCo_2O_4$ showed apparent nanospheres including few pores. Thus, it could be responsible for attaining the high specific capacitance of order 563.0 Fg^{-1} at current density 1 Ag^{-1}. Moreover, the 90.5% capacitance retention was attributed after 5000 cycles [59]. The transmission electron microscopy (TEM) pictures [59] of the same samples were taken in order to provide good agreement between the SEM and

TABLE 12.3
The Important Attributes of Supercapacitors of Perovskite Materials

Perovskites	Methods	Specific Surface Area (m^2/g)	Specific Capacitance	Power Density (Wh/kg)	Cyclic Stability
$LaMnO_{2.97}$ powder [29]	Reverse-phase hydrolysis	10.6	609.8 Fg^{-1} at 2 mV s^{-1} 1 M KOH as electrolyte	220.4 Wh kg^{-1} at 61.2 W kg^{-1}	No data
$LaMnO_3$ powder [30]	Sol–gel	No data	520 Fg^{-1} at 1 Ag^{-1} 0.5 M Na_2SO_4 as electrolyte cycles at 10 A g^{-1}	52.5 Wh kg^{-1} at 1000 W kg^{-1}	Retains about 117% after 7500
$LaMn_{1.1}O_3$ powder [31]	Sol–gel	35.4–45.5	202.1 mA h g^{-1} at 1 Ag^{-1} 1 M KOH as electrolyte	No data	Retains about 75% after 1000 cycles at 3 Ag^{-1}
$La_{0.85}Sr_{0.15}MnO_{2.25}$ powder [32]	Sol–gel	27.90	102 Fg^{-1} at 1 Ag^{-1} 1 M KOH as electrolyte	3.6 Wh kg^{-1} at 120 W kg^{-1}	Retains about 50% after 3000 cycles at 2 Ag^{-1}
$La_{0.5}Ca_{0.5}MnO_{2.811}$ powder [33]	Sol–gel	23.0	170 Fg^{-1} at 1 Ag^{-1} 1 M KOH as electrolyte	7.6 Wh kg^{-1} at 160 W kg^{-1}	Retains 57% after 3000 cycles at 2 Ag^{-1}
Mesoporous spheres of $LaMnO_3$ powder [21]	Sol–gel	16	187 Fg^{-1} at 0.5 Ag^{-1} 1 M KOH as electrolyte	No data	Retains about 43% after 1000 cycles at 3 Ag^{-1}
Flower-like $La_{0.85}Sr_{0.15}MnO_3$ powder [21]	Sol–gel	30	198 Fg^{-1} at 0.5 Ag^{-1} 1 M KOH as electrolyte	No data	Retains about 78% after 1000 cycles at 3 Ag^{-1}
$(La_{0.75}Sr_{0.25})0.95MnO_{3-\delta}$ powder [34]	Commercial procurement	21.13	56 Fg^{-1} at 2 mV s^{-1} 1 M Na_2SO_4 as electrolyte	No data	Retains 98% after 1000 cycles at 5 mA
$La_{0.7}Sr_{0.3}Co_{0.1}Mn_{0.9}O_{3-\delta}$ nanofiber [35]	Electrostatic spinning	27.2	569.1 Fg^{-1} at 2 mV s^{-1} 485 Fg^{-1} at 0.5 Ag^{-1} 1 M KOH as electrolyte	No data	No change after 2000 cycles at 50 mV s^{-1}
$Sr_{0.8}Ba_{0.2}MnO_3$ nanofiber [36]	Electrostatic spinning	No data	446.8 Fg^{-1} at 0.5 Ag^{-1} 0.5 M Na_2SO_4 as electrolyte	37.3 Wh kg^{-1} at 400 W kg^{-1}	Retains 87% after 5000 cycles at 10 Ag^{-1}

(Continued)

TABLE 12.3 (*Continued*)
The Important Attributes of Supercapacitors of Perovskite Materials

Perovskites	Methods	Specific Surface Area (m^2/g)	Specific Capacitance	Power Density (Wh/kg)	Cyclic Stability
$SrRuO_3$ powder [37]	Microwave-assisted citrate method	No data	52.4 Fg^{-1} at 1 Ag^{-1} 1 M KOH as electrolyte	No data	Retains 77.8% after 1000 cycles at 25 Ag^{-1}
$La_{0.7}Sr_{0.3}NiO_{3-\delta}$ nanofiber [38]	Electrostatic spinning	15.46	719.5 Fg^{-1} at 2 Ag^{-1} 1 M Na_2SO_4 as electrolyte	81.4 Wh kg^{-1} at 500 W kg^{-1}	Retains 90% after 2000 cycles at 5 Ag^{-1}
Reduced $LaNiO_3$ powder [39]	Sol–gel	26	230 Fg^{-1} at 1 Ag^{-1} 6 M KOH as electrolyte	No data	3500 cycles at 5 Ag^{-1}
$LaNiO_3$ nanosheet [40]	Sol–gel	181.2	139.2 mA h g^{-1} at 1 Ag^{-1} 6 M KOH as electrolyte	65.84 Wh kg^{-1} at 1.8 kW kg^{-1}	Retains 92.39% after 10,000 cycles at 150 mV s^{-1}
$LaNiO_{2.63}$ powder [41]	Sol–gel	No data	478.7 Fg^{-1} at 0.1 mV s^{-1} 1 M KOH as electrolyte	No data	Retains 94.5% after 15,000 cycles at 10 Ag^{-1}
$SrCo_{0.9}Nb_{0.1}O_{2.418}$ powder [42]	Solid-state reaction	8.63	778.5 Fg^{-1} at 5 Ag^{-1} 6 M KOH as electrolyte	37.5 Wh kg^{-1} at 433.9 W kg^{-1}	Retains 95.7% after 3000 cycles at 5 Ag^{-1}
$Ba_{0.5}Sr_{0.5}Co_{0.8}Fe_{0.2}O_{3-\delta}$ powder [43]	Sol–gel	0.74	610 Fg^{-1} at 1 Ag^{-1} 1 M KOH as electrolyte	No data	Reaches peak capacity around 100 cycles and then sharp decrease
$SrCoO_{3-\delta}$ powder [43]	Sol–gel	1.53	572 Fg^{-1} at 1 Ag^{-1} 1 M KOH as electrolyte	No data	Retains 87.4% after 5000 cycles
$SrCo_{0.9}Mo_{0.1}O_{2.5092}$ powder [44]	Sol–gel	37.4	168.88 mA h g^{-1} at 1 Ag^{-1} 6 M KOH as electrolyte	74.8 Wh kg^{-1} at 734.5 W kg^{-1}	Retains 97.6% after 10,000 cycles at 10 Ag^{-1}
$LaSr_3Fe_3O_{10-\delta}$ powder [45]	Hydrothermal	No data	380 Fg^{-1} at 0.1 Ag^{-1} 6 M KOH as electrolyte	No data	Retains 87.1% after 1000 cycles at 1 Ag^{-1}
$LaFeO_3$ nanotube [46]	Sol–gel	45.09	313.2 Fg^{-1} at 0.8 Ag^{-1} 2 M KOH as electrolyte	No data	Retains 86.1% after 5000 cycles at 100 mV s^{-1}

(*Continued*)

TABLE 12.3 (*Continued*)
The Important Attributes of Supercapacitors of Perovskite Materials

Perovskites	Methods	Specific Surface Area (m^2/g)	Specific Capacitance	Power Density (Wh/kg)	Cyclic Stability
$La_{0.7}Sr_{0.3}FeO_3$ [47]	Electrostatic spinning	27.96	523.2 Fg^{-1} at 0.8 Ag^{-1} 1 M Na_2SO_4 as electrolyte	No data	Retains 83.8% after 5000 cycles at 20 Ag^{-1}
$La_{0.8}Na_{0.2}Fe_{0.8}Mn_{0.2}O_3$ powder [48]	Sol–gel	8.16	56.4 Fg^{-1} at 3 mV s^{-1} 1 M H_2SO_4 as electrolyte	About 1.8 Wh kg^{-1} at 1.42 mA cm^{-2}	Retains 14% after 1000 cycles at 1.42 mA cm^{-2}
$PrBaMn_2O_{5.686}$ powder [49]	Sol–gel	7.83	1252.1 Fg^{-1} at 1 Ag^{-1} 9 M KOH as electrolyte	No data	Retains 90% after 2000 cycles at 1 Ag^{-1}
$(La_{0.75}Sr_{0.25})0.95MnO_{3-\delta}/MnO_2$ composite [50]	Hydrothermal	59.4	437.2 Fg^{-1} at 2 mV s^{-1} 1M Na_2SO_4 as electrolyte	No data	Keeps on increasing in 500 cycles at 10 Ag^{-1}
$LaMnO_3/SiO_2$ porous composite [51]	SBA-15 direct templating	296	100–300 Fg^{-1} at 5 mV s^{-1} in 1 M KOH	No data	No data
$La_{0.85}Sr_{0.15}MnO_3/NiCo_2O_4$ core-shell architecture [52]	Hydrothermal	No data	1341 Fg^{-1} at 0.5 Ag^{-1} 6 M KOH as electrolyte	63.5 Wh kg^{-1} at 900 W kg^{-1}	Retains 200% after 10,000 cycles at 20 Ag^{-1}
$CeO_2/LaMnO_3$ powder [53]	Hydrothermal	28.32	262 Fg^{-1} at 1 Ag^{-1} 1 M Na_2SO_4 as electrolyte	17.2 Wh kg^{-1} at 1015 W kg^{-1}	Retains 92% after 2000 cycles at 6 Ag^{-1}
$LaMnO_3$/Nitrogen-doped reduced graphene oxide [54]	Electrostatic interaction	55	687 Fg^{-1} at 5 mV s^{-1} 1 M KOH as electrolyte	No data	Retains 79% after 2000 cycles at 10 Ag^{-1}
$La_{0.7}Sr_{0.3}CoO_{3-\delta}/Ag$ composite [55]	Solid-state reaction	No data	517.5 Fg^{-1} at 1 mA cm^{-2} 1 M KOH as electrolyte	21.9 mWh cm^{-3} at 90.1 $mWcm^{-3}$	Retains 85.6% after 3000 cycles at 50 mA cm^{-2}
$SrRuO_3$/Reduced graphene oxide [37]	Microwave-assisted citrate method	No data	160 Fg^{-1} at 1 Ag^{-1} 1 M KOH as electrolyte	No data	Retains 90% after 1000 Cycles at 25 Ag^{-1}
CaTiO3/Active carbon [56]	Solid-state reaction	1522	270 Fg^{-1} at 1 Ag^{-1} 6 M KOH as electrolyte	26.3 Wh kg^{-1} at 375 W kg^{-1}	No obvious change up to 5000 cycles at 1Ag^{-1}
$BiFeO_3$ nanowire/reduced graphene oxide [57]	Hydrothermal	No data	928.4 Fg^{-1} at 5 Ag^{-1} 3 M KOH as electrolyte	18.62 Wh kg^{-1} at 950 W kg^{-1}	Retains 87.51% after 1000 cycles at 5 Ag^{-1}

TABLE 12.4
The Important Attributes of Supercapacitors of Spinels, and Their Composite Materials

Spinels and Their Composites	Methods	Surface Area (m^2/g)	Specific Capacitance	Power Density (Wh/kg)	Cyclic Stability
3D hierarchical porous $NiCo_2O_4$/graphene hydrogel/Ni foam electrode [62]	One-step electrodeposition	206	3.84 F cm^{-2} at 2 mA cm^{-2}	65 Wh kg^{-1} and 18.9 kW kg^{-1}	92% retention after 5000 cycles
$NiCo_2O_4$ nanospheres/nitrogen-doped graphene Composites [59]	Hydrothermal route	53.4	563.0 Fg^{-1} at 1 Ag^{-1}	No data	90.5 % retention after 5000 cycles
$MnCo_2O_4$@N-CNF nanofibers [63]	Electrospinning	388.7	871.5 Fg^{-1} at a current density of 0.5 Ag^{-1}	30.26 Wh kg^{-1}	Retention rate of 89.3% after 5000 cycles
$NiCo_2O_4$/C nanorods [60]	Hydrothermal	—	1297 Fg^{-1} at 0.5 Ag^{-1}	7.75 kW kg^{-1} at 27.38 Wh kg^{-1}	Retention of 87.6% after 3000 cycles at 1 Ag^{-1}
$NiFe_2O_4$ powder [64]	Hydrothermal	—	240.9 F/g at the current density of 1 A/g	10.15 Wh/kg at a power density of 140 W/kg	Retention of 128% after 2000 cycles
Fe_2O_3 nanospheres anchored on activated carbon cloth (Fe_2O_3 @ACC) [64]	Hydrothermal	—	2775 mF cm^{-2} in 3 M $LiNO_3$	9.2 mWh cm^{-3} with power density 12 mW cm^{-3}	95% retention after 4000 cycles
Helical CC/$NiCo_2O_4$-N@NiO composites [58]	Hydrothermal	—	709.1 F/g at 2 mA/cm^2	6000 W/kg at 28 Wh/kg	95.2% retention after 10,000 cycles
Helical CC/$NiCo_2O_4$-S@NiO composites [58]	Hydrothermal	—	775.1 F/g at 2 mA/cm^2	5692 W/kg at 20.1 Wh/kg	92.3% retention after 10,000 cycles
$NiCo_2O_4$@$Ni_{0.85}$Se core-shell nanorod arrays on Ni foam [65]	Electro chemical deposition	—	1454 F/g at 1 M KOH	29.3 Wh kg^{-1} at 799 W kg^{-1}	88.5% retention after 10,000 cycles
Hierarchical $NiCo_2O_4$@$NiCo_2S_4$ core/shell nanowires [66]	Ion-exchange route	—	3176 Fg^{-1} at a current density of 2 Ag^{-1}	196.8 Wh kg^{-1} at 752.33 W kg^{-1}	Retain 86.52% at a high current density of 10 Ag^{-1}
$NiCo_2S_4$@PPy//AC-Ni foam) [67]	Hydrothermal & electrochemical	—	1925 Fg^{-1}	86.6 Wh kg^{-1} at a power density of 1046 W kg^{-1}	8000 cycles at a current density of 1.0 A cm^{-2}

TEM studies. In TEM pictures also the clear nanospheres were detected. Recently, it was reported by Dong et al. [60], that the SEM and TEM photos of three-dimensional (3D) porous carbon (OPC) and $NiCo_2O_4$/C (carbon) complexes as shown in reference [60] manifested that the defective structure of carbon allowed to the evolution of nanorods like $NiCo_2O_4$ particles. This in turn attributed the characteristic of offering high specific capacitance. In the case of $NiCo_2O_4$/C nanorods, the high capacitance was recorded to be about 1297 Fg^{-1} at 0.5 Ag^{-1} while the power density was around 7.75 kW kg^{-1} at 27.38 Wh kg^{-1}. The capacitance retention was observed about 87.6% after 3000 cycles at the current density of 1 A/g.

Fu et al. [61], investigated the variation of specific capacitance as a function of $CoFe_2O_4$ (CF) content in crab derived carbon (CDC). In addition, the current density was also varied from 0.5 to 10 A/g. The whole discussion was extracted from capacitive performance of $CoFe_2O_4$/CDC composite as mentioned in reference [61]. From capacitive performance of $CoFe_2O_4$/CDC composite, it was noticed that up to 75% of cobalt ferrite, the specific capacitance values were increased to peak values. In particular, the cobalt ferrite composite performed the specific capacitance about 740 F/g at 0.5 A/g current density (for 75% of CF deposited on CDC). However, the capacitive performance was diminished to lower values at higher values of current density. Likewise, in Table 12.4, different parameters required for SC applications were listed for various materials.

12.3.3 Oxides/Hydroxides/Polymer Composites

Apart from the perovskites and spinels, there were several numbers of oxides, hydroxides, polymers, and their composites providing extensive applications for SCs. The electrochemical storage, and the conversion system can use the below mentioned (in Table 12.5) materials based on their electrochemical performance with respect to different attributes. In view of this, we carried out the material, preparation method, surface morphology, specific capacitance, power density, and cycle life (as listed in Table 12.5). It was obvious from Table 12.5 that the morphology of oxides, hydroxides, polymers, and their composites indicated formation of nanofibers, plates, nanorods, hollow spheres, hollow urchins, clew, lamellar, nanoflowers, layers, nanobelts, nanoribbons, nanotubes, nanoflakes, nanowires, nanowheels, nanoshells, nanoleaflets, hierarchies, helical, nanosheets, hollow rods, caps, etc. However, the different morphologies of these materials (listed in Table 12.5) led to the improvement of attributed for the SC applications rather than the perovskites, and spinels.

As reported by Mukhiya et al. [68], the 3D-cobalt hydroxide wheels/carbon nanofibers (3D-$Co(OH)_2$/CNFs) composites were prepared through electrospinning cum hydrothermal method. Furthermore, the nanofibers, nanoleaflets, and nanowheels were noticed in the surface morphology. It was very much obvious from the SEM pictures of pristine CNFs, field emission scanning electron microscopy (FESEM) pictures of 3D-$Co(OH)_2$/CNFs provided in reference [68]. Especially, in SEM pictures of pristine CNFs, the nanofibers were clearly evidenced for 3D-$Co(OH)_2$/CNFs composite while nanowheels comprising of little bit defect structure were identified in FESEM pictures of 3D-$Co(OH)_2$/CNFs [68]. However, interestingly, the β-$Co(OH)_2$ nanoleaflets were formed within the 200 nm area of the selected spot of present sample. Owing to the limited defective (porous) structure, the capacitive performance was improved much

TABLE 12.5
Important Attributes of Supercapacitors of Oxides/Hydroxides/Polymer Composites

Oxides/Hydroxides/Polymer Composites	Methods	Morphology	Specific Capacitance (F/g)	Power Density (Wh/kg)	Cyclic Stability
3D-cobalt hydroxide wheels/carbon nanofibers (3D $Co(OH)_2$/CNFs) composites [68]	Electrospinning and hydrothermal	Nanofibers, nanoleaflets and nanowheels	1186	60.31	94.6% of retention after 5000 cycles
Seamless integrated C@NiMn-OH-Ni_3S_2/Ni foam [82]	Hydrothermal	—	2521	45.3	~94.6% capacity retention after 10,000 cycles
α-MnO_2 [83]	Hydrothermal	Plate-like, nanorods	72–168	—	—
α-MnO_2 [83]	Hydrothermal	Hollow spheres and urchins	147	—	—
α-MnO_2 [83]	Hydrothermal	Urchin-like	86–152	—	—
α-MnO_2, β-MnO_2 [83]	Hydrothermal	Urchin-like, clew-like	46–120	—	—
Mn_3O_4, MnOOH [83]	Hydrothermal	Cubes and nanowires	~170	—	—
δ-MnO_2 [83]	Hydrothermal	Lamellar	241	—	—
α-MnO_2, γ-MnO_2 [83]	High viscosity process	Rod-shaped	389	—	—
δ-MnO_2 [83]	Room temperature precipitation	Rod-shaped	201	—	—
Cubic MnO_2 (Fd3m) [83]	Low temperature reduction	Nanoflowers	121.5	—	—
Rancieite structure [83]	Low temperature reduction	Layered	17–112	—	—
γ-MnO_2 [83]	Sonochemistry	Spherical particles	118–344	—	—
δ-MnO_2 [83]	Microwave-assisted emulsion	Belt-like	277	—	—
γ-MnO_2 [83]	Sol–gel process	Nanorods	317	—	—

(Continued)

TABLE 12.5 (*Continued*)
Important Attributes of Supercapacitors of Oxides/Hydroxides/Polymer Composites

Oxides/Hydroxides/Polymer Composites	Methods	Morphology	Specific Capacitance (F/g)	Power Density (Wh/kg)	Cyclic Stability
β-MnO_2 [83]	Solution combustion	Plate-like	23–43	71–123	—
MnO_2 NPs/Fe_3O_4 NPs [83]	—	—	21.5	10.2	5000
MnO_2 NPs/AC [83]	—	—	31.0	19.0	5000
δ-MnO_2 nanorods/AC [83]	—	Nanorods	31.0	2	23,000
Fe_3O_4 NPs/AC [76]	—	—	37.9	0.07	500
Amorphous MnO_2 /AC [83]	—	Amorphous	50	0.5–8	100
Amorphous MnO_2 /AC [83]	—	Amorphous	21	16	1,95,000
Hydrous V_2O_5 nanoribbons AC [84]	—	Nanoribbons	64.4	0.07–2	100
NiO nanoflakes /porous carbon [85]	—	Nanoflakes	38	0.01–10	1000
$Ni(OH)_2$/MWNTs/AC [86]	—	Nanotubes	96	1.5	2000
MnO_2/porous carbon V_2O_5 nanowires/MWNTs [87]	—	Nanowires and nanotubes	45	0.075–3.75	100
MnO_2/MWNTs SnO_2/MWNTs [87]	—	Nanotubes	38	143.7	1000
MnO_2/β-FeOOH nanorods [87]	—	Nanorods	51	0.45–2	2000
$LiNi_{0.5}Mn_{1.5}O_4$/AC [83]	—	—	32	—	1000
$LiMn_2O_4$ /MnO_2/MWNTs [83]	—	Nanotubes	60	0.3–2.4	—
$LiCoPO_4$ NPs /carbon nanofoam [84]	—	—	21.9	0.2	1000
$LiMn_2O_4$/AC/$Li_4Ti_5O_{12}$ [83]	—	—	59	4	5000
AC/Li_4Ti_5O12 [76]	—	—	—	—	5000
PFPT/$Li_4Ti_5O_{12}$ NPs [76]	—	—	13	2.5	1500
AC/TiO_2 [76]	—	—	44	0.35	600
MWNTs/TiO_2 nanowires[76]	—	Nanowires	11.5	0.3–1.2	600
MWNTs/Fe_2O_3/MWNTs [76]	—	Nanotubes	80	1	500

(*Continued*)

TABLE 12.5 (*Continued*)
Important Attributes of Supercapacitors of Oxides/Hydroxides/Polymer Composites

Oxides/Hydroxides/Polymer Composites	Methods	Morphology	Specific Capacitance (F/g)	Power Density (Wh/kg)	Cyclic Stability
CNT@Ni_3S_2 1D-hierarchical structures [78]	Rational multi-step transformation route	Hierarchical	514	—	1500
NiO [87]	Microwave method	Nanoflakes	401	—	—
NiO@MnOOH core/shell hierarchies [87]	Electrochemical deposition	Hierarchies	1625.3	—	5000
Nitrogen-doped porous graphitic carbon (NPGC) [88]	Coordination-pyrolysis combination		293	10.5	5000
Polypyrrole@manganese oxide nanosheets@stainless steel yarn (PMS) [89]	—	Nanosheets	181 mF cm^{-2}	10.3 mW cm^{-2}	1000
Nano SnS_2-SnO_2 [90]	Cost-effective solvothermal method		71.4–149.0	1828	3000
NS-$CuMoO_4$/rGO/NF and NS-$CuMoO_4$/NF [91]	Hydrothermal method		1300–2342	3875	4000
$CoNi_2S_4$ core-shells [92]	Electrochemical method	Shell like	425	—	—
$CoNi_2S_4$@polydopamine (PDA) core-shells nanocomposites [92]	Electrochemical method	Shell like	725	—	1000
Multi-hierarchical porous carbon combined with $CoFe_2O_4$ [93]	Novel three-step method	Hierarchical	320.4	4992	20,000 cycles with 91.3% of retention
$NiMoO_4$-PANI core-shell nanocomposite [94]	Chemical polymerization	Shell like	1214		80.7% capacitance retention after 2000 cycles
$NiMoO_4$ nanorods [94]	Solvothermal method	Nanorods	837	23.4	98.6% of original value after 5000 cycles

(*Continued*)

TABLE 12.5 (*Continued*)
Important Attributes of Supercapacitors of Oxides/Hydroxides/Polymer Composites

Oxides/Hydroxides/Polymer Composites	Methods	Morphology	Specific Capacitance (F/g)	Power Density (Wh/kg)	Cyclic Stability
Co_2CrO_4/Co1−xS [94]	Hydrothermal method	Hollow-rod	1580 (max.)	800.5	Remains 88% after 5000 cycles
Co1−xS [94]	Hydrothermal method	—	1341.1	—	—
$NaTi_2(PO_4)_3$ with nitrogen-doped carbon layer (NTP/C) [95]	—	Layer	100	503	100% retention at 5000 cycles
Fe_2MoC [96]	—	—	97.7	21	83.9% capacitance retention after 1000 cycles
Co_3O_4 Films [97]	Electrodeposition method	—	630	—	4000 cycles
Polyaniline(PANI) layer on nanoporous gold(NPG) support (PANI/NPG) [98]	Delloying method	—	6.54 mF/cm^2	1.56 W/cm^2	Charge/discharge rates up to 100 mV/s
ZnNiCo-P [99]	—	—	~958	960	89% of initial specific capacitance after 8000 cycles retained
$Ni(OH)_2$/graphene/$Ni(OH)_2$/Ni foam [100]	Chemical vapor deposition	—	991	750	89.3% retention after 10,000 cycles
Cu_3SbS_4/Ni-5 [100]	Microwave-irradiation process	—	835.24 mAh g^{-1}	6363.63	Retention of about 96.7% even after 1000 cycles
GF-Ni-Au@NiOx cathode [100]	—	—	3.57 F cm^{-2}	334.15 mW cm^{-3}	15,000 cycles capacitance retention of the cathode is 92.1%.
GF-Ni-Au@FeOx anode [100]	—	—	3.34 F cm^{-2}	334.15 mW cm^{-3}	20,000 cycles at the scan rate of 20 mV s^{-1}

(*Continued*)

TABLE 12.5 (*Continued*)
Important Attributes of Supercapacitors of Oxides/Hydroxides/Polymer Composites

Oxides/Hydroxides/Polymer Composites	Methods	Morphology	Specific Capacitance (F/g)	Power Density (Wh/kg)	Cyclic Stability
$Ni(OH)_2$ nanoneedles on N-doped 3D rivet graphene film (N-3DRG) [101]	Chemical vapor deposition (CVD)	Nanoneedles	256.1 mAh g^{-1}	452	71% capacity retention after 10,000 cycles
L-Arginine capped α-$Ni(OH)_2$ microstructures [74]	Microwave irradiation method	Cap and sheet-like	549	601	87.3% after 10,000 cycles
Ni/N-doped PC [102]	—	—	2002.6	800	Retains 91.5% after 5000 cycles
$NiMoO_4$-modified α-MoO_3 Nanobelts [77]	Facile two-step hydrothermal	Nanobelts	1307	37.5	100% retention after 75,000 cycles
PEDOT:Cellulose [103]	—	Film	92	2.01	—
PEDOT:Cellulose [103]	—	Film	91.4	1.98	—
PEDOT:Cellulose [103]	—	Film	89.2	1.77	—
PEDOT:Cellulose [103]	—	Film	87.3	1.64	—

Note: NPs: nanoparticles; AC: activated carbon; PFPT: poly (fluoro phenythiophene); EC: ethylene carbonate; DEC: diethyl carbonate; DMC: dimethyl carbonates; MWNT (multi-wall nanotubes).

and therefore, the recorded specific capacitance was acquired to be 1186 F/g at a current density of 1 A/g. Furthermore, the high power density about 60.31 Wh/kg at power density of 740.8 W/kg was identified. Even, this was still remained to 37 Wh/kg at the power density ~7500 W/kg consisting of significant cycle life. This cyclic stability was around 94.6% of retention after 5000 cycles.

The α-$Ni(OH)_2$ (hydroxide material), and NiO nanosheets were grown using the microwave-assisted liquid phase growth at low atmospheric temperatures. In preparation of nanosheets; FESEM photos of α-$Ni(OH)_2$ [69], the way of performing the reaction was represented. It was seen that within a microwave reactor a 1000 mL, triple-necked flask was inserted. The solution present in the three-necked flask was subjected to rapid microwave heating. As a result, the α-$Ni(OH)_2$ nanosheets were formed, and further heating of α-$Ni(OH)_2$ specimen rendered the formation of nickel oxide nanosheets as shown in FESEM photos of NiO nanosheets [69]. In conclusion, it was found that the instantaneous microwave heating supported to generate nanocrystals. Due to the rapid microwave heating, the desired properties for device applications were enhanced up to larger extent as noticed in the literature [70–73]. Thus, in case of α-$Ni(OH)_2$ specimen, the huge capacitive performance of 4172.5 F/g was recorded at 1 A/g. Moreover, on reaching the current density of 16 A/g, the capacitive performance was further remained constantly at 2680 A/g by offering 98.5 % retention after 2000 cycles. Thus, this compound showed prominent applications for SCs. In support of this work, recently, William et al. [74], prepared the α-$Ni(OH)_2$ with different molarities of additives to the resultant mixture (α-$Ni(OH)_2$) via the microwave irradiation method. That is, 0, 4, 8, 12, and 16 mM. The morphology of all samples exhibited the asymmetric 3D-sheets [74]. Specifically, the morphology in 8 mM-α-$Ni(OH)_2$ revealed the loosely connected 3D-sheets. In addition, the efficient values of specific capacitance, and power densities were noted to be ~549 F/g and 601 Wh/kg respectively. This capacitance was retained about 87.3% even after 10^4 cycles.

Likewise, the MnO_2 (oxide material) thin films associated to porous structures were grown as well using the electrophoretic deposition (EPD). The discussion on EPD method was provided in reference [75]. The pure thin film without varying the formation parameters expressed the existence of highly defective (porous) structured nanofibers in the surface morphology [75]. Afterwards, the morphology was somewhat changed upon manipulating certain formation parameters of charged MnO_2 suspensions [74]. Thus, the remarkable electrochemical performance was achieved for better SCs.

The surface morphology of MoO_3, and $MoO_3/NiMoO_4$ core/shell was shown in published work [76]. It was obvious that the nanobelts were noticed in SEM, and HRTEM pictures of MoO_3, and $MoO_3/NiMoO_4$ samples. The width of these belts was almost 300 to 400 nm, and the length was about 6 to 10 micrometer. For MoO_3, the SEM picture revealed well-defined belts [76], and further, the HRTEM pictures [76] indicated the single nanobelts formation with interplanar distance approximately 0.37 nm (004). In the same fashion, the SEM, and HRTEM pictures of $MoO_3/NiMoO_4$ samples were shown in reference [76]. The corresponding energy dispersive spectrum (EDS) was also provided for elemental confirmation of the samples [76]. It evidenced the presence of Ni, Mo, and O elements in the samples. The outcome of these samples offered the high specific capacitive performance of 1307 F/g while the power density was about 37.5 Wh/kg. Besides, the 100% retention after 75,000 cycles was noticed [76].

Xue et al. [77], synthesized the LaB_6 nanowires for better SC applications. The morphology was shown in reference [77] at different magnifications of 20 micrometer and 2 micrometer, respectively. The electrochemical performance of LaB_6 nanowires suggested that the as prepared samples provided a capacitance of 17.3 mF/cm^{-2} in 1.0 M Na_2SO_4 electrolyte at a current density of 0.1 mA/cm^{-2}. However, the interesting cyclic stability of 10^4 charging/discharging cycles was acquired for this material.

Zhu et al. [78], prepared the 1D hierarchical materials comprising of Ni_3S_2 nanosheets on CNTs backbone with surface area about 64 m^2/g. The charge/discharge performance was checked for as prepared CNT@Ni_3S_2 (I) and NiS nanosheets (II) at a current density of 5.3 Ag^{-1} as mentioned in reference [78]. It was noticed that the capacitance of as prepared CNT@Ni_3S_2 was two times (~500 F/g) greater than NiS nanosheets (~250 F/g). This evidenced a fact that the presence of CNTs induced the specific capacitance value of Ni_3S_2. It can be obtained owing to the readily available charged ions within the nanosheets. Thus, it can allow the improvement of electrical conductivity as well as diffusion kinetics. After reaching 1500 cycles, almost 90% of capacitive retention was attained for CNT@Ni_3S_2. On the other hand, the NiS without CNTs offered approximately 66% retention after the same 1500 cycles. As a whole, it was confirmed that the 1D-hybrid nanostructures such as CNT@Ni_3S_2 can exhibit the outstanding specific capacitance with good cyclic ability.

The small and large size $SnWO_4$ nanoassemblies were prepared through microwave heating method [79]. The current versus voltage (CV) curves were drawn for both the $SnWO_4$ samples by varying the voltage scan rate. It was observed from the cycle voltammogram (CV) curves of small and large size $SnWO_4$ electrodes at 5 mV/s that the small size $SnWO_4$ nanostructure showed the huge value of current of ~4 mA while the large sized $SnWO_4$ nanoassemblies revealed the small value of current of ~1.2 mA at 5 mV/s over the voltage ranging from 0 to 0.55 V. This was seemed to be an interesting in obtaining the high current with respect to the small and large sized $SnWO_4$ assemblies. This kind of trend was happened owing to the quick redox reactions between $Sn^{+2} \leftrightarrow Sn^{+4} + 2e^-$. In general, the capacitance value mainly depends on the redox reactions in compound. That is, the reverse redox reaction can play a key role in adopting the high value of capacitance. In case of $SnWO_4$ assemblies the hopping of two electrons took place between Sn^{+2} & Sn^{+4}. Therefore, it can enhance the electrical conductivity up to the larger extent. Thus, the pseudocapacitance nature was improved much. The CV curves of small- and large-sized $SnWO_4$ assemblies with changing the scan rate from 5–125 mV/s were seen in reported work [79]. Further, the low current was noted at redox peaks for large $SnWO_4$ assemblies. It can be attributed to the decrease of hopping rate of electrons between Sn^{+2} & Sn^{+4}. Thus, the capacitances as well as electrical conductivity parameters were come down to lower values to large-sized $SnWO_4$ assemblies. Furthermore, the variation of specific capacitance as a function of scan rate for small- and large-sized $SnWO_4$ assemblies was analyzed in reference [79]. It can be understood that the capacitive performance was reduced with increase of scan rate from 0.0–140 mV/s. However, in comparison, small-sized $SnWO_4$ assemblies performed better electrochemical ability than the large-sized $SnWO_4$ assemblies. This established a fact that the small size of nanoassemblies can in turn promote the capacitive performance to larger extent. Furthermore, few polymer composites were also revealed the electrochemical performance and the relevant parameters were listed in Table 12.5.

FIGURE 12.5 (a) Trolley bus of Aowei Technology Co. Ltd. Based on $Ni(OH)_2$-AC supercapacitor, (b) schematic of textile based SCs for wearable electronics, (c) fabrication process of inkjet-printed, in-plane SCs on a paper.

In view of the live examples of SCs, specifically, the hybrid supercapacitors were used in the electric buses as shown in Figure 12.5a [80]. This technology was introduced in China for bigger e-buses by Aowei Technology Co., Ltd. (Shanghai, China). Using this technology, one can provide quick charging facility to the big buses during the journey times. Usually, the hybrid configuration of nickel hydroxide-activated carbon ($Ni(OH)_2$-AC) was used in case of trolley bus consisting of the charging time about 90 s, and distance covered is ~8 km at the highest velocity of ~45 km/h [80]. It showed an average velocity of 22 km/h containing the acceleration of 0 to 40 km/h within 16.5 seconds. Moreover, the tram car which can work on electrochemical double-layer capacitors (EDLCs) configuration by possessing the charging time about 30 seconds with the travelling distance of 3–5 km introduced by CSR Co. Ltd. (Chinese). In addition, the knitted and screen-printed supercapacitors on textiles substrates (Figure 12.5b) were used in wearable electronic goods [81]. The prepared supercapacitors revealed the huge capacitance about 51×10^{-2} F/cm^2 with good flexibility for commercial usage. The SCs were used even in inkjet printing technology. This technology can provide the redeveloped thin film patterns by keeping the ink drops to move front on the desired or convenient substrates [81]. These kinds of few patterns were displayed in Figure 12.5c. These patterns will generally have better flexibility with existence for even up to 10^3 bending times. It can provide the remarkable retention even after 10^4 successive cycles [104–106].

12.4 SUMMARY AND CONCLUSIONS

A detailed analysis of various inorganic materials synthesized so far was made for SC applications. The synthesized materials were categorized into three kinds like perovskites, spinels-based materials, and oxides/hydroxides/polymer composites.

It was concluded that the surface morphology played a key role in managing the attributes like capacitive performance, power density, and cycle ability. Moreover, the composite materials with perovskites, spinels, and hydroxides revealed an excellent electrochemical (EC) performance for SC applications. Furthermore, the synthesis technique also influenced the material EC performance for SCs. Especially, the samples comprising of the morphology like nanosheets, nanobelts, nanowires, nanofibers, nanoleaflets, nanorods, nanowheels, nanotubes, helical nanostructures, and in few cases nanospheres exhibited superior specific capacitance, power density, and cyclic stability for SCs. Even the size of the nanomaterials also reinforced to obtain the high magnitudes of above mentioned attributes. The hybrid configuration of nickel hydroxide-activated carbon ($Ni(OH)_2$–AC) was used in case of trolley bus for outstanding charging rate during the long journey distance.

ACKNOWLEDGMENTS

The authors express their sincere thankful to GITAM-Bangalore management and my collaborators for good encouragement throughout my work. Particularly, Dr. D. Baba Basha would like to thank Al'Majmaah, K.S.A, the dean of scientific research at Majmaah University, for supporting this work.

REFERENCES

1. Arunkumar M., Paul A. 2017. Importance of electrode preparation methodologies in supercapacitor applications, *ACS Omega* 2: 8039–8050.
2. Zhang C.J., Nicolosi V. 2019. Graphene and MXene-based transparent conductive electrodes and supercapacitors, *Energy Storage Materials* 16: 102–125.
3. Fic K., Platek A., Piwek J., Frackowiak E. 2018. Sustainable materials for electrochemical capacitors, *Materials Today* 21: 437–454.
4. Lechêne B.P., Cowell M., Pierre A., Evans J.W., Wright P.K., Arias A.C. 2016. Organic solar cells and fully printed super-capacitors optimized for indoor light energy harvesting, *Nano Energy* 26: 631–640.
5. Yang I., Kwon D., Kim M.-S., Jung J.C. 2018. A comparative study of activated carbon aerogel and commercial activated carbons as electrode materials for organic electric double-layer capacitors, *Carbon* 132: 503–511.
6. Du W., Bai Y.-L., Xu J., Zhao H., Zhang L., Li X., Zhang J. 2018. Advanced metal-organic frameworks (MOFs) and their derived electrode materials for supercapacitors, *Journal of Power Sources* 402: 281–295.
7. Yang I., Kwon D., Yoo J., Kim M.-S., Jung J.C. 2019. Design of organic supercapacitors with high performances using pore size controlled active materials, *Current Applied Physics* 19: 89–96.
8. Inagaki M., Kang F., Toyoda M., Konno H. 2014. Carbon materials for electrochemical capacitors, *Advanced Materials Science and Engineering of Carbon*: 237–265. doi:10.1016/b978-0-12-407789-8.00011-9.
9. Inagaki M., Konno H., Tanaike O. 2010. Carbon materials for electrochemical capacitors, *Journal of Power Sources* 195: 7880–7903.
10. Ke Q., Wang J. 2016. Graphene-based materials for supercapacitor electrodes—A review, *Journal of Materiomics* 2: 37–54.
11. Chen T., Dai L. 2013. Carbon nanomaterials for high-performance supercapacitors, *Materials Today* 16: 272–280.

12. Li X., Wei B. 2013. Supercapacitors based on nanostructured carbon, *Nano Energy* 2: 159–173.
13. Javaid A. 2017. 11—Activated carbon fiber for energy storage, activated carbon fiber and textiles, *Woodhead Publishing Series in Textiles*: 281–303. doi:10.1016/B978-0-08-100660-3.00011-0.
14. Libich J., Máca J., Vondrák J., Čech O., Sedlaříková M. 2018. Supercapacitors: Properties and applications, *Journal of Energy Storage* 17: 224–227.
15. Kim B.K., Sy S., Yu A., Zhang J. 2015. Electrochemical Supercapacitors for Energy Storage and Conversion Handbook of Clean Energy Systems. Chichester, UK: John Wiley & Sons, Ltd. pp. 1–25.
16. Muzaffar A., Ahamed M.B., Deshmukh K., Thirumalai J. 2019. A review on recent advances in hybrid supercapacitors: Design, fabrication and applications, *Renewable and Sustainable Energy Reviews* 101: 123–145.
17. Liu L., Feng Y., Wu W. 2019. Recent progress in printed flexible solid-state supercapacitors for portable and wearable energy storage, *Journal of Power Sources* 410–411: 69–77.
18. Kumar N.S., Suvarna R.P. Naidu K.C.B. 2018. Sol-gel synthesized and microwave heated $Pb_{0.8-y}La_yCo_{0.2}TiO_3$ (y = 0.2–0.8) nanoparticles: Structural, morphological and dielectric properties, *Ceramics International* 44: 18189–18199.
19. Naresh U., Kumar R.J., Naidu K.C.B. 2019. Optical, magnetic and ferroelectric properties of $Ba_{0.2}Cu_{0.8-x}La_xFe_2O_4$ (x = 0.2–0.6) nanoparticles, *Ceramics International* 45: 7515–7523.
20. Nan H.-S., Hu X.-Y., Tian H.-W. 2019. Recent advances in perovskite oxides for anion-intercalation supercapacitor: A review, *Materials Science in Semiconductor Processing* 94: 35–50.
21. Wang X.W., Zhu Q.Q., Wang X.E., Zhang H.C., Zhang J.J., Wang L.F. 2016. Structural and electrochemical properties of $La_{0.85}Sr_{0.15}MnO_3$ powder as an electrode material for supercapacitor, *Journal of Alloys and Compounds* 675: 195–200.
22. Kumar N.S., Suvarna R.P., Naidu K.C.B. 2019. Grain and grain boundary conduction mechanism in sol-gel synthesized and microwave heated $Pb_{0.8-y}La_yCo_{0.2}TiO_3$ (y = 0.2 – 0.8) nanofibers, *Materials Chemistry and Physics* 223: 241–248.
23. Lang X., Mo H., Hu X., Tian H. 2017. Supercapacitor performance of perovskite $La_{1-x}Sr_xMnO_3$, *Dalton Transactions* 46: 13720–13730.
24. Raghuram N., Rao T.S., Naidu K.C.B. 2019. Investigations on functional properties of hydrothermally synthesized $Ba_{1-x}Sr_xFe_{12}O_{19}$ (x = 0.0 – 0.8) nanoparticles, *Material Science in Semiconductor Processing* 94: 136–150.
25. Kumar S.N., Suvarna R.P. Naidu K.C.B. 2018. Sol-gel synthesized and microwave heated $Pb_{0.8-y}La_yCo_{0.2}TiO_3$ (y = 0.2–0.8) nanoparticles: Structural, morphological and dielectric properties, *Ceramics International* 44: 18189–18199.
26. Naidu K.C.B., Reddy V.N., Sarmash T.S., Kothandan D., Subbarao T., Kumar N.S. 2019. Structural, morphological, electrical, impedance and ferroelectric properties of BaO-ZnO-TiO_2 ternary system, *Journal of the Australian Ceramic Society* 55: 201–218.
27. Kumar S.N., Suvarna R.P., Naidu K.C.B. Kumar G.R., Ramesh S. 2018. Structural and functional properties of sol-gel synthesized and microwave heated $Pb_{0.8}Co_{0.2-z}La_zTiO_3$ (z = 0.05–0.2) nanoparticles, *Ceramics International* 44: 19408–19420.
28. Sivakumar D., Naidu K.C.B., Nazeer K.P., Rafi M.M., Rameshkumar G., Sathyaseelan B., Killivalavan G., Begam A.A. 2018. Structural characterization and dielectric properties of superparamagnetic iron oxide nanoparticles, *Journal of the Korean Ceramic Society* 55: 230–238.
29. Mefford J.T., Hardin W.G., Dai S., Johnston K.P., Stevenson K.J. 2014. Anion charge storage through oxygen intercalation in $LaMnO_3$ perovskite pseudocapacitor electrodes, *Nature Materials* 13: 726–732.

30. Shafi P.M., Joseph N., Thirumurugan A., Bose A.C. 2018. Enhanced electrochemical performances of agglomeration-free $LaMnO_3$ perovskite nanoparticles and achieving high energy and power densities with symmetric supercapacitor design, *Chemical Engineering Journal* 338: 147–156.
31. Elsiddig Z.A., Xu H., Wang D., Zhang W., Guo X., Zhang Y., Sun Z., Chen J. 2017. Modulating Mn^{4+} ions and oxygen vacancies in nonstoichiometric $LaMnO_3$ perovskite by a facile sol-gel method as high-performance supercapacitor electrodes, *Electrochimica Acta* 253: 422–429.
32. Lang X., Mo H., Hu X., Tian H. 2017. Supercapacitor performance of perovskite $La_{1-x}Sr_xMnO_3$, *Dalton Transactions* 46: 13720–13730.
33. Mo H., Nan H., Lang X., Liu S., Qiao L., Hu X., Tian H. 2018. Influence of calcium doping on performance of $LaMnO_3$ supercapacitors, *Ceramics International* 44: 9733–9741.
34. Lv J., Zhang Y., Lv Z., Huang X., Wang Z., Zhu X., Wei B. 2015. A preliminary study of the pseudo-capacitance features of strontium doped lanthanum manganite, *RSC Advances* 5: 5858–5862.
35. Cao Y., Lin B., Sun Y., Yang H., Zhang X. 2015. Structure, morphology and electrochemical properties of $La_xSr_{1-x}Co_{0.1}Mn_{0.9}O_{3-\delta}$ perovskite nanofibers prepared by electrospinning method, *Journal of Alloys and Compounds* 624: 31–39.
36. George G., Jackson S.L., Luo C.Q., Fang D., Luo D., Hu D., Wen J., Luo Z. 2018. Effect of doping on the performance of high-crystalline $SrMnO_3$ perovskite nanofibers as a supercapacitor electrode, *Ceramics International.* doi:10.1016/j.ceramint.2018.08.313.
37. Galal A., Hassan H.K., Jacob T., Atta N.F. 2018. Enhancing the specific capacitance of $SrRuO_3$ and reduced graphene oxide in $NaNO_3$, H_3PO_4 and KOH electrolytes, *Electrochimica Acta* 260: 738–747.
38. Cao Y., Lin B., Sun Y., Yang H., Zhang X. 2015. Sr-doped lanthanum nickelate nanofibers for high energy density, *Electrochimica Acta* 174: 41–50.
39. Ho K., Wang J. 2017. Hydrazine reduction of $LaNiO_3$ for active materials in supercapacitors, *Journal of the American Ceramic Society* 100: 4629–4637.
40. Li Z., Zhang W., Wang H., Yang B. 2017. Two-dimensional perovskite $LaNiO_3$ nanosheets with hierarchical porous structure for high-rate capacitive energy storage, *Electrochimica Acta* 258: 561–570.
41. Che W., Wei M., Sang Z., Ou Y., Liu Y., Liu J. 2018. Perovskite $LaNiO_{3-\delta}$ oxide as an anion-intercalated pseudocapacitor electrode, *Journal of Alloys and Compounds* 731: 381–388.
42. Zhu L., Liu Y., Su C., Zhou W., Liu M., Shao Z. 2016. Perovskite $SrCo_{0.9}Nb_{0.1}O_{3-\delta}$ as an anion-intercalated electrode material for supercapacitors with ultrahigh volumetric energy density, *Angewandte Chemie International Edition* 55: 9576–9579.
43. Liu Y., Dinh J., Tade M.O., Shao Z. 2016. Design of perovskite oxides as anion-intercalation-type electrodes for supercapacitors cation leaching effect, *ACS Applied Materials & Interfaces* 8: 23774–23783.
44. Sharma R.K., Tomar A.K., Singh G. 2018. Fabrication of Mo-doped strontium cobaltite perovskite hybrid supercapacitor cell with high energy density and excellent cycling life, *ChemSusChem.* doi:10.1002/cssc.201801869.
45. Huang L., Li Q., Zhang G., Zhou X., Shao Z., Zhou W., Cao J. 2017. The preparation of $LaSr_3Fe_3O_{10-\delta}$ and its electrochemical performance, *Journal of Solid State Electrochemistry* 21: 1343–1348.
46. Li Z., Zhang W., Yuan C., Su Y. 2017. Controlled synthesis of perovskite lanthanum ferrite nanotubes with excellent electrochemical properties, *RSC Advances* 7: 12931–12937.

47. Wang W., Lin B., Zhang H., Sun Y., Zhang X., Yang H. 2019. Synthesis, morphology and electrochemical performances of perovskite-type oxide $La_xSr_{1-x}FeO_3$ nanofibers prepared by electrospinning, *Journal of Physics and Chemistry of Solids* 124: 144–150.
48. Rai A., Thakur A.K. 2017. Effect of Na and Mn substitution in perovskite type $LaFeO_3$ for storage device applications, *Ionics* 23: 2863–2869.
49. Liu Y., Wang Z., Veder J.P.M., Xu Z., Zhong Y., Zhou W., Tade M.O., Wang S., Shao Z. 2018. Highly defective layered double perovskite oxide for efficient energy storage via reversible pseudo capacitive oxygen-anion intercalation, *Advanced Energy Materials* 8: 1702604.
50. Lv J., Zhang Y., Lv Z., Huang X., Wang Z., Zhu X., Wei B. 2016. Strontium doped lanthanum manganite manganese dioxide composite electrode for supercapacitor with enhanced rate capability, *Electrochimica Acta* 222: 1585–1591.
51. Piburn G.W., Mefford J.T., Zinni N., Stevenson K.J., Humphrey S.M. 2017. Synthesis and charge storage properties of templated $LaMnO_3$–SiO_2 composite materials, *Dalton Transactions* 46: 977–984.
52. Lang X., Zhang H., Xue X., Li C., Sun X., Liu Z., Nan H., Hu X., Tian H. 2018. Rational design of $La_{0.85}Sr_{0.15}MnO_3@NiCo_2O_4$ core–shell architecture supported on Ni foam for high performance supercapacitors, *Journal Power Sources* 402: 213–220.
53. Nagamuthu S., Vijayakumar S., Ryu K.-S. 2017. Cerium oxide mixed $LaMnO_3$ nanoparticles as the negative electrode for aqueous asymmetric supercapacitor devices, *Materials Chemistry and Physics* 199: 543–551.
54. Elsiddig Z.A., Wang D., Xu H., Zhang W., Zhang T., Zhang P., Tian W., Sun Z., Chen J. 2018. Three-dimensional nitrogen-doped graphene wrapped $LaMnO_3$ nanocomposites as high-performance supercapacitor electrodes, *Journal of Alloys and Compounds* 740: 148–155.
55. Liu P., Liu J., Cheng S., Cai W., Yu F., Zhang Y., Wu P., Liu M. A. 2017. High-performance, electrode for supercapacitors: Silver nanoparticles grown on a porous perovskite-type material $La_{0.7}Sr_{0.3}CoO_{3-\delta}$ substrate, *Chemical Engineering Journal* 328: 1–10.
56. Cao X.L., Ren T.Z., Yuan Z.Y., Bandosz T.J. 2018. $CaTiO_3$ perovskite in the framework of activated carbon and its effect on enhanced electrochemical capacitance, *Electrochimica Acta* 268: 73–81.
57. Moitra D., Anand C., Ghosh B.K., Chandel M., Ghosh N.N. 2018. One-dimensional $BiFeO_3$ nanowire-reduced graphene oxide nanocomposite as excellent supercapacitor electrode material, *ACS Applied Energy Materials* 1: 464–474.
58. Ouyang Y., Huang R., Xia X., Ye H., Jiao X., Wang L., Lei W., Hao Q. 2018. Hierarchical structure electrodes of NiO ultrathin nanosheets anchored to $NiCo_2O_4$ on carbon cloth with excellent cycle stability for asymmetric supercapacitors, *Chemical Engineering Journal*. doi:10.1016/j.cej.2018.08.142.
59. Chang X., Li W., Liu Y., He M., Zheng X., Lv X., Ren Z. 2019. Synthesis and characterization of $NiCo_2O_4$ nanospheres/nitrogen-doped graphene composites with enhanced electrochemical performance, *Journal of Alloys and Compounds*. doi:10.1016/j.jallcom.2019.01.036.
60. Dong K., Wang Z., Sun M., Wang D., Luo S., Liu Y. 2019. Construction of $NiCo_2O_4$ nanorods into 3D porous ultrathin carbon networks for high-performance asymmetric supercapacitors, *Journal of Alloys and Compounds* 783: 1–9.
61. Fu M., Chen W., Ding J., Zhu X., Liu Q. 2019. Biomass waste derived multi-hierarchical porous carbon combined with $CoFe_2O_4$ as advanced electrode materials for supercapacitors, *Journal of Alloys and Compounds*. doi:10.1016/j.jallcom.2018.12.244.
62. Feng H., Gao S., Shi J., Zhang L., Peng Z., Cao S. 2019. Construction of 3D hierarchical porous $NiCo_2O_4$/graphene hydrogel/Ni foam electrode for high-performance supercapacitor, *Electrochimica Acta*. doi:10.1016/j.electacta.2018.12.177.

63. Cai N., Fu J., Chan V., Liu M., Chen W., Wang J., Zeng H., Yu F. 2019. $MnCo_2O_4$@ nitrogen-doped carbon nanofiber composites with meso-microporous structure for high performance symmetric supercapacitors, *Journal of Alloys and Compounds*. doi:10.1016/j.jallcom.2018.12.044.
64. Gao X., Wang W., Bi J., Chen Y., Hao X., Sun X., Zhang J. 2018. Morphology controllable preparation of $NiFe_2O_4$ as high performance electrode material for supercapacitor, *Electrochimica Acta*. doi:10.1016/j.electacta.2018.11.054.
65. Li J., Wang Y., Xu W., Wang Y., Zhang B., Luo S., Zhou X., Zhang C., Gu X., Hu C. 2018. Porous Fe_2O_3 nanospheres anchored on activated carbon cloth for high-performance symmetric supercapacitors, *Nano Energy*. doi:10.1016/j.nanoen.2018.12.061.
66. Sui Y., Ye A., Qi J., Wei F., He Y., Meng Q., Ren Y., Sun Z. 2018. Construction of $NiCo_2O_4$@$Ni_{0.85}Se$ core-shell nanorod arrays on Ni foam as advanced materials for an asymmetric supercapacitor, *Journal of Alloys and Compounds*. doi:10.1016/j.jallcom.2018.10.354.
67. Wang X., Shi B., Huang F., Fang Y., Rong F., Que R. 2018. Fabrication of hierarchical $NiCo_2O_4$@$NiCo_2S_4$ core/shell nanowire arrays by an ion-exchange route and application to asymmetric supercapacitors, *Journal of Alloys and Compounds*. doi:10.1016/j.jallcom.2018.07.074.
68. Xinming Wu, Lian Meng, Qiguan Wang, Wenzhi Zhang, Yan Wang. 2019. A novel inorganic-conductive polymer core-sheath nanowire arrays as bendable electrode for advanced electrochemical energy storage, *Chemical Engineering Journal* 358: 1464–1470.
69. Mukhiya, T., Dahal, B., Ojha, G. P., Kang, D., Kim, T., Chae, S.-H. ... Kim, H. Y. 2019. Engineering nanohaired 3D cobalt hydroxide wheels in electrospun carbon nanofibers for high-performance supercapacitors, *Chemical Engineering Journal*, 361:1225–1234.
70. Zhu Y.Q., Cao C., Tao S., Chu W., Wu Z., Li Y. 2014. Ultrathin Nickel hydroxide and oxide nanosheets: Synthesis, characterizations and excellent supercapacitor performances, *Scientific Reports* 4: 5787. doi:10.1038/srep05787.
71. Kumar N.S., Suvarna R.P., Naidu K.C.B. 2018. Multiferroic nature of microwave-processed and sol-gel synthesized nano $Pb_{1-x}Co_xTiO_3$ ($x = 0.2–0.8$) ceramics, *Crystal Research and Technology* 53: 1800139. doi:10.1002/crat.201800139.
72. Naidu C.B., Kumar N.S., Kumar G.R., Kumar S.N. 2018. Temperature and frequency dependence of complex impedance parameters of microwave sintered NiMg ferrites, *Journal of the Australian Ceramic Society*. doi:10.1007/s41779-018-0260-x.
73. Naidu K.C.B., Madhuri W. 2018. Ceramic nanoparticle synthesis at lower temperatures for LTCC and MMIC technology, *IEEE Transactions on Magnetics* 54. doi:10.1109/TMAG.2018.2855663.
74. Kumar N.S., Suvarna R.P., Naidu K.C.B. 2018. Structural and ferroelectric properties of microwave heated lead cobalt titanate nanoparticles synthesized by sol–gel technique, *Journal of Materials Science: Materials in Electronics* 29: 4738–4742.
75. William J.J., Babu I.M., Muralidharan G. 2019. Microwave assisted fabrication of l-Arginine capped α-$Ni(OH)_2$ microstructures as an electrode material for high performance hybrid supercapacitors, *Materials Chemistry and Physics*. doi:10.1016/j.matchemphys.2018.12.044.
76. Wei W., Cui X., Chen W., Ivey D.G. 2011. Manganese oxide-based materials as electrochemical supercapacitor electrodes, *Chemical Society Reviews* 40: 1697–1721.
77. Zhang X., Wei L., Guo X. 2018. Ultrathin mesoporous $NiMoO_4$-modified MoO_3 core/shell nanostructures: Enhanced capacitive storage and cycling performance for supercapacitors, *Chemical Engineering Journal* 353: 615–625.
78. Xue Q., Tian Y., Deng S., Huang Y., Zhu M., Pei Z. et al. 2018. LaB_6 nanowires for supercapacitors, *Materials Today Energy* 10: 28–33.

79. Zhu T., Wu H.B., Wang Y., Xu R., Lou X.W. 2012. Formation of 1D hierarchical structures composed of Ni_3S_2 nanosheets on CNTs backbone for supercapacitors and photocatalytic H_2 production, *Advanced Energy Materials* 2: 1497–1502.
80. Ede S.R. Kundu S. 2015. Microwave synthesis of $SnWO_4$ nanoassemblies on DNA scaffold: A novel material for high performance supercapacitor and as catalyst for butanol oxidation, *ACS Sustainable Chemistry & Engineering* 3: 2321–2336.
81. Muzaffar A., Ahamed M.B., Deshmukh K., Thirumalai J. 2019. A review on recent advances in hybrid supercapacitors: Design, fabrication and applications, *Renewable and Sustainable Energy Reviews* 101: 123–145.
82. Liu L., Feng Y., Wu W. 2019. Recent progress in printed flexible solid-state supercapacitors for portable and wearable energy storage, *Journal of Power Sources* 410–411: 69–77.
83. Xu T., Li G., Yang X., Guo Z., Zhao L. 2019. Design of the seamless integrated C@NiMn-OHNi$_3$S$_2$/Ni foam advanced electrode for supercapacitors, *Chemical Engineering Journal*. doi:10.1016/j.cej.2019.01.083.
84. Zhao X., Sanchez B.M., Dobson P.J., Grant P.S. 2011. The role of nanomaterials in redox-based supercapacitors for next generation energy storage devices, *Nanoscale* 3: 839–855.
85. Huang Y., Liang J., Chen Y. 2012. An overview of the applications of graphene-based materials in supercapacitors, *Small* 8: 1805–1834.
86. Vijayakumar S., Nagamuthu S., Muralidharan G. 2013. Supercapacitor studies on NiO nanoflakes synthesized through microwave route, *ACS Applied Materials & Interfaces*. doi:10.1021/am400012h.
87. Wang H., Dai H. 2013. Strongly coupled inorganic–nano-carbon hybrid materials for energy storage, *Chemical Society Reviews* 42: 3088–3113. doi:10.1039/c2cs35307e.
88. Sun J., Li W., Zhang B., Li G., Jiang L., Chen Z., Zou R., Hu J.. 2014. 3D core/shell hierarchies of MnOOH ultrathin nanosheets grown on NiO nanosheet arrays for high-performance supercapacitors, *Nano Energy* 4: 56–64.
89. Sun L., Tian C., Fu Y., Yang Y., Yin J., Wang L., Fu H. 2013. Nitrogen-doped porous graphitic carbon as an excellent electrode material for advanced supercapacitors, *Chemistry–A European Journal* 19: 1–12.
90. Yang C., Liu Q., Zang L., Qiu J., Wang X., Wei C. et al. 2019. High-performance yarn supercapacitor based on metal–inorganic–organic hybrid electrode for wearable electronics, *Advanced Electronic Materials* 5: 1800435.
91. Asen P., Haghighi M., Shahrokhian S., Taghavinia N. 2019. One step synthesis of SnS_2-SnO_2 nano-heterostructured as an electrode material for supercapacitor applications, *Journal of Alloys and Compounds* 782: 38–50.
92. Bahmani F., Kazemi S.H., Kazemi H., Kiani M.A., Feizabadi S.S.Y. 2019. Nanocomposite of copper–molybdenum–oxide nanosheets with graphene as high-performance materials for supercapacitors, *Journal of Alloys and Compounds*. doi:10.1016/j.jallcom.2018.12.353.
93. Ding X., Zhu J., Hu G., Zhang S. 2018. Core–shell structured $CoNi_2S_4$@polydopamine nanocomposites as advanced electrode materials for supercapacitors, *Ionics*. doi:10.1007/s11581-018-2798-6.
94. Fu M., Chen W., Ding J., Zhu X., Liu Q. 2019. Biomass waste derived multi-hierarchical porous carbon combined with $CoFe_2O_4$ as advanced electrode materials for supercapacitors, *Journal of Alloys and Compounds*. doi:10.1016/j.jallcom.2018.12.244.
95. Gao H., Wu F., Wang X., Hao C., Ge C. 2018. Preparation of $NiMoO_4$-PANI core-shell nanocomposite for the high-performance all-solid-state asymmetric supercapacitor, *International Journal of Hydrogen Energy* 43(39): 18349–18362.
96. Ge J., Wu J., Ye B., Fan L., Jia J. 2018. Hollow rod-like hybrid Co_2CrO_4/$Co_{1-x}S$ for high-performance asymmetric supercapacitor, *Journal of Materials Science: Materials in Electronics*. doi:10.1007/s10854-018-0373-6.

97. Guo X., Luan Z., Lu Y., Fu L., Gai L. 2019. $NaTi_2(PO_4)_3$/C|| carbon package asymmetric flexible supercapacitors with the positive material recycled from spent Zn-Mn dry batteries, *Journal of Alloys and Compounds.* doi:10.1016/j.jallcom.2018.12.163.
98. Hao X., Bi J., Wang W., Chen Y., Gao X., Sun X., Zhang J. 2018. Bimetallic carbide Fe_2MoC as electrode material for high-performance capacitive energy storage, *Ceramics International.* doi:10.1016/j.ceramint.2018.08.297.
99. Kalyani M., Emerson R.N. 2018. Electrodeposition of nano crystalline cobalt oxide on porous copper electrode for supercapacitor, *Journal of Materials Science: Materials in Electronics.* doi:10.1007/s10854-018-0389-y.
100. Lee K.U., Byun J.Y., Shin H.J., Kim S.H. 2018. A high-performance supercapacitor based on polyaniline-nanoporous gold, *Journal of Alloys and Compounds.* doi:10.1016/j.jallcom.2018.11.022.
101. Li J., Liu Z., Zhang Q., Cheng Y., Zhao B., Dai S. et al. 2018. Anion and cation substitution in transition-metal oxides nanosheets for high-performance hybrid supercapacitors, *Nano Energy.* doi:10.1016/j.nanoen.2018.12.011.
102. Liu S., Yin Y., San Hui K., Hui K.N., Lee S.C., Jun S.C. 2018. Nickel hydroxide/chemical vapor deposition-grown graphene/nickel hydroxide/nickel foam hybrid electrode for high performance supercapacitors, *Electrochimica Acta.* doi:10.1016/j.electacta.2018.11.070.
103. Mariappan V.K., Krishnamoorthy K., Pazhamalai P., Sahoo S., Nardekar S.S., Kim S.-J. 2018. Nanostructured ternary metal chalcogenide-based binder-free electrodes for high energy density asymmetric supercapacitors, *Nano Energy.* doi:10.1016/j.nanoen.2018.12.031.
104. Wen S., Qin K., Liu P., Zhao N., Shi C., Ma L., Liu E. 2019. Ultrafine $Ni(OH)_2$ nanoneedles on N-doped 3D rivet graphene film for high-performance asymmetric supercapacitor, *Journal of Alloys and Compounds.* doi:10.1016/j.jallcom.2018.12.347.
105. Yang H.X., Zhao D.L., Meng W.J., Zhao M., Duan Y.J., Han X.Y., Tian X.M. 2019. Nickel nanoparticles incorporated into N-doped porous carbon derived from N-containing nickel-MOF for high-performance supercapacitors, *Journal of Alloys and Compounds.* doi:10.1016/j.jallcom.2018.12.259.
106. Brooke R., Edberg J., Sawatdee A., Grimoldi A., Åhlin J., Gustafsson G. et al. 2019. Supercapacitors on demand: All-printed energy storage devices with adaptable design, *Flexible and Printed Electronics.* doi:10.1088/2058-8585/aafc4f.

13 Molybdenum Based Materials for Supercapacitors Beyond TMDs

Swapnil S. Karade and Deepak P. Dubal

CONTENTS

13.1 Introduction 239
13.2 Basic Properties of MoS_2 240
13.3 Synthesis Routes 241
13.4 MoS_2 Based Supercapacitors 242
13.5 MoS_2-Nanocarbon Composite for Supercapacitors 243
13.6 MoS_2-Conducting Polymer Composite Based Supercapacitors 245
13.7 Other Composites of MoS_2 248
13.8 Conclusion and Forward Look 252
References 252

13.1 INTRODUCTION

The energy source is a prime requirement of the recent advanced smart grid digital technological devices which made our life easy. However, thinking about endangered condition of natural energy sources such as coal, fossil fuel, crude oil, etc. concluded that there is need to find some artificial substitution of energy generation and storage devices with significant efficiency. Renewable energy consisting of solar energy and wind energy are in demand for energy production, but limitations arise due to non-availability of the sunlight at night or during a rainy season, which is out of our control. However, it is dynamic to find more resourceful ways to store these types of energy effectively because they are frequently sporadic in nature. In this regard, much of the recent research efforts are dedicated towards the improvement of new nonstructural materials with simple, cheap, effective and efficient outlook for the promising application in the field of energy storage system. The novel materials with excellent electrochemical properties such as huge stability, effective surface morphology and least diffusion path are greatly influential factors on electrochemical performance [1,2]. Moreover, simple, inexpensive and eco-friendly synthesis methods are producing a significant approach towards materials progresses at commercial level.

In the last decade, transition metal dichalcogenides (TMDs) have been utilized at large extend for the energy storage and conversion applications. Several transition metal sulfides such as VS_2, NiS, MnS, FeS, MoS_2, CuS and CoS are synthesized using simple a chemical route and utilized for supercapacitor electrode. On the basis of layered structure with effective surface morphology, the metal sulfides attained the identical height with respect to traditional materials in the field of advanced materials development technology. Through the wide applications of solar cell, supercapacitor, lithium/sodium ion batteries, gas sensing and photocatalysis metal sulfides have shown their growing footprints towards nanostructural development.

Loh and coworkers have firstly considered the MoS_2 nanowall films and reported their capacitive behavior in 2007 until that scientists did not think about TMDs for the application of supercapacitors [3]. Though, after 2007, a significant number of articles have been issued through the researchers on MoS_2 and MoS_2 composite- (MoS_2/carbon and MoS_2/conducting polymer) based electrodes for supercapacitor applications [4]. The synthesis methods of MoS_2 are most of all high-temperature solution-based or they contain exfoliation of bulk MoS_2 crystals trough top-down approach, forming high crystalline MoS_2 materials [4,5]. Generally, highly crystalline phase-oriented pseudocapacitive materials exhibits poor specific capacitance (Cs) whereas amorphous phase with low crystallinity shows a higher specific capacitance [6]. This phenomenon has been verified by examining the amorphous and crystalline phase of NiO along with its bulk and surface contribution in the capacitance [7]. Moreover, Lu et al. tried to find the proof with experimental verification and interestingly they obtained three-fold increment in capacitance for amorphous phase of NiO over crystalline phase. In recent, Zhang et al. have developed a core shell structure of crystalline Ni_3S_4 and amorphous MoS_2, which revealed improved capacitive performance than bare Ni_3S_4 [8].

In this chapter, we cover sufficient literature, which has been highlighted in order to understand the properties and application of molybdenum sulfide (MoS_2) in supercapacitors as electrode material. Also, the discussion has extended by introducing composite structural and morphological development of MoS_2 material through carbon, polymer and other inorganic-based nanomaterials. Hopefully, this chapter visualizes the open path for the development of MoS_2 materials towards energy storage application.

13.2 BASIC PROPERTIES OF MoS_2

Since the last two decades, a considerable amount of work towards optical, physical, chemical and electrical properties of molybdenum for suitable application point of view has been accomplished [9–16]. According to experimental survey it is known that, the molybdenum sulfide mainly available in three structural polytypes such as 2H-MoS_2, 3R-MoS_2, and 1T-MoS_2. Generally, MoS_2 possess extreme anisotropy in terms of electrical, optical and physical properties as it is a one of the TMDs. The electric and thermal conductivity of the layered compounds are the higher order of magnitude along the direction parallel to the basal plane, which are a smaller magnitude in the direction perpendicular to the basal plane [10]. The layered hexagonal crystal structure in MoS_2 is a result of strong covalent bonding and weak van der Waals forces between Mo and S in layered plane and S-Mo-S sandwiched layer [9,16]. The gap between the S-Mo-S sandwiched layers is approximately 3.49 Å which is examined by using X-ray diffraction (XRD) analysis [10,17,18]. The difference in electronegativity between the Mo and S leads to covalent bond within the S-Mo-S. The sulfur anions take up valence electrons from molybdenum cations to acquire II^- oxidation state and oxidation state of molybdenum cations become IV^+ [10]. However, to the formation of hexagonal unit cell of 2H-MoS_2 in layered structure, every Mo atom becomes coordinate with six sulfur atoms whereas every S atom contribute their coordination with three Mo atoms. On the other hand, the arrangements of molybdenum and sulfur atoms in hexagonal sheets are along a direction perpendicular to the basal plane [10]. The trigonal prismatic coordination of Mo atoms gives rise to the six equivalent cylindrical bond functions with a combination of the 4d, 5s, and 5p orbitals [16,18–20]. Pauling and Hultgren have described this type of orbital grouping as d4sp hybridization [21,22]. Four valence electrons of the molybdenum atom primarily utilized to form the 4d orbital, which are strongly responsible for sulfur bonding and the rest of two-valence electrons exist in non-bonding orbitals. Coordination to three molybdenum atoms is attained by each sulfur atom through the hybridization of 3p and 3d orbitals. The van der Waal bonding between S-Mo-S sandwiched later is a result of the interaction between saturate sulfur 3s subshell, which extends perpendicular to the basal plane [10,16]. Thus, the extension of the bulk structure along the c-direction upon intercalation is allowed by weak interlayer van-der Waal bonding.

13.3 SYNTHESIS ROUTES

Nanostructured transition metal dichalcogenides (TMDs) have been verified to expose a significant electrochemical performance because of their unique shape, size and structure. So far, various methods have been recognized for the synthesis of nanostructured TMDs to be used in supercapacitor. Each synthetic method has its own advantages and drawbacks. Even under considering the limitations, it is possible to develop structural and morphological properties of the materials by using every synthetic method.

Low temperature hydrothermal method is one of the most general process and also frequently used for the synthesis of nanostructured MoS_2 material with possible to control specific size and morphology of the material. In this technique, the reaction is carried out in air tight container using water soluble metal salts within temperature limit of 200°C. In the meantime, a high pressure will self-develop and is associated to the reaction temperature, the percentage of the liquid filled, and any dissolved salts. On the other hand, solvothermal is also firmly used for the synthesis of MoS_2 compound. This technique is similar to hydrothermal method where the water is substituted by an organic solvent. Solvothermal is one of a promising technique for the synthesis of composite structures. Apart from these, several other synthesis methods also have reported for the synthesis of MoS_2 compound such as chemical bath deposition, microwave heating, successive ionic layer adsorption and reaction and electrodeposition. These techniques of synthesis are also simple, cost effective and easily controllable which is also much effective for the controlling size and morphology of the compound.

13.4 MoS_2 BASED SUPERCAPACITORS

Considering numerous advances in its distinctive chemical and physical characteristics, nanostructured MoS_2 is a cheering candidate for the application of energy storage and conversion [23]. The MoS_2 has been applied in the form of various nanostructures for the application of supercapacitors [23–25]. In this regard, we have successfully prepared homogeneous ultrathin MoS_2 nanoflakes at room temperature chemical route and successfully utilized as a supercapacitor electrode. The prepared MoS_2 ultrathin nanoflakes thin film electrode revealed superior electrochemical performance with a maximum specific capacitance of 576 F/g at 5 mV/s and significant capacity retention of 82% after 3000 cycles [26]. The significant electrochemical parameter achieved by hydrothermally prepared MoS_2 nanoflake array on copper foil along with fundamental charge and discharge process has explained in this report with schematic illustration for hexagonal MoS_2 crystal structure [27]. In addition, solvothermally prepared MoS_2 on nickel foam electrode has reported for the symmetric supercapacitor device. The fabricated symmetric device of MoS_2@Ni foam electrode shows 39 F/g specific capacitance with excellent long-term stability of 81% after 10,000 cycle [28]. Furthermore, neoteric approach by using magnetron sputtering method has developed to produce high efficient thin film-based electrode of MoS_2 for supercapacitor application. The electrode exhibited a significant specific capacitance of 330 F/g at 10 mV/s and also obtained excellent yield in specific capacitance of 97% after 5000 cycles [29]. Recently, Pujari et al. have reported a thin film of MoS_2 by using low-temperature soft-chemical-bath deposition. The direct growth of MoS_2 on stainless steel substrate revealed a significant specific capacitance of 180 F/g at 5 mV/s and offers 23 Wh/kg specific energy at a power density of 2100 W/kg [30]. Krishnamoorthy at co-workers has established a ball milling methodology to achieving few-layered MoS_2 via bulk MoS_2 exfoliation. The electrode exhibits remarkable specific capacitance, energy density and cyclic stability [31]. Further, a zinc ion-assisted hydrothermal route has followed to prepare hierarchical MoS_2 microspheres which demonstrates excellent pseudocapacitive properties and offers remarkable specific capacitance, rate capability and cyclic stability [32]. Islam and co-workers have reported beyond one step of previous one by assembling coin cell supercapacitor using vertically oriented MoS_2 nanosheets on plasma pyrolyzed cellulose filter paper [33]. The organic electrolyte used in this work which offers extended potential window of 2.5 V along with significant capacitance of 48 mF/cm^2. Likewise, there are many hydrothermal routes that can be suitable for the synthesis of MoS_2 in order to achieve favorable electrode morphologies such as hierarchical hollow nanospheres [34,35], flower-like MoS_2 nanostructures [25], mesoporous MoS_2 nanostructures [36], and MoS_2 nanosheets [24,37].

According to the above estimated results, MoS_2 possesses potential ability to develop further up to the device extend in the field of supercapacitor. But, by considering a layered structure of MoS_2, a small space between the stacked MoS_2 would not in favor of electrochemical applicability, which further results into a poor and inefficient charge transport. An electrode developed with an enlarged interlayer distance of MoS_2 layers by material/composite as one of the key constituents to overcome this limitation. In this regard, growing research is available on MoS_2 composites with a carbon, polymers and other electrochemical active materials.

13.5 MoS_2-NANOCARBON COMPOSITE FOR SUPERCAPACITORS

To improve the device storage efficiency there is much needed to work on the development of a multifaceted material structure [38]. The significant energy storage capacity has shown by MoS_2 nanosheets, which makes them potentially suitable candidate for supercapacitors electrode. But, their excellent electrochemical properties cannot be effectively utilized due to the low conductivity of MoS_2. In addition, synthesis of single or multi-layered MoS_2 nanosheets is a difficult task, but it would offer the high surface area, which is extremely valuable property of the material useful for energy storage [39]. We are quite aware about the contribution of carbon-based materials in the field of energy storage due to its physical, chemical, thermal, and mechanical properties along with its eco-friendly nature. The well-known carbon-based materials like activated carbon [40], carbon spheres [41,42] graphene aerogels [43], microporous carbon derived from metal-organic framework [44], graphene [45,46] and carbon nanotubes [39,47] possess good electrical conductivity and widely studied as a supercapacitor electrode. Additionally, carbon-based materials are much familiar because they offer high utilized surface area with π-π stacking and strong electrostatic interaction for the interfacial conjugation of supporting nanomaterials [48]. Additionally, they deliver supplementary access for ion/electron transfer which further resulted into reduced charge transfer resistance. Rather low capacitance due to non-faradaic storage behavior even its high surface area limits its extended potential ability [49]. On the basis of above favorable properties, enormous efforts have been taken to designing materials conjugation of carbon and MoS_2-based composite for supercapacitor [50]. Huang et al. reported a facile hydrothermal route for the synthesis of carbon aerogel incorporated flower-like MoS_2 supported by L-cysteine. The MoS_2/carbon aerogel hybrid revealed a significant specific capacitance of 260 F/g at 1 A/g current density and also shows only 8% degradation in capacitance after 1500 cycles [51]. Interestingly, Xu et al. prepared a tubular KF@MoS_2/rGO composite by using facile hydrothermal method. The aerogel of KF@MoS_2/rGO shows a typical wrinkled and crosslinking structure with a uniform dispersion of tubular KF@MoS_2 and presented in Figure 13.1. The specific capacitance reached at 347 F/g for the composite electrode and further used for the fabrication of asymmetric supercapacitor device with KF@MnO_2 positive electrode [52].

Recently, Thakur et al. reported MoS_2 flake integrated with boron and carbon doped carbon. The composition ratio of BCN/MoS_2-11 reveals a superior electrochemical performance with excellent cyclic stability of 90% after 5000 cycles [53]. Sun and co-workers have reported a uniform MoS_2 nanolayer with sulphur vacancy on carbon nanotube networks. The composite electrode of MoS_{2-x}@CNT shows enhanced GCD profile by obtaining highest specific capacitance of 512 F/g and superior cyclic stability increment by 6% after 2000 cycles [54]. A novel aligned CNT/MoS_2 hybrid nanostructure has been prepared to effectively combine the advantages of the CNT and MoS_2 with remarkable electrochemical performances. The fibrous supercapacitor shows significant capacitance of 135 F/cm^2 and long-term cycle life of 92% after 10,000 cycles [39]. Further, the significant electrochemical performance has reported by synthesizing coaxial yarn electrodes composed of activated carbon fibers tows and MoS_2 nanosheets (ACFTs/MoS_2). The predictable enhancement

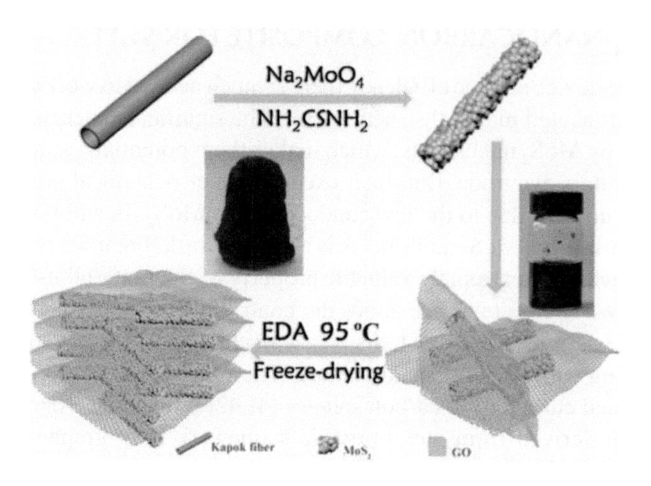

FIGURE 13.1 Schematic illustration for fabrication of KF@MoS_2/rGO. (With kind permission from Springer Science+Business Media: *J. Mater. Sci.*, 53, 2018, 11659, Xu, W. et al.)

in the electrochemical performance was obtained for ACFTs/MoS_2 composite by obtaining maximum gravimetric capacitance of 308 F/g at a lower scan rate of 5 mV/s and superior cyclic retention more than 97% after 6000 cycles [40]. The challenge of achieving a maximum energy density without sacrificing the rate capability was successfully tackled through the development of fibers by integrating MoS_2 and graphene nanosheets into well-aligned multiwalled carbon nanotubes (MoS_2-graphene/MWCNTs fibers). The developed MoS_2-graphene/MWCNT fibers have perfectly utilized by fabricating an asymmetric supercapacitor which further obtaining extended potential limit of 1.4 V with excellent retention in coulombic efficiency, significant rate capability, improved energy density and excellent cycling stability [45].

Yang et al. have reported a facile hydrothermal reduction approach fort the synthesis of MoS_2/graphene aerogel composites. The as-prepared MoS_2/graphene aerogel composite were examined through electrochemical testing for supercapacitor electrode, and revealed a maximum specific capacitance of 268 F/g at lower current density of 0.5 A/g, along with that a high rate capability (88% capacitance retention after 15 A/g current density), and non-degradable cyclic stability with 7% loss in capacitance after 1000 cycles [43]. The highly effective and efficient hydrothermal technique has followed in order to the synthesis of complex structure of MoS_2-modified carbon nanospheres in two different morphologies of flower and spherical shaped. The as-prepared electrode is showing a pseudocapacitive behavior in aqueous electrolyte. The flower structured MoS_2/CNT-based electrode exhibited a 231 F/g specific capacitance with an efficient energy density of 26 Wh/kg by maintaining power density of 6443 W/kg [41]. Weng et al. have proposed a three step scheme for the complexed structural synthesis of MoS_2 nanosheets wrapped with MOF-derived microporous carbons (MoS_2@MPC). The complex composite electrode of MoS_2@MPC delivered a significant specific capacitance and long-term cyclic stability along with excellent rate performance because of mutual participation of combined structural portion under the electrochemical charge-discharge

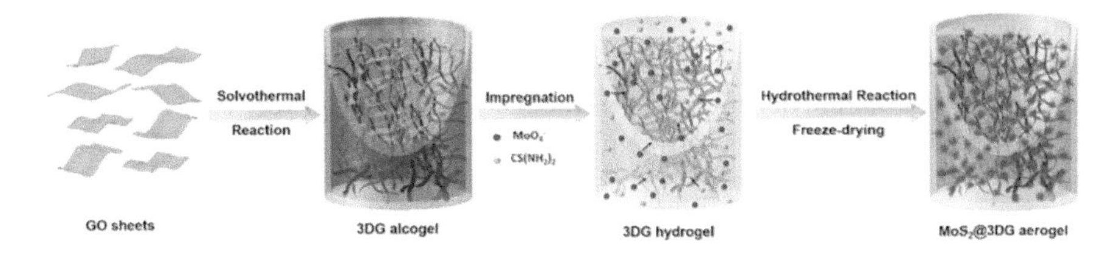

FIGURE 13.2 Schematic of fabrication of the MoS_2@3DG hybrid structure. (Reprinted with permission from Zhang, F. et al., *ACS Appl. Mater. Interfaces*, 8, 4691, 2016. Copyright 2016 American Chemical Society.)

process [44]. Masikhwa et al. developed a MoS_2/graphene foam with different graphene foam loading. The composite electrode of MoS_2/graphene foam revealed a maximum specific capacitance of around 400 F/g at a scan rate of 5 mV/s. Also, MoS_2/graphene foam electrode was successfully utilized by fabricating asymmetric device with an activated carbon derived from expanded graphite (AEG) as suitable negative electrode and it offers a significant specific capacitance and cyclic stability [55]. An in-situ hydrothermal approach has reported by Murugan et al. to synthesize the graphene-MoS_2 nanocomposites. The cyclic voltammetry study of graphene-MoS_2 nanocomposites electrode shown an improved specific capacitance (334 F/g at 10 mV/s) with significant retention by 10% degradation in stability after 500 cycles [56]. The hierarchical structure of graphene-wrapped CNT@MoS_2 was successfully developed by Sun and co-workers. The CNT@MoS_2 electrode revealed an improved specific capacitance of 498 F/g at a current density of 1 A/g with a high rate performance of 56.5% after 10 A/g and showing an excellent retention by losing only 4% capacitance after 10,000 cycles [57]. Zhang et al. have developed an effective composite structure of MoS_2 with spheres of mesoporous carbon (MoS_2/MCS). A hybrid electrode of MoS_2/MCS offers maximum specific capacitance of 411 F/g at minimum current density of 1 A/g and retained at 272 F/g after 10 A/g current which implies a good rate capability. Subsequently, capacitance retained greater than 93% after 1000 cycles implies the MoS_2/MCS is a promising candidate for future energy storage development [58]. Additionally, Zhang et al. developed a MoS_2@3DG aerogel by using simple solution-based process. The schematic of hybrid structure formation process is depicted in Figure 13.2. The hybrid structure is utilized to fabricate hybrid supercapacitor device which further resulted into wide potential and high energy supercapacitor device [59].

13.6 MoS_2-CONDUCTING POLYMER COMPOSITE BASED SUPERCAPACITORS

Among the list of all pseudocapacitive materials, conducting polymers are well known and high efficient class due to their easy synthesis, cost effectiveness, and non-hazardous. However, conducting polymers does not sustain long term stability due to the faradaic charge storage mechanism mechanical will be resulted into subsequent degradation of the polymer [60]. Therefore, it is worth to be hybridized with organic and/or inorganic compounds to improve the performance and stability [61,62]. By considering this, nanocomposites of PANI with MoS_2 have developed and successfully utilized as an electrode material for a supercapacitor with much attractive electrochemical outcomes [63]. Wang et al. reported an intercalated composites of poly(3,4-ethylenedioxythiophene) and MoS_2 though very common method of in-situ polymerization by using ammonium persulfate as an oxidizing agent. A significant enhancement was obtained in specific capacitance and obtaining the value of 405 F/g along with only 10% performance degradation after 1000 cycles [64].

There are some commonly used conduction polymers such as polyaniline (PANI), polypyrrole (PPy) and poly(3,4-ethylenedioxythiophene) (PEDOT) in order to making composites with MoS_2 for the development of improved electrode material for supercapacitor [65]. In this orientation, Huang et al. reported a facile route to synthesize PANI/graphene analogue MoS_2 composite by an in-situ polymerization. A maximum specific capacitance of 575 F/g was obtained at 1 A/g current density and less than 2% decrement in specific capacitance after 500 cycles at 1 A/g which further denotes a PANI/MoS_2 composite electrode is a promising candidate for the application of supercapacitor [62]. Towards the forward step, Yang et al. developed a nanostructure of carbon shell-coated PANI grown on 1T MoS_2 monolayers (MoS_2/PANI@C). As prepared composite electrode reveals an improved specific capacitance value of 678 F/g at1 mV/s scan rate and remarkable cyclic retention of 80% was obtained after 10,000 cycles. The composite electrode shows an outstanding enhancement than the bare electrodes of MoS_2 [61]. Sha et al. have reported a two-step strategy in order to synthesize a ternary composite of a MoS_2/PANI/graphene for high efficient supercapacitor. The composite electrode exhibits a high specific capacitance of 618 F/g at low current of 1 A/g and retained a 476 F/g at very high current of 20 A/g implying an excellent rate capability along with extraordinary cyclic stability [66]. Ren et al. developed an efficient hybrid supercapacitor electrode of 3D tubular MoS_2/PANI (as shown in Figure 13.3) and optimize the effect of different amounts of PANI concentration. The well optimized hybrid electrode with 60% loading concentration of PANI not only reveals an excellent specific capacitance of 552 F/g at lower current density, but also retained 82% after very high current density of 30 A/g [67]. Ansari et al. proposed an efficient nanocomposite by combining a mechanically exfoliated MoS_2 sheet and PANI. This nanocomposite revealed significant enhancement in specific capacitance of 510 F/g [68]. Wang and co-workers have also published a binary nanocomposite of MoS_2/PANI by using chemical exfoliation for supercapacitor application. The outstanding electrode performance was obtained by reached at maximum specific capacitance value of 560 F/g for 1 A/g current density [69]. Fu et al. have presented a novel composite of PANI/MoS_2 with an effective morphology of spongia-shaped nanospheres by using polyvinylpyrrolidone addition. The as prepared composite electrode demonstrates enhanced specific capacitance of 605 F/g at 1 A/g with an effective energy density of 53.78 Wh/kg at power density of 400 W/kg along with 88% capacitance retention after 1000 cycles [70]. Thakur et al. constructed a facile and efficient-hybrid composite of PANI/CNT/MoS_2 ternary compound, which shown their promising electrochemical behavior with improved specific capacitance 350 F/g at 1A/g and obtaining 68% retention after 2000 cycles at 10 A/g [71]. Chang et al. developed a decorative growth of MoS_2 nanoflakes on one dimensional PPy nanotubes as shown in Figure 13.4. This nanocomposite exhibits an effective specific capacitance of 307 F/g at a current density of 1 A/g, with outstanding retention of 96% in specific capacitance after 1000 cycles [72]. Furthermore, Chen et al. designed a

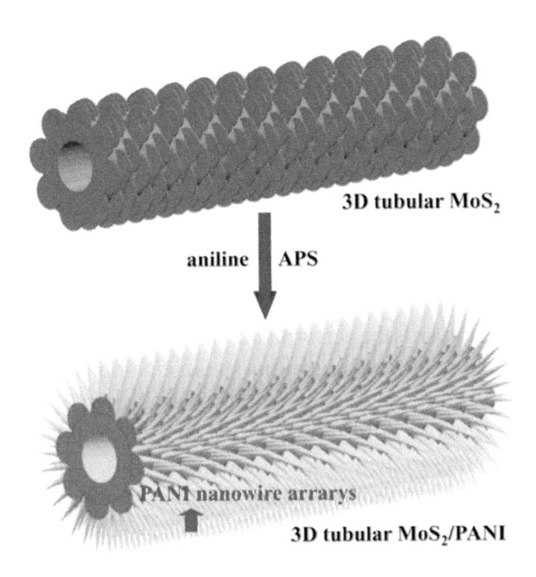

FIGURE 13.3 Formation schematic illustration of the 3D tubular MoS_2/PANI. (Reprinted with permission from Ren, L. et al., *ACS Appl. Mater. Interfaces*, 7, 28294, 2015. Copyright 2015 American Chemical Society.)

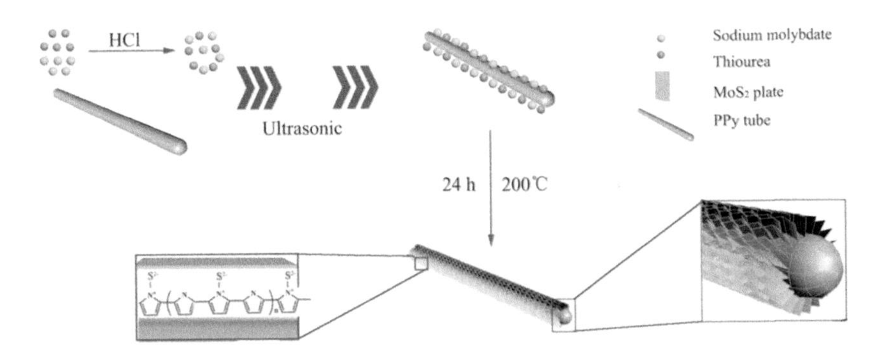

FIGURE 13.4 Illustration of the preparation of MoS_2/PPy nanocomposites. (With kind permission from Springer Science+Business Media: *J. Mater. Sci.: Mater. Electron.*, 28, 2017, 1777, Chang, C. et al.)

tubular MoS_2/PPy nanocomposites and examined as an efficient electrode material by delivering an excellent capacitance with remarkable cyclic stability [73]. An advanced nanocomposite of MoS_2-PEDOT have synthesized by Alamro and co-workers for supercapacitor electrode. The composite electrode shown a promising electrochemical behavior with a specific capacitance of 452 F/g [74]. Recently, Chao, et al. have reported hierarchical nanoflowers assembled from MoS_2/PANI sandwiched nanosheets for high performance supercapacitor electrode. The MoS_2/PANI composite electrode exhibited 456 F/g specific capacitance with significant 86% cyclic stability even after 16,000 cycles [75]. Interestingly, the pronounced electrophoretic deposition technique has developed for the synthesis of PANI/MoS_2 hybrid electrode. The as prepared PANI/MoS_2 electrode showing an enhanced specific capacitance of 812 F/g [76]. Moreover, the in-situ polymerization method has successfully utilized for the synthesis of MoS_2/PPy nanocomposite. The specific capacitance reaches up to 700 F/g at 10 mV/s scan rate with significant rate capability after high scan rate and high current density [77].

13.7 OTHER COMPOSITES OF MoS_2

The efforts are devoted towards the performance improvement of the electrode materials. In this regard core-shell nanostructure with a suitable composition could have a better choice to improve not only the conductivity but also charge storage ability of the electrode material for energy storage devices. To this, material like nickel sulfide is an attractive candidate owing to their several oxidation states, excellent electrochemical performance, scalable, earth abundant, and non-toxic nature. But, due to their low electrical conductivity and performance degradation of rate capability limit their practical application in daily use [78,79]. Recently, promising research efforts have started by researchers and tried to it with MoS_2 and in order to achieve remarkable enhancement in the electrochemical properties. In this orientation, Wang and co-workers have introduced a simple, green and low-cost solution approach to design a core shell structure of Ni_3S_2@MoS_2 nanoarray. The designed core shell Ni_3S_2@MoS_2 electrode exhibited an enhanced electrochemical performance by obtaining 848 F/g specific capacitance which is remarkable improvement than bare Ni_3S_2 electrode [80]. The facile hydrothermal method has used to synthesize hierarchical double core-shell structure of carbon@Ni_3S_2@MoS_2. The electrode of C@Ni_3S_2@MoS_2 nanorods offers an improved specific capacitance (1544 F/g at 2 A/g) along with remarkable retention by losing 7% performance form initial capacitance after 2000 cycles [78]. Moreover, the effective four-step process has developed for the synthesis of Ni_3S_4-MoS_2 heterostructure by a four-step process. The maximum specific capacitance of 985 F/g was observed at current density of 1 A/g with excellent rate capability and cyclic stability [79]. Additionally, Huang et al. has synthesized Ni_3S_4@MoS_2 core-shell intervened carbon fiber paper. The nanocomposite exhibited an outstanding supercapacitor performance through specific capacitance (1296 F/g), rate capability (58%) and stability (96%@5000 cycles) [81]. Furthermore, the hydrothermal reaction has generated for the synthesis of hybrid Bi_2S_3/MoS_2 nanosheet array on Mo foil for supercapacitor application. The heterostructure of Bi_2S_3/MoS_2 revealed an improved capacitance in terms of electrode area (1.45 F/cm^2 at 10 mA/cm^2 areal current density) with a significant retention of 96% after 1000 cycles which was around 10 times more that of the bare MoS_2 electrode [82]. Recently, the composite electrode of Ag nanowires with MoS_2 nanosheets has reported for high rate micro-supercapacitor. The fabricated Ag/MoS_2 electrode exhibited superior electrochemical parameters with 23 mF/cm^2 capacitance and 96% retention after 20,000 cycles [83]. Huang et al. reported a phase controlled Ni_3S_2@MoS_2 structure though one-step process. The significant electrochemical performance was achieved with maximum specific capacitance of 1418 F/g and 76% cyclic retention after 1250 cycles [84]. Wang et al. have reported in situ growth of MoS_2/CoS_2 nanotubes array by sign hydrothermal method. The enhanced specific capacitance of 145 mF/cm^2 is achieved by MoS_2/CoS_2 hybrid electrode with 93% stability after 1000 cycles [85].

In progressive way of MoS_2 composites, 3D core-shell nanostructure of uniform CeO_2 hollow spheres as core decorated with ultrathin MoS_2 nanosheets as shell has developed by using wet chemical method. A symmetric supercapacitor cell of core shell nanocomposite is depicted a superior electrochemical performance with noteworthy outcomes in terms of specific capacitance, energy density and long term stability [86]. Wang et al. have used a two-step chemical route by using hydrothermal followed by chemical precipitation to synthesize a hierarchical MoS_2/Mn_3O_4 nanoparticles. Also, after 2000 cycles there is 70% cyclic retention was achieved for MoS_2/Mn_3O_4 at 1 A/g which is two-fold higher than bare MoS_2 electrode [87]. Gen et al. have explored a multilayer metallic MoS_2-H_2O system (M-MoS_2-H_2O) for supercapacitor which has further obtained a high specific capacitance 380 F/g at lower scan rate of 5 mV/s and still retained at 105 F/g after 10 V/s scan rate [88]. Sarno et al. synthesized a physically exfoliated graphene (PHG) supported MoS_2/Fe_3O_4 nanocomposite. The MoS_2/Fe_3O_4/PHG nanocomposite reveals an excellent specific capacitance (830 F/g) at a current density of 1 A/g. Also insignificant change in the capacitance after long-term cyclic test (about 96%@2200 cycles) made it a promising candidate for supercapacitor [89]. Wang et al. utilized the layered MoS_2 nanosheets in order to design a hybrid hererostructure with a porous and high-efficient transition metal oxides like Fe_2O_3, NiO and Co_3O_4. Among these, the heterostructure of MoS_2/NiO revealed an outstanding specific capacitance of 1080 F/g at 1 A/g current density by charge-discharge profile and retention stability of 102% after 9000 cycles at 2 A/g [90]. An Ag nanoparticle puts their footprints towards an efficient supercapacitor electrode. In this regard, Wu et al. have utilized it to the formation of nanocomposite. They have used a self-assembly method for the preparation of AgNP/MoS_2 nanocomposite. The hybrid AgNP/MoS_2 electrode presented an excellent electrochemical performance [91]. Liang et al. have developed a novel scheme of laser ablation in liquids and aging-induced induced phase transformation for the synthesis of MoS_2/Co_3O_4 composite structure. The MoS_2/Co_3O_4 composites electrode demonstrated a battery-type behavior with maximum capacity of 69 mAh/g at 0.5 A/g and retains the capacity of 87% after 500 cycles which is much improved than those of bare MoS_2 and Co_3O_4 electrodes [92]. Beside, Su et al. presented the scheme for growing 3D hierarchical CoS_2/MoS_2 nanocomposite on carbon cloth as an efficient supercapacitor electrode. The CoS_2/MoS_2 nanocomposite electrode delivered a specific capacitance of 406 F/g and 5% degradation of the initial capacitance after very long-term cycles (10,000) [93]. Designing of MoS_2-graphene carbon nitride (MoS_2-g-C_3N_4) heterostructure was performed for the application of supercapacitor. The heterostructure MoS_2-g-C_3N_4 electrode delivered a significant specific capacitance of 241 F/g with better cyclic stability [94]. Yang et al. have prepared a heterocomposite of nickel sulfides/MoS_2 on carbon nanotubes. The composite electrode was further utilized for the fabrication of asymmetric supercapacitor which further revealed a significant specific capacitance of 108 F/g at 0.5 A/g and retained the capacitance without any loss after 10,000 cycles [95]. Some of the other MoS_2-based compound and their composites are summarized in Table 13.1 along with their electrochemical performance [96–116].

TABLE 13.1
Some of the Highlighted Molybdenum Based Compound and Composites for Supercapacitor Application

Material	Method	Morphology	Electrolyte	Specific Capacitance (F/g)	Stability	References
MoS_2	DC magnetron sputtering	Nanoworms	1M Na_2SO_4	589	—	[96]
MoS_2	Hydrothermal	Nanosheets	1M KCl	138	93%@1000	[97]
MoS_2	Solvothermal	Petal like	1M KCl	811	83%@1000	[98]
MoS_2	Hydrothermal	Nanoparticle	3M KOH	1531	82%@3000	[99]
MoS_2	Electrochemical	Nanosheet	6M KOH	118	92%@5000	[100]
MoS_2/graphene	Chemical reaction	Hybrid nanosheets	1M KOH	756	88%@10,000	[101]
MoS_2@ mesoporous hollow carbon spheres	Hydrothermal	Core shell	1M Na_2SO_4	613	90%@2000	[102]
MoS_2/rGO	Hydrothermal	Particles wrapped on rGO nanosheets	1M KOH	1022	—	[103]
MoS_2/3DG	Hydrothermal	Flower-like	1M Na_2SO_4	410	80%@10,000	[104]
MoS_2/MWCNTs	CBD	Nanoflakes over MWCNTs	0.5M Na_2SO_4	592	94%@2000	[105]
MoS_2/PANI	In-situ oxidative polymerization	Core/shell microspheres	1M H_2SO_4	633	86%@1000	[106]
PPY/MoS_2	Hydrothermal	3D multilayer structure	1M H_2SO_4	895	98%@10,000	[107]
PANI/MoS_2	Template method with hydrothermal reaction	Nanorods covered with flower like structure	1M HCl	603	87%@6000	[108]
MoS_2/PPY	Hydrothermal and oxidative polymerization	Aggregated nanosheets	1M KCl	182	80%@1000	[109]

(Continued)

TABLE 13.1 (*Continued*)
Some of the Highlighted Molybdenum Based Compound and Composites for Supercapacitor Application

Material	Method	Morphology	Electrolyte	Specific Capacitance (F/g)	Stability	References
Bi_2S_3/MoS_2	Hydrothermal	Flower-like	6M KOH	3040	92%@5000	[110]
MoS_2/SiC	Hydrothermal	Nanoflowers decorated on nanowires	1M Na_2SO_4	200	96%@2000	[111]
$NiFe_2O_4/MoS_2$	Solution reaction and freeze-dried process	Nanosheets and nanoparticles	1M KOH	506	91%@3000	[112]
$NiCo_2S_4@MoS_2$	Multi-step treatment process	Rhombic dodecahedral shape coated with nanosheets	6M KOH	860	—	[113]
PANI/graphene/MoS_2	In situ oxidative polymerization	Interlayer structure encapsulated with nanorods	1M H_2SO_4	699	—	[114]
MoS_2/MoO_3/PPY	In situ oxidation polymerization	Nanoparticle-decorated nanosheets	1M KCl	352	105%@2000	[115]
α-Fe_2O_3/MoS_2	Hydrothermal	Sheets supported nanoparticles	3M KOH	266	84%@2000	[116]

13.8 CONCLUSION AND FORWARD LOOK

The recent progresses in the growth of MoS_2 and MoS_2-based composites with high performance in various electrochemical domains are discussed in this chapter. Molybdenum sulfide is the most promising candidate in TMDs due to their significant structural, electrical and morphological properties over metal oxides. However, their poor electrical conductivity than carbon-based materials have impended their utility for their use in commercial appliances, so diverse nanostructures and composites have been developed for MoS_2 to overcome the limitations. In this regard, carbon, polymer and other metallic compound-based composites of MoS_2 shows significant enhancement in the electrochemical performance through storage ability and cyclic stability. But for the future attention of the researchers should be focused on to the solving electrochemical functioning mechanism, develop the MoS_2 electrode structural stability, material phase, and electrochemical cycle stability.

REFERENCES

1. D. P. Dubal, O. Ayyad, V. Ruiz and P. Gomez-Romero, *Chem. Soc. Rev.*, 2015, 44, 1777.
2. D. P. Dubal, N. R. Chodankar, D. H. Kim and P. Gomez-Romero, *Chem. Soc. Rev.*, 2018, 47, 2065.
3. J. M. Soon and K. P. Loh, *Electrochem. Solid-State Lett.*, 2007, 10, 250.
4. X. Chia, A. Y. S. Eng, A. Ambrosi, S. M. Tan and M. Pumera, *Chem. Rev.*, 2015, 115, 11941–11966.
5. C. N. R. Rao, K. Gopalakrishnan and U. Maitra, *ACS Appl. Mater. Interfaces*, 2015, 7, 7809.
6. Q. Lu, J. G. Chen and J. Q. Xiao, *Angew. Chem. Int. Ed.*, 2013, 52, 1882.
7. Q. Lu, Z. J. Mellinger, W. Wang, W. Li, Y. Chen, J. G. Chen and J. Q. Xiao, *ChemSusChem*, 2010, 3, 1367.
8. Y. Zhang, W. Sun, X. Rui, B. Li, H. T. Tan, G. Guo, S. Madhavi, Y. Zong and Q. Yan, *Small*, 2015, 11, 3694.
9. R. G. Dickinson and L. Pauling, *J. Am. Chem. Soc.*, 1923, 45, 1466.
10. R. E. Bell and R. E. Herfert, *J. Am. Chem. Soc.*, 1957, 79(13), 3351.
11. R. J. Traill, *Can. Mineral.*, 1963, 7, 524.
12. J. C. Wildervanck and F. Jellinek, *Z. Anorg. Allg. Chem.*, 1964, 328, 309.
13. Y. Takeuchi and W. Nowacki, *Schweiz. Mineral. Petrogr. Mitt.*, 1964, 44, 105.
14. F. E. Wickman and D. K. Smith, *Am. Mineral.*, 1970, 55, 1843.
15. J. A. Wilson and A. D. Yoffe, *Adv. Phys.*, 1969, 18, 193.
16. G. A. N. Connell, J. A. Wilson and A. D. Yoffe, *J. Phys. Chem. Solids*, 1969, 30, 287.
17. R. J. J. Newberry, *Am. Mineral.*, 1979, 64, 758.
18. L. F. Mattheiss, *Phys. Rev. B: Condens. Matter Mater. Phys.*, 1973, 8, 3719.
19. P. D. Fleischauer, J. R. Lince, P. A. Bertrand and R. Bauer, *Langmuir*, 1989, 5, 1009.
20. S. V. Didziulis, J. R. Lince, D. K. Shuh, T. D. Durbin and J. A. Yarmoff, *J. Electron Spectrosc. Relat. Phenom.*, 1992, 60, 175.
21. L. Pauling, *J. Am. Chem. Soc.*, 1931, 53, 1367.
22. R. Hultgren, *Phys. Rev.*, 1932, 40, 891.
23. G. Zhang, H. Liu, J. Qu and J. Li, *Energy Environ. Sci.*, 2016, 9, 1190.
24. K. J. Huang, J. Z. Zhang, G. W. Shi and Y. M. Liu, *Electrochim. Acta*, 2014, 132, 397.
25. X. Wang, J. Ding, S. Yao, X. Wu, Q. Feng, Z. Wang and B. Geng, *J. Mater. Chem.* A, 2014, 2, 15958.
26. S. S. Karade, D. P. Dubal and B. R. Sankapal, *RSC Adv.*, 2016, 6, 39159.

27. H. Gong, F. Zheng, Z. Li, Y. Li, P. Hu, Y. Gong, S. Song, F. Zhan and Q. Zhen, *Electrochim. Acta*, 2017, 227, 101.
28. R. K. Mishra, M. Krishnaih, S. Y. Kim, A. K. Kushwaha and S. H. *Jin, Mater. Lett.*, 2019, 236, 167.
29. N. Choudhary, M. Patel, Y. H. Ho, N. B. Dahotre, W. Lee, J. Y. Hwang and W. Choi, *J. Mater. Chem. A*, 2015, 3, 24049.
30. R. B. Pujari, A. C. Lokhande, A. R. Shelke, J. H. Kim and C. D. Lokhande, *J. Colloid Interface Sci.*, 2017, 496, 1.
31. K. Krishnamoorthy, P. Pazhamalai, G. K. Veerasubramani and S. J. Kim, *J. Power Sources*, 2016, 321, 112.
32. L. Xu, L. Ma, T. Rujiralai, X. Zhou, S. Wu and M. Liu, *RSC Adv.*, 2017, 7, 33937.
33. N. Islam, S. Wang, J. Warzywoda, Z. Fan, *J. Power Sources*, 2018, 400, 277.
34. L. Wang, Y. Ma, M. Yang and Y. Qi, *Electrochim. Acta*, 2015, 186, 391.
35. H. Xiao, S. Wang, S. Zhang, Y. Wang, Q. Xu, W. Hu, Y. Zhou, Z. Wang, C. An and J. Zhang, *Mater. Chem. Phys.*, 2017, 192, 100.
36. A. Ramadoss, T. Kim, G. S. Kim and S. J. Kim, *New J. Chem.*, 2014, 38, 2379.
37. Z. Wu, B. Li, Y. Xue, J. Li, Y. Zhang and F. Gao, *J. Mater. Chem. A*, 2015, 3, 19445.
38. S. S. Karade and B. R. Sankapal, *ACS Sustainable Chem. Eng.*, 2018, 6, 15072.
39. Y. Luo, Y. Zhang, Y. Zhao, X. Fang, J. Ren, W. Weng Y. Jiang et al., *J. Mater. Chem. A*, 2015, 3, 17553.
40. L. Gao, X. Li, X. Li, J. Cheng, B. Wang, Z. Wang and C. Li, *RSC Adv.*, 2016, 6, 57190.
41. T. N. Y. Khawula, K. Raju, P. J. Franklyn, L. Sigalas and K. I. Ozoemena, *J. Mater. Chem. A*, 2016, 4, 6411.
42. L. Q. Fan, G. J. Liu, C. Y. Zhang, J. H. Wu and Y. L. Wei, *Int. J. Hydrogen Energy*, 2015, 40, 10150.
43. M. H. Yang, J. M. Jeong, Y. S. Huh and B. J. Choi, *Compos. Sci. Technol.*, 2015, 121, 123.
44. Q. Weng, X. Wang, X. Wang, C. Zhang, X. Jiang, Y. Bando and D. Grolberg, *J. Mater. Chem. A*, 2015, 3, 3097.
45. G. Sun, X. Zhang, R. Lin, J. Yang, H. Zhang and P. Chen, *Angew. Chem. Int. Ed.*, 2015, 54, 4651.
46. N. Li, T. Lv, Y. Yao, H. Li, K. Liu and T. Chen, *J. Mater. Chem. A*, 2017, 5, 3267.
47. A. Liu, H. Lv, H. Liu, Q. Li and H. Zhao, *J. Mater. Sci.: Mater. Electron.*, 2017, 28, 8452.
48. G. Z. Chen, *Prog. Nat. Sci. Mater. Int.*, 2013, 23, 245.
49. S. S. Karade and B. R. Sankapal, *J. Electroanal. Chem.*, 2017, 802, 131.
50. K. J. Huang, L. Wang, J. Z. Zhang, L. L. Wang and Y. P. Mo, *Energy*, 2014, 67, 234.
51. K. J. Huang, L. Wang, J. Z. Zhang and K. Xing, *J. Electroanal. Chem.*, 2015, 752, 33.
52. W. Xu, B. Mu1 and A. Wang, *J. Mater. Sci.*, 2018, 53, 11659.
53. A. K. Thakur, M. Majumder, R. B. Choudhary and S. B. Singh, *J. Power Sources*, 2018, 402, 163.
54. P. Sun, R. Wang, Q. Wang, H. Wang and X. Wang, *App. Surf. Sci.*, 2019, 475, 793.
55. T. M. Masikhwa, M. J. Madito, A. Bello, J. K. Dangbegnonb and N. Manyala, *J. Colloid Interface Sci.*, 2017, 488, 155.
56. M. Murugan, R. M. Kumar, A. Alsalme, A. Alghamdi and R. Jayavel, *J. Nanosci. Nanotechnol.*, 2017, 17, 5469.
57. T. Sun, X. Liu, Z. Li, L. Ma, J. Wang and S. Yang, *New J. Chem.*, 2017, 41, 7142.
58. S. Zhang, R. Hu, P. Dai, X. Yu, Z. Ding, M. Wu, G. Li, Y. Ma and C. Tu, *Appl. Surf. Sci.*, 2017, 396, 994.
59. F. Zhang, Y. Tang, H. Liu, H. Ji, C. Jiang, J. Zhang, X. Zhang and C. S. Lee, *ACS Appl. Mater. Interfaces*, 2016, 8, 4691.
60. W. J. Zhang, and K. J. Huang, *Inorg. Chem. Front.*, 2017, 4, 1602.

61. C. Yang, Z. Chen, I. Shakir, Y. Xu and H. Lu, *Nano Res.*, 2016, 9, 951.
62. K. J. Huang, L. Wang, Y. J. Liu, H. B. Wang, Y. M. Liu and L. L. Wang, *Electrochim. Acta*, 2013, 109, 587.
63. K. Gopalakrishnan, S. Sultan, A. Govindaraj and C. N. R. Rao, *Nano Energy*, 2015, 12, 52.
64. J. Wang, Z. Wu, H. Yin, W. Li and Y. Jiang, *RSC Adv.*, 2014, 4, 56926.
65. J. Wang, Z. Wu, K. Hu, X. Chen and H. Yin, *J. Alloys Compd.*, 2015, 619, 38.
66. C. Sha, B. Lu, H. Mao, J. Cheng, X. Pan, J. Lu and Z. Ye, *Carbon*, 2016, 99, 26.
67. L. Ren, G. Zhang, Z. Yan, L. Kang, H. Xu, F. Shi, Z. Lei and Z. H. Liu, *ACS Appl. Mater. Interfaces*, 2015, 7, 28294.
68. S. A. Ansari, H. Fouad, S. G. Ansari, M. P. Sk and M. H. Cho, *J. Colloid Interface Sci.*, 2017, 504, 276.
69. J. Wang, Q. Luo, C. Luo, H. Lin, R. Qi, N. Zhong and H. Peng, *J. Solid State Electrochem.*, 2017, 1.
70. G. Fu, L. Ma, M. Gan, X. Zhang, M. Jin, Y. Lei, P. Yang and M. Yan, *Synth. Met.*, 2017, 224, 36.
71. A. K. Thakur, A. B. Deshmukh, R. B. Choudhary, I. Karbhal, M. Majumder and M. V. Shelke, *Mater. Sci. Eng. B*, 2017, 223, 24.
72. C. Chang, X. Yang, S. Xiang, H. Que and M. Li, *J. Mater. Sci.: Mater. Electron.*, 2017, 28, 1777.
73. Y. Chen, W. Ma, K. Cai, X. Yang and C. Huang, *Electrochim. Acta*, 2017, 246, 615.
74. T. Alamro and M. K. Ram, *Electrochim. Acta*, 2017, 235, 623.
75. J. Chao, J. Deng, W. Zhou, J. Liu, R. Hu, L. Yang, M. Zhu, and O. G. Schmidt, *Electrochimica. Acta*, 2017, 243, 98.
76. M. S. Nam, U. Patil, B. Park, H. B. Sim and S. C. Jun, *RSC Adv.*, 2016, 6, 101592.
77. H. Tang, J. Wang, H. Yin, H. Zhao, D. Wang and Z. Tang, *Adv. Mater.*, 2015, 27, 1117.
78. L. Li, H. Yang, J. Yang, L. Zhang, J. Miao, Y. Zhang, C. Sun, W. Huang, X. Dong and B. Liu, *J. Mater. Chem. A*, 2016, 4, 1319.
79. W. Luo, G. Zhang, Y. Cui, Y. Sun, Q. Qin, J. Zhang and W. Zheng, *J. Mater. Chem. A*, 2017, 5, 11278.
80. J. Wang, D. Chao, J. Liu, L. Li, L. Lai, J. Lin and Z. Shen, *Nano Energy*, 2014, 7, 151.
81. F. Huang, A. Yan, Y. Sui, F. Wei, J. Qi, Q. Meng and Y. He, *J. Mater. Sci.: Mater. Electron.*, 2017, 1.
82. Y. Ma, Y. Jia, L. Wang, M. Yang, Y. Bi and Y. Qi, *J. Power Sources*, 2017, 342, 921.
83. J. Li, Q. Shi, Y. Shao, C. Hou, Y. Li, Q. Zhang, and H. Wang, *Energy Storage Mater.*, 2019, 16, 212.
84. L. Huang, H. Hou, B. Liu, K. Zeinu, X. Zhu, X. Yuan, X. He, L. Wu, J. Hu, and J. Yang, *Appl. Surf. Sci.*, 2017, 425, 879.
85. L. Wang, X. Zhang, Y. Ma, M. Yang, and Y. Qi, *J. Phys. Chem. C*, 2017, 121, 9089.
86. N. Li, H. Zhao, Y. Zhang, Z. Liu, X. Gong and Y. Du, *CrystEngComm*, 2016, 18, 4158.
87. M. Wang, H. Fei, P. Zhang and L. Yin, *Electrochim. Acta*, 2016, 209, 389.
88. X. Geng, Y. Zhang, Y. Han, J. Li, L. Yang, M. Benamara, L. Chen and H. Zhu, *Nano Lett.*, 2017, 17, 1825.
89. M. Sarno and A. Troisi, *J. Nanosci. Nanotechnol.*, 2017, 17, 3735.
90. K. Wang, J. Yang, J. Zhu, L. Li, Y. Liu, C. Zhang and T. Liu, *J. Mater. Chem. A*, 2017, 5, 11236.
91. X. Wu, X. Yan, Y. Dai, J. Wang, J. Wang and X. Cheng, *Mater. Lett.*, 2015, 152, 128.
92. D. Liang, Z. Tian, J. Liu, Y. Ye, S. Wu, Y. Cai and C. Liang, *Electrochim. Acta*, 2015, 182, 376.
93. C. Su, J. Xiang, F. Wen, L. Song, C. Mu, D. Xu, C. Hao and Z. Liu, *Electrochim. Acta*, 2016, 212, 941.

94. S. A. Ansari and M. H. Cho, *Sci. Rep.*, 2017, 7, 43055.
95. X. Yang, L. Zhao and J. Lian, *J. Power Sources*, 2017, 343, 373.
96. A. Sanger, V. K. Malik and R. Chandra, *Inter. J. Hydrogen Energy*, 2018, 43, 11141.
97. L. Wang, Y. Ma, M. Yang and Y. Qi, *Appl. Surf. Sci.*, 2017, 396, 1466.
98. R. K. Mishra, S. Manivannan, K. Kim, H. I. Kwon and S. H. Jin, *Curr. Appl. Phys.*, 2018, 18, 345.
99. C. Nagaraju, C. V. V. Muralee Gopi, J. W. Ahn and H. J. Kim, *New J. Chem.*, 2018, 42, 12357.
100. A. Ejigu, I. A. Kinloch, E. Prestat and R. A. W. Dryfe, *J. Mater. Chem. A*, 2017, 5, 11316.
101. D. Vikramana, K. Karuppasamy, S. Hussainb, A. Kathalingamd and A. Sanmugame, J. Jung, H. S. Kim, *Composites Part B*, 2019, 161, 555.
102. L. Zheng, T. Xing, Y. Ouyang, Y. Wang and X. Wang, *Electrochica. Acta*, 2019, 298, 630.
103. L. Hongtao, X. Zichen, Z. Lina, Z. Zhiqiang and X. Li, *Chem. Phys. Lett.*, 2019, 716, 6.
104. T. Sun, Z. Li, X. Liu, L. Ma, J. Wang and S. Yang, *J. Power Sources*, 2016, 331, 180.
105. S. S. Karade, D. P. Dubal, and B. R. Sankapal, *ChemistrySelect*, 2017, 2, 10405.
106. X. Zhang, L. Ma, M. Gan, G. Fu, M. Jin and Y. Zhai, *Appl. Surf. Sci.*, 2018, 460, 48.
107. M. Lian, X. Wu, Q. Wang, W. Zhang and Y. Wang, *Ceramics Inter.*, 2017, 43, 9877.
108. H. Wang, L. Ma, M. Gan and T. Zhou, *J. Alloys Comp.*, 2017, 699, 176.
109. C. C. Tu, P. W. Peng and L. Y. Lin, *Appl. Surf. Sci.*, 2018, 444, 789.
110. L. Fang, Y. Qiu, T. Zhai, F. Wang, M. Lan, K. Huang and Q. Jing, *Colloid. Surf. A*, 2017, 535, 41.
111. A. Meng, Z. Yang, Z. Li, X. Yuan and J. Zhao, *J. Alloys Comp.*, 2018, 746, 93.
112. Y. Zhao, L. Xu, J. Yan, W. Yan, C. Wu, J. Lian, Y. Huang et al., *J. Alloys Comp.* 2017, 726, 608.
113. X. Z. Song, F. F. Sun, Y. L. Meng, Z. W. Wang, Q. F. Su and Z. Tan, *New J. Chem.*, 2019, 43, 3601.
114. S. Palsaniya, H. B. Nemade and A. K. Dasmahapatra, *Polymer*, 2018, 150, 150.
115. C. W. Chu, F. N. I. Sari, J. C. R. Ken and J. M. Ting, *Appl. Surf. Sci.*, 2018, 462, 526.
116. X. Yang, H. Sun, L. Zhang, L. Zhao, J. Lian and Q. Jiang, *Sci. Rep.*, 2016, 6, 31591.

14 Iron-Based Electrode Materials for an Efficient Supercapacitor

Leonardo Vivas, Rafael Friere, Javier Enriquez, Rajesh Kumar, and Dinesh Pratap Singh

CONTENTS

14.1 Introduction ... 257
14.2 Electrochemical Behavior of Iron-Based Electrodes ... 259
14.3 Iron-Based Materials ... 262
14.3.1 Fe_2O_3-Based Materials for Supercapacitors ... 262
14.3.2 FeOOH-Based Materials for Supercapacitors ... 266
14.3.3 Fe_3O_4-Based Materials for Supercapacitors ... 268
14.3.4 FeS and FeS_2-Based Materials for Supercapacitors ... 270
14.4 Conclusion and Perspective ... 272
Acknowledgements ... 272
References ... 272

14.1 INTRODUCTION

A strong need to develop new technologies in order to take advantage of ecofriendly energies such as hydro, biomass, wind and solar, have encouraged the development of efficient devices for the energy storage, providing a solution to the irregular rate of electricity generated by these resources [1]. Currently, lithium-ion batteries of high energy density (100–250 Wh/kg) are the most promising candidate for these applications, but their low power density, poor cyclic stability, health risk, high cost due to limited resources [2,3] restrict their usability. These drawbacks impelled researchers to think about new materials and devices to meet the current needs, driven by human development day by day [4].

Supercapacitors are energy storage devices with high rate capability, pulse power supply, long cycle life, simple principles, fast dynamics of charge propagation and low maintenance cost [5]. These devices as a mean of electrical energy storage have its own pros and cons. Advantages of supercapacitors are higher cycle stability with very low capacity retention, rapid charging-discharging and high specific power. A major disadvantage of electrochemical capacitors in comparison with batteries is a lower energy density [6]. The energy density of practical supercapacitor cells is limited to 10 Wh/kg, whereas the energy density of common lead acid batteries is of 30–40 Wh/kg. The two most important examples of electrochemical technology for energy storage are batteries and supercapacitors. The consumer electronics are mostly based on lithium batteries because they have a high energy density (≈180 Wh/kg) [3]. However, the use of batteries has given rise to serious safety issues such as low mobility of electrons and ions generate a series of resistive losses and hence heat production at high power, which results into dendrites [7]. This type of failure has occurred in cars manufactured by Tesla and in the Dreamliner airplane of the company Boeing [8].

The important elements with which the supercapacitors are manufactured are the electrodes, which can be made of carbon materials, conducting polymers and metal oxides [9]. Supercapacitors are mainly based on two processes: electrostatic or faradic. If the electrodes are immersed in an electrolyte, the storage of the charge will occur at the interface between the electrolyte and the electrodes made of the active material. That is why the surface area of the active material plays a very important role. It has been observed that using an active material that increases the surface area by 1000 times results an increase of 100,000 times in energy density and specific capacitance and increases faradic reactions in comparison to a normal capacitor. Thus, the technologies are at a point to move forward from capacitors that store microfarads or millifarads to new devices, called supercapacitors, that can store the energies up to thousands of farads [3].

Supercapacitors can be classified into two classes, which depend on the mechanism with which the load is stored: Electrochemical double-layer capacitors and asymmetric supercapacitors. Electrochemical double-layer capacitors electrostatically store charges on the interface of the electrodes/electrolyte [10,11]. Asymmetric supercapacitors store charge via faradic process which involves the transfer of charge between electrode and electrolyte [12].

Asymmetric supercapacitors have shown great interest because they show high specific capacitance and higher energy density, up to 100 times greater than double-layer capacitors [13–16]. The use of transition metal oxides and transition metal hydroxides for electrodes in asymmetric capacitors have been extensively studied, as their crystalline structures jointly with valance states of the metal ions allow fast redox reactions, that upfront these materials in the list of more attractive and promising [17–19] candidates.

Recently various materials have been explored that show high capacity to be used in pseudo capacitors. Among all, few extensively studied materials for electrodes are MnO_2, RuO_2, $Ni(OH)_2$, or electrically conducting polymers, but these materials show a stable and wide working window, only in positive potentials. Therefore, they can only be used as positive electrodes. Similarly, research interest has also been made on the materials that show a stable working window in negative potentials. Some of these are Fe-based oxides/hydroxides, such as Fe_2O_3, $NiFe_2O_4$, $CoFe_2O_4$, $MnFe_2O_4$, $CuFe_2O_4$, Fe_3O_4, FeOOH, $BiFeO_3$, etc. as shown in Figure 14.1 [20].

Due to its striking properties the iron-based materials show a great premise to be utilized as an asymmetric supercapacitor. Few of them are: (1) multiple state of valence states, Fe^0, Fe^{2+}, Fe^{3+} etc., for its active redox reactions, which ensure a high specific capacitance [21,22]; (2) these Fe-based materials can be defined as green materials because they are environmentally friendly when purchased with other

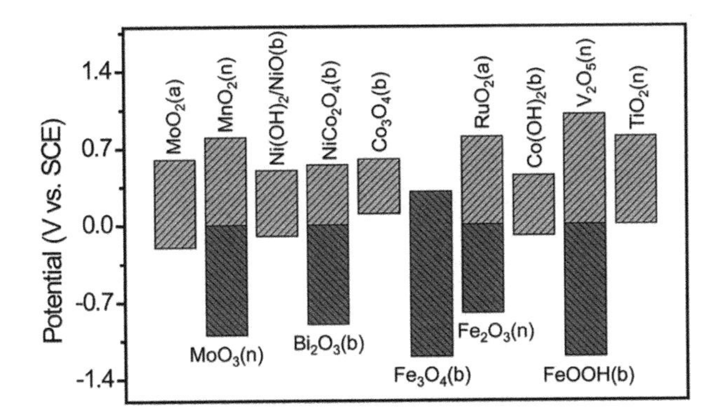

FIGURE 14.1 The working potential windows of various electrode materials in aqueous electrolyte. The acid-base property of electrolyte is marked inside the braces, a for acid, b for basic and n for neutral. (From Zeng, Y. et al.: Iron-based supercapacitor electrodes: Advances and challenges. *Advanced Energy Materials*. 2016. 6. 1601053. Copyright John Wiley & Sons. Reproduced with permission.)

oxides or transition metal hydroxides [23,24]; and (3) iron is low cost, because it is abundant in the earth. Therefore, all these promising characteristics make iron oxides/hydroxides of great interest for widespread industrial applications [20,25,26].

14.2 ELECTROCHEMICAL BEHAVIOR OF IRON-BASED ELECTRODES

In general, there are no major differences between cyclic voltammetry (CV) behavior of iron oxides and hydroxides in both neutral (e.g. Na_2SO_4) and alkaline (e.g. KOH) aqueous electrolytes irrespective of different crystal forms and compositions but its pseudocapacitive behavior and faradic reactions that are observed in the charging and discharging are not clearly understood yet. Several attempts have been made to explain it, starting from the fact that the CV behavior of the electrodes depends on the potential window used. Many studies have been carried out using a potential window from 0 to −0.8 V, in which rectangular CV curves are observed without any clear redox peaks (Figure 14.2) [27]. It was concluded that in this region there is a redox reaction between Fe^{3+} and Fe^{2+}, which explains the pseudocapacitive behavior as reviewed by Q. Xia et al. [27]. Chen et al. [22], analyzed FeOOH electrodes by in situ XAS, where Li_2SO_4 as an electrolyte and a potential window of −0.8–0 V was utilized. The XAS results showed the evidence of a reversibility between Fe^{3+} and Fe^{2+} and changing the potential window from 0 to values below −1 V the shape of the CV curve is different, and the intensity of the redox peaks decreased (Figure 14.2a). R. Li et al. [28] reinforced the results, where they examined the Fe_3O_4 electrode in a KOH electrolyte discharged to 1.3 V by X-ray diffraction (XRD) and X-ray photoelectron spectroscopy (XPS). In the study the formation of metallic Fe^0 was clearly observed which indicate a pronounced redox reaction between Fe^{3+} and Fe^0 in a greater potential range.

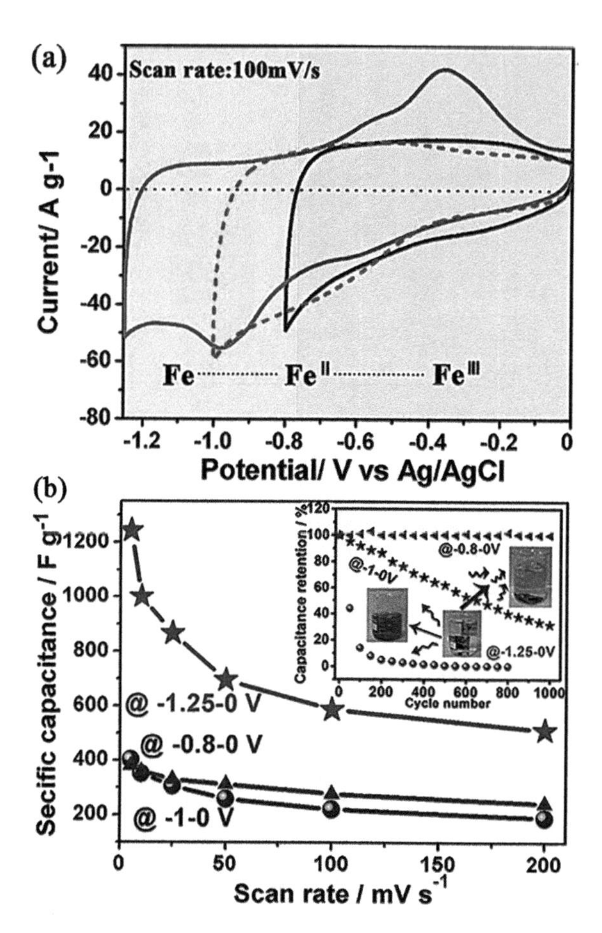

FIGURE 14.2 (a) Typical CV curves of a FeOOH/graphene composite electrode in different potential windows at a scan rate of 100 mVs^{-1}. (b) Rate performance and cycle performance of a FeOOH/graphene composite electrode in different potential windows. (From Xia, Q. et al.: Nanostructured iron oxide/hydroxide-based electrode materials for supercapacitors. *ChemNanoMat*. 2016. 2. 588–600. Copyright John Wiley & Sons. Reproduced with permission.)

K. A. Owusu et al. [29], analyzed the CV curves for the electrode α-Fe_2O_3 at different cycles, in an electrolyte of 2M KOH and in a potential window of between −1.2 and 0 V. They observed a pair of faradaic peaks at −0.66 and −1.05 V for the first cycle. The intensity of these peaks gradually reduced during the first ten cycles and became stable afterwards, which suggested the change in crystalline structure during the first few cycles (Figure 14.3). The ex situ XRD, X-ray photoelectron spectroscopy (XPS), SEM and TEM analysis confirmed the transformation of α-Fe_2O_3 into FeOOH after ten cycles. The α-Fe_2O_3 phase was not recovered in the subsequent discharge process, instead, a mixture of FeOOH and $Fe(OH)_2$ was obtained. The feasible transformation reaction and charge-storage mechanism was given as follow:

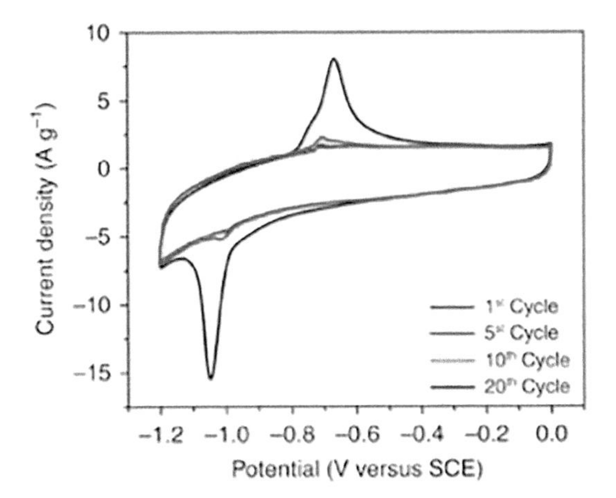

FIGURE 14.3 Cycle Voltammetry curve of FeOOH nanoparticles. (Reprinted by permission from Macmillan Publishers Ltd. *Nat. Commun.*, Owusu, K.A. et al., 2017, Copyright 2017.)

During first ten cycles:

$$Fe_2O_3 + H_2O \rightarrow 2FeOOH$$

Subsequent discharge process:

$$FeOOH + H_2O + e^- \rightarrow Fe(OH)_2 + OH^-$$

Subsequent charge process:

$$Fe(OH)_2 + OH^- \rightarrow FeOOH + H_2O + e^-$$

This explanation has also been given for the Redox processes for Ni–Fe alkaline aqueous batteries. T. Zhai et. al. [30] reported anode oxidation reactions Fe-based oxide/hydroxide in alkaline electrolytes as follows.

$$Fe + 2OH^- \leftrightarrow Fe(OH)_2 + 2e^-$$

$$3Fe(OH)_2 + 2OH^- \leftrightarrow Fe_3O_4 + 4H_2O + 2e^-$$

On/and

$$Fe(OH)_2 + OH^- \leftrightarrow FeOOH + H_2O + e^-$$

These equations suggested that cations of electrode materials also play an important role in the redox reactions. But still we lack the clear understanding of the redox processes of iron oxides and iron hydroxides.

14.3 IRON-BASED MATERIALS

The electrochemical properties of electrode materials are influenced by their morphology, composition, crystalline phases, defects, etc. Iron-based materials also show a direct correlation of their electrochemical behavior with their structural characteristics [20,30,31]. That is why the synthesis methods used to obtain iron-based materials have a very role in the performances. In the literature many synthesis methods are reported based on electrochemical deposition [32], sol-gel [33], thermal decomposition [34], hydrothermal/chemical/solvothermal method [35,36], microwave-assisted methods [37], aerogel [38] etc. for the controlled growth iron oxides/hydroxides nanostructures [27]. By means of which various Fe-based nanostructures with desirable morphology, for example, nanorods, nanosheets, nanotubes, nanowires, nanoflowers etc. are achieved. These materials considerably increase the surface area due to formation of nanostructures and shorten the diffusion length for electron and ion transport, which make them attractive for the supercapacitor devices [20].

This chapter deals the latest work done on the use of iron-based nanostructured materials such as of Fe_2O_3, FeOOH, Fe_3O_4, FeS and FeS_2 as an electrode for supercapacitors as described below.

14.3.1 Fe_2O_3-Based Materials for Supercapacitors

In general, Fe-based materials have received considerable interest due to their high performance as electrodes for supercapacitors. Among all Fe-based materials, emphasis has been given on the study of Fe_2O_3-based materials. The Fe_2O_3, exists in four different crystal structures, α-Fe_2O_3, β-Fe_2O_3, γ-Fe_2O_3, and ε-Fe_2O_3 [39] but among all hematite (α-Fe_2O_3), due to its most thermodynamically stable phase, abundance in the earth's crust, environmental benign and high theoretical specific capacitance [20,40,41], has been extensively studied.

The main challenge in the development of supercapacitor devices is the efficient use of materials to build a device with high performance. For this it is necessary to develop a highly advanced electrode material and the design or architecture for devices where these electrodes will be utilized.

Usually, the construction of symmetrical supercapacitors is done by placing two electrodes of the same material, one in front of the other, in which, one will serve as positive electrode and the other as negative electrode. Y. Wang and others [36] fabricated asymmetric supercapacitor, where each electrode was composed of porous Fe_2O_3 nanospheres anchored in a layer of activated carbon (Fe_2O_3@ACC). The symmetric supercapacitor assembled by two pieces of Fe_2O_3@ACC electrodes gave 1.8 V operating voltage in 3 M $LiNO_3$ aqueous electrolyte, as can be seen in the Figure 14.4. This supercapacitor showed an energy density of 9.2 mWh/cm^3 and the power density of 12 mW/cm^3 in an electrolyte of 3 M $LiNO_3$. The capacity retention obtained was 95% after 4000 cycles and a specific capacitance of 1565 $\mu F/cm^3$ was recorded.

Similarly, A. M. Khattak et al. [42] showed the performance of a solid-state symmetrical supercapacitor built with a hybrid Fe_2O_3/graphene aerogel electrode.

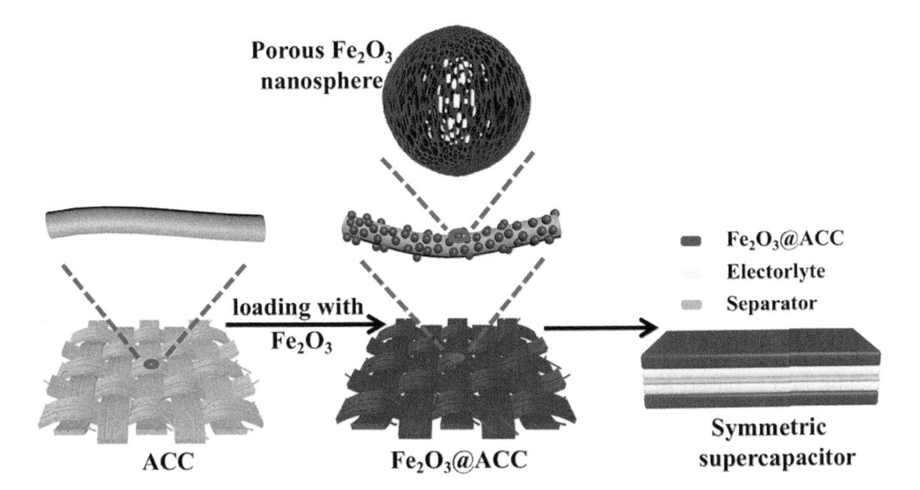

FIGURE 14.4 Schematic illustration of the Fe_2O_3@ACC electrode and the structure of symmetric supercapacitor. (Reprinted from *Nano Energy*, 57, Wang, Y. et al., Porous Fe_2O_3 nanospheres anchored on activated carbon cloth for high-performance symmetric supercapacitors, 379–387, Copyright 2018, with permission from Elsevier.)

They synthesized the Fe_2O_3/graphene aerogel using the in situ hydrothermal method. The electrochemical property of the synthesized material was tested in a first stage using a three-electrode system. Thus, they observed well-defined redox peaks corresponding to the redox reactions of Fe $2p_{3/2}$. A high specific capacitance of 1045.3 F/g was achieved when the current density was 0.4 A/g.

For the manufacture of the device, the first step was to cover the nickel foam with Fe_2O_3/graphene aerogel to make the electrodes and then submerge it in to PVA-KOH gel electrolyte and further a separator was inserted between these electrodes to make a symmetric supercapacitor. This device gave an energy density of 9.8 Wh/kg and a power density of 90.1 W/kg with a voltage window of 0 and 0.8 V. They also concluded that the electrochemical performance of the device was suitable for different angles of flexion providing a specific capacitance of 440 F/g. The proposed device retained 90% of the capacitance up to 2200 cycles, which indicates a good cyclic stability of the device.

Similarly, J. W. Park et al. [43] were able to synthesize hybrid nanoparticles such as Janus, α-Fe_2O_3/poly (3,4-ethylenedioxythiophene) (PEDOT) using sonochemical methods of crystallization assisted by liquid-liquid diffusion and polymerization by vapor deposition. This material showed a large surface area (376.4 m^2/g), high conductivity values (up to 120 S/cm) and remarkable electrochemical characteristics. They manufactured a symmetrical solid-state supercapacitor, using Fe_2O_3/PEDOT for the two electrodes and then immersed in a hydrogel electrolyte based on H_2SO_4-PVA. This device demonstrated a high specific capacitance of 258.8 F/g in potential range of 2.0 V, an excellent energy density of 136.3 Wh/Kg and a power density of 10.526 W/kg. Around 92% of capacity retention was observed even after 1000 cycles.

An interesting supercapacitor device was assembled by L. F. Chen et al. [25] by a simple method of using ferric oxides on carbon (CC/Fe_2O_3) as positive and negative electrode ($CC/Fe_2O_3//CC/Fe_2O_3$), with two sheets of titanium as current collectors, a liquid solution of, 2.0 M Li_2SO_4 as electrolyte and the filter paper as separator intercalated by the two electrodes. When working with a potential window of 0–2 V the device displayed an energy density of 11.0 mWh/cm^3 and power density of 1543.7 mW/cm^3. In addition, from the cyclic voltammetry curves a specific capacitance of 1428 F/cm^2 was obtained with a capacity retention of 84.65%. They showed that the manufactured device can be scaled, with few changes in specific capacitance, energy density and power when switching from devices with 1.5 cm^2 electrodes to devices with 100 cm^2 electrodes. Figure 14.5 shows how a supercapacitor manufactured with a 100.00 cm^2 electrode that is charged at a constant current of 0.25 A–2.0 V, can illuminate 50 red light emitting diodes (LEDs) for 6 minutes.

The Fe_2O_3-based materials were also utilized to design and construct the asymmetric supercapacitors, which include two electrodes with different materials. Among the possible electrode combinations, both electrodes can have different forms or mechanisms of charge storage: e.g. capacitive and/or Faradaic battery type. These devices are called hybrid supercapacitors. These supercapacitors also consist of two electrodes with different materials, each. These supercapacitor devices currently attracted a huge attention as an attractive class of supercapacitors, due to its expected and reported high energy and power density as compared to symmetrical and asymmetrical supercapacitors.

There are many asymmetric supercapacitors reported, based on Fe_2O_3 combined with other materials to obtain efficient devices. M. Serrapede et al. [38] synthesized, Fe_2O_3 decorated reduced graphene oxide (rGO/Fe_2O_3) by using a green surfactant extracted from olive leaves and by the method of hydrothermal. The rGO/Fe_2O_3

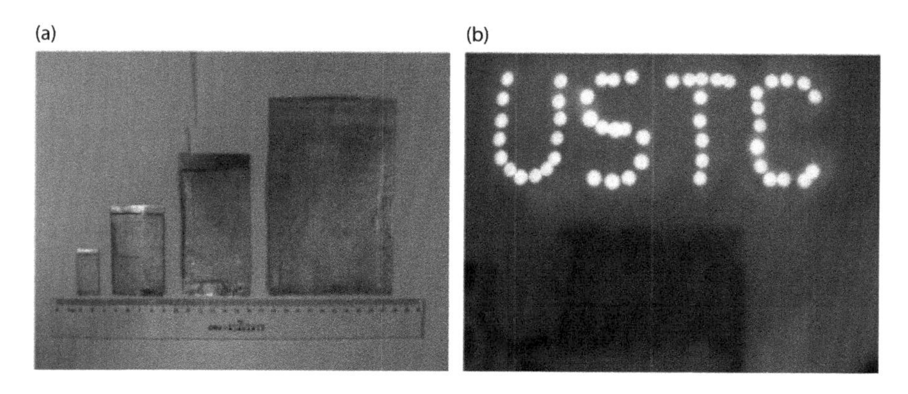

FIGURE 14.5 (a) Photograph of the supercapacitors with CC/Fe_2O_3 electrode materials of different sizes from 6.00 cm^2 to 100.00 cm^2 (1.80 × 3.33 cm^2, 4.00 × 6.25 cm^2, 5.50 × 9.09 cm^2, and 8.00 × 12.50 cm^2, thickness of the electrode materials and the separator: 0.42 cm). (b) Digital image of 50 red light-emitting diodes (LED) connected in parallel lighted by the CC/Fe_2O_3-based supercapacitor device with an electrode material area of 10,000 cm^2 (electrode materials: 8.00 × 12.50 cm^2, thickness of the electrode materials and the separator: 0.42 cm). (Reprinted from *Nano Energy*, 9, Chen, L.F. et al., In situ hydrothermal growth of ferric oxides on carbon cloth for low-cost and scalable high-energy-density supercapacitors, 345–354, Copyright 2014, with permission from Elsevier.)

aerogel exhibited very good pseudocapacitive performance and further this material was used in combination with MnOx nanostructured electrodes, to synthesize a flexible supercapacitor device. The asymmetric device that was manufactured showed that it can reach capacitance values above 50 F/g in 2 M KOH and values above 20 F/g when using the gel electrolyte (2 M KOH + PVA).

Likewise, Y. Dong et al. [44] manufactured a quasi-solid state asymmetric super capacitor assembled by using MnO_2 nanosheets as a cathode and α-Fe_2O_3@C nanowire arrays as anode materials, PVA/Na_2SO_4 electrolyte gel and the separator. The reported specific capacitance of the α-Fe_2O_3@C//MnO_2 supercapacitor was ~55.125 F/g at a current density of 0.75 A/g. The evaluation of the devices made with α-Fe_2O_3@C//MnO_2 to 4 A/g showed that after 10,000 cycles there was a retention capacity of 82%. The Ragone plots of the α-Fe_2O_3@C//MnO_2 device as shown in Figure 14.6, where it is compared with data from other works for devices with different compositions, also shows the device in operation. The assembled supercapacitor device exhibited an energy density of 30.625 and 11.944 Wh/Kg at current densities of 0.75 and 5 A/g respectively.

A similar arrangement was shown by Y. Liu et al. [45] where they used MnO_2/rGO and Fe_2O_3/rGO to manufacture a device with asymmetric configuration. The device or asymmetric supercapacitor showed a rectangular cyclic voltammetry curve when measured in a voltage window of 0–1.8 V and a specific capacitance of 152 mF/cm^2. The shape of the charge and discharge curves were almost triangular, thus showing that it has a good columbine efficiency. The reported capacitance retention was 90% after 1000 cycles.

The Fe_2O_3 was also used by H. Jiang et al. [46] as a negative electrode, V_2O_5 as a positive electrode and PVA/Na_2SO_4 gel as electrolyte for the fabrication of all-solid-state asymmetric supercapacitor. By utilizing electrospinning followed by calcination highly porous Fe_2O_3 and V_2O_5 nanofibers were obtained. From the cyclic voltammetry curve, for device, in a potential window of 0–1.8 V, at various scan rate of 2, 5, 10, 20, 30 and 50 mV/s, the specific capacitance was 48.1, 42.4, 39.5 and 36.5 F/g, respectively. In addition, these devices showed high energy densities of 32.2 and 16.4 Wh/kg with

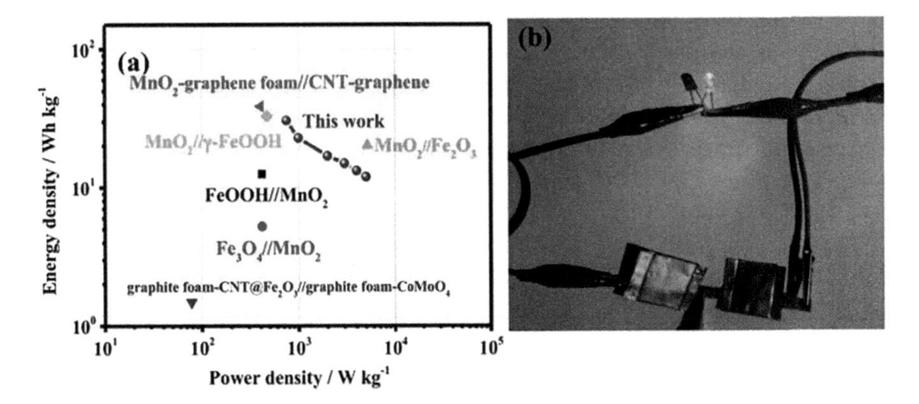

FIGURE 14.6 (a) Ragone plots of ASC. (b) Digital image of LED lighted by the α-Fe_2O_3@C//MnO_2 ASC device. (Reprinted with permission from Dong, Y. et al., *Nanomaterials*, 8, 487, 2018. MDPI Publishing Group.)

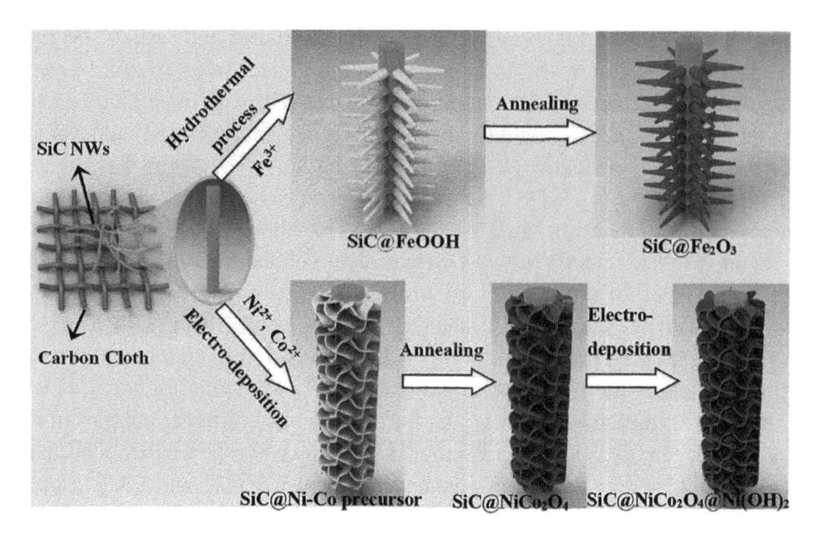

FIGURE 14.7 Schematic illustration for the synthesis procedures of the SiC NWs @Fe_2O_3 NNAs negative electrode and NWs@$NiCo_2O_4$@$Ni(OH)_2$ HNAs positive electrode on carbon cloth. (From Zhao, J. et al.: A High-Energy Density Asymmetric Supercapacitor Based on Fe_2O_3 Nanoneedle Arrays and $NiCo_2O_4$/$Ni(OH)_2$ Hybrid Nanosheet Arrays Grown on SiC Nanowire Networks as Free-Standing Advanced Electrodes. *Advance Energy Materials*. 2018. Copyright John Wiley & Sons. Reproduced with permission.)

power densities of 128.7 and 3285 W/kg respectively. When evaluating cycle stability, the supercapacitor maintained the specified capacitance ~97% of the initial value after 2000 cycles when measured with a scan rate of 100 mV/s.

J. Zhao et al. [47] recently proposed a new material, formed by aligned and trimmed Fe_2O_3 nano-needles (Fe_2O_3 NNAs) together with mesoporous structures (nanosheets) of $NiCo_2O_4$/$Ni(OH)_2$ and ($NiCo_2O_4$/$Ni(OH)_2$ HNAs) within SiC nanothread structures (SiC NW). This arrangement resulted in a material with high conductivity and surface area besides highly resistant to oxidation and corrosion. Figure 14.7 schematically illustrates the method of synthesis of negative electrodes of SiC NWs @ Fe_2O_3 NNAs and positive electrodes of SiC NWs@$NiCo_2O_4$@$Ni(OH)_2$ HNAs on carbon cloth and then manufactured an asymmetric supercapacitor, with 2 mol/L KOH solution as an electrolyte and commercially available filter paper as a separator. This assembly provided a specific capacitance of 242 F/g at a working current of 4 A/g and a very high specific energy density of 103 Wh/kg at 3.5 kW/kg with a capacitance retention of 86.6% after 5000 cycles.

There are many reports on the materials based on Fe_2O_3 as a negative electrode that show high performances. Table 14.1 summarizes some of the latest work where Fe_2O_3 is a protagonist in the capacitive properties of the devices.

14.3.2 FeOOH-Based Materials for Supercapacitors

Since 2008 FeOOH has emerged as a material with excellent capacitive properties. it was Wei-Hong Jin et al. [48] who first studied it as an anode material for supercapacitors and built a hybrid supercapacitor by utilizing MnO_2 as an anode and FeOOH as a

TABLE 14.1
A Summary of the Fe_2O_3 and its Composites According to Preparation, Electrolyte, Window of Potential, Specific Capacitance and Cycling Performance

Electrode	Preparation	Electrolyte	Window of Potential	Specific Capacitance	Cycling Performance	References
Fe_2O_3/CNTs	Assisted microwave	Na_2SO_4	0–1 V	204 F/g	87% at 10,000 cycles	[37]
α-Fe_2O_3/graphene	Hydrothermal	1 M KOH	0–1 V	1051 F/g in 1 A/g	85.4% at 1000 cycles	[35]
rGO/Fe_2O_3	Aerogel	2 M KOH	–1.5–0.2 V	330 F/g	75% at 2500 cycles	[36]
Fe_2O_3@ACC	Hydrothermal	3 M $LiNO_3$	–0.8–0 V	2775 μF/cm^2	95% at 4000 cycles	[38]
α-Fe_2O_3/rGO	Hydrothermal	2 M KOH	0.15–0.55 V	621.3 F/g	77% at 14,000 cycles	[56]
Fe_2O_3@CF	Electrodepositing	Gel	–0.8–0 V	42 F/g	96% at 5000 cycles	[57]
α-Fe_2O_3 thin film	Deposition	Na_2SO_3	–1.2–0 V	960 F/g	—	[58]
Fe_2O_3@C	Hydrothermal	1 M KOH	0–0.6 V	288 F/g	72.3% at 2000 cycles	[59]
α-Fe_2O_3@Ag microboxes	Mof as precursor and self-template	1 M Na_2SO_3	–1–0 V	123 F/g	80% at 2000 cycles	[60]
Fe_2O_3 nanospheres	Hydrothermal	2 M KOH	0–1.5 V	—	82.6% at 5000 cycles	[61]

cathode with a wide voltage range (0–1.85 V) in Li_2SO_4 electrolyte solution. It delivered an energy density of 12 Whk/g and a power density of 3700 W/kg [20,48].

Recently there are various reports in which FeOOH plays an important role in the design and implementation of electrodes for asymmetric supercapacitors. For example Q. Liao [49] synthesized an anode based on carbon fabric (CF), vertically aligned graphene nanosheets (VAGN) and FeOOH nanorods (CF/VAGN/FeOOH) using the deposition method, together with a CF/Co_3O_4 cathode. The solid-state supercapacitor assembled with CF/VAGN/FeOOOH as anode and CF/Co_3O_4 as cathode, in addition PVA/KOH gel was used as electrolyte and a paper filter as separator, to avoid contact between the two electrodes. From the charge and discharge curves they obtained specific capacitance values of 83.6, 71.3, 56.3, 49.8 and 54.4 F/g when averaged at current densities of 1.9, 3.9, 7.8, 15.6 and 31.3 A/g, respectively. The asymmetric super capacitor showed that it could reach energy density values of 20, 21.8, 22.5, 28.6 and 33.5 Wh/kg with power densities of 26.6, 13.3, 6.6, 3.3 and 1.6 kW/kg, respectively. Figure 14.8 shows the excellent results of the study of the asymmetric supercapacitor. It is worth mentioning the Ragone graph where the energy and power density of this work is compared with that of other works, the percentage of capacitance retention after 3000 cycles and the device in operation.

T. Nguyen et al. [50], reported the preparation by electrodeposition of lepidocrocite γ-FeOOH and amorphous Ni–Mn hydroxide arranged over carbon nanofoam to manufactured hybrid FeOOH-CNFP||Ni–Mn hydroxide-CNFP supercapacitors that work with alkaline electrolytes. This device demonstrated a high energy density of 1515 mW h/cm^2 and a power density of 9 mW/cm^2 with a capacity retention of 95% even after 10,000 cycles which showed excellent cyclic stability of the proposed device.

The importance of FeOOH-based materials lies in the fact that they function as negative electrodes with a high capacitance, comparable to the capacitances of positive electrodes, which have been extensively studied. R. Chen et al. [51] extensively studied the synthesis of α-FeOOH and β-FeOOH and its characterization as a good candidate to be used as a negative electrode in supercapacitors devices. They developed a new method to synthesize α-FeOOH and β-FeOOH through liquid-liquid interface (PELLI) methods, for electrode materials. Using PELLI allowed them to achieve a final product of non-agglomerated powders, which made it easier for them to mix them efficiently with multi-walled carbon nanotubes (MWCNT) resulting in better electrolyte penetration to the particle surface. The results for negative electrodes of α-FeOOH-MWCNT showed a specific capacitance of 5.86 F/cm^2 and low impedance, which was comparable to positive electrodes; among many other reports.

14.3.3 Fe_3O_4-Based Materials for Supercapacitors

Fe_3O_4 has also been extensively studied, due to its interesting characteristics as it has, a high electrical conductivity approximately 200 S/cm, high specific capacitance, abundance in nature and friendly to the environment. Recent review has displayed that nanostructures of Fe_3O_4, with a very large surface area can exhibit exceptional capacitance. There are many synthesis methods reported for the synthesis of Fe_3O_4

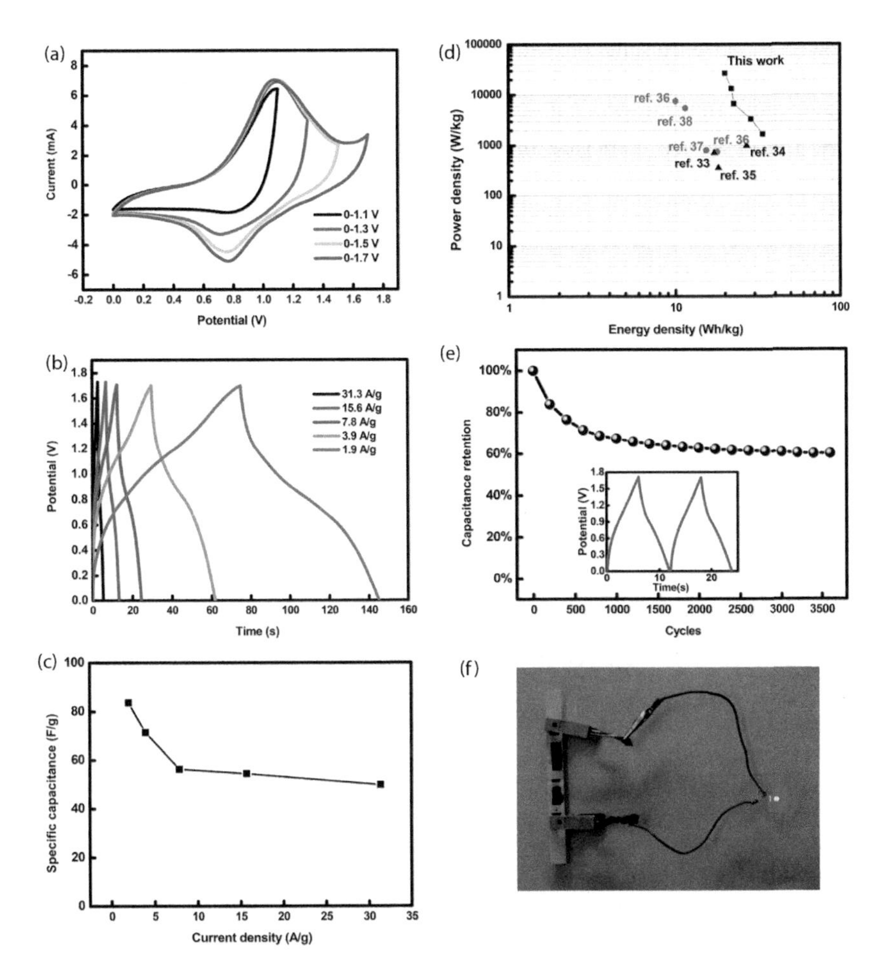

FIGURE 14.8 (a) CV curves of the asymmetric supercapacitor at various potential windows and (b) galvanostatic charging/discharging curves of the asymmetric supercapacitor at various current densities. (c) Specific capacitance vs. current densities for the asymmetric supercapacitor and (d) Ragone plots of the asymmetric supercapacitor. The values of other reported supercapacitors are added for comparison. (e) Cycling stability of the asymmetric supercapacitor; the inset shows the galvanostatic charge/discharge curves of the last 2 cycles. (f) Two asymmetric supercapacitor units are connected to light a red LED. (Liao, Q. and Wang, C., *CrystEngComm.*, 21, 662–672, 2019. Reproduced by permission of The Royal Society of Chemistry.)

nanostructures and among them, few most used methods are, hydrothermal, sol-gel, electroplating and electrospinning to obtain a well-defined morphology and architecture. [20].

Fe_3O_4, in combination with other materials, has also improved its properties. A. Kumar et al. [52] utilized Fe_3O_4 nanoparticles attached to the surface of carbon nanotubes (CNT-Fe_3O_4) as a negative electrode for asymmetric supercapacitor. The device was assembled by intercalating a polypropylene mesh separator between the positive electrode (CNT-Mn_3O_4) and the negative electrode (CNT-Fe_3O_4) soaked in electrolyte, followed by an encapsulation with polyethylene. This arrangement provided a specific capacitance of 135.2 F/g at 10 mV/s scan rate and a potential window of 0–1.8 V. Exceptionally 100% of its initial capacitance was retained even after 15000 cycles.

W. Liu et al. [53] reported in situ decorated Fe_3O_4 nanospheres on graphene through a facile solvothermal procedure, with which they manufactured a graphene/MnO_2//graphene/Fe_3O_4 asymmetric supercapacitor device, which displayed an energy density of 87.6 Wh/kg and superior cycling stability with 93.1% capacity retention after 10,000 cycles.

The use of Fe_3O_4 as a negative electrode in supercapacitor devices is also studied a lot. Recent literature survey revealed that this material shows very high performance when utilized as a negative electrode. Supercapacitors made with C-Fe_3O_4 electrodes revealed a specific capacitance of 526 F/g up to 98% capacitance retention in 1000 work cycles [54]. When combined with another material based on Fe negative electrodes of the type Fe_2O_3-Fe_3O_4/N-rGO exhibited a specific capacitance of 111.95 F/g with the capacity retention of 92% after 10000 cycles for asymmetric supercapacitors manufactured with this electrode [55].

14.3.4 FeS and FeS_2-Based Materials for Supercapacitors

Due to their outstanding electrochemical properties iron sulfides have demonstrated a great potential to be utilized as electrode materials for supercapacitors. Various efficient supercapacitors are realized by utilized these sulfides as electrode besides of iron oxides and iron hydroxide.

S. S. Karade [62] manufactured a high performance and flexible hybrid supercapacitor with which they were able to activate an array of 21 LEDs. The device consisted of FeS electrodes and through electrochemical measurements the achieved specific capacitance was 65.17 mF/cm^3, operating in a potential range of 2 V. They also reported a high energy density of 2.56 Wh/gr with a capacity retention of 91% after 1000 cycles.

In combination with reduced graphite oxide the iron sulfide was also utilized as an electrode for efficient supercapacitor. C. Zhao [63] demonstrated a supercapacitor made with electrode of RGO/FeS which showed a specific capacitance of 300 F/g and retention if capacitance up to 97.5% even after 2000 cycles. Further an asymmetric super capacitor using RGO/FeS as negative electrode and $Ni(OH)_2$ as positive electrode was also derived by them which exhibited a specific capacitance of 62.02 F/g and an energy density of 27.91 Wh/kg.

Y. Wang et al. [64] showed how FeS_2 together with double nitrogen and sulfur-doped graphite (FeS_2/NSG) can show an outstanding electrochemical performance (in three-electrode array measurements) with a specific capacitance of 528.7 F/g measured at 1 A/g and a capacity retention of 89% after 10000 charge/discharge cycles.

By combining FeS_2 nanosheets with Fe_2O_3 nanospheres Y. Zhong et al. [65] proposed a more efficient supercapacitors instead of using only Fe_2O_3 as electrode. The supercapacitor based on this electrode exhibited a high specific capacitance of 255 F/g with a capacitance retention of 90% after 5000 cycles.

Various other relevant reports based on the FeS and FeS_2 as an electrode material are summarized in the Table 14.2.

TABLE 14.2
Summary of Results Shown for Electrodes Made with FeS or FeS_2 for Supercapacitor

Electrode	Preparation	Electrolyte	Potential Range	Specific Capacitance	Cycling Performance (retention of Capacitance)	References
FeS/NF	Hydrothermal	2 M KOH	−0,4–0,8 V	2007,61 F/g	52,27% at 3000 cycles	[62]
FeSN-C	Hydrothermal	6 M KOH	−0,9–0,3 V	1320,3 F/cm^3	100% at 50,000 cycles	[67]
Pyrite FeS_2 (Device)	Solvothermal	3,5 M KOH	0–0,9 V	58 F/g	79,8% at 5000 cycles	[68]
FeS_2-CNF	Microwave irradiation	30 WT% KOH	−1,2–0 V	612 F/g	97% at 2000 cycles	[69]
FeS_2/CoS_2 (device)	Microwave assisted	PAAS/KOH	0–1,2 V	75 F/g	91% at 5000 cycles	[70]
FeS/RGO/FeS@Fe and $Ni(OH)_2$ (Device)	Hydrothermal	2 M KOH	0–2 V	75,72 F/g	90% at 3000 cycles	[71]
FeS_2/NSG	Rapid microwave	6 M KOH	−0,6–0 V	528,7 F/g	89% at 10,000 cycles	[65]

14.4 CONCLUSION AND PERSPECTIVE

In summary, iron-based materials are promising materials for negative electrodes of asymmetric supercapacitors using aqueous electrolytes because of their good electrochemical behavior, environmental benign, abundance and cost effectiveness. In this chapter we have reviewed some of the latest work showing the use of iron-based materials as various electrodes for supercapacitors. Usually the presence of other materials, such as graphite, metal oxides, polymers etc. make it more promising for the consideration as a potential candidate for future use in the manufacture of low-cost devices for the commercialization.

The various studies on the electrochemical behavior of iron-based electrodes show its better dependency with a window of potential and excellent behavior in the charging/discharging cycles, which helps to maintain the capacitance for long cycles.

It is important to continue working on nanostructures based on iron, as an active material, that can increase the surface area and shorten the diffusion trajectory of the ions in order to increase the capacitance. With which we could improve the electrical conductivity, that is a limitation of these materials with respect to the storage capacity. To study the importance of the different crystalline structures that form the materials based on iron and their role in the electrochemical properties is very important when designing electrodes for supercapacitors. There is still much work to be done before using these materials in commercial devices, but for its abundance, price and relationship with the environment is a strong motive to innovate and continuing research on this material for future hopes and perspectives.

ACKNOWLEDGEMENTS

The author LV and DPS acknowledge with gratitude the financial support from Millennium Institute for Research in Optics (MIRO) Chile and FONDECYT REGULAR 1151527.

REFERENCES

1. C. Liu, F. Li, L.-P. Ma, H.-M. Cheng, Advanced materials for energy storage, *Adv. Mater.* 22 (2010) E28–E62. doi:10.1002/adma.200903328.
2. Y. Ma, H. Chang, M. Zhang, Y. Chen, Graphene-based materials for lithium-ion hybrid supercapacitors, *Adv. Mater.* 27 (2015) 5296–5308. doi:10.1002/adma.201501622.
3. Y. Shao, M.F. El-Kady, J. Sun, Y. Li, Q. Zhang, M. Zhu, H. Wang, B. Dunn, R.B. Kaner, Design and mechanisms of asymmetric supercapacitors, *Chem. Rev.* 118 (2018) 9233–9280. doi:10.1021/acs.chemrev.8b00252.
4. J.W. Long, T. Brousse, D. Bélanger, Electrochemical capacitors: Fundamentals to applications, *J. Electrochem. Soc.* 162 (2015) Y3–Y3. doi:10.1149/2.0261505jes.
5. C. Liu, Z. Yu, D. Neff, A. Zhamu, B.Z. Jang, Graphene-based supercapacitor with an ultrahigh energy density, *Nano Lett.* 10 (2010) 4863–4868. doi:10.1021/nl102661q.
6. C. Zhao, W. Zheng, A review for aqueous electrochemical supercapacitors, *Front. Energy Res.* 3 (2015). doi:10.3389/fenrg.2015.00023.
7. T.M. Bandhauer, S. Garimella, T.F. Fuller, A critical review of thermal issues in lithium-ion batteries, *J. Electrochem. Soc.* 158 (2011) R1. doi:10.1149/1.3515880.

8. V.V.N. Obreja, Supercapacitors specialities: Materials review, In 2014. pp. 98–120. doi:10.1063/1.4878482.
9. J. Jiang, Y. Zhang, P. Nie, G. Xu, M. Shi, J. Wang, Y. Wu, R. Fu, H. Dou, X. Zhang, Progress of nanostructured electrode materials for supercapacitors, *Adv. Sustain. Syst.* 2 (2018) 1700110. doi:10.1002/adsu.201700110.
10. S. Zheng, Z.-S. Wu, S. Wang, H. Xiao, F. Zhou, C. Sun, X. Bao, H.-M. Cheng, Graphene-based materials for high-voltage and high-energy asymmetric supercapacitors, *Energy Storage Mater.* 6 (2017) 70–97. doi:10.1016/j.ensm.2016.10.003.
11. Y. Zhai, Y. Dou, D. Zhao, P.F. Fulvio, R.T. Mayes, S. Dai, Carbon materials for chemical capacitive energy storage, *Adv. Mater.* 23 (2011) 4828–4850. doi:10.1002/adma.201100984.
12. S.-M. Chen, R. Ramachandran, V. Mani, R. Saraswathi, Recent advancements in electrode materials for the high-performance electrochemical supercapacitors: A review, 2014. www.electrochemsci.org (Accessed November 27, 2018).
13. F. Wang, S. Xiao, Y. Hou, C. Hu, L. Liu, Y. Wu, Electrode materials for aqueous asymmetric supercapacitors, *RSC Adv.* 3 (2013) 13059. doi:10.1039/c3ra23466e.
14. S.R.C. Vivekchand, C.S. Rout, K.S. Subrahmanyam, A. Govindaraj, C.N.R. Rao, Graphene-based electrochemical supercapacitors, *J. Chem. Sci.* 120 (2008) 9–13. doi:10.1007/s12039-008-0002-7.
15. L.L. Zhang, X. Zhao, M.D. Stoller, Y. Zhu, H. Ji, S. Murali, Y. Wu, S. Perales, B. Clevenger, R.S. Ruoff, Highly conductive and porous activated reduced graphene oxide films for high-power supercapacitors, *Nano Lett.* 12 (2012) 1806–1812. doi:10.1021/nl203903z.
16. Z. Bo, Z. Wen, H. Kim, G. Lu, K. Yu, J. Chen, One-step fabrication and capacitive behavior of electrochemical double layer capacitor electrodes using vertically-oriented graphene directly grown on metal, *Carbon N.Y.* 50 (2012) 4379–4387. doi:10.1016/J.CARBON.2012.05.014.
17. M. Down, S.J. Rowley-Neale, G. Smith, C.E. Banks, Fabrication of graphene oxide supercapacitor devices, *ACS Appl. Energy Mater.* (2018). doi:10.1021/acsaem.7b00164.
18. Y. Zhu, S. Murali, M.D. Stoller, K.J. Ganesh, W. Cai, P.J. Ferreira, A. Pirkle et al., Carbon-based supercapacitors produced by activation of graphene, *Science* 322(6037) (2011) 1537–1541. doi:10.1126/science.1200770.
19. J. Zhu, A.S. Childress, M. Karakaya, S. Dandeliya, A. Srivastava, Y. Lin, A.M. Rao, R. Podila, Defect-engineered graphene for high-energy- and high-power-density supercapacitor devices, *Adv. Mater.* (2016) 7185–7192. doi:10.1002/adma.201602028.
20. Y. Zeng, Y. Meng, X. Lu, Y. Tong, M. Yu, P. Fang, Iron-based supercapacitor electrodes: Advances and challenges, *Adv. Energy Mater.* 6 (2016) 1601053. doi:10.1002/aenm.201601053.
21. J.S. Lee, D.H. Shin, J. Jun, C. Lee, J. Jang, Fe_3O_4/carbon hybrid nanoparticle electrodes for high-capacity electrochemical capacitors, *ChemSusChem.* 7 (2014) 1676–1683. doi:10.1002/cssc.201301188.
22. Y.-C. Chen, Y.-G. Lin, Y.-K. Hsu, S.-C. Yen, K.-H. Chen, L.-C. Chen, Novel iron oxyhydroxide lepidocrocite nanosheet as ultrahigh power density anode material for asymmetric supercapacitors, *Small.* 10 (2014) 3803–3810. doi:10.1002/smll.201400597.
23. T. Liu, Y. Ling, Y. Yang, L. Finn, E. Collazo, T. Zhai, Y. Tong, Y. Li, Investigation of hematite nanorod–nanoflake morphological transformation and the application of ultrathin nanoflakes for electrochemical devices, *Nano Energy.* 12 (2015) 169–177. doi:10.1016/J.NANOEN.2014.12.023.
24. M.-S. Wu, R.-H. Lee, Electrochemical growth of iron oxide thin films with nanorods and nanosheets for capacitors, *J. Electrochem. Soc.* 156 (2009) A737. doi:10.1149/1.3160547.

25. L.-F. Chen, Z.-Y. Yu, X. Ma, Z.-Y. Li, S.-H. Yu, In situ hydrothermal growth of ferric oxides on carbon cloth for low-cost and scalable high-energy-density supercapacitors, *Nano Energy.* 9 (2014) 345–354. doi:10.1016/J.NANOEN.2014.07.021.
26. S. Shivakumara, T.R. Penki, N. Munichandraiah, Synthesis and characterization of porous flowerlike-Fe_2O_3 nanostructures for supercapacitor application, *ECS Electrochem. Lett.* 2 (2013) A60–A62. doi:10.1149/2.002307eel.
27. Q. Xia, M. Xu, H. Xia, J. Xie, Nanostructured iron oxide/hydroxide-based electrode materials for supercapacitors, *ChemNanoMat.* 2 (2016) 588–600. doi:10.1002/cnma.201600110.
28. R. Li, Y. Wang, C. Zhou, C. Wang, X. Ba, Y. Li, X. Huang, J. Liu, Carbon-stabilized high-capacity ferroferric oxide nanorod array for flexible solid-state alkaline battery-supercapacitor hybrid device with high environmental suitability, *Adv. Funct. Mater.* 25 (2015) 5384–5394. doi:10.1002/adfm.201502265.
29. K.A. Owusu, L. Qu, J. Li, Z. Wang, K. Zhao, C. Yang, K.M. Hercule, C. Lin, C. Shi, Q. Wei, L. Zhou, L. Mai, Low-crystalline iron oxide hydroxide nanoparticle anode for high-performance supercapacitors, *Nat. Commun.* 8 (2017) 14264. doi:10.1038/ncomms14264.
30. T. Zhai, S. Xie, M. Yu, P. Fang, C. Liang, X. Lu, Y. Tong, Oxygen vacancies enhancing capacitive properties of MnO_2 nanorods for wearable asymmetric supercapacitors, *Nano Energy.* 8 (2014) 255–263. doi:10.1016/j.nanoen.2014.06.013.
31. Y. Zeng, Y. Han, Y. Zhao, Y. Zeng, M. Yu, Y. Liu, H. Tang, Y. Tong, X. Lu, Advanced ti-doped Fe_2O_3 @PEDOT core/shell anode for high-energy asymmetric supercapacitors, *Adv. Energy Mater.* 5 (2015) 1402176. doi:10.1002/aenm.201402176.
32. M. Aghazadeh, M.R. Ganjali, Evaluation of supercapacitive and magnetic properties of Fe_3O_4 nano-particles electrochemically doped with dysprosium cations: Development of a novel iron-based electrode, *Ceram. Int.* 44 (2018) 520–529. doi:10.1016/J.CERAMINT.2017.09.206.
33. M.P. Kumar, L.M. Lathika, A.P. Mohanachandran, R.B. Rakhi, A high-performance flexible supercapacitor anode based on polyaniline/Fe_3O_4 composite@carbon cloth, *ChemistrySelect.* 3 (2018) 3234–3240. doi:10.1002/slct.201800305.
34. S. Yu, V.M. Hong Ng, F. Wang, Z. Xiao, C. Li, L.B. Kong, W. Que, K. Zhou, Synthesis and application of iron-based nanomaterials as anodes of lithium-ion batteries and supercapacitors, *J. Mater. Chem. A.* 6 (2018) 9332–9367. doi:10.1039/C8TA01683F.
35. Y. Gao, T. Wang, Y. Tan, P. Liu, D. Jia, D. Wu, Reduced graphene oxide-coated mulberry-shaped α-Fe_2O_3 nanoparticles composite as high performance electrode material for supercapacitors, *J. Alloys Compd.* 738 (2017) 89–96. doi:10.1016/j.jallcom.2017.12.131.
36. Y. Wang, C. Hu, X. Zhou, J. Li, C. Zhang, Y. Wang, W. Xu, X. Gu, B. Zhang, S. Luo, Porous Fe_2O_3 nanospheres anchored on activated carbon cloth for high-performance symmetric supercapacitors, *Nano Energy.* 57 (2018) 379–387. doi:10.1016/j.nanoen.2018.12.061.
37. L. Yue, S. Zhang, H. Zhao, M. Wang, J. Mi, Y. Feng, D. Wang, Microwave-assisted one-pot synthesis of Fe_2O_3/CNTs composite as supercapacitor electrode materials, *J. Alloys Compd.* 765 (2018) 1263–1266. doi:10.1016/j.jallcom.2018.06.283.
38. M. Serrapede, A. Rafique, M. Fontana, A. Zine, P. Rivolo, S. Bianco, L. Chetibi, E. Tresso, A. Lamberti, Fiber-shaped asymmetric supercapacitor exploiting rGO/Fe_2O_3 aerogel and electrodeposited MnOx nanosheets on carbon fibers, *Carbon N. Y.* 144 (2019) 91–100. doi:10.1016/j.carbon.2018.12.002.
39. B. Xu, M. Zheng, H. Tang, Z. Chen, Y. Chi, L. Wang, L. Zhang, Y. Chen, H. Pang, Iron oxide-based nanomaterials for supercapacitors, *Nanotechnology.* (2019). doi:10.1088/1361-6528/ab009f.
40. M. Zhu, Y. Wang, D. Meng, X. Qin, G. Diao, Hydrothermal synthesis of hematite nanoparticles and their electrochemical properties, *J. Phys. Chem. C.* 116 (2012) 16276–16285. doi:10.1021/jp304041m.

41. L. Liu, J. Lang, P. Zhang, B. Hu, X. Yan, Facile synthesis of FE_2O_3 NANO-DOTS@ nitrogen-doped graphene for supercapacitor electrode with ultralong cycle life in koh electrolyte, *ACS Appl. Mater. Interfaces.* 8 (2016) 9335–9344. doi:10.1021/acsami.6b00225.
42. A.M. Khattak, H. Yin, Z.A. Ghazi, B. Liang, A. Iqbal, N.A. Khan, Y. Gao, L. Li, Z. Tang, Three dimensional iron oxide/graphene aerogel hybrids as all-solid-state flexible supercapacitor electrodes, *RSC Adv.* 6 (2016) 58994–59000. doi:10.1039/c6ra11106h.
43. J.W. Park, W. Na, J. Jang, Hierarchical core/shell Janus-type α-Fe_2O_3/PEDOT nanoparticles for high performance flexible energy storage devices, *J. Mater. Chem. A.* 4 (2016) 8263–8271. doi:10.1039/c6ta01369d.
44. Y. Dong, L. Xing, K. Chen, X. Wu, Porous α-Fe_2O_3@C nanowire arrays as flexible supercapacitors electrode materials with excellent electrochemical performances, *Nanomaterials.* 8 (2018) 487. doi:10.3390/nano8070487.
45. Y. Liu, D. Luo, K. Shi, X. Michaud, I. Zhitomirsky, Asymmetric supercapacitor based on MnO_2 and Fe_2O_3 nanotube active materials and graphene current collectors, *Nano-Structures and Nano-Objects.* 15 (2018) 98–106. doi:10.1016/j.nanoso.2017.08.010.
46. H. Jiang, H. Niu, X. Yang, Z. Sun, F. Li, Q. Wang, F. Qu, Flexible Fe_2O_3 and V_2O_5 nanofibers as binder-free electrodes for high-performance all-solid-state asymmetric supercapacitors, *Chem. A Eur. J.* 24 (2018) 10683–10688. doi:10.1002/chem.201800461.
47. J. Zhao, Z. Li, X. Yuan, Z. Yang, M. Zhang, A. Meng, Q. Li, A High-energy density asymmetric supercapacitor based on Fe_2O_3 nanoneedle arrays and $NiCo_2O_4/Ni(OH)_2$ hybrid nanosheet arrays grown on sic nanowire networks as free-standing advanced electrodes, *Adv. Energy Mater.* 8 (2018) 1–14. doi:10.1002/aenm.201702787.
48. W.H. Jin, G.T. Cao, J.Y. Sun, Hybrid supercapacitor based on MnO_2 and columned FeOOH using Li_2SO_4 electrolyte solution, *J. Power Sources.* 175 (2008) 686–691. doi:10.1016/j.jpowsour.2007.08.115.
49. Q. Liao, C. Wang, Amorphous FeOOH nanorods and Co_3O_4 nanoflakes as binder-free electrodes for high-performance all-solid-state asymmetric supercapacitors, *CrystEngComm.* 21 (2019) 662–672. doi:10.1039/c8ce01639a.
50. T. Nguyen, M. Fátima Montemor, γ-FeOOH and amorphous Ni-Mn hydroxide on carbon nanofoam paper electrodes for hybrid supercapacitors, *J. Mater. Chem. A.* 6 (2018) 2612–2624. doi:10.1039/c7ta05582j.
51. R. Chen, I.K. Puri, I. Zhitomirsky, High areal capacitance of FeOOH-carbon nanotube negative electrodes for asymmetric supercapacitors, *Ceram. Int.* 44 (2018) 18007–18015. doi:10.1016/j.ceramint.2018.07.002.
52. A. Kumar, D. Sarkar, S. Mukherjee, S. Patil, D.D. Sarma, A. Shukla, Realizing an asymmetric supercapacitor employing carbon nanotubes anchored to Mn_3O_4 Cathode and Fe_3O_4 anode, *ACS Appl. Mater. Interfaces.* 10 (2018) 42484–42493. doi:10.1021/acsami.8b16639.
53. W. Liu, K. Cheng, S. Sheng, D. Cao, K. Ye, J. Yan, G. Wang, K. Zhu, Fe_3O_4 nanospheres in situ decorated graphene as high-performance anode for asymmetric supercapacitor with impressive energy density, *J. Colloid Interface Sci.* 536 (2018) 235–244. doi:10.1016/j.jcis.2018.10.060.
54. M.M. Vadiyar, X. Liu, Z. Ye, Utilizing waste thermocol sheets and rusted iron wires to fabricate carbon–Fe_3O_4 nanocomposite-based supercapacitors: Turning wastes into value-added materials, *ChemSusChem.* 11 (2018) 2410–2420. doi:10.1002/cssc.201800852.
55. S. Mallick, P.P. Jana, C.R. Raj, Asymmetric supercapacitor based on chemically coupled hybrid material of Fe_2O_3-Fe_3O_4 heterostructure and nitrogen-doped reduced graphene oxide, *ChemElectroChem.* 5 (2018) 2348–2356. doi:10.1002/celc.201800521.

56. S. Zhu, X. Zou, Y. Zhou, Y. Zeng, Y. Long, Z. Yuan, Q. Wu, M. Li, Y. Wang, B. Xiang, Hydrothermal synthesis of graphene-encapsulated 2D circular nanoplates of α-Fe_2O_3 towards enhanced electrochemical performance for supercapacitor, *J. Alloys Compd.* 775 (2019) 63–71. doi:10.1016/j.jallcom.2018.10.085.
57. P. Zhao, N. Wang, W. Hu, S. Komarneni, Anode electrodeposition of 3D mesoporous Fe_2O_3 nanosheets on carbon fabric for flexible solid-state asymmetric supercapacitor, *Ceram. Int.* (2019). doi:10.1016/j.ceramint.2019.02.101.
58. S.N. Khatavkar, S.D. Sartale, α-Fe_2O_3 thin film on stainless steel mesh: A flexible electrode for supercapacitor, *Mater. Chem. Phys.* 225 (2019) 284–291. doi:10.1016/j.matchemphys.2018.12.079.
59. M. Zhu, J. Kan, J. Pan, W. Tong, Q. Chen, J. Wang, S. Li, One-pot hydrothermal fabrication of A-Fe_2O_3@C nanocomposites for electrochemical energy storage, *J. Energy Chem.* (2019) 1–8. doi:10.1016/j.jechem.2017.09.021.
60. Z. Yu, X. Zhang, L. Wei, X. Guo, MOF-derived porous hollow α-Fe_2O_3 microboxes modified by silver nanoclusters for enhanced pseudocapacitive storage, *Appl. Surf. Sci.* 463 (2019) 616–625. doi:10.1016/j.apsusc.2018.08.262.
61. X. Yun, J. Li, Z. Luo, J. Tang, Y. Zhu, Advanced aqueous energy storage devices based on flower-like nanosheets-assembled $Ni_{0.8}5Se$ microspheres and porous Fe_2O_3 nanospheres, *Electrochim. Acta.* 302 (2019) 449–458. doi:10.1016/j.electacta.2019.02.038.
62. S.S. Karade, P. Dwivedi, S. Majumder, B. Pandit, B.R. Sankapal, First report on a FeS-based 2 V operating flexible solid-state symmetric supercapacitor device, Sustain. *Energy Fuels.* 1 (2017) 1366–1375. doi:10.1039/C7SE00165G.
63. C. Zhao, X. Shao, Z. Zhu, C. Zhao, X. Qian, One-pot hydrothermal synthesis of RGO/FeS composite on Fe foil for high performance supercapacitors, *Electrochim. Acta.* 246 (2017) 497–506. doi:10.1016/j.electacta.2017.06.090.
64. Y. Wang, M. Zhang, T. Ma, D. Pan, Y. Li, J. Xie, S. Shao, A high-performance flexible supercapacitor electrode material based on nano-flowers-like FeS_2/NSG hybrid nanocomposites, *Mater. Lett.* 218 (2018) 10–13. doi:10.1016/j.matlet.2018.01.135.
65. Y. Zhong, J. Liu, Z. Lu, H. Xia, Hierarchical FeS_2 nanosheet@Fe_2O_3 nanosphere heterostructure as promising electrode material for supercapacitors, *Mater. Lett.* 166 (2016) 223–226. doi:10.1016/J.MATLET.2015.12.092.
66. K.D. Ikkurthi, S. Srinivasa Rao, M. Jagadeesh, A.E. Reddy, T. Anitha, H.-J. Kim, Synthesis of nanostructured metal sulfides *via* a hydrothermal method and their use as an electrode material for supercapacitors, *New J. Chem.* 42 (2018) 19183–19192. doi:10.1039/C8NJ04358B.
67. X. Dong, H. Jin, R. Wang, J. Zhang, X. Feng, C. Yan, S. Chen, S. Wang, J. Wang, J. Lu, High volumetric capacitance, ultralong life supercapacitors enabled by waxberry-derived hierarchical porous carbon materials, *Adv. Energy Mater.* 8 (2018) 1702695. doi:10.1002/aenm.201702695.
68. S. Venkateshalu, P. Goban Kumar, P. Kollu, S.K. Jeong, A.N. Grace, Solvothermal synthesis and electrochemical properties of phase pure pyrite FeS_2 for supercapacitor applications, *Electrochim. Acta.* 290 (2018) 378–389. doi:10.1016/j.electacta.2018.09.027.
69. V. Sridhar, H. Park, Carbon nanofiber linked FeS_2 mesoporous nano-alloys as high capacity anodes for lithium-ion batteries and supercapacitors, *J. Alloys Compd.* 732 (2018) 799–805. doi:10.1016/j.jallcom.2017.10.252.
70. Z. Sun, X. Yang, H. Lin, F. Zhang, Q. Wang, F. Qu, Bifunctional iron disulfide nanoellipsoids for high energy density supercapacitor and electrocatalytic oxygen evolution applications, *Inorg. Chem. Front.* 6 (2019) 659–670. doi:10.1039/C8QI01230J.
71. X. Shao, Z. Zhu, C. Zhao, C. Zhao, X. Qian, Hierarchical FeS/RGO/FeS@Fe foil as high-performance negative electrode for asymmetric supercapacitors, *Inorg. Chem. Front.* 5 (2018) 1912–1922. doi:10.1039/C8QI00227D.

15 Metal-Organic Frameworks for Supercapacitors

Mustapha Mohammed Bello, Anam Asghar, and Abdul Aziz Abdul Raman

CONTENTS

15.1 Introduction..277
15.2 Supercapacitors..278
15.2.1 Fundamentals and Mechanisms of Energy Storage in Supercapacitors..280
15.2.1.1 Electrochemical Double-Layer Capacitors (EDLCs)...280
15.2.1.2 Pseudocapacitors..282
15.2.1.3 Hybrid Supercapacitors..283
15.3 Metal-Organic Framework..284
15.4 Applications of MOFs and MOFs-Based Materials in Supercapacitors.....286
15.4.1 Pristine and Modified MOFs..286
15.4.2 MOF-Based Composites..289
15.4.2.1 MOF-GO..290
15.4.2.2 MOF-CNT..291
15.4.2.3 MOF-Conducting Polymer..292
15.4.3 MOF-Derived Materials..293
15.4.3.1 MOF-Derived Porous Carbon..293
15.4.3.2 MOF-Derived Metal Oxides..296
15.5 Conclusion..296
References..297

15.1 INTRODUCTION

The unsustainability of conventional energy sources and their contribution to global warming have necessitated the search for clean and renewable energy. A significant research interest has been directed at developing energy from renewable sources such as solar and wind. However, key challenges in utilizing these clean technologies are the conversion and storage of the generated energy. There has been increasing interest to utilize electrochemical technologies such as supercapacitors for energy conversion and storage (Zheng et al., 2018a). Supercapacitors have attracted wide interests as energy storage devices due to their excellent power density, fast charge-discharge cycles and durability (Jiao et al., 2019, 2016; Luo et al., 2013). Thus, supercapacitors have found many applications in power tools, electronics and energy storage (Jabeen et al., 2016; Zheng et al., 2018a).

Supercapacitors are generally charge-storage devices consisting of two electrodes made up of high surface area porous materials, an electrolyte and a separator. The presence of electrode/electrolyte interface induces the formation of a high surface area electric double-layer interface, where charges can be stored and separated. This improved structure of supercapacitors leads to a better energy storage compared to conventional dielectric capacitors (Kim et al., 2014). The function of supercapacitors is therefore considered as a bridge between the conventional dielectric capacitors, which have high power output and batteries/fuel cells, which have high-energy storage (Lu et al., 2012).

The efficacy of supercapacitors depends largely upon the electrode materials. Therefore, selection of suitable cathode materials is a critical factor and represents a significant research direction in the development of supercapacitors. The most important criteria for electrode materials include high pseudo-capacitance and stability (Qi et al., 2018). Among the different possible materials for electrode, metal-organic frameworks (MOFs) have received a wide attention. MOFs are coordination polymers, comprising ligands and metal atoms that are joined through coordination bonds. MOFs are porous materials, which offer high surface areas, controllable structures and tunable pore sizes (Wang et al., 2016). Since the pioneering work on MOFs by Yaghi and Li (1995), a significant research interest has been directed towards the use of MOFs for various applications, including their potential use as electrode materials for supercapacitors. The excellent features of MOFs could be exploited to develop high performance supercapacitors (Bahaa et al., 2019; Li et al., 2018b). However, pristine MOFs face challenges due to their poor conductivity and chemical instability in conventional electrolytes (Qu et al., 2016; Xu et al., 2017a). Consequently, the use of pristine MOFs in supercapacitors is mostly limited to precursors and templates for other advanced materials such as porous carbon and metal oxides (Li et al., 2019a; Yang et al., 2018).

A significant research effort has been devoted to enhancing the performance of MOFs as electrode materials in supercapacitors. These include the modification of MOFs using metallic and nonmetallic materials, and the development of MOFs-based hybrids as supercapacitor electrodes. Figure 15.1 shows different MOF-based materials that have been investigated as potential materials for supercapacitors. Here, we review the advances in the utilization of MOFs and MOFs-based materials in supercapacitors. In particular, we focus on the recent advances in supercapacitors, development of electrode using MOFs and MOFs-based composites. In addition, the challenges facing the use of MOFs in supercapacitors are highlighted and perspectives are offered for future research directions.

15.2 SUPERCAPACITORS

Supercapacitors are electrical components that bridge the performance gap between the high-energy density derived from batteries and the high-power density provided by dielectric capacitors (Zhou et al., 2018). The development of this technology can be traced back to a simple invention made by Leyden Jar in 1746. The invention was based

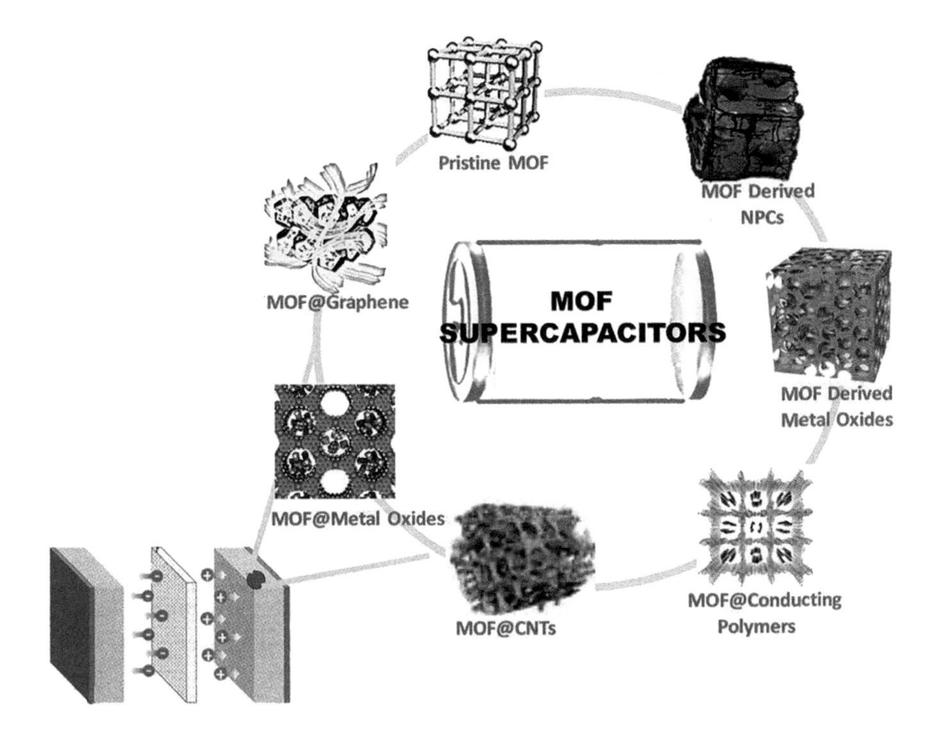

FIGURE 15.1 MOF-based materials for supercapacitors. (Reprinted from *Coord. Chem. Rev.* 369, Sundriyal, S. et al. Metal-organic frameworks and their composites as efficient electrodes for supercapacitor applications, 15–38, Copyright 2018, with permission from Elsevier.)

on a simple setup comprising glass vessel and metal foils. The glass jar acted as dielectric while metal foils acted as electrodes (Poonam et al., 2019). The first electrolytic capacitor was introduced in 1920s. In the middle of twentieth century, the first supercapacitor (electrochemical double-layer capacitor; EDLC) with a maximum capacitance of ~1 F was patented by General Electric (GE) using activated charcoal as the plates (Libich et al., 2018). This supercapacitor had a limited scope due to its high resistance and poor energy density. In order to prompt its practical application, Maxwell Laboratories (USA) commercialized various EDLC supercapacitors with a low equivalent series resistance (ESR) and a maximum capacity of 1 kF (Libich et al., 2018). With these continuous developments, recent supercapacitors possess higher specific capacity (over several thousand F), high current density, and excellent cyclic stability.

Supercapacitors have received enormous attention due to high specific capacitance, high power density, fast charge/discharge rate and being almost maintenance free (Kandalkar et al., 2010; Zhang and Zhao, 2009). The development of supercapacitors has received tremendous interests and several categories have been developed to fulfill the current demands for several applications. Generally, two categories of supercapacitors are available: (i) non-faradic supercapacitors and (ii) faradic supercapacitors, which includes EDLCs, pseudocapacitors (PCs) and hybrid

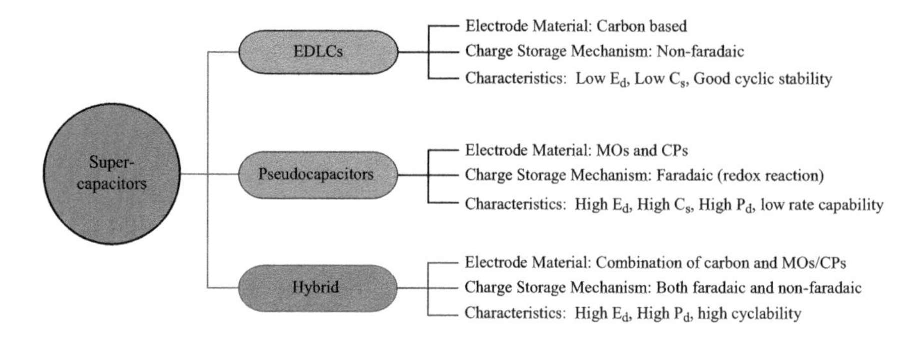

FIGURE 15.2 Comparison of supercapacitors.

supercapacitors (HSCs) (Du et al., 2018; Liu and Peng, 2017). Figure 15.2 compares and summarizes few fundamental features of supercapacitors (EDLCs, PCs and HSCs) based on energy storage mechanism and electrode materials.

This classification is based on different charge storage mechanisms. For instance, in EDLCs charge adsorption/separation takes place at the electrode-electrolyte interface while fast and reversible faradic reactions take place in pseudocapacitors. However, both faradic and non-faradic processes are involved in the charge storage mechanism of hybrid supercapacitors (Poonam et al., 2019; Wang et al., 2013; Zhang et al., 2017). In spite of these advancements, the energy density (E_d) for these types of supercapacitors are still low, which limit their utilization in cases where high-energy capacity is required. Therefore, continuous efforts are in progress to improve the energy density and capacitance of supercapacitors. These include the development of high capacitance electrode material and electrolytes with wide potential window (Béguin et al., 2014; Liu et al., 2019; Qiu et al., 2015; Xia et al., 2014; Xiong et al., 2019). The combination of carbon and different metal oxides, also called metal-organic frameworks (MOFs), offers excellent alternatives due to their high specific surface area, large internal pore volumes, redox activities of metal oxides and the potential to increase the overall cell voltage (Hua et al., 2019b; Mahmoud and Tan, 2018; Pham et al., 2019; Sonai Muthu and Gopalan, 2019). Therefore, in addition to increasing the overall cell voltage, efforts to increase the specific capacitance and the overall performance of supercapacitors are required. The first approach of amplifying cell voltage is linked with hybrid capacitors while the later criterion is associated with pseudocapacitive electrodes and EDLC. It is therefore pertinent to improve the performance of supercapacitors and their utilization for commercial applications.

15.2.1 Fundamentals and Mechanisms of Energy Storage in Supercapacitors

15.2.1.1 Electrochemical Double-Layer Capacitors (EDLCs)

Electrochemical double-layer capacitors (EDLCs), using carbon as electrodes, have attained considerable attention as one of the energy storage devices (Li et al., 2018). The use of carbons results in high specific capacitance, short charging time, high-power density and long life cycles (Li et al., 2018b; Liu et al., 2019; Soneda, 2013). In comparison with secondary batteries or ceramic/aluminum electrolytic capacitors that work via conventional electrochemical reactions, EDLCs perform better due their intrinsic features such as high-power density and good cyclic stability. The difference between these two types of devices is the different charge storage mechanism. In the case of supercapacitors, charge is stored on the electrode surface while it is stored in bulk in the case of batteries.

For EDLCs, the capacitance is associated with the accumulation of charges at the electrode-electrolyte interface. Non-faradaic redox reactions are involved and pure physical charge accumulation i.e. adsorption of ions at the surface of polarized electrodes, as shown in Figure 15.3, plays a fundamental role in determining the specific power and cyclic stability of the charge-discharge process (Han et al., 2018b; Simon et al., 2017). EDLCs resemble parallel-plate capacitors and therefore can be described by Equation 15.1 (Muzaffar et al., 2019).

$$C = \frac{\varepsilon_o \, \varepsilon_\gamma A}{d} \tag{15.1}$$

where ε_γ is the dielectric constant of the electrolyte, ε_o is the permeability factor, A is the contacting surface area of electrode with electrolyte and d is the effective thickness of the electric double layer. In addition, the accumulation of the charge (Q) on the electrode can be determined using Equation 15.2 (Simon et al., 2017).

$$Q = C \cdot \Delta \tag{15.2}$$

Equation 15.2 is valid when electrodes are immersed paralleled into an electrolyte and the applied potential difference of ΔV exists between the surfaces. Here, the adsorption/accumulation of the charges will be fully reversible and the maximum voltage will be controlled by the redox potential of the electrolyte or electrode

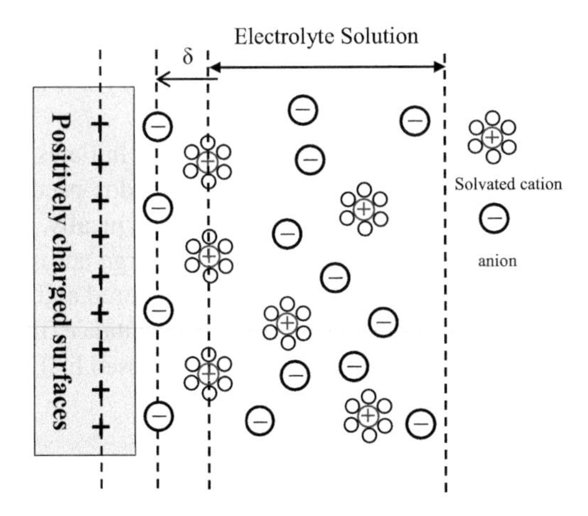

FIGURE 15.3 Negative polarization of carbon electrode.

material (Xu et al., 2007). Therefore, EDLC can be considered a series combination of two capacitors i.e. a negative electrode associated with cation in solution and a positive electrode accompanied with anions in the solution (Conway, 1999). The charge of the double electrochemical layer during negative polarization of carbon electrode is shown in Figure 15.3 (Zhang and Zhao, 2009).

As Equation 15.1 shows, the capacitance of EDLCs is influenced by parameters such as the effective thickness (d) and specific surface area (A) of the electrode. The distance between electrolyte and electrode for EDLC is much smaller than that in conventional batteries, leading to the higher capacitance of the former. Thus, the factor that will primarily be considered in obtaining high capacitance is the carbon/electrolyte interfacial surface area. Thus, utilization of materials with high specific surface area could enhance the capacitance of supercapacitors (Conway, 1999; Zhang and Zhao, 2009). However, in most cases, the high specific surface area has not brought the expected high capacitance as not all pores are involved in the reaction (Barbieri et al., 2005). Therefore, efforts are in progress to develop carbon-based materials with controlled pore size distribution and specific capacitance. Subsequently, materials like carbon nanotubes (Pan et al., 2010) graphene (Ke and Wang, 2016), exfoliated carbon fibers (Soneda, 2013) and other ordered porous carbon materials (Li et al., 2018) were developed. Other options include carbon surfaces doped with heteroatoms (such as oxygen, nitrogen, sulfur) and functionalized with metal oxides to induce faradaic redox reactions (Wang et al., 2018).

15.2.1.2 Pseudocapacitors

The capacitance of pseudocapacitors is due to the rapid faradaic reduction-oxidation reactions occurring at or near the plane of electrode (Liu et al., 2009). The redox reactions involved are fast but have slow kinetics than that of pure physical charge adsorption as the later does not involve any ionic diffusion or charge transfer phenomenon. Therefore, the faradaic redox reactions result in low power density and structural instability (Wang et al., 2018). In this regard, the limited ion conductivity and thus the pseudocapacitance can significantly be enhanced by inducing redox reaction on the surface or interior of the electrode materials (Zhai et al., 2017). Pseudocapacitance is non-electrostatic in nature and usually designates a capacitive material with linear dependence of the charge stored on the width of the potential window (Xu et al., 2007).

The charge storage mechanism in pseudocapacitors initiates with several reactions mechanisms such as underpotential deposition, redox pseudocapacitance and intercalation pseudocapacitance (Zhang et al., 2013). Typically, pseudocapacitance occurs at the electrode surface with the passage of charge across the double layer. In this case, the capacitance is linked with the charge stored and the varying potential (Simon et al., 2017). Thus, considering pseudocapacitance, the linear change in the stored charge (ΔQ) with potential (ΔV) can be expressed by Equation 15.3.

$$\Delta Q = C \times \Delta V \tag{15.3}$$

Here, pseudocapacitance (C) can be expressed as the derivative of charge stored (ΔQ) and potential (ΔV) (Equation 15.4) (Muzaffar et al., 2019).

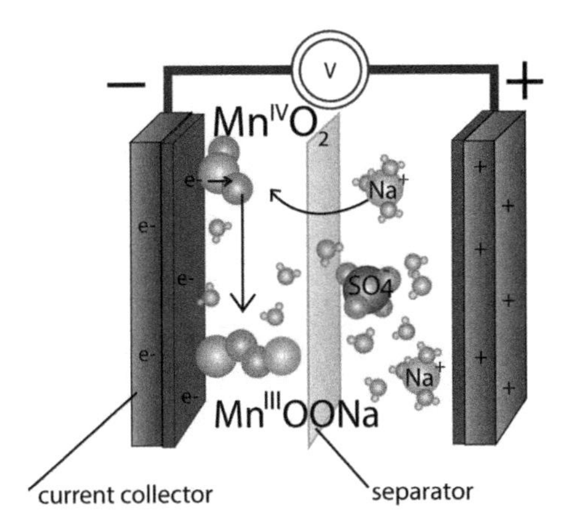

FIGURE 15.4 Charge storage mechanism in pseudocapacitor. (Jost, K. et al., *J. Mater. Chem. A*, 2, 10776–10787, 2014. Reproduced by permission of The Royal Society of Chemistry.)

$$C = \frac{d\,(\Delta Q)}{d\,(\Delta V)} \tag{15.4}$$

As explained earlier, the charge storage in pseudocapacitive materials involves fast and reversible redox reaction (Figure 15.4). Most of the pseudocapacitive materials tested to date are electronic conductive polymers and metal oxides (Liu et al., 2019). These include ruthenium oxides (Frackowiak and Béguin, 2001), vanadium nitride (Choi et al., 2006), manganese oxide (Zhang et al., 2008), polyaniline (PANI) (Fan et al., 2007) carbon-based hetero-atoms (Jeong et al., 2011; Lee et al., 2011), transition metals (Hulicova et al., 2005) and nano-porous or exfoliated carbons (Lota et al., 2005). The capacitance values for most of these materials range from 100 to 1500 F/g.

Since faradaic reactions involve charge storage, the energy density of pseudocapacitive material is higher than EDLC. However, the corresponding power densities are lower because of the slower kinetics (Simon et al., 2017). Therefore, the value of pseudocapacitance can further be improved by using nano-structured materials as electrodes. Such types of materials are known for high surface area and decreased diffusion lengths.

15.2.1.3 Hybrid Supercapacitors

The concept of hybrid supercapacitors was induced with an opportunity to increase energy density of conventional EDLC to a range of 20–30 Wh/kg (Burke et al., 2000). Coupling the advantages of high cycling rates of carbon negative electrodes with those of faradaic positive electrodes, the charge storage mechanism in hybrid supercapacitor is directed by the combined principles of EDLC and pseudocapacitors (Figure 15.5) (Muzaffar et al., 2019). Hybrid supercapacitors combine the high energy density of batteries and high-power density from supercapacitors (Liu et al., 2017). This combination leads to overcoming the shortcomings of combined elements and delivers a higher capacitance.

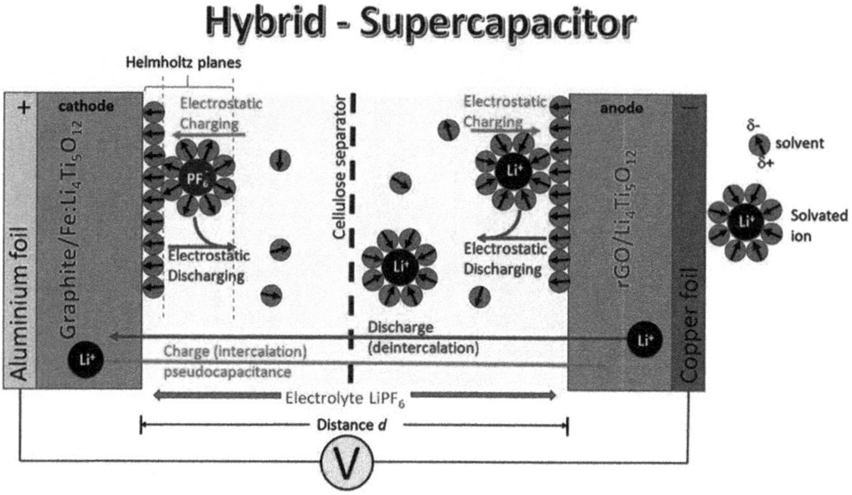

FIGURE 15.5 Depiction of charge storage mechanism in hybrid capacitors. (Repp, S. et al., *Nanoscale*, 10, 1877–1884, 2018. Reproduced by permission of The Royal Society of Chemistry.)

Hybrid supercapacitors, depending upon the configuration, are classified as: (i) symmetric and (ii) asymmetric. The configuration comprising two similar EDLC and pseudocapacitive electrodes is classified as symmetric hybrid supercapacitors while a combination of two dissimilar electrodes forms asymmetric hybrid supercapacitor (Mastragostino et al., 2002; Wang et al., 2006). Commercial supercapacitors are asymmetrical in configuration and mostly utilize polymer electrodes, which involve a series of redox reactions to store energy (Gómez-Romero et al., 2010).

One of the limitations of hybrid supercapacitor is internal resistance, which is due to ESR, a combination of electrolyte resistance, current collectors and electrodes (Khomenko et al., 2008; Park et al., 2002). Therefore, the overall performance of hybrid supercapacitor depends on the selection of both electrode and electrolyte materials. It is essential to make careful selection of electrolyte or electrode material for the improvement of overall performance of hybrid supercapacitor.

15.3 METAL-ORGANIC FRAMEWORK

Metal organic frameworks (MOFs) are coordination compounds constructed from periodic metallic centers coordinated by organic linkers, which have recently attracted a significant attention due to their excellent properties (Burtch et al., 2014; Yang and Xu, 2017).MOFs have excellent properties, including tunable pore size and large surface area (Choi et al., 2008). Consequently, MOFs can be manipulated and used in specific applications, including supercapacitor electrodes (Eddaoudi et al., 2001; Fang et al., 2005).

MOFs are conventionally synthesized through hydrothermal methods, using autoclave at higher temperature for prolonged hours (Ni and Masel, 2006;

Tranchemontagne et al., 2008). Similar to conventional heating, microwaves irradiation has also been used for the preparation of MOFs (Choi et al., 2008; Ni and Masel, 2006). Microwave heating can result in fast crystallization, phase selectivity and controlled morphology. The electrochemical method was first reported by Muller et al. (2007). Other methods include mechano-chemical synthesis and ultrasound induced synthesis methods (Rademann et al., 2010; Yuan et al., 2010). Ultrasound treatment has proven efficient because it is simple, fast and energy efficient (Ando, 1984; Gilvaldo et al., 2016). Ultrasound irradiation, based on acoustic cavitation, is an effective method for the modification of different surface structures. The hot spots generated by ultrasound cavitation hold sufficient energy to activate or split larger particles into smaller ones and increase the porosity of solid surfaces (Suslick, 1990).

The above-mentioned synthesis techniques are impractical for industrial scale applications, necessitating the development of alternative synthetic routes for large-scale applications. In this regard, Gilvaldo et al. (2016) explored a more practical synthesis approach through sonoelectrochemistry. Sonoelectrochemistry seems efficient and practical as it is widely applied for applications such as electrodeposition/electroplating and nano-particles synthesis. Figure 15.6 depicts a schematic for the synthesis of MOF through sonoelectrochemical method.

Due to the excellent properties of MOFs, there is growing interest in their applications in supercapacitors (Díaz et al., 2012; Hu et al., 2010; Liu et al., 2010; Yang et al., 2014; Yu et al., 2015; Zhang et al., 2015). However, poor conductivity and chemical instability still remain major challenges in the application of MOFs in supercapacitors. Thus, a great effort is devoted to address these challenges. In this

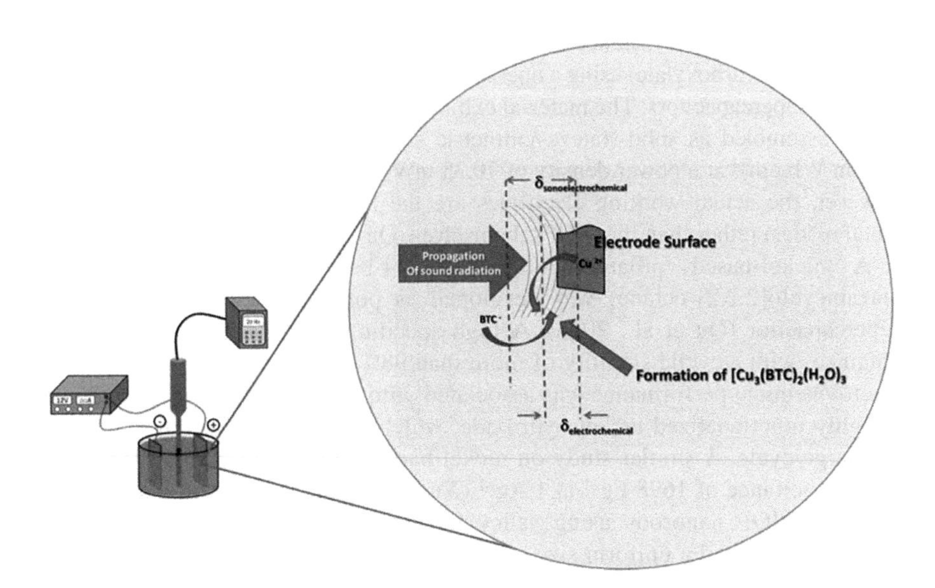

FIGURE 15.6 A schematic of a typical sono-electrochemical synthesis of MOF. (Reprinted from *Synth. Met.*, 220, Givaldo, et al., 369–373, copyright 2016, with permission from Elsevier.)

regard, many studies have reported the functionalization of MOFs to improve their conductivity and stability, development of MOF-based composites and MOF-derived advanced materials.

15.4 APPLICATIONS OF MOFs AND MOFs-BASED MATERIALS IN SUPERCAPACITORS

One of the challenges in the development of supercapacitors is the selection of appropriate cathode materials. MOFs have received wide attentions as potential electrode materials for supercapacitors. The key considerations in the use of MOFs-based materials include specific capacitance, energy and power densities as well as the cycle stability of the supercapacitors. Table 15.1 summarized some recent studies on the use of MOF-based materials as electrode in supercapacitors, highlighting the materials compositions and the resulting specific capacitance, energy/power density and stability of the supercapacitors. These applications are further discussed under pristine and modified MOFs, MOF-based composites and MOFs-derived materials.

15.4.1 Pristine and Modified MOFs

Many studies have reported the applications of pristine MOFs as electrode materials in supercapacitors. However, pristine MOFs are characterized by low conductivity and poor chemical stability, hindering their applications as electrode materials (Dai et al., 2018). Due to their chemical instability, most pristine MOFs are not stable in traditional electrolytes, which are characterized by acidic or alkaline conditions. Despite these challenges, some studies have reported the application of pure MOFs in supercapacitors. For instance, Yan et al. (2018) prepared a vanadium MOFs (V-O-1,4-benzenedicarboxylate) using a one-step solvothermal strategy for potential application in supercapacitors. The material exhibited a specific capacitance of 572.1 Fg^{-1}. When assembled as solid-state asymmetric supercapacitors, a specific energy of 6.72 m Wh cm^{-3} at a power density of 70.35 mW cm^{-3} was obtained. In most cases, however, the actual working electrodes are the MOF-derived materials (such as metal oxides) rather than the MOFs themselves (Qu et al., 2016; Xu et al., 2017a).

A nickel-based pillared MOFs (Ni/1,4-benzenedicarboxylic acid/1,4-diazabicyclo[2.2.2]-octane) was developed as potential electrode materials in supercapacitor (Qu et al., 2016). A high specific capacitance of 552 Fg^{-1} was obtained, with a cyclic stability of more than 98% after 16,000 cycles. This high electrochemical performance was associated with the conversion of the MOFs to a highly functionalized nickel hydroxide, which remained stable during charge-discharge cycle. A similar study on nickel-based MOF resulted in a higher specific capacitance of 1698 Fg^{-1} at 1 Ag^{-1} (Xue et al., 2016). The authors produced Ni-based MOF nanorods using salicylate ions as organic linkers. The MOF nanorods exhibited a uniform size nano structure (average length of 900 nm and diameter of 60 nm) and a high specific capacitance. When the Ni-based MOF nanorods were assembled with graphene in an asymmetric supercapacitor, an excellent specific capacitance (166 Fg^{-1}) was achieved.

TABLE 15.1
Applications of MOF-Based Materials as Supercapacitor Electrodes

MOF Description	Specific Surface Area (m^2/g)	Scan Rate/ Current Density	Specific Capacitance (F/g)	Energy Density (Wh/kg)	Power Density (W/kg)	Cycle Stability (%)	References
Ni-MOF/AC	758.24	1 A/g	1453.5	56.85	480	86.67	Yue et al. (2019)
$NiCoSe_2$: Ni-Co-ZIF 67 + Se	232.8	1 A/g	300.2	—		100	Sun et al. (2019b)
Zn-MOF	1234	1 A/g	322	70	—	95.8	Osman et al. (2019)
$CuCo_2S_4$ nanosheet	132.92	3 mA cm^{-2}	409.2 mA h/g	89.6	663	94.2	Bahaa et al. (2019)
Ni/P/N/C-MOF		1 A/g	2887.87			90	Yu et al. (2019)
MnOOH/NiA-LDH	110.17	1 A/g	1331.11	26.89	800	—	Hua et al. (2019a)
$Ni(OH)_2$@ZnCoS-nanosheet	—	3 mA cm^{-2}	2730	75	0.4	82	Sun et al. (2019a)
Zn-pephenylenediamine MOF		1 A/g	—	62.8	4500	96	Kannangara et al. (2019a)
Ni-MOF	347	1 A/g	250.6 mA h/g	48.1	1064.7	92	Rawool et al. (2019)
ZIF-67-based $Co_3O_4/Co(OH)_2$ hybrids	151.6	1 A/g	184.9 mA h/g	37.6	47	91	Lee and Jang (2019)
ZnO@MOF@PANI)	—	1 A/g	340.7	—	—	82.5	Zhu et al. (2019)
Al-MOF-derived RAPC	1886	1 A/g	332	—	—	97.8	Jiang et al. (2019)
Ni@CoNi-MOF	88.1	0.5 A/cm^3	100 F/cm^3	31.3 mWh/cm^3	3796.4 mW/cm^3	100	Hong et al. (2019)
Ni-MOF	—	1 mA cm^{-2}	684.4 mF cm^{-2}	61.3	900	—	Jiao et al. (2019)

(Continued)

TABLE 15.1 (*Continued*)
Applications of MOF-Based Materials as Supercapacitor Electrodes

MOF Description	Specific Surface Area (m^2/g)	Scan Rate/ Current Density	Specific Capacitance (F/g)	Energy Density (Wh/kg)	Power Density (W/kg)	Cycle Stability (%)	References
rGO@ZIF- 67@ NiAl-LDHs/AC	—	1 A/g	2291.6	91.4	875	—	Guo et al. (2019)
Ni-pPDA//GC	20.98	1 A/g	184.7	57.5	5980	80	Kannangara et al. (2019b)
PO-based MOFs@rGO	13.8	0.5 A/g	178	20.1	9071	94	Tan et al. (2019)
Cu-MOF-based porous carbon	2491	0.5 A/g	260.5	18.38	350	91.1	Li et al. (2019)
Ni/Co-MOF-rGO	—	1 A/g	860	72.8	850	91.6	Rahmanifar et al. (2018)
Zn-MOF-derived porous AC	2314.9	1 A/g	325	—	71.2	98.8%	Osman et al. (2018)
Cu_7S_4/C	—	0.5 A/g	321.9	—	—	78.1	Han et al. (2018a)
V-MOF	116.8	0.5 A/g	572.1	6.72 mWh cm^{-3}	70.35 mW cm^{-3}	92.8	Yan et al. (2018)
CC@CoO@S-Co_3O_4	24.7	1 mA cm^{-2}	1013 mF cm^{-2}	0.71 mWh cm^{-3}	21.3 mW cm^{-3}	87.9	Dai et al. (2018)
MOF-derived Co_9S_8@C composite	62.95	1 A/g	1887	58	1000	90	Sun et al. (2018)
nHKUST-1-derived $Cu_7S_{4/}C$		0.5 A/g	321.9			78.1	Han et al. (2018)
$Ni_xCo_{3-x}O_4$/CNTs	—	1 A/g	668	23.65	800.15	93.75	Xue et al. (2018)
CC@CoO@S-Co_3O_4	—	1 mA cm^{-2}	1013 mF cm^{-2}	0.71 mWh cm^{-3}	21.3 mW cm^{-3}	87.9	Dai et al. (2018)
Ni-MOF	295.7	1 A/g	123.4 mA h/g	55.8	7000	—	Jiao et al. (2016)
rGO/HKUST-1	1241	1 A/g	385	42	3100	98	Srimuk et al. (2015)

Abbreviations: LDHS: Layered double hydroxides; RAPC: Rod-like porous activated carbon; *p*-PDA: *p*-phenylenediamine; GC: Graphite carbon; PO: Polyoxometallate

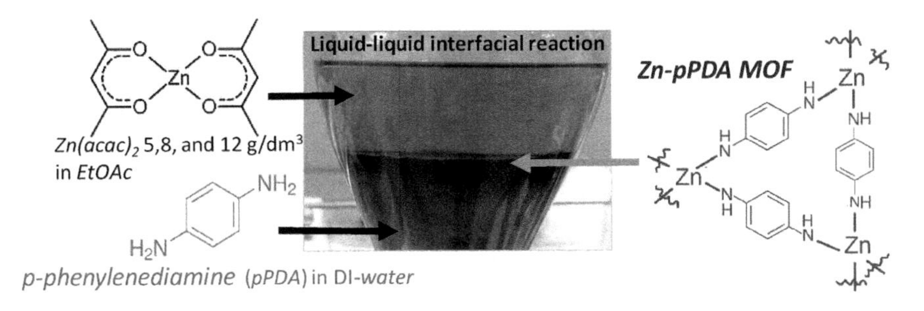

FIGURE 15.7 Synthesis of pPDA through liquid-liquid interfacial reaction and the chemical structure of the synthesized MOF. (Reproduce from *Electrochim. Acta* 297, Kannangara, et al., 145–154, Copyright 2019a, with permission from Elsevier)

The electrochemical performance of MOFs depends on the precursor materials as well as the preparation method. Kannangara et al. (2019a) synthesized Zn-pephenylenediamine (Zn-pPDA) MOFs through liquid-liquid interfacial reactions for use in supercapacitors. The synthesis method involves the use of an organometallic molecule, $Zn(acac)_2$, and an organic ligand, pPDA to produce the MOF complexes (Figure 15.7). When the MOFs were applied as an electrode in an asymmetric supercapacitor, an energy density of 62.8 Wh/kg at a power density of 4500 W/kg was achieved. The material also exhibited an excellent cyclic stability (96%). However, a residual $Zn(acac)_2$ was observed to leach into the Zn-pPDA, reducing the electrical conductivity of the electrode.

The poor chemical stability of pristine MOFs hinders their performance as electrode materials in supercapacitors. In an attempt to increase their stability and the performance of supercapacitors, many studies have reported the applications of modified MOFs. Yue et al. (2019) modified a Ni-based MOF using nonmetallic elements. The procedure involved hydrothermal synthesis of the MOF and the subsequent modification of the MOF using urea, dihydrogen phosphate and thioacetamide to obtain N-modified Ni-MOF, P-modified Ni-MOF and S-modified Ni-MOF respectively. These modifications resulted in the formation of more stable framework structures and more active sites. In particular, the S-modified MOF delivered a specific capacitance of 1453.5 Fg^{-1} and 89.23% cyclic stability after 5000 cycles. When applied in a supercapacitor, a specific energy of 56.85 Wh kg^{-1} at a specific power of 480 W kg^{-1} was achieved.

15.4.2 MOF-Based Composites

The modification of MOFs using metallic and nonmetallic materials improves their stability and electrochemical performance. However, using MOFs as the only active materials still faces challenges of poor conductivity and instability. These challenges could be addressed by functionalizing MOFs with conducting materials such as GO, CNTs and conducting polymers (Xu et al., 2017; Zhu et al., 2019). The composites of MOFs and these conductive materials offer a viable approach to develop supercapacitors with excellent performance and stability. Consequently, recent studies on MOFs in supercapacitors are largely on MOFs-based composites, including MOF-GO, MOF-CNT, and MOF-Conducting polymers.

15.4.2.1 MOF-GO

Graphene-based carbons exhibit high specific areas and excellent conductivity, which are advantageous in energy storage devices such as supercapacitors. Graphene oxide (GO) has been utilized to produce MOF-based composite with high conductivity and excellent electrochemical properties for applications in supercapacitors and batteries (Wei et al., 2017). The stability and conductivity of pristine MOFs are generally enhanced with the incorporation of GO to into the MOF.

MOF-GO composites have been explored as electrode materials in supercapacitors. The results of previous studies indicated that the electrochemical properties and chemical stability of the MOF-GO are superior to those of the pristine MOF. Srimuk et al. (2015) synthesized a hybrid of rGO and HKUST-1 and applied it in supercapacitors. The hybrid exhibited excellent textural properties and delivered an excellent specific capacitance (385 Fg^{-1}), which is higher than that of pristine HKUST-1 (0.5 Fg^{-1}). When coated on a carbon fiber paper, the composite delivered a power density of 3100 W kg^{-1} and an energy density of 42 Wh kg^{-1}. In addition, the rGO-HKUST-1 supercapacitor exhibited a higher cyclic stability (about 98.5%) than rGO supercapacitor (85%) over 4000 cycles.

The use of a dual Ni/Co-MOF-rGO composite as a supercapacitor electrode was recently reported by Rahmanifar et al. (2018). The composite was prepared through a one-pot co-synthesis of the MOFs and rGO and delivered a specific capacitance of 860 Fg^{-1} at 1.0 Ag^{-1}. When assembled to form an asymmetric supercapacitor of activated carbon//Ni/Co-MOF-rGO, a specific energy of 72.8 Wh kg^{-1} and a specific power of 42,500 W kg^{-1} were obtained, with a good cycle stability (91.6% over 6000 cycles). The excellent electrochemical properties of the Ni/Co-MOF-rGO, which are better than those reported by Srimuk et al. (2015), are largely due to the intrinsic synergy of the dual metal MOFs and the rGO.

Yu et al. (2017) prepared hybrids of MOF-GO (HKUST-1/GO, ZIF-67/GO and ZIF-8/GO) for supercapacitor electrodes. The preparation method of the hybrids involves mixing the GO with metal ion sources, formation of the layer-by-layer metal ion-GO hybrid and a reaction with an organic linker to produce MOF-GO hybrid. A further step was introduce to convert the MOF-GO into layered double hydroxide (LDH) or porous carbon-GO hybrids. When used in an asymmetric supercapacitor, it delivered a specific capacitance of 142.1 Fg^{-1} and a specific energy of 50.5 Wh kg^{-1} at a specific power of 853.3 W kg^{-1}. The hybrid maintained 81.2% of its capacitance after 3000 cycles at 5 Ag^{-1}.

Xu et al. (2017) synthesized 3D GO/MOF composites through a facile method and subsequently used them as precursors to developed rGO/MOF-derived aerogels. The preparation method (Figure 15.8) involves mixing GO and MOF to produce GO/MOF hydrogel, which is then subjected to a freeze-drying to produce GO/MOF aerogel. The GO-MOF was then converted to rGO/MOF aerogel through annealing. The resulting composite showed high specific capacitances of 869.2 and 289.6 Fg^{-1} at current densities of 1 and 20 Ag^{-1} respectively.

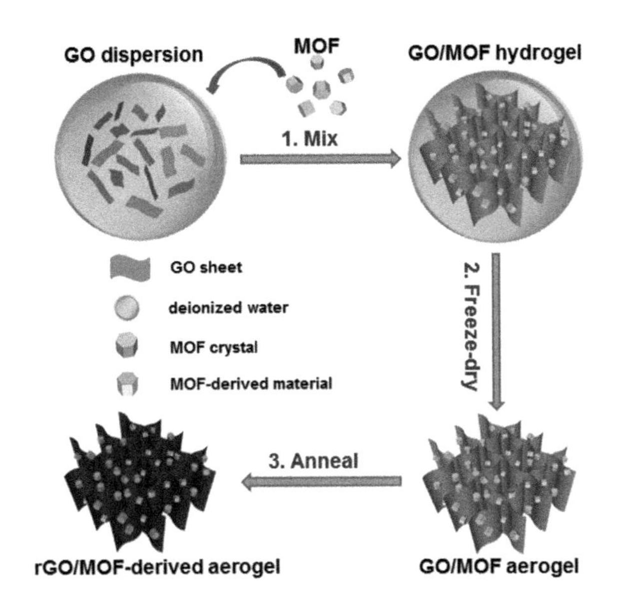

FIGURE 15.8 preparation method of GO/MOF composites for supercapacitor applications. (Reprinted with permission from Xu, M.W. et al., *J. Solid State Electrochem.*, 11, 372–377, 2007, Copyright 2007 American Chemical Society.)

15.4.2.2 MOF-CNT

Combining CNTs with MOFs is another strategy towards enhancing their performance in supercapacitors. CNTs have excellent conductivity and tunable structure that can be exploited to developed MOF-CNT composites, resulting in enhanced capacitance and energy/power densities in supercapacitors. The enhanced performance of the composites is due to the porous nature of the MOFs and conductivity of the CNTs, resulting in a synergistic effect in the supercapacitors.

Xue et al. (2018) prepared porous $Ni_xCo_{3-x}O_4$/CNTs nanocomposites for supercapacitors electrode. Figure 15.9 depicts the synthesis method, which involves the in-situ growth of precursor (ZIF-67) onto CNTs, with a subsequent annealing of the product. The hybrid exhibited a specific capacitance of 668 Fg^{-1} and delivered energy and power densities of 23.56 Wh kg^{-1} and 800.15 W kg^{-1} respectively. In addition, ~94% cycle stability over 3000 cycles was obtained.

Wen et al. (2015) developed a Ni-MOF/CNT composite for use as a supercapacitor electrode. The synergy of the excellent conductivity of CNTs and the intrinsic feature of the Ni-MOF resulted in a composite with an excellent specific capacitance of 1765 Fg^{-1}. When applied as an electrode in asymmetric supercapacitor, the composite delivered a specific energy and power of 36.6 Wh kg^{-1} and 480 W kg^{-1} respectively. Similarly, Fu et al. (2016) prepared an MOF composite consisting of a zirconium-based MOF, CNT and poly(3,4-ethylenedioxythiophene)-GO. The synthesis involved coating the MOF onto the CNT, followed by a deposition of the PEDOT-GO, resulting in an electrode with excellent electrochemical properties.

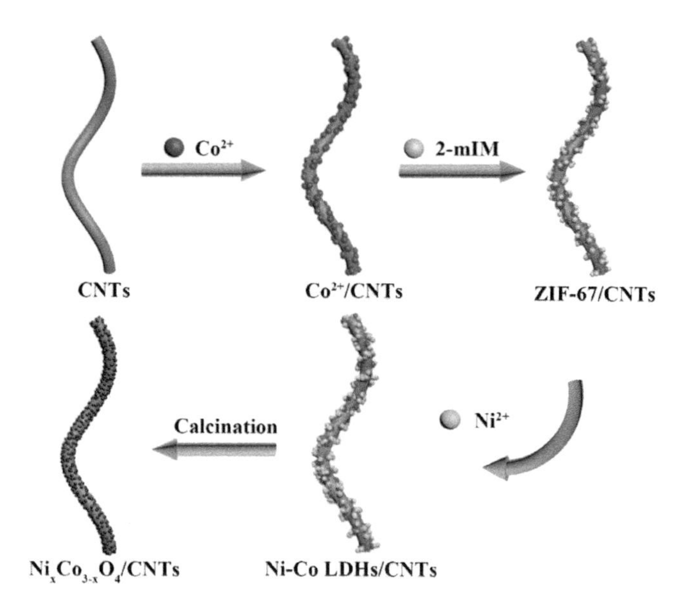

FIGURE 15.9 Preparation of MOF-CNT composite using ZIF-67. (Reproduced from *J. Colloid Interface Sci.*, 530, Xue, et al., 233–242, Copyright 2018, with permission from Elsevier.)

15.4.2.3 MOF-Conducting Polymer

Conducting polymers can be used to functionalize MOFs, improving their conductivity and other electrochemical properties. The conducting polymers provide linkages between isolated MOFs crystals, thereby enhancing the electron transportation through the crystals (Qi et al., 2018). The enhancement in the structural and electrochemical properties of the resultant composite depends on the MOF precursor, type and deposition method of the polymer (Zhu et al., 2019). Thus, the selection of polymer and functionalization method are important considerations for developing effective MOF-conducting polymer composites. Conducting polymers, such as polyaniline and polypyrrole, have been used to functionalize MOFs for effective applications in supercapacitors.

Many studies have reported the use of polyaniline to functionalized MOFs for supercapacitor electrodes. Wang et al. (2015) prepared a cobalt-based MOF (ZIF-67) and then introduced a polyaniline to interweave the MOF crystals, forming a flexible conductive electrode. Figure 15.10 depicts the scheme for electrolyte and electron conduction in both MOF and MOF-polyaniline composite. The presence of polyaniline in the composite provides a conduction path for the electron, thus enhancing the conductivity of the composite. Clearly, the polyaniline assisted in improving the electrochemical properties of the MOF.

The capacitance achieved by the MOF-polyaniline described above is much higher than MOF-polypyrrole as reported by Qi et al. (2018). In their study, Qi and co-workers prepared an MOF-polypyrrole composite. The composite achieved a

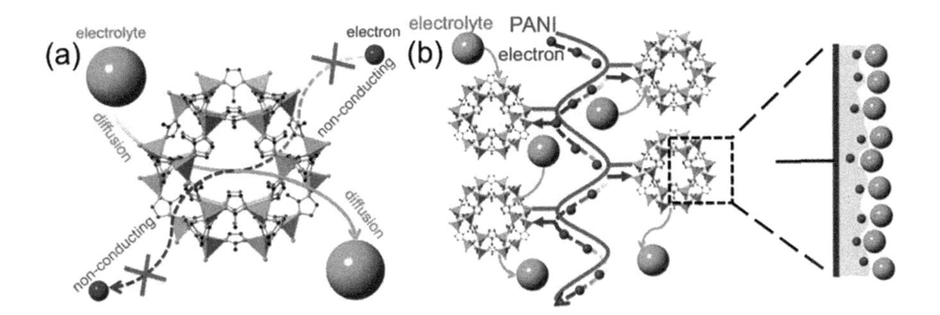

FIGURE 15.10 (a) Pristine MOF without conduction path (b) MOF-PANI showing the conduction the PANI. The presence of PANI creates a conduction path in the composite. (Reprinted with permission from Wang, L. et al., *J. Am. Chem. Soc.*, 137, 4920–4923, 2015. Copyright 2015 American Chemical Society.)

specific capacitance of 206 mF cm^{-2} when used as an electrode in an all-solid-state fabric supercapacitor. This higher performance could be largely due to the superior properties of polyaniline compared to polypyrrole (Blinova et al., 2007; Kim et al., 2006). As highlighted previously, the enhancement in the conductivity of the MOF composite would depend, among other things, on the type of polymer.

Liu et al. (2018) utilize polypyrrole to functionalize MOF (ZIF-67) for application in supercapacitors. The polypyrrole improved the electrochemical properties of the MOF, delivering specific capacitance of 180.2 mF cm^{-2}, which is lower than that reported by Qi et al. (2018). However, the composite exhibited an excellent cycling stability of 100% after 40,000 cycles.

15.4.3 MOF-Derived Materials

MOFs can be utilized as precursors to produce excellent advanced materials such as porous carbon and metal oxides that can be used as electrode in supercapacitors. The structure of MOF can be easily tuned into a highly porous carbon/metal oxide through thermal treatment of the MOF precursors under inert gas/air condition (Zhang et al., 2019). MOFs-derived porous carbon and metal oxides are highly ordered three-dimensional structures with unique electrochemical properties that can be utilized in supercapacitors (Bao et al., 2016; Yu et al., 2017). MOF-derived porous carbon and metal oxides represent one of the major research focus in the applications of MOF-based materials in supercapacitors.

15.4.3.1 MOF-Derived Porous Carbon

MOF-derived porous carbons are generated from high-temperature carbonization of MOFs, resulting in porous activated carbons with excellent controllable pore size, large pore volume and excellent conductivity (Bao et al., 2016). The excellent properties of this class of activated carbon are superior to those of activated carbons produced through conventional methods. Consequently, a significant research interest is accorded to the development of MOF-derived porous activated carbon and their applications as electrode in supercapacitors.

Yan et al. (2014) prepared porous carbon through carbonization (at 1073 K) of three different MOFs (Cu-MOF, Zn-MOF and Al-MOF). Although the specific surface areas of the produced porous carbon were relatively low (Cu-MOF carbon = 50 $m^2\ g^{-1}$, Zn-MOF = 420 $m^2\ g^{-1}$ and Al-MOF carbon = 1103 $m^2\ g^{-1}$), they exhibited good specific capacitance (Cu-MOF carbon = 82.9 Ag^{-1}, Zn-MOF = 87 Ag^{-1} and Al-MOF carbon = 173.6 Ag^{-1}).

Jiang et al. (2019) prepared a hierarchical porous rod-like activated carbon using aluminum-based MOF as a starting material. The procedure involved the synthesis of Al-MOF through direct precipitation of $AlCl_3$ and deprotonated organic ligand, sodium 1,3,5-benzene-tricarboxylate (TBC), followed by the carbonization and chemical activation (using KOH and KNO_3) of the Al-based MOF. This procedure is depicted in Figure 15.11. The derived carbon exhibited high specific surface area (1886 $m^2\ g^{-1}$), pore volume (1.29 $cm^3\ g^{-1}$) high specific capacitance (332 Fg^{-1}) and capacitive stability of 97.8% after 100,000 cycles.

Osman et al. (2019) prepared a porous-like amorphous carbon from Zn-based fluorinated MOF. In this case, the procedure involved the preparation of the Zn-MOF thorough a solvothermal method using 2,3,5,6–tetrafluorobenzenedicarboxylate (TFBDC) as organic linker. This was followed by the carbonization of the Zn-TFBDC at 750°C under nitrogen environment. The prepared material had a high specific surface area (1234 $m^2\ g^{-1}$) and excellent electrochemical properties. Under

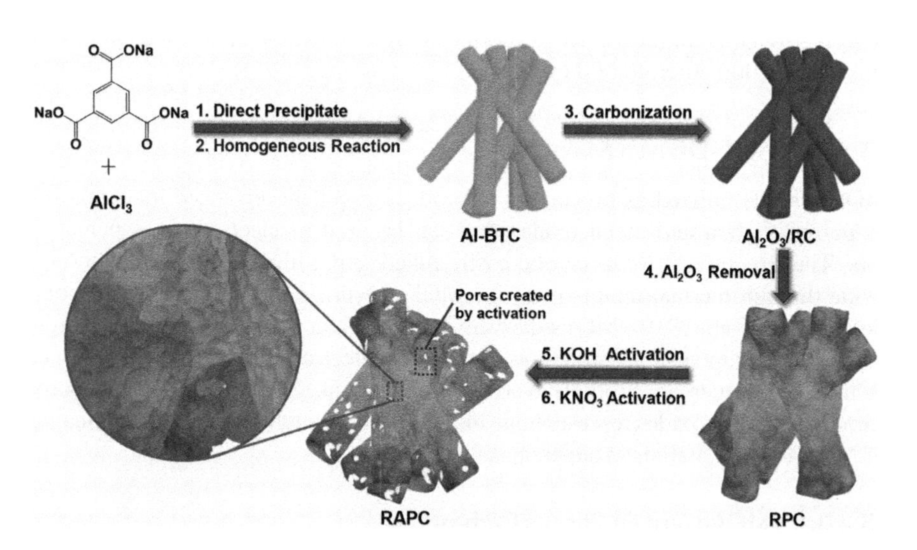

FIGURE 15.11 Schematic illustration of the preparation of rod-like porous activated carbon (RPAC) derived from aluminum-benzene- tricarboxylate (Al-TBC) MOF for application in supercapacitors. (Reprinted from *J. Power Sources*, 409, Jiang, W., et al., 13–23, Copyright 2019, with permission from Elsevier.)

electrochemical test, the porous carbon exhibited specific capacitance and specific energy density of 322 Fg^{-1} and 70 Wh kg^{-1} at a current density of 1 Ag^{-1}, with a cycling stability of 95.8% after 100,000 cycles. These results are comparable to those obtained by Jiang et al. (2019) as discussed above. However, an earlier study by Yu et al. (2017) reported a Zn-MOF-derived porous carbon with a higher specific surface area (2618.7 m^2 g^{-1}) but lower specific capacitance (150.8 Fg^{-1}), energy density (17.37 Wh kg^{-1}) and cyclic stability (94.8% after 10,000 cycles).

Li et al. (2019) prepared a porous carbon using a 2D Cu-MOF as a precursor. The derived porous carbon exhibited high specific surface area of 2491 m^2 g^{-1}, a pore volume of 1.50 cm^3 g^{-1} and a specific capacitance of 260.5 Fg^{-1}. The supercapacitor delivered a specific energy of 18.38 Wh Kg^{-1} and specific power of 350 W kg^{-1}, with a cyclic stability of 91.1% after 5000 cycles.

Although porous carbons are mostly generated starting with pristine or modified MOFs, composites of MOFs and other excellent materials such as GO have also been utilized. For example, Tan et al. (2019) synthesized porous carbon using a composite of polyoxometallate-based MOF/rGO composites as precursors. The synthesis procedure, which is based on hydrothermal treatment and calcination, is depicted in Figure 15.12. The derived carbon exhibited a specific capacitance of 178 Fg^{-1},

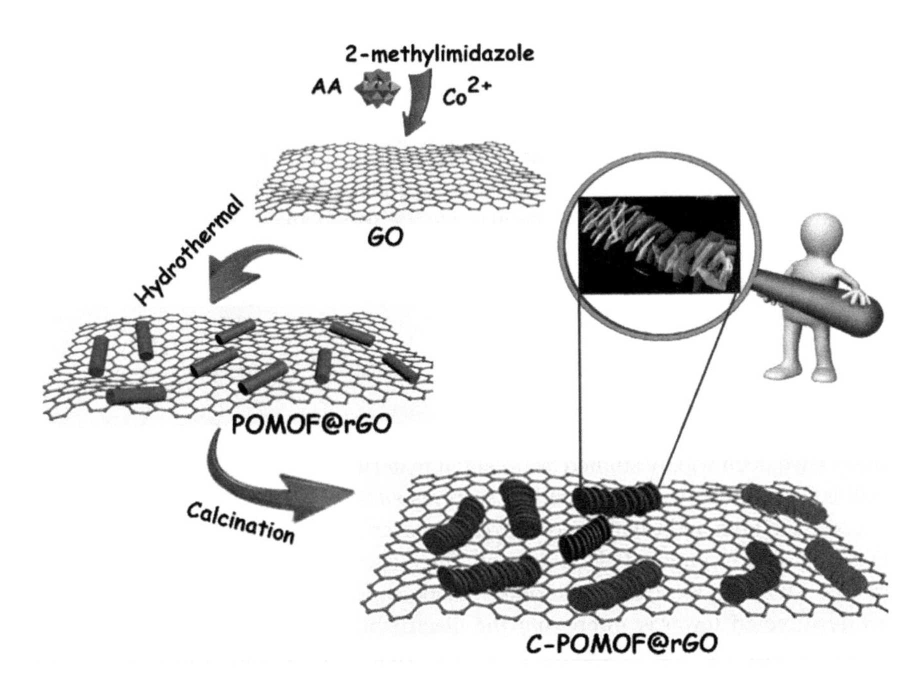

FIGURE 15.12 Preparation of porous carbon from polyoxometallate/MOF/rGO (POMOF@rGO). The procedure involves a hydrothermal preparation of POMOF@rGO, followed by its calcination to obtain the C-POMOF@rGO. (Reprinted from Tan, L., et al. *J. Electroanal. Chem.*, 836, 112–117.)

delivering a specific energy and power of 20.1 Wh kg^{-1} and 9071 W kg^{-1} respectively when applied in an asymmetric supercapacitor. In addition, a cycling stability of 94% was achieved over 5000 cycles.

15.4.3.2 MOF-Derived Metal Oxides

MOFs can be utilized as precursors to produce metal oxides with excellent properties for application in supercapacitors. The metal oxides are obtained by a careful thermal treatment of MOFs, generating a highly nanostructured materials. These nanoparticles are then grown on a flexible carbon fiber, generating electrode with excellent electrical conductivity, high capacitance and structural stability (Yang et al., 2018). The derived metal oxide depends on the metal present in the precursor MOF. Lee and Jang (2019) utilized ZIF-67 as a precursor to developed a hybrid layered double hydroxide consisting of Co_3O_4 distributed on a hexagonal $Co(OH)_2$ plate. The hybrid delivered a high specific capacitance of 184.9 mAh g^{-1}. When applied in an all-solid state supercapacitor, specific energy and specific power of 37.6 and 47 kW kg^{-1} were respectively obtained.

Besides single metal oxides, binary and ternary metal oxides can be derived from MOFs. The synergy of different metals results in higher electrochemical performance compared to a single metal. For example, Lim et al. (2018) derived cobalt-zinc bimetallic oxide using a zinc-cobalt MOF as a precursor. The derived bimetallic oxides delivered a specific capacity that is 300% more than that of a single-metal MOF-derived oxide. Additionally, the cyclic stability of the bimetallic oxide was much higher than single metal oxide. Indeed, the electrochemical performance can further be enhanced using a higher number of metals. Zheng et al. (2018b) synthesized a Ni-Co-Mn layer double hydroxide nanoflake using bimetallic imidazolate MOF as a precursor. The ternary metal oxides delivered an excellent specific capacitance of 2012.5 Fg^{-1} at 1 Ag^{-1} compared to that delivered by Ni-Co oxide (1266.2 Fg^{-1}). The cyclic stability of the ternary metal oxide (75.0%) was also higher than that of the binary oxides (41.8%).

15.5 CONCLUSION

MOFs have been widely studied as potential materials for supercapacitor electrodes. This is largely due to their excellent features such as porous structure, high specific surface area and tunable pore size. However, poor conductivity and chemical instability in conventional electrolyte are the major challenges in the application of pristine MOFs as electrode materials in supercapacitors. Thus, recent studies are mostly directed towards improving the electrochemical properties of the MOFs. In this regard, the use of modified MOFs, MOFs-based composites and MOFs-derived materials in supercapacitors have been reported. The composite/hybrid of MOFs-GO, MOFs-CNTs, MOFs-conducting polymer, MOFs-derived porous AC and metal oxides are particularly pursued. These materials have resulted in supercapacitors with excellent specific capacitance, energy and power densities and cyclic stability. Despite the noted progress in the use of MOFs-based materials in supercapacitors, it is necessary to address some technical challenges for their successful utilization. The challenges of instability and poor conductivity of MOF have been somewhat addressed by the incorporation of other materials such as carbon-based materials and conducting polymers. However, there is no clear information on the mechanisms of the composite formation and the interaction between MOF and the functionalizing materials. This information is necessary for effective design and optimization of electrodes and the supercapacitors. Numerous preparation methods for both MOF precursors and MOF-based composites have been reported in the literature, each resulting in materials with different properties. The absence of an established common preparation method makes it difficult to make meaningful comparisons in the reported performance of the electrodes. Thus, there is a need for refine and common synthetic methods. Since the application of MOF-based materials in supercapacitors is still at laboratory stage, the existing technical challenges are expected to be addressed through a continuous research in the field.

REFERENCES

Ando, T., 1984. Sonochemical switching of reaction pathways in solid-liquid two-phase reactions. *J. Chem. Soc. Chem. Commun.* 7, 439–440.

Bahaa, A., Balamurugan, J., Kim, N.H., Lee, J.H., 2019. Metal organic framework derived hierarchical copper cobalt sulfide nanosheet arrays for high-performance solid-state asymmetric supercapacitors. *J. Mater. Chem. A* 7, 8620–8632. https://doi.org/10.1039/c9ta00265k.

Bao, W., Mondal, A.K., Xu, J., Wang, C., Su, D., Wang, G., 2016. 3D hybrid-porous carbon derived from carbonization of metal organic frameworks for high performance supercapacitors. *J. Power Sources* 325, 286–291. https://doi.org/10.1016/j.jpowsour.2016.06.037.

Barbieri, O., Hahn, M., Herzog, A., Kötz, R., 2005. Capacitance limits of high surface area activated carbons for double layer capacitors. *Carbon N.Y.* 43, 1303–1310. https://doi.org/10.1016/j.carbon.2005.01.001.

Béguin, F., Presser, V., Balducci, A., Frackowiak, E., 2014. Carbons and electrolytes for advanced supercapacitors. *Adv. Mater.* 26, 2219–2251. https://doi.org/10.1002/adma.201304137.

Blinova, N. V., Stejskal, J., Trchová, M., Prokeš, J., Omastová, M., 2007. Polyaniline and polypyrrole: A comparative study of the preparation. *Eur. Polym. J.* 43, 2331–2341. https://doi.org/10.1016/j.eurpolymj.2007.03.045.

Burke, L.D., Collins, J.A., Horgan, M.A., Hurley, L.M., O'Mullane, A.P., 2000. Importance of the active states of surface atoms with regard to the electrocatalytic behaviour of metal electrodes in aqueous media. *Electrochim. Acta* 45, 4127–4134. https://doi.org/10.1016/S0013-4686(00)00532-6.

Burtch, N.C., Jasuja, H., Walton, K.S., 2014. Water stability and adsorption in metal—Organic frameworks. *Chem. Rev.* 114, 10575–10612. https://doi.org/10.1021/cr5002589.

Choi, D., Blomgren, G.E., Kumta, P.N., 2006. Fast and reversible surface redox reaction in nanocrystalline vanadium nitride supercapacitors. *Adv. Mater.* 18, 1178–1182. https://doi.org/10.1002/adma.200502471.

Choi, J., Son, W., Kim, J., Ahn, W., 2008. Microporous and mesoporous materials metal—Organic framework MOF-5 prepared by microwave heating: Factors to be considered. *Microporous Mesoporous Mater.* 116, 727–731. https://doi.org/10.1016/j.micromeso.2008.04.033.

Conway, B.E., 1999. *Electrochemical Supercapacitors Scientific Fundamental and Technological Applications*. Kluwer Academic/Plenum Publishers, New York.

Dai, S., Yuan, Y., Yu, J., Tang, J., Zhou, J., Tang, W., 2018. Metal–organic framework-templated synthesis of sulfur-doped core–sheath nanoarrays and nanoporous carbon for flexible all-solid-state asymmetric supercapacitors. *Nanoscale* 10, 15454–15461. https://doi.org/10.1039/c8nr03743d.

Díaz, R., Orcajo, M.G., Botas, J.A., Calleja, G., Palma, J., 2012. Co8-MOF-5 as electrode for supercapacitors. *Mater. Lett.* 68, 126–128. https://doi.org/10.1016/j.matlet.2011.10.046.

Du, W., Bai, Y.L., Xu, J., Zhao, H., Zhang, L., Li, X., Zhang, J., 2018. Advanced metal-organic frameworks (MOFs) and their derived electrode materials for supercapacitors. *J. Power Sources* 402, 281–295. https://doi.org/10.1016/j.jpowsour.2018.09.023.

Eddaoudi, M., Moler, D.B., Li, H., Chen, B., Reineke, T.M., Keeffe, M.O., Yaghi, O.M., 2001. Modular chemistry: Secondary building units as a basis for the design of highly porous and robust metal—Organic carboxylate frameworks. *Acc. Chem. Res.* 34, 319–330.

Fan, L.Z., Hu, Y.S., Maier, J., Adelhelm, P., Smarsly, B., Antonietti, M., 2007. High electroactivity of polyaniline in supercapacitors by using a hierarchically porous carbon monolith as a support. *Adv. Funct. Mater.* 17, 3083–3087. https://doi.org/10.1002/adfm.200700518.

Fang, Q., Zhu, G., Xue, M., Sun, J., Wei, Y., Qiu, S., Xu, R., 2005. A metal–organic framework with the zeolite MTN topology containing large cages of volume 2.5 nm^3. *Angew. Chem. Int.* ed. 6, 3845–3848. https://doi.org/10.1002/anie.200462260.

Frackowiak, E., Béguin, F., 2001. Carbon materials for the electrochemical storage of energy in capacitors. *Carbon N.Y.* 39, 937–950. https://doi.org/10.1016/S0008-6223(00)00183-4.

Fu, D., Zhou, H., Zhang, X.M., Han, G., Chang, Y., Li, H., 2016. Flexible solid–state supercapacitor of metal–organic framework coated on carbon nanotube film interconnected by electrochemically-codeposited PEDOT-GO. *ChemistrySelect* 1, 285–289. https://doi.org/10.1002/slct.201600084.

Gilvaldo, G., Silva, C.S., Ribeiro, R.T., Ronconi, C.M., Barros, B.S., Neves, J.L., Alves, S., 2016. Sonoelectrochemical synthesis of metal-organic frameworks. *Synth. Met.* 220, 369–373. https://doi.org/10.1016/j.synthmet.2016.07.003.

Gómez-Romero, P., Ayyad, O., Suárez-Guevara, J., Muñoz-Rojas, D., 2010. Hybrid organic-inorganic materials: From child's play to energy applications. *J. Solid State Electrochem.* 14, 1939–1945. https://doi.org/10.1007/s10008-010-1076-y.

Guo, D., Song, X., Tan, L., Ma, H., Sun, W., Pang, H., Zhang, L., Wang, X., 2019. A facile dissolved and reassembled strategy towards sandwich-like rGO@NiCoAl-LDHs with excellent supercapacitor performance. *Chem. Eng. J.* 356, 955–963. https://doi.org/10.1016/j.cej.2018.09.101.

Han, Y., Fan, H., Qi, X., Li, L., Liu, Y., Li, X., Meng, L., 2018a. Metal-organic framework-derived carbon coated copper sulfide nanocomposites as a battery-type electrode for electrochemical capacitors. *Mater. Lett.* 236, 131–134. https://doi.org/10.1016/j.matlet.2018.10.093.

Han, Z.J., Huang, C., Meysami, S.S., Piche, D., Seo, D.H., Pineda, S., Murdock, A.T., Bruce, P.S., Grant, P.S., Grobert, N., 2018b. High-frequency supercapacitors based on doped carbon nanostructures. *Carbon N.Y.* 126, 305–312. https://doi.org/10.1016/j.carbon.2017.10.014.

Hong, M., Zhou, C., Xu, S., Ye, X., Yang, Z., Zhang, L., Zhou, Z., Hu, N., Zhang, Y., 2019. Bi-metal organic framework nanosheets assembled on nickel wire films for volumetric-energy-dense supercapacitors. *J. Power Sources* 423, 80–89. https://doi.org/10.1016/j.jpowsour.2019.03.059.

Hu, J., Wang, H., Gao, Q., Guo, H., 2010. Porous carbons prepared by using metal—organic framework as the precursor for supercapacitors. *Carbon N.Y.* 48, 3599–3606. https://doi.org/10.1016/j.carbon.2010.06.008.

Hua, X., Mao, C.J., Chen, J.S., Chen, P.P., Zhang, C.F., 2019a. Facile synthesis of new-type MnOOH/NiAl-layered double hydroxide nanocomposite for high-performance supercapacitor. *J. Alloys Compd.* 777, 749–758. https://doi.org/10.1016/j.jallcom.2018.11.005.

Hua, Y., Li, X., Chen, C., Pang, H., 2019b. Cobalt based metal-organic frameworks and their derivatives for electrochemical energy conversion and storage. *Chem. Eng. J.* 370, 37–59. https://doi.org/10.1016/j.cej.2019.03.163.

Hulicova, D., Yamashita, J., Soneda, Y., Hatori, H., Kodama, M., 2005. Supercapacitors prepared from melamine-based carbon. *Chem. Mater.* 17, 1241–1247. https://doi.org/10.1021/cm049337g.

Jabeen, N., Xia, Q., Savilov, S. V., Aldoshin, S.M., Yu, Y., Xia, H., 2016. Enhanced pseudo-capacitive performance of α-MnO_2 by cation preinsertion. *ACS Appl. Mater. Inter.* 8, 33732–33740. https://doi.org/10.1021/acsami.6b12518.

Jeong, H.M., Lee, J.W., Shin, W.H., Choi, Y.J., Shin, H.J., Kang, J.K., Choi, J.W., 2011. Nitrogen-doped graphene for high-performance ultracapacitors and the importance of nitrogen-doped sites at basal planes. *Nano Lett.* 11, 2472–2477. https://doi.org/10.1021/nl2009058.

Jiang, W., Pan, J., Liu, X., 2019. A novel rod-like porous carbon with ordered hierarchical pore structure prepared from Al-based metal-organic framework without template as greatly enhanced performance for supercapacitor. *J. Power Sources* 409, 13–23. https://doi.org/10.1016/j.jpowsour.2018.10.086.

Jiao, Y., Hong, W., Li, P., Wang, L., Chen, G., 2019. Metal-organic framework derived Ni/NiO micro-particles with subtle lattice distortions for high-performance electrocatalyst and supercapacitor. *Appl. Catal. B Environ.* 244, 732–739. https://doi.org/10.1016/j.apcatb.2018.11.035.

Jiao, Y., Pei, J., Yan, C., Chen, D., Hu, Y., Chen, G., 2016. Layered nickel metal-organic framework for high performance alkaline battery-supercapacitor hybrid devices. *J. Mater. Chem. A* 4, 13344–13351. https://doi.org/10.1039/c6ta05384j.

Jost, K., Dion, G., Gogotsi, Y., 2014. Textile energy storage in perspective. *J. Mater. Chem. A* 2, 10776–10787. https://doi.org/10.1039/c4ta00203b.

Kandalkar, S.G., Dhawale, D.S., Kim, C.K., Lokhande, C.D., 2010. Chemical synthesis of cobalt oxide thin film electrode for supercapacitor application. *Synth. Met.* 160, 1299–1302. https://doi.org/10.1016/j.synthmet.2010.04.003.

Kannangara, Y.Y., Rathnayake, U.A., Song, J.K., 2019a. Redox active multi-layered Zn-pPDA MOFs as high-performance supercapacitor electrode material. *Electrochim. Acta* 297, 145–154. https://doi.org/10.1016/j.electacta.2018.11.186.

Kannangara, Y.Y., Rathnayake, U.A., Song, J.K., 2019b. Hybrid supercapacitors based on metal organic frameworks using p-phenylenediamine building block. *Chem. Eng. J.* 361, 1235–1244. https://doi.org/10.1016/j.cej.2018.12.173.

Ke, Q., Wang, J., 2016. Graphene-based materials for supercapacitor electrodes—A review. *J. Mater.* 2, 37–54. https://doi.org/10.1016/j.jmat.2016.01.001.

Khomenko, V., Raymundo-Piñero, E., Béguin, F., 2008. High-energy density graphite/AC capacitor in organic electrolyte. *J. Power Sources* 177, 643–651. https://doi.org/10.1016/j.jpowsour.2007.11.101.

Kim, B.K., Sy, S., Yu, A., Zhang, J., 2014. Electrochemical supercapacitors for energy storage and conversion. *Handbook of Clean Energy Systems*. https://doi.org/10.1002/9781118991978.hces112.

Kim, J., Deshpande, S.D., Yun, S., Li, Q., 2006. A comparative study of conductive polypyrrole and polyaniline coatings on electro-active papers. *Polym. J.* 38, 659–668. https://doi.org/10.1295/polymj.pj2005185.

Lee, G., Jang, J., 2019. High-performance hybrid supercapacitors based on novel Co_3O_4/$Co(OH)_2$ hybrids synthesized with various-sized metal-organic framework templates. *J. Power Sources* 423, 115–124. https://doi.org/10.1016/j.jpowsour.2019.03.065.

Lee, Y.H., Lee, Y.F., Chang, K.H., Hu, C.C., 2011. Synthesis of N-doped carbon nanosheets from collagen for electrochemical energy storage/conversion systems. *Electrochem. Commun.* 13, 50–53. https://doi.org/10.1016/j.elecom.2010.11.010.

Li, G., Gao, X., Wang, K., Cheng, Z., 2018a. Porous carbon nanospheres with high EDLC capacitance. *Diam. Relat. Mater.* 88, 12–17. https://doi.org/10.1016/j.diamond.2018.06.010.

Li, X., Zhang, J., Bai, Y.-L., Zhao, H., Xu, J., Du, W., Zhang, L., 2018b. Advanced metal-organic frameworks (MOFs) and their derived electrode materials for supercapacitors. *J. Power Sources* 402, 281–295. https://doi.org/10.1016/j.jpowsour.2018.09.023.

Li, L., Liu, Y., Han, Y., Qi, X., Li, X., Fan, H., Meng, L., 2019a. Metal-organic framework-derived carbon coated copper sulfide nanocomposites as a battery-type electrode for electrochemical capacitors. *Mater. Lett.* 236, 131–134. https://doi.org/10.1016/j.matlet.2018.10.093.

Li, Z.-X., Yang, B.-L., Zou, K.-Y., Kong, L., Yue, M.-L., Duan, H.-H., 2019b. Novel porous carbon nanosheet derived from a 2D Cu-MOF: Ultrahigh porosity and excellent performances in the supercapacitor cell. *Carbon N.Y.* 144, 540–548. https://doi.org/10.1016/j.carbon.2018.12.061.

Libich, J., Máca, J., Vondrák, J., Čech, O., Sedlaříková, M., 2018. Supercapacitors: Properties and applications. *J. Energy Storage* 17, 224–227. https://doi.org/10.1016/j.est.2018.03.012.

Lim, G.J.H., Liu, X., Guan, C., Wang, J., 2018. Co/Zn bimetallic oxides derived from metal organic frameworks for high performance electrochemical energy storage. *Electrochim. Acta* 291, 177–187. https://doi.org/10.1016/j.electacta.2018.08.105.

Liu, B., Shioyama, H., Jiang, H., Zhang, X., Xu, Q., 2010. Metal-organic framework (MOF) as a template for syntheses of nanoporous carbons as electrode materials for supercapacitor. *Carbon N.Y.* 48, 456–463. https://doi.org/10.1016/j.carbon.2009.09.061.

Liu, C.F., Liu, Y.C., Yi, T.Y., Hu, C.C., 2019. Carbon materials for high-voltage supercapacitors. *Carbon N.Y.* 145, 529–548. https://doi.org/10.1016/j.carbon.2018.12.009.

Liu, L.Y., Zhang, X., Li, H.X., Liu, B., Lang, J.W., Kong, L. Bin, Yan, X. Bin, 2017. Synthesis of Co–Ni oxide microflowers as a superior anode for hybrid supercapacitors with ultralong cycle life. *Chinese Chem. Lett.* 28, 206–212. https://doi.org/10.1016/j.cclet.2016.07.027.

Liu, Y., Li, K., Wang, J., Sun, G., Sun, C., 2009. Preparation of spherical activated carbon with hierarchical porous texture. *J. Mater. Sci.* 44, 4750–4753. https://doi.org/10.1007/s10853-009-3710-6.

Liu, Y., Peng, X., 2017. Recent advances of supercapacitors based on two-dimensional materials. *Appl. Mater. Today* 7, 1–12. https://doi.org/10.1016/j.apmt.2017.01.004.

Liu, Y., Xu, N., Chen, W., Wang, X., Sun, C., Su, Z., 2018. Supercapacitor with high cycling stability through electrochemical deposition of metal-organic frameworks/polypyrrole positive electrode. *Dalt. Trans.* 47, 13472–13478. https://doi.org/10.1039/c8dt02740d.

Lota, G., Grzyb, B., Machnikowska, H., Machnikowski, J., Frackowiak, E., 2005. Effect of nitrogen in carbon electrode on the supercapacitor performance. *Chem. Phys. Lett.* 404, 53–58. https://doi.org/10.1016/j.cplett.2005.01.074.

Lu, X.F., Li, G.R., Tong, Y.X., 2012. A review of negative electrode materials for electrochemical supercapacitors. *Chem. Soc. Rev.* 41, 797–828. https://doi.org/10.1007/s11431-015-5931-z.

Luo, J., Jang, H.D., Huang, J., 2013. Effect of sheet morphology on the scalability of graphene-based ultracapacitors. *ACS Nano* 7, 1464–1471. https://doi.org/10.1021/nn3052378.

Mahmoud, M.E., Tan, J.-C., 2018. Metal-organic framework based composite, In. *Comprehensive Composite Materials II.* Amsterdam, the Netherlands: Elsevier, pp. 239–261. https://doi.org/10.1016/B978-0-12-803581-8.09974-4.

Mastragostino, M., Arbizzani, C., Soavi, F., 2002. Conducting polymers as electrode materials for supercapacitors. *Solid State Ion.* 148, 493–498.

Muller, U., Putter, H., Hesse, M., Schubert, M., Wessel, H., Huff, J., Guzmann, M., 2007. Method for electrochemical production of a crystalline porous metal organic skeleton material. US 2007/0227898 A1.

Muzaffar, A., Ahamed, M.B., Deshmukh, K., Thirumalai, J., 2019. A review on recent advances in hybrid supercapacitors: Design, fabrication and applications. *Renew. Sustain. Energy Rev.* 101, 123–145. https://doi.org/10.1016/j.rser.2018.10.026.

Naoi, K., Naoi, W., Aoyagi, S., Miyamato, J.-I., Kamino, T., 2013. New generation "Nanohybrid Supercapacitor." *Acc. Chem. Res.* 46, 1075–1083.

Ni, Z., Masel, R.I., 2006. Rapid production of metal—Organic frameworks via microwave-assisted solvothermal synthesis. *J. AM. Chem. Soc.* 128, 12394–12395. https://doi.org/10.1021/ja0635231.

Osman, S., Senthil, R.A., Pan, J., Li, W., 2018. Highly activated porous carbon with 3D microspherical structure and hierarchical pores as greatly enhanced cathode material for high-performance supercapacitors. *J. Power Sources* 391, 162–169. https://doi.org/10.1016/j.jpowsour.2018.04.081.

Osman, S., Senthil, R.A., Pan, J., Sun, Y., 2019. A novel coral structured porous-like amorphous carbon derived from zinc-based fluorinated metal-organic framework as superior cathode material for high performance supercapacitors. *J. Power Sources* 414, 401–411. https://doi.org/10.1016/j.jpowsour.2019.01.026.

Pan, H., Li, J., Feng, Y.P., 2010. Carbon nanotubes for supercapacitor. *Nanoscale Res. Lett.* 5, 654–668. https://doi.org/10.1007/s11671-009-9508-2.

Park, J.H., Park, O.O., Shin, K.H., Jin, C.S., Kim, J.H., 2002. An electrochemical capacitor based on a $Ni(OH)_2$ activated carbon composite electrode. *Electrochem. Solid-State Lett.* 5, H7. https://doi.org/10.1149/1.1432245.

Pham, T.N., Park, D., Lee, Y., Kim, I.T., Hur, J., Oh, Y.K., Lee, Y.C., 2019. Combination-based nanomaterial designs in single and double dimensions for improved electrodes in lithium ion-batteries and faradaic supercapacitors. *J. Energy Chem.* 119–146. https://doi.org/10.1016/j.jechem.2018.12.014.

Poonam, Sharma, K., Arora, A., Tripathi, S.K., 2019. Review of supercapacitors: Materials and devices. *J. Energy Storage* 21, 801–825. https://doi.org/10.1016/j.est.2019.01.010.

Qi, K., Hou, R., Zaman, S., Qiu, Y., Xia, B.Y., Duan, H., 2018. Construction of metal-organic framework/conductive polymer hybrid for all-solid-state fabric supercapacitor. *ACS Appl. Mater. Interfaces* 10, 18021–18028. https://doi.org/10.1021/acsami.8b05802.

Qiu, H.J., Liu, L., Mu, Y.P., Zhang, H.J., Wang, Y., 2015. Designed synthesis of cobalt-oxide-based nanomaterials for superior electrochemical energy storage devices. *Nano Res.* 8, 321–339. https://doi.org/10.1007/s12274-014-0589-6.

Qu, C., Jiao, Y., Zhao, B., Chen, D., Zou, R., Walton, K.S., Liu, M., 2016. Nickel-based pillared MOFs for high-performance supercapacitors: Design, synthesis and stability study. *Nano Energy* 26, 66–73. https://doi.org/10.1016/j.nanoen.2016.04.003.

Rademann, K., Klimakow, M., Klobes, P., Th, A.F., 2010. Mechanochemical synthesis of metal—Organic frameworks: A fast and facile approach toward quantitative yields and high specific surface areas. *Chem. Mater.* 22, 5216–5221. https://doi.org/10.1021/cm1012119.

Rahmanifar, M.S., Hesari, H., Noori, A., Yaser, M., Morsali, A., Mousavi, M.F., 2018. A dual Ni/Co-MOF-reduced graphene oxide nanocomposite as a high performance supercapacitor electrode material. *Electrochim. Acta* 275, 76–86. https://doi.org/10.1016/j.electacta.2018.04.130.

Rawool, C.R., Karna, S.P., Srivastava, A.K., 2019. Enhancing the supercapacitive performance of Nickel based metal organic framework-carbon nanofibers composite by changing the ligands. *Electrochim. Acta* 294, 345–356. https://doi.org/10.1016/j.electacta.2018.10.093.

Repp, S., Harputlu, E., Gurgen, S., Castellano, M., Kremer, N., Pompe, N., Wörner et al., 2018. Synergetic effects of Fe^{3+} doped spinel $Li_4Ti_5O_{12}$ nanoparticles on reduced graphene oxide for high surface electrode hybrid supercapacitors. *Nanoscale* 10, 1877–1884. https://doi.org/10.1039/c7nr08190a.

Simon, P., Brousse, T., Favier, F., 2017. Supercapacitors based on carbon nanomaterials. In. *Supercapacitors Based on Carbon or Pseudocapacitive Materials*. London, UK: ISTE Ltd and John Wiley & Sons, pp. 1–25.

Sonai Muthu, N., Gopalan, M., 2019. Mesoporous nickel sulphide nanostructures for enhanced supercapacitor performance. *Appl. Surf. Sci.* 480, 186–198. https://doi.org/10.1016/j.apsusc.2019.02.250.

Soneda, Y., 2013. Carbons for supercapacitors, *Handbook of Advanced Ceramics: Materials, Applications, Processing, and Properties*. 2nd Ed. San Diego, CA: Elsevier. https://doi.org/10.1016/B978-0-12-385469-8.00013-7.

Srimuk, P., Luanwuthi, S., Krittayavathananon, A., Sawangphruk, M., 2015. Solid-type supercapacitor of reduced graphene oxide-metal organic framework composite coated on carbon fiber paper. *Electrochim. Acta* 157, 69–77. https://doi.org/10.1016/j.electacta.2015.01.082.

Sun, S., Luo, J., Qian, Y., Jin, Y., Liu, Y., Qiu, Y., Li, X., Fang, C., Han, J., Huang, Y., 2018. Metal–organic framework derived honeycomb Co9S8@C composites for high-performance supercapacitors. *Adv. Energy Mater.* 8, 1–9. https://doi.org/10.1002/aenm.201801080.

Sun, W., Du, Y., Wu, G., Gao, G., Zhu, H., Shen, J., Zhang, K., Cao, G., 2019a. Constructing metallic zinc–cobalt sulfide hierarchical core–shell nanosheet arrays derived from 2D metal–organic-frameworks for flexible asymmetric supercapacitors with ultrahigh specific capacitance and performance. *J. Mater. Chem.* A 7, 7138–7150. https://doi.org/10.1039/c8ta12153b.

Sun, Z., Qi, J., Meng, Q., Sui, Y., Miao, Y., Zhang, D., Wei, F., Ren, Y., He, Y., 2019b. Polyhedral NiCoSe2 synthesized via selenization of metal-organic framework for supercapacitors. *Mater. Lett.* 242, 42–46. https://doi.org/10.1016/j.matlet.2019.01.096.

Sundriyal, S., Kaur, H., Bhardwaj, S.K., Mishra, S., Kim, K.H., Deep, A., 2018. Metal-organic frameworks and their composites as efficient electrodes for supercapacitor applications. *Coord. Chem. Rev.* 369, 15–38. https://doi.org/10.1016/j.ccr.2018.04.018.

Suslick, K.S., 1990. Sonochemistry. *Science* 247(4949), 1439–1445.

Tan, L., Guo, D., Liu, J., Song, X., Liu, Q., Chen, R., Wang, J., 2019. In-situ calcination of polyoxometallate-based metal organic framework/reduced graphene oxide composites towards supercapacitor electrode with enhanced performance. *J. Electroanal. Chem.* 836, 112–117. https://doi.org/10.1016/j.jelechem.2019.01.051.

Tranchemontagne, D.J., Hunt, J.R., Yaghi, O.M., 2008. Room temperature synthesis of metal-organic frameworks: MOF-5, MOF-74, *Tetrahedron* 64, 8553–8557. https://doi.org/10.1016/j.tet.2008.06.036.

Wang, C., Sun, P., Qu, G., Yin, J., Xu, X., 2018. Nickel/cobalt based materials for supercapacitors. *Chinese Chem. Lett.* 29, 1731–1740. https://doi.org/10.1016/j.cclet.2018.12.005.

Wang, F., Xiao, S., Hou, Y., Hu, C., Liu, L., Wu, Y., 2013. Electrode materials for aqueous asymmetric supercapacitors. *RSC Adv.* 3, 13059–13084. https://doi.org/10.1039/c3ra23466e.

Wang, L., Feng, X., Ren, L., Piao, Q., Zhong, J., Wang, Y., Li, H., Chen, Y., Wang, B., 2015. Flexible solid-state supercapacitor based on a metal-organic framework interwoven by electrochemically-deposited PANI. *J. Am. Chem. Soc.* 137, 4920–4923. https://doi.org/10.1021/jacs.5b01613.

Wang, L., Han, Y., Feng, X., Zhou, J., Qi, P., Wang, B., 2016. Metal-organic frameworks for energy storage: Batteries and supercapacitors. *Coord. Chem. Rev.* 307, 361–381. https://doi.org/10.1016/j.ccr.2015.09.002.

Wang, Y., Yu, L., Xia, Y., 2006. Electrochemical capacitance performance of hybrid supercapacitors based on $Ni(OH)_2$ carbon nanotube composites and activated carbon. *J. Electrochem. Soc.* 153, A743. https://doi.org/10.1149/1.2171833.

Wei, T., Zhang, M., Wu, P., Tang, Y.J., Li, S.L., Shen, F.C., Wang, X.L., Zhou, X.P., Lan, Y.Q., 2017. POM-based metal-organic framework/reduced graphene oxide nanocomposites with hybrid behavior of battery-supercapacitor for superior lithium storage. *Nano Energy* 34, 205–214. https://doi.org/10.1016/j.nanoen.2017.02.028.

Wen, P., Gong, P., Sun, J., Wang, J., Yang, S., 2015. Design and synthesis of Ni-MOF/CNT composites and rGO/carbon nitride composites for an asymmetric supercapacitor with high energy and power density. *J. Mater. Chem. A* 3, 13874–13883. https://doi.org/10.1039/c5ta02461g.

Xia, X., Chao, D., Fan, Z., Guan, C., Cao, X., Zhang, H., Fan, H.J., 2014. A new type of porous graphite foams and their integrated composites with oxide/polymer core/shell nanowires for supercapacitors: Structural design, fabrication, and full supercapacitor demonstrations. *Nano Lett.* 14, 1651–1658. https://doi.org/10.1021/nl5001778.

Xiong, C., Li, B., Lin, X., Liu, H., Xu, Y., Mao, J., Duan, C., Li, T., Ni, Y., 2019. The recent progress on three-dimensional porous graphene-based hybrid structure for supercapacitor. *Compos. Part B Eng.* 165, 10–46. https://doi.org/10.1016/j.compositesb.2018.11.085.

Xu, M.W., Bao, S.J., Li, H.L., 2007. Synthesis and characterization of mesoporous nickel oxide for electrochemical capacitor. *J. Solid State Electrochem.* 11, 372–377. https://doi.org/10.1007/s10008-006-0155-6.

Xu, Q., Zou, R., Liang, Z., Guo, W., Qu, C., 2017a. Pristine metal-organic frameworks and their composites for energy storage and conversion. *Adv. Mater.* 30, 1702891. https://doi.org/10.1002/adma.201702891.

Xu, X., Shi, W., Li, P., Ye, S., Ye, C., Ye, H., Lu, T., Zheng, A., Zhu, J., Xu, L., Zhong, M., Cao, X., 2017b. Facile fabrication of three-dimensional graphene and metal-organic framework composites and their derivatives for flexible all-solid-state supercapacitors. *Chem. Mater.* 29, 6058–6065. https://doi.org/10.1021/acs.chemmater.7b01947.

Xue, B., Li, K., Gu, S., Lu, J., 2018. Zeolitic imidazolate frameworks (ZIFs)-derived $NixCo_3–xO_4$/CNTs nanocomposites with enhanced electrochemical performance for supercapacitor. *J. Colloid Interface Sci.* 530, 233–242. https://doi.org/10.1016/j.jcis.2018.06.077.

Xue, Y., Wang, C., Yang, C., Chen, Z., Xu, J., Cao, J., 2016. Facile synthesis of novel metal-organic nickel hydroxide nanorods for high performance supercapacitor. *Electrochim. Acta* 211, 595–602. https://doi.org/10.1016/j.electacta.2016.06.090.

Yaghi, O.M., Li, H., 1995. Hydrothermal synthesis of a metal-organic framework containing large rectangular channels. *J. Am. Chem. Soc.* 117, 10401–10402. https://doi.org/10.1021/ja00146a033.

Yan, X., Li, X., Yan, Z., Komarneni, S., 2014. Porous carbons prepared by direct carbonization of MOFs for supercapacitors. *Appl. Surf. Sci.* 308, 306–310. https://doi.org/10.1016/j.apsusc.2014.04.160.

Yan, Y., Luo, Y., Ma, J., Li, B., Xue, H., Pang, H., 2018. Facile synthesis of vanadium metal-organic frameworks for high-performance supercapacitors. *Small* 14, 1–8. https://doi.org/10.1002/smll.201801815.

Yang, J., Zheng, C., Xiong, P., Li, Y., Wei, M., 2014. supercapacitive performance. *J. Mater. Chem. A Mater. Energy Sustain.* 2, 19005–19010. https://doi.org/10.1039/C4TA04346D.

Yang, Q., Liu, Y., Xiao, L., Yan, M., Bai, H., Zhu, F., Lei, Y., Shi, W., 2018. Self-templated transformation of MOFs into layered double hydroxide nanoarrays with selectively formed Co9S8 for high-performance asymmetric supercapacitors. *Chem. Eng. J.* 354, 716–726. https://doi.org/10.1016/j.cej.2018.08.091.

Yang, Q., Xu, Q., 2017. Metal—Organic frameworks meet metal nanoparticles: Synergistic effect for. *Chem. Soc. Rev.* 46, 4774–4808. https://doi.org/10.1039/C6CS00724D.

Yu, D., Ge, L., Wei, X., Wu, B., Ran, J., Wang, H., Xu, T., 2017. A general route to the synthesis of layer-by-layer structured metal organic framework/graphene oxide hybrid films for high-performance supercapacitor electrodes. *J. Mater. Chem. A* 5, 16865–16872. https://doi.org/10.1039/c7ta04074a.

Yu, F., Wang, T., Wen, Z., Wang, H., 2017a. High performance all-solid-state symmetric supercapacitor based on porous carbon made from a metal-organic framework compound. *J. Power Sources* 364, 9–15. https://doi.org/10.1016/j.jpowsour.2017.08.013.

Yu, F., Xiong, X., Zhou, L.Y., Li, J.L., Liang, J.Y., Hu, S.Q., Lu, W.T., Li, B., Zhou, H.C., 2019. Hierarchical nickel/phosphorus/nitrogen/carbon composites templated by one metal-organic framework as highly efficient supercapacitor electrode materials. *J. Mater. Chem. A* 7, 2875–2883. https://doi.org/10.1039/c8ta11568k.

Yu, H., Xu, D., Xu, Q., 2015. Dual template effect of supercritical CO2 in ionic liquid to fabricate highly mesoporous cobalt metal-organic framework. *Chem. Commun.* 51, 13197–13200. https://doi.org/10.1039/C5CC04009D.

Yuan, W., Garay, L., Pichon, A., Clowes, R., Wood, C.D., Cooper, I., James, S.L., 2010. Study of the mechanochemical formation and resulting properties of an archetypal MOF: Cu^3 (BTC) 2 (BTC ¼ 1,3,5-benzenetricarboxylate). *CrystEngrComm* 12, 4063–4065. https://doi.org/10.1039/c0ce00486c.

Yue, L., Guo, H., Wang, X., Sun, T., Liu, H., Li, Q., Xu, M., Yang, Y., Yang, W., 2019. Non-metallic element modified metal-organic frameworks as high-performance electrodes for all-solid-state asymmetric supercapacitors. *J. Colloid Interface Sci.* 539, 370–378. https://doi.org/10.1016/j.jcis.2018.12.079.

Zhai, T., Wan, L., Sun, S., Chen, Q., Sun, J., Xia, Q., Xia, H., 2017. Phosphate Ion Functionalized Co_3O_4 ultrathin nanosheets with greatly improved surface reactivity for high performance pseudocapacitors. *Adv. Mater.* 29, 1–8. https://doi.org/10.1002/adma.201604167.

Zhang, D., Shi, H., Zhang, R., Zhang, Z., Wang, N., Li, J., Yuan, B., Bai, H., Zhang, J., 2015. RSC advances quick synthesis of zeolitic imidazolate framework micro fl owers with enhanced supercapacitor and electrocatalytic performances. *RSC Adv.* 5, 58772–58776. https://doi.org/10.1039/C5RA08226A.

Zhang, F., Zhang, T., Yang, X., Zhang, L., Leng, K., Huang, Y., Chen, Y., 2013. A high-performance supercapacitor-battery hybrid energy storage device based on graphene-enhanced electrode materials with ultrahigh energy density. *Energy Environ. Sci.* 6, 1623–1632. https://doi.org/10.1039/c3ee40509e.

Zhang, H., Cao, G., Wang, Z., Yang, Y., Shi, Z., Gu, Z., 2008. Growth of manganese oxide nanoflowers on vertically-aligned carbon nanotube arrays for high-rate electrochemical capacitive energy storage. *Nano Lett.* 8, 2664–2668. https://doi.org/10.1021/nl800925j.

Zhang, H., Nai, J., Yu, L., Lou, X.W. (David), 2017. Metal-organic-framework-based materials as platforms for renewable energy and environmental applications. *Joule.* 1, 77–107. https://doi.org/10.1016/j.joule.2017.08.008.

Zhang, L.L., Zhao, X.S., 2009. Carbon-based materials as supercapacitor electrodes. *Chem. Soc. Rev.* 38, 2520–2531. https://doi.org/10.1039/b813846j.

Zhang, Y., Wang, T., Wang, Y., Wang, Y., Wu, L., Sun, Y., Zhou, X., Hou, W., Du, Y., Zhong, W., 2019. Metal organic frameworks derived hierarchical hollow Ni 0.85 SelP composites for high-performance hybrid supercapacitor and efficient hydrogen evolution. *Electrochim. Acta* 303, 94–104. https://doi.org/10.1016/j.electacta.2019.02.069.

Zheng, S., Xue, H., Pang, H., 2018a. Supercapacitors based on metal coordination materials. *Coord. Chem. Rev.* 373, 2–21. https://doi.org/10.1016/j.ccr.2017.07.002.

Zheng, X., Han, X., Zhao, X., Qi, J., Ma, Q., Tao, K., Han, L., 2018b. Construction of Ni-Co-Mn layered double hydroxide nanoflakes assembled hollow nanocages from bimetallic imidazolate frameworks for supercapacitors. *Mater. Res. Bull.* 106, 243–249. https://doi.org/10.1016/j.materresbull.2018.06.005.

Zhou, L., Li, C., Liu, X., Zhu, Y., Wu, Y., van Ree, T., 2018. Metal oxides in supercapacitors. In. *Metal Oxides in Energy Technologies*. Amsterdam, the Netherlands: Elsevier. https://doi.org/10.1016/b978-0-12-811167-3.00007-9.

Zhu, C., He, Y., Liu, Y., Kazantseva, N., Saha, P., Cheng, Q., 2019. ZnO@MOF@PANI core-shell nanoarrays on carbon cloth for high-performance supercapacitor electrodes. *J. Energy Chem*. 35, 124–131. https://doi.org/10.1016/j.jechem.2018.11.006.

16 Amino Acid-Assisted Inorganic Materials for Supercapacitors

M. Hamed Misbah, Amr A. Essawy, Rawya Ramadan, E.F. ElAgammy, and Maged El-Kemary

CONTENTS

16.1 Introduction 307
16.2 Lysine Amino Acid 308
16.3 Histidine Amino Acid 311
16.4 Arginine Amino Acid 312
16.5 Cysteine Amino Acid 314
16.6 Aspartic Amino Acid 315
16.7 Glycine Amino Acid 317
16.8 Guanidine Amino Acid 318
16.9 Glutamic Amino Acid 319
References 320

16.1 INTRODUCTION

Due to the significance of energy in the development of human society, energy storage has received the researcher attention. Among these storage systems are the supercapacitors that are considered as the supreme auspicious green energy storage systems [1–6]. In supercapacitors, the electrode pile electrons; meanwhile, the discharge course is assisted by the electrochemical reactions that are provoking an electron drift over an outward path. Thus, the charge and discharge efficiency directly depends on the proficiency of the supercapacitor electrodes. In this regard, different active materials have been applied as anode or cathode, which are categorized into metal oxides, conducting polymers and carbon resources [1–11]. In contrast to the conventional synthesis of the electrode materials, nanotechnology provides effective ways for developing electrode nanomaterials characterized by their superior power density and high retention efficiency (R %) after a high number charge-discharge cycles (CDCs) [4,12]. Thus, nanomaterials have been exploited for constructing electrodes with a high capacitance efficiency promised for supercapacitor applications as compared to their bulk equivalents. This enhancement stems from the great surface area accompanying the nanomaterials, thus enabling proficient ion carriage with a high energy-transformation course [13]. However, the nanomaterials tend to aggregate because of their inherent high specific area, and in consequence their cycle life constancy is reduced [14]. In order to overcome this aggregation problem, different methodologies-based organic molecules have been developed to generate nanostructured materials with supreme specific properties. One of those methodologies that relays on soft-approaches along with the microwave- and hydrothermal-assisted synthesis methods is the green chemistry [6,15–20]. In this technique, different biomolecules are employed to control the nucleation and growth processes, thus generating a well-generated nanocrystals with precise crystallite size, shape and dispersity [5,19–23]. Among those biomolecules, amino acids are characterized by their abundance and low-cost, as well as they show extraordinary properties with the ability to form hierarchical nanostructures [5,19,20,23–28]. They have been employed to overcome the different synthetic limitations with other methods. Usually, the amino acids have been used as (i) directing agents for self-assembling nanostructure materials, (ii) particle reducer and (iii) surface modifier as oxygen, nitrogen and sulfur dopants in the assembly of nanomaterials. Furthermore, they can be applied rather than the organic and inorganic surfactants in assisting the hydrolysis of the precursor materials. This can result in the generation of different hierarchical structures with precious control in size and morphology, as a result of their carboxyl, amino and thiol functional groups (Figure 16.1).

16.2 LYSINE AMINO ACID

As the lysine amino acid includes functional groups such as amine on its side chain, it has been employed as a chelating agent to support the generation of different nanostructures. Lysine has assisted the Stöber technique to fabricate finite nanospheres of formaldehyde/resorcinol and silica materials [19,20]. By means of this strategy, different nanoparticles with precise control of size, morphology and disparsity can be obtained. For instant, lysine amino acid has been used as catalyst to generate polymer nanospheres of resorcinol/formaldehyde, with size ranging from 30 to 650 nm, as a function of pH value [19,20]. Furthermore, carbonization process of the nanoparticles results in the generation of mesoporous nanospheres of large surface area; with approximately 300–400 m^2/g. Due to its redox property with p-type semiconductive, nickel oxide (NiO) can achieve a high capacitance performance with theoretical value of 2573–2584 F/g [3]. Different nanostructures of nickel oxide such as nanoflakes, nanowire and nanoflowers have been controlled by various hydrothermal- and microwave-assisted solution-based preparation techniques [3,5]. In the synthesized electrode materials by these strategies, the electron/ion transport into porous network can be improved by including lysine amino acids as a template to assist the hydrothermal process, thus well formation of hierarchical nickel oxide micro-flower-like spheres [3,5,29,30]. This well generation is attributed to the slow release rate of the hydroxyl (OH^-) ions during the hydrolysis process (Equation 16.1). Thus, the lysine amino acid has an implication on the controlling the formation of nickel hydroxide ($Ni(OH)_2$), (Equation 16.2), and therefore the nucleation and growth of the nickel oxide (NiO) flower-like microspheres.

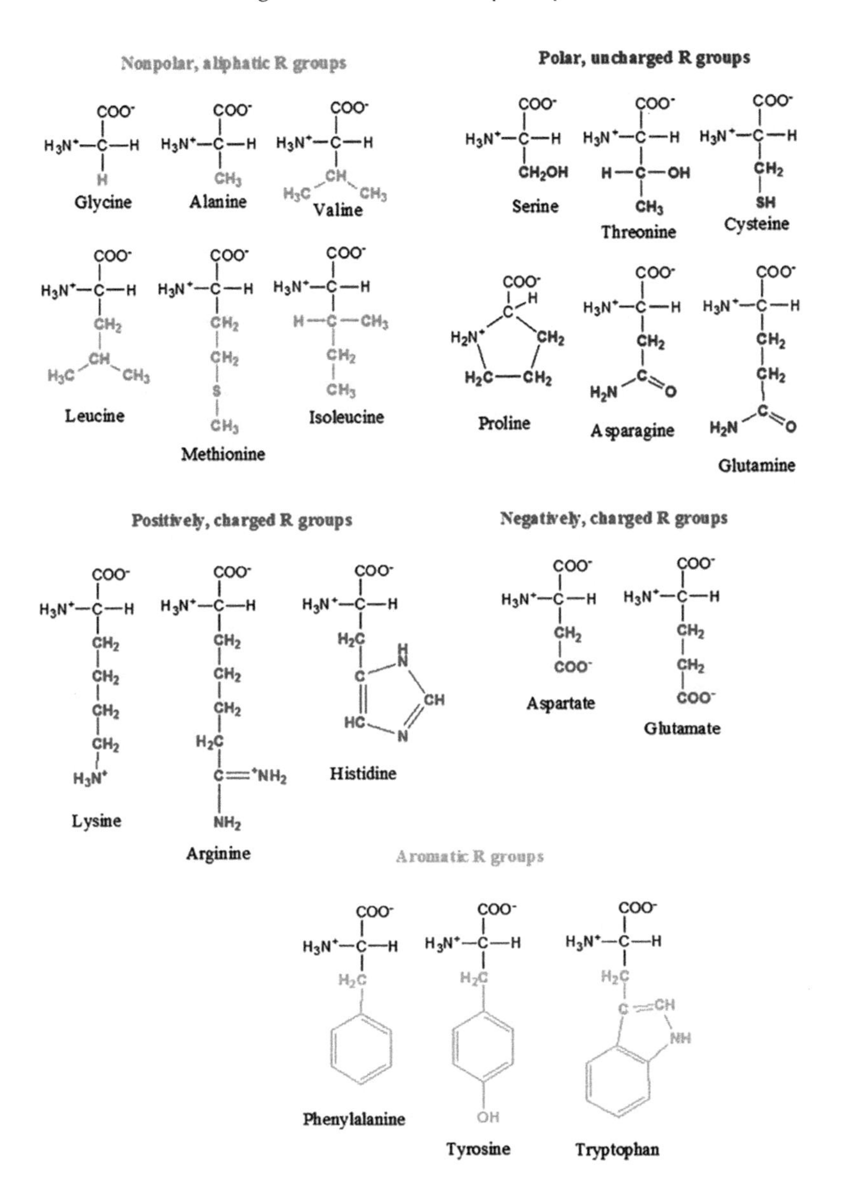

FIGURE 16.1 The essential 20 amino acids.

$$H_2N(CH_2)_4CH(NH_2)COOH + H_2O \rightarrow [H_3N(CH_2)_4CH(NH_2)COOH]^+ + OH^- \quad (16.1)$$

$$Ni^{2+} + 2OH^- \leftrightarrow Ni(OH)_2 \quad (16.2)$$

These microspheres have been delivered about 324 F/g at 2 A/g, with R value about 83% after 1000 CDCs at 20 A/g in 6 M of KOH [5]. A key parameter to increasing the capacitance of the electroactive nickel structures is to control the morphologies and pore size for transporting the hydroxyl groups into the electrode matrix. In this mesoporous nickel oxide, a faradaic redox process can be resulted by the hopping transition between Ni(II) and Ni(III), as following [3,5,29,30]:

$$NiO + 2\,OH^{-} \leftrightarrow NiO\,OH + e^{-} \tag{16.3}$$

Likewise, the lysine amino acid has been employed for controlling the formation of urchin cobalt hydroxide (β-Co (OH)) arrangements composed of nanowires as building blocks [7]. This well generation is stemmed from the synergistic impact of chloride (Cl^{-}) ions and the lysine amino acids to the formed β-cobalt hydroxide arrays. This hierarchical porous structure can result in delivering a good capacitance with about 421 and 370 F/g at 1.33 and 5.33 A/g, respectively, with 3.6% of degradation in capacitance value after 1000 CDCs. Furthermore, the chemical composition influences the capacitance performance [3,5,29]. For example, the nickel hydroxyde/cobalt hydroxide nanocomposite can exhibit an improved capacitance behavior attributed to the hopping between the redox pairs Co^{2+}/Co^{3+} and Ni^{2+}/Ni^{3+} [31,32]. In this regard, the amino acids have been employed to improve the specific capacitance delivered by those composites including the Co and Ni elements. For example, the introduction strategy of the CoO_4 to NiO microspheres for designing an effective electrode material of NiO-CoO_4 flower-like microsphere has been arisen in delivering the capacitance 1988.6 and 1491.4 F/g at 1 and 15 A/g, respectively [29]. This superior capacitance performance is ascribed to the improved specific area 43.4–51.8 m^2/g and the pore size 5.4–8.5 nm for the NiO-CoO_4 flower-like microsphere structure. Thus, the specific capacitance delivered by the heterostructures composed of various metal components can be improved with orders higher than all individual ingredients [6,29,31]. A further example of this enhancement is the synthesis of face-centered cubic spinel $Li_4Mn_5O_{12}$ having a flake-like structure that has been synthesized by lysine amino acid. In this case, the lysine acid has been employed to act as a complex agent and pH stabilizer during the synthesis methods [6]. This cubic structure can provide a capacitance of 168 F/g at the potential range 0–1.4 V, 5 mV/s, in 1 M of Li_2SO_4, with R value of 98.2% after 100 CDCs.

Aqueous assembly based on the amphiphilic molecules is a fascinating technique due to their superior structural organization and morphological miscellany [10,11]. These amphiphilic molecules play a key factor in generating hierarchical nanostructures, by engaging the conventional inorganic acids and bases as catalysts to initiating the cross-linking and formation of the polymer frameworks. In this regard, scalable synthesis of hierarchical ordered nanostructures of carbon materials have been developed by means of amino acids. For instance, the formation of the hierarchical carbon monolith can be well-controlled in the presence of lysine with a crack-free property, 600 m^2/g and macropore volume of 3.52 cm^3/g [10]. Along with the aqueous assembly approach of the organic molecules, they could be employed as nitrogen-dopant to generate nitrogen-doped nanostructures for the supercapacitors applications [12]. Among those nitrogen precursors, urea has been applied as a nitrogen dopant for the graphene oxide. The resulted nitrogen-doped heterostructures exhibit 405 F/g at 1 A/g, with 12.3% of degradation in capacitance value after

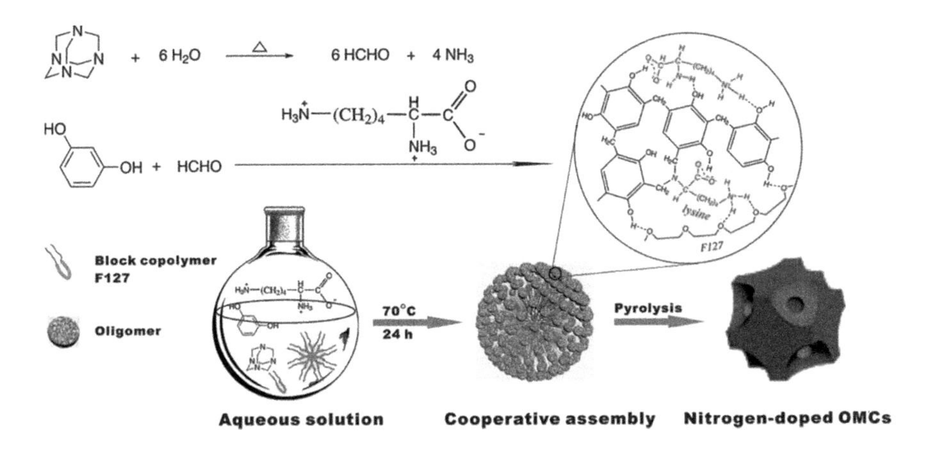

FIGURE 16.2 Lysine amino acid-assisted the generation of cubic nitrogen-doped ordered mesoporous carbon. (From Liu, D. et al., *J. Power Sources*, 321, 143–154, 2016.)

5000 CDC at 4 A/g. However, the generated materials display disordered mesostructures that hinder their employing in the energy storing [33]. To overwhelm the aforementioned limitation, lysine amino acid has been employed to produce mesoporous carbons rich with the nitrogen atoms and highly regular-pore arrays [10,11]. For instance, lysine has been exploited to act as catalyst and assembly promoter in combination with the structural directing Pluronic F127 as a triblock copolymer for generating the ordered mesoporous carbons (OMCs). This mesostructure is composed of highly regular pore arrays (Figure 16.2) and fully interconnected macropores that is in consequence resulted in attaining a mesoporous carbon rich with the nitrogen atom and superior surface area of 2422 m^2/g. Thus, the capacitance can be improved and deliver a value of 186 F/g at 0.25 A/g.

16.3 HISTIDINE AMINO ACID

Histidine amino acid is characterized by their carboxyl, amine and imidazole groups found in its molecular structures. These functional groups have been exploited in controlling the formation of different nanostructures for supercapacitor applications [34]. For example, recombinant elastin-like polypeptides containing hexa-histidine has been mixed with the metal salts Fe^{3+}/Co^{2+} to control the formation of magnetite (Fe_3O_4) and tricobalt tetroxide (Co_3O_4) nanoparticle with 5 nm. After the annealing process, the polypeptide containing hexa-histidine amino acids has been degraded into a porous microsphere of carbon that embedded with the magnetite and tricobalt oxide nanoparticles. This has been resulted in accomplishing a robust sodiation/desodiation reaction that is desired for sodium ion batteries [35]. For the carbon-encapsulated hematite and tricobalt tetroxide microspheres; a charge capacitance of 657 and 583 mAh/g, respectively, at 0.1 A/g can be delivered. Furthermore, histidine has been employed as a nitrogen source for doping the mesoporous graphene-like walls, thus developing nanostructures with a high capacity performance [36]. The protonation of histidine followed by

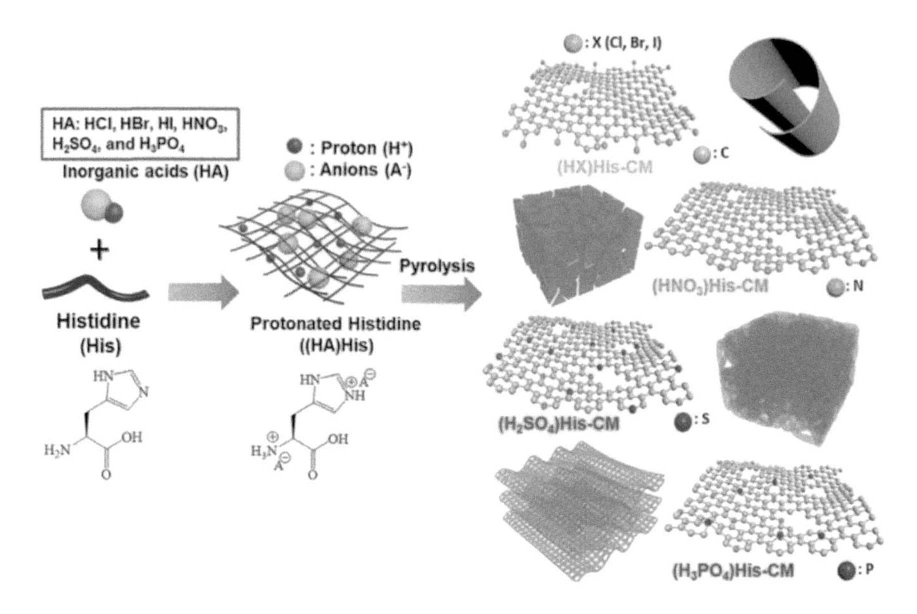

FIGURE 16.3 Protonated histidine amino acid-assisted the production of carbon materials. (From Tran, T.-N. et al., *J. Ind. Eng. Chem.*, 2019.)

pyrolysis is an effective way for generating hierarchical nanosheet-like structure. These nanosheets are promised for the supercapacitor applications, especially if a sup-additional heteroatom-dopant such as P and S atoms is introduced along with the original nitrogen doping (Figure 16.3) [37]. For example, carbon material doped with the sup-additional atom P ((H_3PO_4) histidine-carbon) displays a superior capacitance of 152 and 144 F/g in three- and two-electrode configurations, respectively, at 1.0 A/g, with 12% of degradation in capacitance value after 5000 CDCs, at 5.0 A/g.

16.4 ARGININE AMINO ACID

Arginine amino acid has been used as hydrolysis-controlling and stabilizer agent for generating metal hydroxides such as the two-dimensional α-cobalt(II) hydroxide via hydrothermal process [38] and the three-dimensional-connected nanosheets of α-nickel hydroxide via microwave method [25]. This controlling is assigned to the functional groups such as guandio, amino and carboxyl in the chemical structure of arginine amino acid. The α-cobalt hydroxide nanosheets [38] have a good performance capacitance of 427 F/g at 5.33 A/g. This capacitance value is reduced to about 414 F/g after 2000 CDCs at 0.3–0.45 V. In addition to the aforementioned roles, arginine amino acid has been applied as complexing agent to generate double layer hydroxide of Co/Al, forming porous flake-like structure [39]. The morphology, dispersion and size of these nanosheets depend on the pH value of the solution, since irregular nanosheets of hexagonal platelets are formed with a size of about 150 × 400 nm at pH 9; monodispersed nanosheets with a size of about 200 × 400 nm are formed at pH 11; uniform nanosheets with size of about 250 × 400 nm are generated at pH 13. This tuning ability is attributed to the coordination bonds between the arginine molecule and Co^{2+}/Al^{3+} ions, as well as to stabilization by the hydrogen bonding between nitrogen/oxygen and hydrogen atoms. Thus, the hydrotalcite formation can be taken place in the presence of the hydroxyl ions. Finally, the coordination and hydrogen bonds are destroyed by heating process; meanwhile the CO_3^{2-} ions produced by the hydrolysis of urea have been inserted into layers of hydroxide to form Co/Al layers. The generated structure has been displayed an improved electrochemical capacitance of 1158 F/g. Moreover, the produced mesopores participate in the quick transfer of electrolyte ions, thus delivering a fast redox reaction lucrative for the electrochemical supercapacitors. Furthermore, the nanosheets can be connected to form three dimensional structure via microwave method with the addition of arginine (Figure 16.4) [25]. The α-nickel hydroxide sheets display a capacitance of about 549 C/g at 2 A/g with 12.7% of degradation in capacitance value after 10,000 CDCs. Also, the three dimensional of α-nickel hydroxide/polyurethane structures can provide an energy density of about 57 Wh K/g at 601 W K/g.

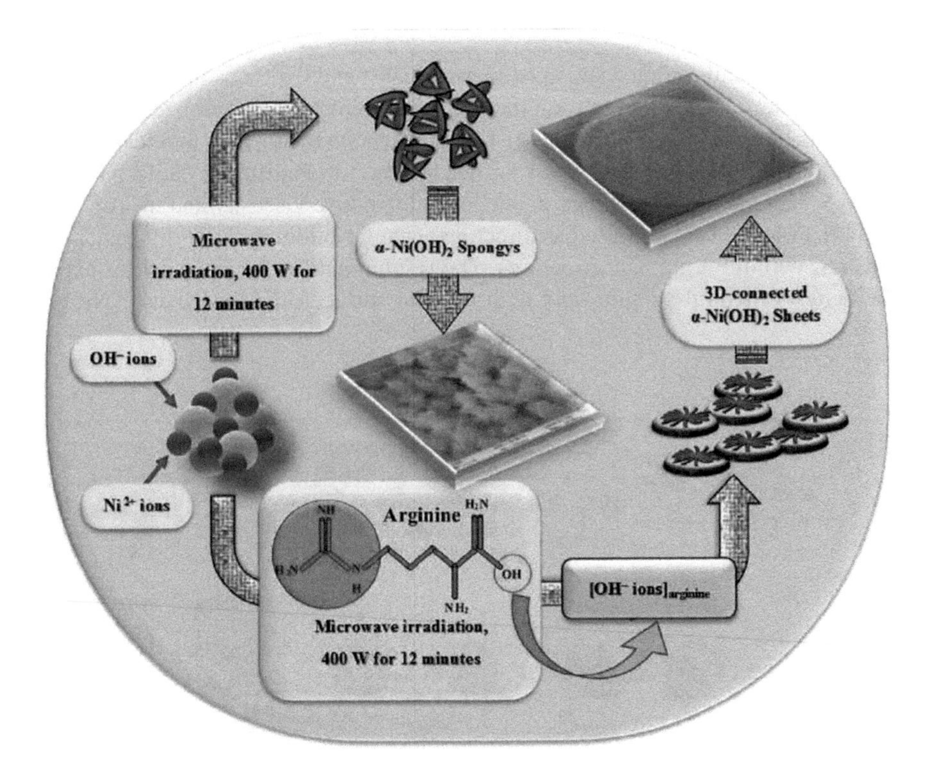

FIGURE 16.4 Arginine amino acid-assisted synthesis of 3D-connected α-$Ni(OH)_2$ sheets. (From William, J.J. et al., *Mater. Chem. Phys.*, 224, 357–368, 2019.)

The MnO_2 nanostructures have been generated by the reduction of $KMnO_4$ not only by using arginine amino acid, but also by using the glycine amino acid, and glutamic amino acid at the room temperature [40]. The generated nanorods with size of about 200 nm have exhibited a capacitance of about 250, 218 and 212 F/g for MnO_2-Arginine, MnO_2-Glycine and MnO_2-Glutamic, respectively, at 1 mA/cm.

16.5 CYSTEINE AMINO ACID

Due to its sulfur atom, the cysteine amino acid has assisted with the synthesis of different metal sulfide nanostructures, such as nickel sulfides, zinc sulfides, molybdenum sulfides, and indium sulfides via hydrothermal technique [41–43]. This generation ability stems from the strong tendency of the cysteine –SH to coordinate with the metal ions, thus forming metal-ions-cysteine complexes. These complexes can interact with each other to generate nuclei of nickel disulfide, which in consequence have been self-assembled into uniform aggregates [41]. Thus, hollow spheres of nickel sulfide has been produced after the calcination process of the nickel disulfide nuclei in air under different temperatures as shown in Figure 16.5h. The capacitance of the nickel sulfide and nickel disulfide electrodes is about 1643, 1076 F/g, respectively, at 1 A/g. This capacitance has been retained at 455 F/g and 368 F/g after 1000 CDCs.

Furthermore, the complexation between the Ni^{2+} and Co^{2+} ions in mixed aqueous solution of tertiary butanol to form the $NiCo_2S_4$ heterostructure has been performed by means of cysteine amino acid [44]. Cysteine can be hydrolyzed to release the S^{2-} ions with slow rate, and therefore a sufficient time for controlling the formation of $NiCo_2S_4$ nanoflakes. In this regard, due to its distinctive viscosity and hydrophobicity [45], cysteine has been used as directing agent for building three-dimensional of $NiCo_2S_4$ with ultrathin nanostructures. The resultant $NiCo_2S_4$ nanoparticles exhibit the better structural stability and high specific surface area than that of $NiCo_2S_4$

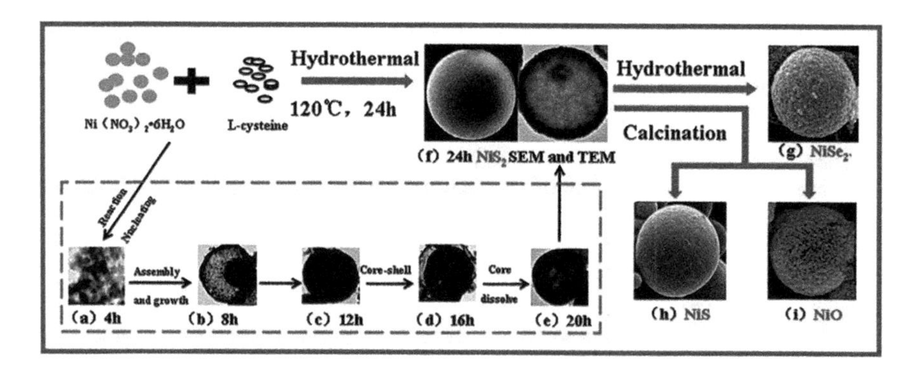

FIGURE 16.5 Cysteine amino acid-assisted the morphology evolution of NiS_2, NiS, 190 NiO and $NiSe_2$. (From Lu, M et al., *J. Mater. Chem. A*, 5, 3621–3627, 2017.)

fabricated by the conventional two-step method. By this way, the flower molybdenum disulfide with a diameter lower than 200 nm can be achieved by the sulfurization method [46]:

$$HS\ CH_2CH\ NH_2COOH + H_2O \rightarrow CH_3CO\ COOH + NH_3 + H_2S \quad (16.4)$$

$$4\ MoO_4^{2-} + 9\ H_2S + 6\ CH_3CO\ COOH \rightarrow 4\ MoS_2 + SO_4^{2-} + 6\ CH_3CO\ COO^- + 12\ H_2O \quad (16.5)$$

Molybdenum disulfide electrode exhibits a capacitance of about 129.2 F/g at 1 A/g, with R value of about 85.1% after 500 CDCs. This specific capacitance has been enhanced by the addition of graphene sheets, where cysteine has been employed as a sulfurization agent. The graphene sheets inhibit the growth of layered molybdenum disulfide crystals via the hydrothermal process [46]. Li et al. revealed that the complexation between the functional group found on the graphene sheets improving the well-controlled growth of molybdenum disulfide [47]. This in consequence has resulted in generating a three-dimensional sphere-like structure (molybdenum disulfide: graphene ratio = 1:2), 1100 mAh/g at 100 mA/g, where the capacity decreases to 900 mAh/g after 1000 CDCs. Similarly, the generated SnS_2/SnO_2 combination by the microwave-supported cysteine amino acid results in an improved capacitance behavior when it is used as anode material in lithium ion batteries. This capacitance performance depends on the molar ratio of $SnCl_4$:cysteine. At 1:6 molar ratio of $SnCl_4$:cysteine, the primary capacity is about 593 mAh/g [48]. Cysteine has been employed as a sulfurization agent for the preparation of cobalt sulfide nanowires, with a capacitance of about 508 F/g [49]. As well as, Bi_2S_3 flower-like structure with well-aligned nanorods has been generated by the assistance of cysteine amino acid, with a capacitance of about 142 (mA h)/g [42].

Cysteine amino acid has been employed to reduce the graphene sheets via its thiol group [50]. Thus, the reduced graphene exhibits a high dispersion behavior with enhanced capacity and cycle-life for charge-discharge process. For example, the composite of reduced graphene oxide/nickel sulfides nanospheres can be produced via hydrothermal technique [51]. In this case, the cysteine is not used only as a reducing agent but also as a sulfur donor, and linker to generate ultrafine nanocomposites of nickel sulfides and graphene sheets. This composite displays 1169 and 761 F/g at 5 and 50 A/g, respectively. These superior capacitances are stemmed from the high conductivity of the reduced graphene oxide, the connection between the nickel sulfide and graphene sheets, and the micropores within the nickel sulfide nanospheres.

16.6 ASPARTIC AMINO ACID

Due to its carboxylic functional group, aspartic acid has been used to control the formation of copper oxide (CuO) nanoparticles of about 6 nm [27], and the generation of uniform nanowires of Cu-based porous organization polymers with length of about 200 nm [52]. These nanowires exhibit an improved capacitance of 367 F/g with 6% of degradation in capacitance value after 1000 CDCs compared to the electrode composed of the copper oxide nanotube. Furthermore, the $Co_2Fe(CN)_6$ particles have been formed by aspartic acid as a reducing agent [53], where the pH value has a remarkable effect in chelating the aspartic acid to hinder the metal ions agglomerations. This nanostructure revealed a large capacitance of 758.86 F/g at 2 mV/s in 1 M of KOH. Also, aspartic acid has been used to assist the formation of ultra-high reduced graphene aerogels via hydrothermal treatment [54]. The strength of this aerogel matrix is enhanced by the addition of Fe^{3+} ions as a cross-linker at pH 9. This strength is stemmed from the coordination bonding between the Fe^{3+} ions or carboxylic and the reduced sheets of the graphene oxide. The later matrix exhibits 276.4 F/g at 0.5 A/g, with 11.8% degradation in capacitance value after 5000 CDCs.

The functional groups of the glutamic, alanine and arginine amino acids have been exploited to control the formation of the iron oxide (Fe_2O_3) nanoparticles [55]. Furthermore, they have been employed as N-doping and morphology-assisting to generate the iron oxide/nitrogen-doped graphene hydrogels (Figure 16.6). In this

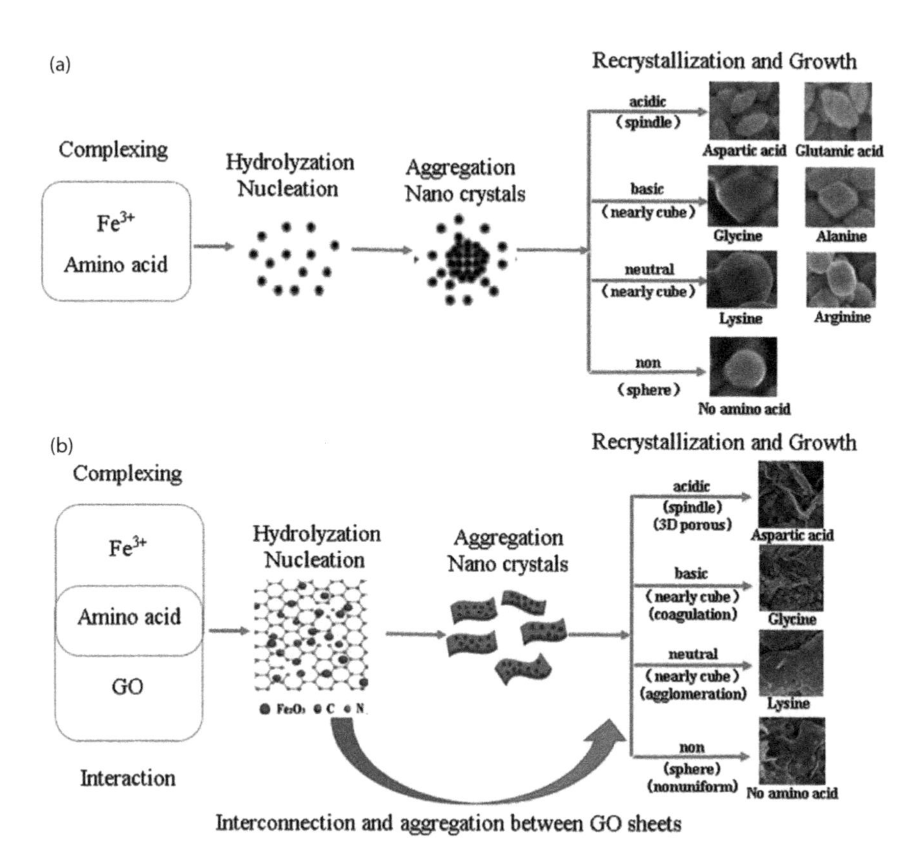

FIGURE 16.6 Different amino acids-assisted the preparation of (a) Fe_2O_3, and (b) Fe_2O_3/nanographene composites. (From Wu, D. et al., *Electrochim. Acta*, 283, 1858–1870, 2018.)

case, the effect of the amino acid acidity on the morphological properties of iron oxide has been clarified. Thus, the iron oxide prepared by glutamic exhibit a mulberry-liked structure (the iron oxide nanoparticles have a length and width of 130–170 and 56 nm, respectively), while the iron oxide nanoparticles prepared by the alanine arginine acids show a nearly cube-like structure with size of about 420–540 nm. These well-generated morphologies with the precise size are different from the spherical one of bare iron oxide nanoparticles of about 200 nm. The regulation of the iron oxide nanoparticles by the nanographene as a template in the presence of amino acid has been resulted in generating the same morphologies of the iron oxide nanoparticles but with smaller size. Thus, the size of iron oxide nanoparticles is reduced in the case of iron oxide-nanographene oxide, to about 40–70 nm. The iron oxide/nanographene-aspartic combination has been exhibited a cross-linked mesoporous matrix of pore size with a few μm. Furthermore, iron oxide/graphene nanocomposites prepared by glycine and lysine have been generated including folded and wrinkled graphene sheets, thus indicating the occurrence of severe aggregation in the glycine and lysine amino acids-assisted the synthesis procedures. The key parameter in controlling this morphology is the isoelectric point of the amino acid [56]. For instant, the aspartic acid can bear a negative charge above its isoelectric point and therefore control the formation of dispersed graphene sheets. In contrast, the glycine and lysine amino acid bear a positive charge below their isoelectric point, and therefore aggregated graphene nanosheets are produced. Furthermore, the adsorption of amino acid molecules on the surface of the inorganic nanoparticles such as iron oxide plays an important role in controlling their growth and morphologies [57,58]. Therefore, the amino acids assist the hydrothermal synthesis for developing nanocomposites with an improved capacitance performance than that of the iron oxide/nitrogen-doped graphene from a conventional nitrogen sources [59–63]. Iron oxide/nanographene composites controlled by the aspartic acid have been demonstrated the best capacity retention, which can be attributed to the iron oxide nanoparticle accompanied by the active sites on their surfaces.

16.7 GLYCINE AMINO ACID

Glycine can form coordination bonds with metal ions such as iron, cobalt, indium and nickel through its functional groups [64–66]. The glycine amino acid has been employed as a template or doping agent, as well as an organic ligand to develop different nanomaterials at mild condition [67,68], efficient for supercapacitor applications [64,65,68]. Glycine amino acid has been used to prepare the non-uniform spherical nanoparticles Na_4Ni_3 $(PO_4)_2$ P_2O_7 with a porous structure by combustion synthesis technique with [65]. Coating this Na_4Ni_3 $(PO_4)_2$ P_2O_7 nanostructures with carbon has been resulted in improving the pseudo-capacitance to about 1094 F/g, and excellent cycling stability in 2 M of NaOH. This high capacitance is attributed to the redox pair of Ni^{2+}/Ni^{3+} and carbon concentration accessible to enhancing the conductivity. Furthermore, the spinal $CoFe_2O_4$ nanocrystalline of spherical structures with 40–80 nm has been developed by employing the glycine and aspartic acids as a template and directing agent [69].

Glycine amino acid has been exploited to develop uniform small crystals of iron oxide nanoparticles to be applied as an anode material for lithium-ion-batteries [16,70]. This controlling stems from the complex formation between the glycine and iron ions, thus controlling the nucleation process at early stage of the reaction. Although iron oxide has been verified to be competent for high lithium storage, the delivered capacitance has been improved by its deposition onto highly uniform and ordered multimodal porous carbon [16]. The resultant combination exhibits a high specific area and unique hierarchical porosities, promised for developing electrode materials in the supercapacitor applications. Remarkably, glycine amino acid provides a crucial effect in controlling the morphology and distribution of iron oxide nanoparticles along with the ordered mesoporous carbons via hydrothermal treatment [71]. Thus, well-dispersed iron oxide nanoparticles of a cube-like structure sustained on the ordered porous carbon exhibit a significant capacitance performance. For example, 10 wt% Fe_2O_3/ordered multimodal porous carbons exhibit a capacitance of 294 F/g at 1.5 A/g, which is about two and four times greater than that of ordered mesoporous and Fe_2O_3, respectively.

Sheet-like structures of cobalt-glycine complex have been prepared in the presence of glycine amino acid and cobalt nitrate hexahydrate via hydrothermal treatment [72]. This material exhibits a high pseudo-capacitive performance with interesting cycle stability as a result of its amorphous structure and hierarchical porosity. When this material is examined in 6 M KOH as an electrode, it has a discharge capacitance of 2511.6 and 1564.8 F/g at of 1 and 20 A/g, respectively, with 0.5% degradation in the capacitance value after 2000 CDCs at 10 A/g. The nickel oxide urchin-like nanoporous of spherical structures have been well controlled by the glycine amino acid via hydrothermal method [73]. These nanospheres has been displayed a specific capacitance of 1027 mA h/g in lithium-ion-batteries. The nanostructure of Co_xNi_{1-x}Al-layered-triple hydroxides have been prepared by using the glycine amino acid as a chelating agent via hydrothermal treatment [64]. At a molar ratio Co:Ni of 3:2, 1375 F/g at 0.5 A/g can be achieved with R value of 93.3% after 1000 CDCs at 2A/g.

16.8 GUANIDINE AMINO ACID

Guanidine hydrochloride has been applied to control the one-dimensional hollow cobalt oxide (Co_3O_4) nanotubes with a diameter size of 200–300 nm, by the hydrothermal technique via triggered carbon as a nanotemplate [74]. The slow rate release of hydroxyl ions results in controlling the reaction rate, which is the key parameter for achieving the desired nucleation and growth processes (Figure 16.7). In this case, the template can induce the primary crystallites formations or ultrafine nanoparticles. According to the Kirkendall effect [75], and at slow heating rate, a hollow nanotube structure can be well-generated. The generated nanotubes has been exhibited a rough surface as it has been built from cobalt oxide nanoparticles. They have been delivered a superior capacitance about 1006 F/g at 1 A/g, with 9% degradation in capacitance value after 1000 CDCs.

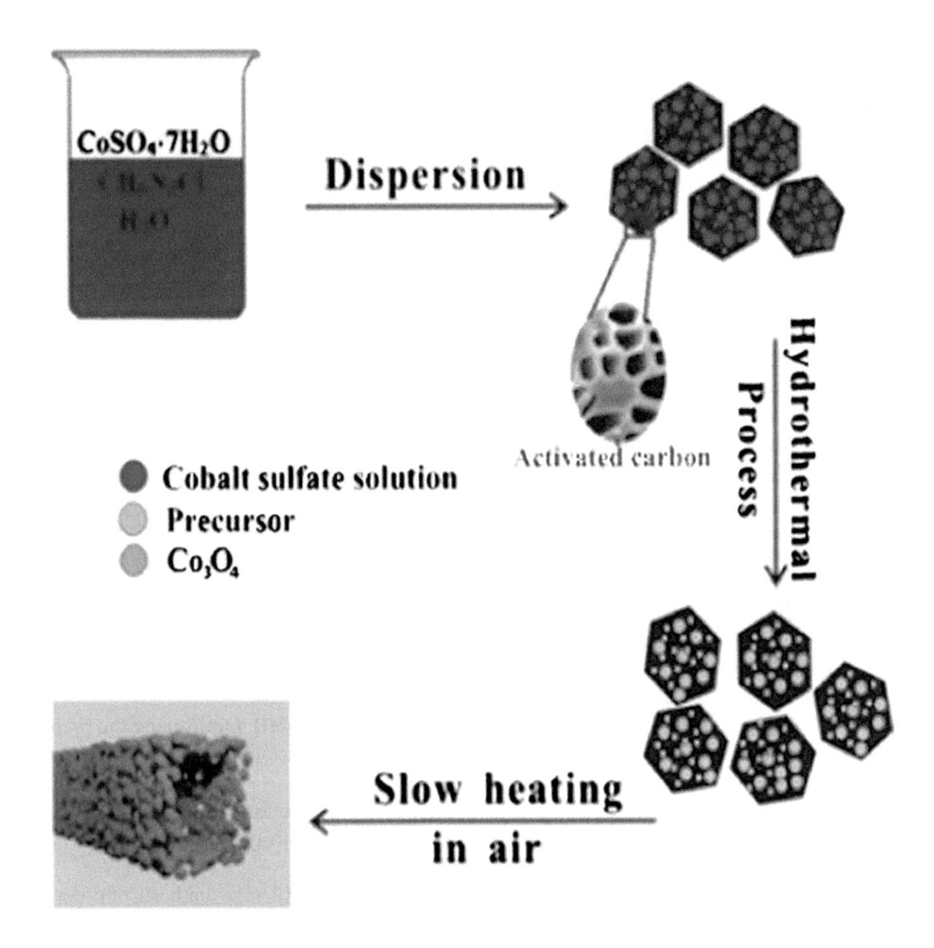

FIGURE 16.7 Guanidine amino acid-assisted synthesis process of 1D hierarchical hollow Co_3O_4 nanotubes. (From Yao, M. et al., *J. Alloy Compd.*, 644, 721–728, 2015.)

16.9 GLUTAMIC AMINO ACID

Glutamic acid has been exploited to form the β-nickel hydroxide sheets with 300–500 nm via hydrothermal treatment [76]. These nanosheets have been displayed the superior capacitance 2537.4 and 2290.0 F/g at 1 and 10 A/g, respectively, with 22.4% degradation in capacitance value after 3000 CDCs. Due to coordination bonding between the functional groups of the amino acids and zinc cations, glutamic acid has been used as a carbon precursor to prepare highly microporous carbons [77–79]. These carbons are rich with nitrogen and have exhibited the superior specific area 1203 m^2/g. When these carbons has been examined in 6 M of KOH, they delivered a capacitance about 217 F/g at 0.5 A/g and 9% degradation in capacitance value after 30000 CDCs.

REFERENCES

1. Xia X, Zhang Y, Chao D, Guan C, Zhang Y, Li L, et al. Solution synthesis of metal oxides for electrochemical energy storage applications. *Nanoscale* 2014;6:5008–5048.
2. Fihri A, Sougrat R, Rakhi RB, Rahal R, Cha D, Hedhili MN, et al. Nanoroses of nickel oxides: Synthesis, electron tomography study, and application in CO oxidation and energy storage. *ChemSusChem* 2012;5:1241–1248.
3. Khairy M, El-Safty SA. Mesoporous NiO nanoarchitectures for electrochemical energy storage: Influence of size, porosity, and morphology. *RSC Advances* 2013;3:23801–23809.
4. Khairy M, Ayoub HA, Banks CE. Large-scale production of CdO/Cd $(OH)_2$ nanocomposites for non-enzyme sensing and supercapacitor applications. *RSC Advances* 2018;8:921–930.
5. Wu Q, Liu Y, Hu Z. Flower-like NiO microspheres prepared by facile method as supercapacitor electrodes. *Journal of Solid State Electrochemistry* 2013;17:1711–1716.
6. Zhao Y, Lai Q, Zeng H, Hao Y, Lin Z. Li_4 Mn_5 O_{12} prepared using l-lysine as additive and its electrochemical performance. *Ionics* 2013;19:1483–1487.
7. Yuan C, Zhang X, Hou L, Shen L, Li D, Zhang F, et al. Lysine-assisted hydrothermal synthesis of urchin-like ordered arrays of mesoporous Co $(OH)_2$ nanowires and their application in electrochemical capacitors. *Journal of Materials Chemistry* 2010;20:10809–10816.
8. Bhattacharjya D, Yu J-S. Activated carbon made from cow dung as electrode material for electrochemical double layer capacitor. *Journal of Power Sources* 2014;262:224–231.
9. Fathy M, Gomaa A, Taher FA, El-Fass MM, Kashyout AE-HB. Optimizing the preparation parameters of GO and rGO for large-scale production. *Journal of Materials Science* 2016;51:5664–5675.
10. Hao G-P, Li W-C, Wang S, Wang G-H, Qi L, Lu A-H. Lysine-assisted rapid synthesis of crack-free hierarchical carbon monoliths with a hexagonal array of mesopores. *Carbon* 2011;49:3762–3772.
11. Liu D, Zeng C, Qu D, Tang H, Li Y, Su B-L, et al. Highly efficient synthesis of ordered nitrogen-doped mesoporous carbons with tunable properties and its application in high performance supercapacitors. *Journal of Power Sources* 2016;321:143–154.
12. Elessawy NA, El Nady J, Wazeer W, Kashyout A. Development of high-performance supercapacitor based on a novel controllable green synthesis for 3D nitrogen doped graphene. *Scientific Reports* 2019;9:1129.
13. Yu G, Hu L, Liu N, Wang H, Vosgueritchian M, Yang Y, et al. Enhancing the supercapacitor performance of graphene/MnO_2 nanostructured electrodes by conductive wrapping. *Nano Letters* 2011;11:4438–4442.
14. White JL, Baruch MF, Pander III JE, Hu Y, Fortmeyer IC, Park JE, et al. Light-driven heterogeneous reduction of carbon dioxide: Photocatalysts and photoelectrodes. *Chemical Reviews* 2015;115:12888–12935.
15. Guan D, Gao Z, Yang W, Wang J, Yuan Y, Wang B, et al. Hydrothermal synthesis of carbon nanotube/cubic Fe_3O_4 nanocomposite for enhanced performance supercapacitor electrode material. *Materials Science and Engineering: B* 2013;178:736–743.
16. Chen J, Huang K, Liu S. Hydrothermal preparation of octadecahedron Fe_3O_4 thin film for use in an electrochemical supercapacitor. *Electrochimica Acta* 2009;55:1–5.
17. Shoueir K, El-Sheshtawy H, Misbah M, El-Hosainy H, El-Mehasseb I, El-Kemary M. Fenton-like nanocatalyst for photodegradation of methylene blue under visible light activated by hybrid green DNSA@ Chitosan@ MnFe2O4. *Carbohydrate Polymers* 2018;197:17–28.

18. Salahuddin NA, El-Kemary M, Ibrahim EM. Synthesis and characterization of ZnO nanotubes by hydrothermal method. *International Journal of Scientific and Research Publications* 2015;5:1–4.
19. Liu J, Qiao SZ, Liu H, Chen J, Orpe A, Zhao D, et al. Extension of the Stöber method to the preparation of monodisperse resorcinol–formaldehyde resin polymer and carbon spheres. *Angewandte Chemie* 2011;123:6069–6073.
20. Dong Y-R, Nishiyama N, Egashira Y, Ueyama K. Basic amid acid-assisted synthesis of resorcinol–formaldehyde polymer and carbon nanospheres. *Industrial & Engineering Chemistry Research* 2008;47:4712–4716.
21. Misbah MH, Espanol M, Quintanilla L, Ginebra M, Rodríguez-Cabello JC. Formation of calcium phosphate nanostructures under the influence of self-assembling hybrid elastin-like-statherin recombinamers. *RSC Advances* 2016;6:31225–31234.
22. Misbah MH, Santos M, Quintanilla L, Günter C, Alonso M, Taubert A, et al. Recombinant DNA technology and click chemistry: A powerful combination for generating a hybrid elastin-like-statherin hydrogel to control calcium phosphate mineralization. *Beilstein Journal of Nanotechnology* 2017;8:772–783.
23. Hu X, Sun A, Kang W, Zhou Q. Strategies and knowledge gaps for improving nanomaterial biocompatibility. *Environment International* 2017;102:177–189.
24. Misbah MH, Quintanilla L, Alonso M, Rodríguez-Cabello JC. Evolution of amphiphilic elastin-like co-recombinamer morphologies from micelles to a lyotropic hydrogel. *Polymer* 2015;81:37–44.
25. William JJ, Babu IM, Muralidharan G. Microwave assisted fabrication of l-Arginine capped α-Ni $(OH)_2$ microstructures as an electrode material for high performance hybrid supercapacitors. *Materials Chemistry and Physics* 2019;224:357–368.
26. Rahimi-Nasrabadi M, Ahmadi F, Eghbali-Arani M. Novel route to synthesize nanocrystalline nickel titanate in the presence of amino acids as a capping agent. *Journal of Materials Science: Materials in Electronics* 2016;27:11873–11878.
27. El-Trass A, ElShamy H, El-Mehasseb I, El-Kemary M. CuO nanoparticles: Synthesis, characterization, optical properties and interaction with amino acids. *Applied Surface Science* 2012;258:2997–3001.
28. Wei W, Chen W, Ding L, Cui S, Mi L. Construction of hierarchical three-dimensional interspersed flower-like nickel hydroxide for asymmetric supercapacitors. *Nano Research* 2017;10:3726–3742.
29. Liu P, Hu Z, Liu Y, Yao M, Zhang Q, Xu Z. Novel three-dimensional hierarchical flower-like NiO–Co_3O_4 composites as high-performance electrode materials for supercapacitors. *International Journal of Electrochemical Science* 2014;9:7986–7996.
30. Xia X, Tu J, Zhang J, Wang X, Zhang W, Huang H. Electrochromic properties of porous NiO thin films prepared by a chemical bath deposition. *Solar Energy Materials and Solar Cells* 2008;92:628–633.
31. Li M, Xu S, Zhu Y, Yang P, Wang L, Chu PK. Heterostructured Ni $(OH)_2$–Co $(OH)_2$ composites on 3D ordered Ni–Co nanoparticles fabricated on microchannel plates for advanced miniature supercapacitor. *Journal of Alloys and Compounds* 2014;589:364–371.
32. Essawy AA, Nassar AM, Arafa WA. A novel photocatalytic system consists of Co (II) complex@ ZnO exhibits potent antimicrobial activity and efficient solar-induced wastewater remediation. *Solar Energy* 2018;170:388–397.
33. Zhao D, Wan Y, Zhou W. *Ordered Mesoporous Materials*. Hoboken, NJ: John Wiley & Sons; 2012.
34. Gao X, Chen Z, Yao Y, Zhou M, Liu Y, Wang J, et al. Direct heating amino acids with silica: A universal solvent-free assembly approach to highly nitrogen-doped mesoporous carbon materials. *Advanced Functional Materials* 2016;26:6649–6661.

35. Zhou Y, Sun W, Rui X, Zhou Y, Ng WJ, Yan Q, et al. Biochemistry-derived porous carbon-encapsulated metal oxide nanocrystals for enhanced sodium storage. *Nano Energy* 2016;21:71–79.
36. Yao Y, Chen Z, Zhang A, Zhu J, Wei X, Guo J, et al. Surface-coating synthesis of nitrogen-doped inverse opal carbon materials with ultrathin micro/mesoporous graphene-like walls for oxygen reduction and supercapacitors. *Journal of Materials Chemistry A* 2017;5:25237–25248.
37. Tran T-N, Kim HJ, Samdani JS, Hwang JY, Ku B-C, Lee JK, et al. A facile in-situ activation of protonated histidine-derived porous carbon for electrochemical capacitive energy storage. *Journal of Industrial and Engineering Chemistry* 2019;73:316–327.
38. Hou L, Yuan C, Yang L, Shen L, Zhang F, Zhang X. Biomolecule-assisted hydrothermal approach towards synthesis of ultra-thin nanoporous α-Co $(OH)_2$ mesocrystal nanosheets for electrochemical capacitors. *CrystEngComm* 2011;13:6130–6135.
39. Lin W, Yu W, Hu Z, Ouyang W, Shao X, Li R, et al. Superior performance asymmetric supercapacitors based on flake-like Co/Al hydrotalcite and graphene. *Electrochimica Acta* 2014;143:331–339.
40. Aswathy R, Munaiah Y, Ragupathy P. Unveiling the charge storage mechanism of layered and tunnel structures of manganese oxides as electrodes for supercapacitors. *Journal of The Electrochemical Society* 2016;163:A1460–A1468.
41. Lu M, Yuan X-P, Guan X-H, Wang G-S. Synthesis of nickel chalcogenide hollow spheres using an L-cysteine-assisted hydrothermal process for efficient supercapacitor electrodes. *Journal of Materials Chemistry A* 2017;5:3621–3627.
42. Zhang B, Ye X, Hou W, Zhao Y, Xie Y. Biomolecule-assisted synthesis and electrochemical hydrogen storage of Bi_2S_3 flowerlike patterns with well-aligned nanorods. *The Journal of Physical Chemistry B* 2006;110:8978–8985.
43. Jiang J, Yu R, Yi R, Qin W, Qiu G, Liu X. Biomolecule-assisted synthesis of flower-like NiS microcrystals via a hydrothermal process. *Journal of Alloys and Compounds* 2010;493:529–534.
44. Wang H, Yang Y, Zhou X, Li R, Li Z. $NiCo_2S_4$/tryptophan-functionalized graphene quantum dot nanohybrids for high-performance supercapacitors. *New Journal of Chemistry* 2017;41:1110–1118.
45. Tao Y, Ruiyi L, Zaijun L, Junkang L, Guangli W, Zhiquo G. A free template strategy for the fabrication of nickel/cobalt double hydroxide microspheres with tunable nanostructure and morphology for high performance supercapacitors. *RSC Advances* 2013;3:19416–19422.
46. Chang K, Chen W. L-cysteine-assisted synthesis of layered MoS_2/graphene composites with excellent electrochemical performances for lithium ion batteries. *ACS Nano* 2011;5:4720–4728.
47. Li Y, Wang H, Xie L, Liang Y, Hong G, Dai H. MoS_2 nanoparticles grown on graphene: An advanced catalyst for the hydrogen evolution reaction. *Journal of the American Chemical Society* 2011;133:7296–7299.
48. Chang K, Chen W-X, Li H, Li H. Microwave-assisted synthesis of SnS_2/SnO_2 composites by l-cysteine and their electrochemical performances when used as anode materials of Li-ion batteries. *Electrochimica Acta* 2011;56:2856–2861.
49. Bao S-J, Li CM, Guo C-X, Qiao Y. Biomolecule-assisted synthesis of cobalt sulfide nanowires for application in supercapacitors. *Journal of Power Sources* 2008;180:676–681.
50. Zhang D, Zhang X, Chen Y, Wang C, Ma Y, Dong H, et al. Supercapacitor electrodes with especially high rate capability and cyclability based on a novel Pt nanosphere and cysteine-generated graphene. *Physical Chemistry Chemical Physics* 2012;14:10899–10903.

51. Xing Z, Chu Q, Ren X, Tian J, Asiri AM, Alamry KA, et al. Biomolecule-assisted synthesis of nickel sulfides/reduced graphene oxide nanocomposites as electrode materials for supercapacitors. *Electrochemistry Communications* 2013;32:9–13.
52. Wu M-K, Zhou J-J, Yi F-Y, Chen C, Li Y-L, Li Q, et al. High-performance supercapacitors of Cu-based porous coordination polymer nanowires and the derived porous CuO nanotubes. *Dalton Transactions* 2017;46:16821–16827.
53. Shanmugavani A, Kalpana D, Selvan RK. Electrochemical properties of $CoFe_2O_4$ nanoparticles as negative and $Co(OH)_2$ and $Co_2Fe(CN)_6$ as positive electrodes for supercapacitors. *Materials Research Bulletin* 2015;71:133–141.
54. Wang B, Kang Y, Shen T-Z, Song J-K, Park HS, Kim J-H. Ultralight and compressible mussel-inspired dopamine-conjugated poly (aspartic acid)/Fe^{3+}-multifunctionalized graphene aerogel. *Journal of Materials Science* 2018;53:16484–16499.
55. Wu D, Liu P, Wang T, Chen X, Yang L, Jia D. Amino acid-assisted synthesis of Fe_2O_3/nitrogen doped graphene hydrogels as high performance electrode material. *Electrochimica Acta* 2018;283:1858–1870.
56. Wang T, Wang L, Wu D, Xia W, Zhao H, Jia D. Hydrothermal synthesis of nitrogen-doped graphene hydrogels using amino acids with different acidities as doping agents. *Journal of Materials Chemistry A* 2014;2:8352–8361.
57. Hatzor A, Weiss P. Molecular rulers for scaling down nanostructures. *Science* 2001;291:1019–1020.
58. Cao H, Qian X, Wang C, Ma X, Yin J, Zhu Z. High symmetric 18-facet polyhedron nanocrystals of Cu_7S_4 with a hollow nanocage. *Journal of the American Chemical Society* 2005;127:16024–16025.
59. Guan C, Liu J, Wang Y, Mao L, Fan Z, Shen Z, et al. Iron oxide-decorated carbon for supercapacitor anodes with ultrahigh energy density and outstanding cycling stability. *ACS Nano* 2015;9:5198–5207.
60. Zhao C, Shao X, Zhang Y, Qian X. Fe_2O_3/reduced graphene oxide/Fe_3O_4 composite in situ grown on Fe foil for high-performance supercapacitors. *ACS Applied Materials & Interfaces* 2016;8:30133–30142.
61. Lu X, Zeng Y, Yu M, Zhai T, Liang C, Xie S, et al. Oxygen-deficient hematite nanorods as high-performance and novel negative electrodes for flexible asymmetric supercapacitors. *Advanced Materials* 2014;26:3148–3155.
62. El-Khodary SA, El-Enany GM, El-Okr M, Ibrahim M. Modified iron doped polyaniline/sulfonated carbon nanotubes for all symmetric solid-state supercapacitor. *Synthetic Metals* 2017;233:41–51.
63. Tabrizi AG, Arsalani N, Mohammadi A, Namazi H, Ghadimi LS, Ahadzadeh I. Facile synthesis of a $MnFe_2O_4$/rGO nanocomposite for an ultra-stable symmetric supercapacitor. *New Journal of Chemistry* 2017;41:4974–4984.
64. Zhang F, Jiang J, Yuan C, Hao L, Shen L, Zhang L, et al. Glycine-assisted hydrothermal synthesis of nanostructured Co_xNi_{1-x}–Al layered triple hydroxides as electrode materials for high-performance supercapacitors. *Journal of Solid State Electrochemistry* 2012;16:1933–1940.
65. Senthilkumar B, Ananya G, Ramaprabhu S. Synthesis of carbon coated nano-Na_4Ni_3 (PO_4) $2P_2O_7$ as a novel cathode material for hybrid supercapacitors. *Electrochimica Acta* 2015;169:447–455.
66. Alharbi S, El-Sheshtawy H. Glycine capped SnO_2 nanoparticles: Synthesis, photophysical properties and photodegradation efficiency. *Nanoscience and Nanotechnology Letters* 2017;9:266–271.
67. Görbitz CH. Crystal structures of amino acids: From bond lengths in glycine to metal complexes and high-pressure polymorphs. *Crystallography Reviews* 2015;21:160–212.

68. Wei C, Pang H, Cheng C, Zhao J, Li P, Zhang Y. Mesoporous 3D ZnO–NiO architectures for high-performance supercapacitor electrode materials. *CrystEngComm* 2014;16:4169–4175.
69. Shanmugavani A, Selvan RK, Layek S, Vasylechko L, Sanjeeviraja C. Influence of pH and fuels on the combustion synthesis, structural, morphological, electrical and magnetic properties of $CoFe_2O_4$ nanoparticles. *Materials Research Bulletin* 2015;71:122–132.
70. Chaudhari NK, Kim M-S, Bae T-S, Yu J-S. Hematite (α-Fe_2O_3) nanoparticles on vulcan carbon as an ultrahigh capacity anode material in lithium ion battery. *Electrochimica Acta* 2013;114:60–67.
71. Chaudhari NK, Chaudhari S, Yu JS. Cube-like α-Fe_2O_3 Supported on ordered multimodal porous carbon as high performance electrode material for supercapacitors. *ChemSusChem* 2014;7:3102–3111.
72. Hou X, Chen L, Xu H, Zhang Q, Zhao C, Xuan L, et al. Engineering of two-dimensional cobalt-glycine complex thin sheets of vertically aligned nanosheet basic building blocks for high performance supercapacitor electrode materials. *Electrochimica Acta* 2016;210:462–473.
73. Mondal AK, Su D, Wang Y, Chen S, Liu Q, Wang G. Microwave hydrothermal synthesis of urchin-like NiO nanospheres as electrode materials for lithium-ion batteries and supercapacitors with enhanced electrochemical performances. *Journal of Alloys and Compounds* 2014;582:522–527.
74. Yao M, Hu Z, Xu Z, Liu Y. Template synthesis of 1D hierarchical hollow Co_3O_4 nanotubes as high performance supercapacitor materials. *Journal of Alloys and Compounds* 2015;644:721–728.
75. Yin Y, Rioux RM, Erdonmez CK, Hughes S, Somorjai GA, Alivisatos AP. Formation of hollow nanocrystals through the nanoscale kirkendall effect. *Science* 2004;304:711–714.
76. Hu X, Liu S, Li C, Huang J, Luv J, Xu P, et al. Facile and environmentally friendly synthesis of ultrathin nickel hydroxide nanosheets with excellent supercapacitor performances. *Nanoscale* 2016;8:11797–11802.
77. Dong X-L, Lu A-H, He B, Li W-C. Highly microporous carbons derived from a complex of glutamic acid and zinc chloride for use in supercapacitors. *Journal of Power Sources* 2016;327:535–542.
78. Brede F, Mandel K, Schneider M, Sextl G, Müller-Buschbaum K. Mechanochemical surface functionalisation of superparamagnetic microparticles with in situ formed crystalline metal-complexes: A fast novel core–shell particle formation method. *Chemical Communications* 2015;51:8687–8690.
79. Cheng F, Li W-C, Zhu J-N, Zhang W-P, Lu A-H. Designed synthesis of nitrogen-rich carbon wrapped Sn nanoparticles hybrid anode via in-situ growth of crystalline ZIF-8 on a binary metal oxide. *Nano Energy* 2016;19:486–494.

17 Co-Based Materials for Supercapacitors

N. Suresh Kumar, R. Padma Suvarna, S. Ramesh, D. Baba Basha, and K. Chandra Babu Naidu

CONTENTS

17.1 Introduction 325
17.2 History of Supercapacitors 326
17.3 Types of Supercapacitors 326
17.4 Cobalt Based Supercapacitors 328
References 335

17.1 INTRODUCTION

Numerous techniques have seen the light to satisfy the demands of global energy with less environmental damage. Improvement of technology for the storage and production of electrical energy is one of the best strategies to meet this [1–5]. Specifically, this requires better technology for supercapacitors and advanced batteries for storage of electrical energy. Because of processability, abundance and electrochemical potential, carbon was in use for a long time for such technology [6]. In addition, the existence of graphene and its immense properties have triggered common interest in the material as well as in electrochemical energy systems [7,8]. A high-capacity capacitor with capacitance larger than other capacitors such as electrolyte capacitors and rechargeable batteries etc., is called as supercapacitor [9–11]. These supercapacitors are also called Gold cap or ultracapacitors [12]. Due to high energy densities and longer cyclic life spans, these ultracapacitors are used in various devices such as elevators, cars, buses, cranes etc. for rapid charge or discharge cycles (short-term energy storage) instead of long-term energy storage (batteries and electrolytic and electrostatic capacitors) [13–15]. In supercapacitors the energy storage mainly depends on the contact surface of the electrode materials with the electrolyte. Although these supercapacitors have good properties, still it is a long way to be exploited in a wide range [16]. One of the important factors of the ultracapacitors is the electrode materials that are related to power density and energy density. Figure 17.1 shows the power density vs. energy density curve of various charge storage devices. Till now huge contributions have been revealed to increase the performance of supercapacitors [17] like increasing the surface area of the materials and searching for new electrode materials like MnO_2, graphene, polyaniline, carbon nanotubes etc. [18]. Predominantly, some battery type materials have been proposed and served as ultracapacitor materials. Generally,

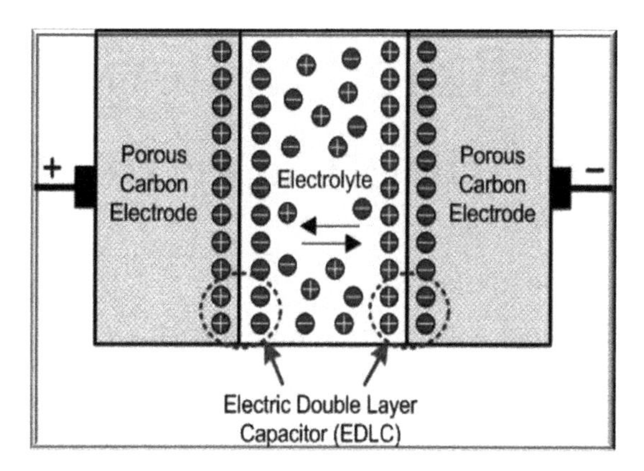

FIGURE 17.1 Power density vs. energy density curve.

materials with a low price, more natural resources and environment friendly have received substantial research interest for supercapacitor electrode materials such as cobalt oxide, cobalt hexoxide etc. [19,20]. This chapter concentrates on cobalt-based supercapacitors and the challenges in the commercialization of such materials.

17.2 HISTORY OF SUPERCAPACITORS

The low-voltage electrolytic capacitor was the first capacitor developed by Becker in 1957 using carbon electrodes. The pores in the carbon stored energy in the form of charges [21,22]. The next version of this component "electrical energy storage apparatus" was first developed at SOHIO (Standard Oil of Ohio) in 1966 [23]. This was marketed as supercapacitors in 1971 by NEC for providing backup power for computer memories [24]. In 1970 Donald L. Boos patented an electrochemical capacitor using activated carbon as electrodes. The term supercapacitor was coined by Brian Evans Conway in 1999 with an intension of explaining increased (super) capacitance. According to Conway, surface redox reaction associated with faradaic charge transfer between ions and electrodes leads to an increased electrical charge storage. Supercapacitor became a successful energy source for memory backup applications. 1980s witnessed supercapacitors with further increased capacitance due to advancement in electrode material. In 1982 PRI developed a supercapacitor with low internal resistance for military applications and marketed as "PRI Ultracapacitors." In 1992 Maxwell Laboratories called ultracapacitors as Boost capacitors. In 1994 David A Evans developed an Electrolytic-Hybrid Electrochemical capacitor, which combines the features of electrolytic and electrochemical capacitors [25].

17.3 TYPES OF SUPERCAPACITORS

Supercapacitors exhibits economic power and energy densities and stores more energy than the conventional capacitors. Based on their charge storage mechanism, supercapacitors are categorized as electrochemical double-layer capacitors (EDLCs), pseudocapacitors (PC) and hybrid supercapacitors (HSC). Figure 17.2 represents the

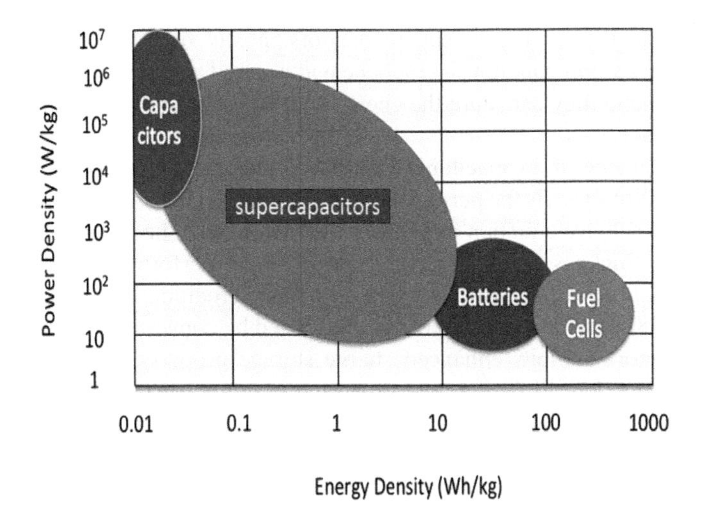

FIGURE 17.2 Classification of supercapacitors.

classification of supercapacitors. In 1957, H.I. Becker successfully demonstrated the operation of EDLC with porous carbon electrode with aqueous electrolyte as stated by electrical double-layer theory of Helmholtz model. Figure 17.3 denotes the simple EDLC with porous carbon electrode. These capacitors are also called as non-faradaic supercapacitors. From Figure 17.3 it is clear that the EDLCs are similar to parallel plate capacitors [26,27] in which the capacitance is originated from the charge which is accumulated at the interface of electrode and electrolyte.

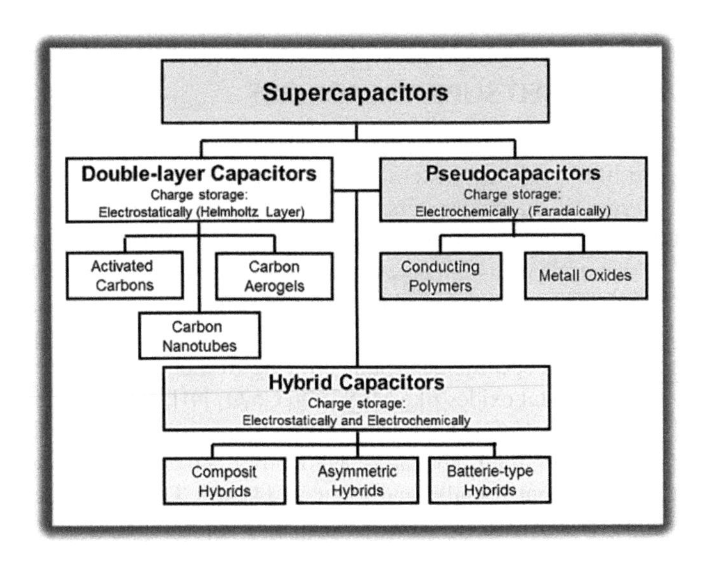

FIGURE 17.3 Electrochemical double-layer (EDLC) supercapacitor.

In addition, in these types of supercapacitors mostly high surface carbon-based materials like carbon nanotubes, graphene and ordered porous carbon etc., are used as electrodes and they can store the energy by fast ion adsorbing at electrode-electrolyte interface.

The capacitance of a capacitor is can be written as $C = \varepsilon_0\varepsilon_r/dA$, where ε_r is the relative permittivity, ε_o is the permittivity of free space, d is the thickness and A is the area. From the above equation it is obvious that in EDLC the capacitance is greatly influenced by thickness and surface area. However, the performance of these carbon-based materials can be improved by doping with heteroatoms like oxygen, nitrogen, sulfur, etc. [28]. Pseudocapacitance was introduced by convey et al. in the year 1975. Pseudocapacitors exhibits enhanced charge storage property than EDLC because these capacitors depend on the reversible faradic redox (reduction-oxidation) reactions which happen near or at the surface of the electrode materials. Furthermore, the presence of chemical reaction helps to improve the capacitance of the pseudocapacitors. Comparing to cyclic life and charge storage capacity, pseudocapacitors falls intermediate between battery and supercapacitor. Three distinct mechanisms which contribute towards the pseudocapacitance are underpotential deposition, redox, and intercalation [29].

In addition to these (EDLCs and pseudocapacitors) there is other type of supercapacitors which exhibits high charge capacity like batteries and high-power density like supercapacitors. These supercapacitors are named as hybrid supercapacitors for example Li-ion capacitors [30]. Actually Li-ion capacitors are combination of two technologies (electrostatic and electrochemical) i.e. they use both faradaic and non-faradaic mechanism in energy storage. So, hybrid capacitors have one electrostatic electrode and one electrochemical electrode, because of this they show high energy density than supercapacitor and more cyclic life than batteries. Due to all these reasons hybrid supercapacitors are contemplated as promising aspirants for next generation energy storage devices.

17.4 COBALT BASED SUPERCAPACITORS

One of the prominent energy storage systems with high energy density and having long cyclic life is supercapacitor. Various types of supercapacitors are already mentioned in types of supercapacitors section. Nevertheless, for acquiring supercapacitors with high energy and power densities researchers concentrated on searching new electrode materials with high potential and capacitance range. In addition to this they also focused on synthesizing new transition metal oxides-like MnO_2, Co_2O_3, etc. and enhancing the surface area of carbon-based materials [31–37]. Compared to all, metal oxides have high specific capacitance owing to their electrochemical behavior [38,39]. Various metal oxides like MnS [40], Co_3O_4 [41], $NiCo_2S_4$ [42], ZnO [43], MnO_2 [44], MoS_2 [45], $NiCo_2O_4$ [46] have been extensively used as electrodes in supercapacitors due to their abundant availability, faradaic property, excellent capacitance, environment friendly and low cost [47,48]. This chapter concentrates on Co-based materials as electrodes for supercapacitors applications. Owing to admirable reversible capacity arising from redox reaction and high specific capacitance, rutheniumoxide (RuO_2) is considered as most prominent candidate metal oxide (M-O) electrode material for supercapacitor [49,50]. Nevertheless, the usage of ruthenium is limited for commercial application owing to its toxic nature, high cost and low porosity. Among other metal oxides, cobalt-based materials such as Co_3O_4 have engrossed considerable attentions due to their low cost, environmental friendliness, good electrochemical performance, controllable size and shape, tunable structural properties, etc. Many reports are available on the synthesis of cobalt oxide structures like nano tubes, nano wire (one dimension) [51–54], nano sheets, nano flakes (two dimension) [55–57], nano flowers and nano bowls (three dimension) [58] on different substrates. Following Table 17.1 exhibits some of the cobalt based materials for supercapacitor applications.

Tao et al. [59] examined the properties of a new type of cobalt-based electrode material for supercapacitor applications. They synthesized the cobalt sulphide (CoS_x) by chemical precipitation technique. The investigation on electrochemical properties revealed that CoS_x material have specific capacitance 474 Fg^{-1} at 5 mA current density, also the material shows 369 Fg^{-1} at 50 mA current density. Kandalkar et al. prepared the Cobalt–nickel thin films via chemical bath deposition route as the electrode materials for supercapacitors. Studies on structure and morphology of these films using XRD and SEM revealed irregular-shaped nano platelets as shown in Figure 17.2 of the published work [60]. These thin films show maximum specific capacitance of the order of 324 Fg^{-1}. Continuing this, Wei et al. [61] synthesized enhanced superconductor electrode material by sol gel route. They prepared nickel cobaltite aerogels and their studies revealed that the specific capacitance value enormously increased in the prepared aerogels compared to others like RuO_2. Reported highest value of 1580 Fg^{-1} and they retain 91% capacitance after 2000 cycles. Long cyclic stability, enhanced capacity and high specific surface area of the materials make these aerogels as prominent candidates for the next generation energy storage systems.

Chen et al. [62], used hydrothermal synthesis route to prepare shape controlled Co-based nano cubes, discs and flowers and compared the results with lithium storage properties. Figure 5 of the reported work [62] exhibits the XRD, SEM and TEM images which confirm the cubes, discs and flower pattern of Co_3O_4 at 300°C. The obtained results are compared with the lithium storages and existing graphite materials and concluded that in all cases (cubes, discs and flower) these materials exhibit high capacity even after 100 cycles. They published that the prepared materials exhibited high specific capacity (1350 Fg^{-1}). The sample with high specific area exhibited low capacitance value 700 Fg^{-1}. Wang et al. [63], prepared nickel cobaltite nano wires by hydrothermal and thermal decomposition Process and reported that nickel cobaltite nanowires are most promising candidates for the electrochemical supercapacitors with high capacity (946 Fg^{-1}) and admirable cyclic stability (retains 81% after 3000 cycles). Also exhibited that the co-existence of cobalt with nickel enhanced the capacity and optimize the power density (5000 Wkg^{-1}) as well as energy density (19.72 Wh k^{-1}).

Wang et al. [64] prepared nickel cobalt oxide ($Ni_{0.37}Co_{0.63}(OH)_2$) by chemical precipitation method to serve as supercapacitor electrode material. They described that the prepared materials exhibit non-crystalline structure which affords large specific capacitance of 1840 Fg^{-1}. Figure 2 in the reference [64] displays the SEM images of

TABLE 17.1
Important Attributes of Co-Based Materials for Supercapacitors

Material	Method of Synthesis	Specific Surface Area (m^2g^{-1})	Electrolyte	Specific Capacitance	Energy Density	Retention	References
$Ni_{0.37}Co_{0.63}(OH)_2$	Chemical precipitation method	No data	1.0 M NaOH+0.5 M Na_2SO_4	1840 Fg^{-1}	No data	90% retention after 500 cycles at 1 mV s^{-1}	[69]
(Co_3O_4)-MnO_2-NiO Ternary 1D Hybrid Nanotubes	Electrochemical deposition	No data	1M KOH	2525 Fg^{-1}	108.8 Wh/kg at 12.2 A/g	Retains 80% after 5700 cycles at 2 mA cm^{-2}	[75]
Fe-Co-Ni ternary oxide (FCNO) nanowires	Sol-gel	No data	1 M KOH	61.58 mF·cm^{-2} at 0.1 mA cm^{-2}	16.74 μWh·cm^{-2} at 69.94 μWh·cm^{-2}	Retains 90.2% after 4000 cycles at 2 mA cm^{-2}	[76]
Cobalt–nickel (Co–Ni) thin films	Chemical bath deposition	No data	2 M KOH	324 Fg^{-1}	No data	No data	[60]
CoS_x powders	Chemical precipitation method	No data	6 M KOH	474 Fg^{-1} at 5 mA current density	No data	No data	[59]
$NiCo_2O_4$ (Nickel cobaltite) aerogels	Sol-gel	123	Propylene oxide	1580 Fg^{-1}	No data	91% retention after 2000 cycles	[61]
Co_3O_4 nanocubes, discs and flowers	Hydrothermal	91.4	1.0 M NaOH	1350 Fg^{-1}	No data	More than 75% retention after 100 cycles	[62]

(Continued)

TABLE 17.1 (*Continued*)
Important Attributes of Co-Based Materials for Supercapacitors

Material	Method of Synthesis	Specific Surface Area (m^2g^{-1})	Electrolyte	Specific Capacitance	Energy Density	Retention	References
Nickel cobaltite nanowires	Hydrothermal and thermal-decomposition process	No data	1 M KOH	946 Fg^{-1}	19.72 Wh kg^{-1} at 5000 W kg^{-1}	81% retention after 3000 cycles	[63]
Co_3O_4 nanowires with brush like morphology	Substrate-assisted solvothermal route	75.94	1 M KOH	911 Fg^{-1}	69.7 Wh kg^{-1} at 81 Wh kg^{-1}	Retains 94% after 5000 cycles	[65]
$Ni_{1/3}Co_{1/3}Mn_{1/3}(OH)_2$ composite	Chemical precipitation technique	268.5	1 M NAOH	1840 Fg^{-1} at 1mV s^{-1}	No data	85% retention after 50 cycles	[66]
Ni-Co LDH hybrid film	Hydrothermal co-deposition method	No data	1 M HCL	2682 Fg^{-1} at 3 Ag^{-1}	77.3 Wh kg^{-1} at 623 W kg^{-1}	82% retention after 5000 cycles	[67]
$La_{0.7}Sr_{0.3}Co_{0.1}Mn_{0.9}O_{3-\delta}$ nanofiber	Electrostatic spinning	27.2	1 M KOH	569.1 Fg^{-1} at 2 mV s^{-1} 485 Fg^{-1} at 1 Ag^{-1}	No data	No change after 2000 cycles at 50 mV s^{-1}	[69]
$NiCo_2S_4$ hollow spheres	Solvothermal technique	53.9	6 M KOH	1036 Fg^{-1} at 1 Ag^{-1}	42.3 Wh kg^{-1} at 476 W kg^{-1}	87% retention after 2000 cycles	[70]
$Ba_{0.5}Sr_{0.5}Co_{0.8}Fe_{0.2}O_{3-\delta}$ (BSCF)	EDTA-citrate complexing (EDTA-CA) technique	63.05	1 M KOH	610 Fg^{-1}	No data	92% retention after 3000 cycles	[71]

(*Continued*)

TABLE 17.1 *(Continued)*
Important Attributes of Co-Based Materials for Supercapacitors

Material	Method of Synthesis	Specific Surface Area (m^2g^{-1})	Electrolyte	Specific Capacitance	Energy Density	Retention	References
$SrCoO_{3-\delta}$ (SC)	EDTA-citrate complexing (EDTA-CA) technique	24.13	1 M KOH	572 Fg^{-1}	27.5 Wh kg^{-1} at 750 W kg^{-1}	Retains 88% after 5000 cycles	[71]
$La_{0.7}Sr_{0.3}CoO_{3-\delta}$/Ag composite	Solid-state reaction	No data	1 M KOH	517.46 Fg^{-1} at 1 mA cm^{-2}	21.88 mWh cm^{-3} at 90.11 $mWcm^{-3}$	85.6% retention after 3000 cycles at 50 mA cm^{-2}	[72]
$SrCo_{0.9}Mo_{0.1}O_{2.5092}$ powder	Sol–gel technique	37.39	6 M KOH	168.90 mA h g^{-1} at 1 Ag^{-1}	74.79 Wh kg^{-1} at 734.48 W kg^{-1}	97.6% retention after 10,000 cycles at 10 Ag^{-1}	[78]
$La_{0.85}Sr_{0.15}MnO_3$/ $NiCo_2O_4$ core-shell architecture	Hydrothermal method	No data	6 M KOH	1341.02 Fg^{-1} at 0.5 Ag^{-1}	63.47 Wh kg^{-1} at 900 W kg^{-1}	200% retention after 10,000 cycles at 20 Ag^{-1}	[79]
Co_xO_y/CNFs embedded with Ag	Electrospinning route	No data	4 M KOH	698 Fg^{-1} at 1 Ag^{-1}	No data	81.1% retention after 5000 cycles at 5 Ag^{-1}	[73]
CoS/rGO nano particles	Hydrothermal process	No data	No data	813 Fg^{-1} at 0.5 Ag^{-1}	No data	91.2% retention after 1000 cycles	[74]
Co_9S_8@$NiCo_2O_4$ nano brushes	Multistep route	No data	3 M KOH	1966 Fg^{-1} at 1 Ag^{-1}	86 Wh kg^{-1} at 792 W kg^{-1}	Retains 91.2% after 1000 cycles	[77]

the prepared materials and it confirms the amorphous nature of the $Ni_{0.37}Co_{0.63}(OH)_2$ materials. Figure 4 [64] represents the potential vs. current density at various potentials. From these curves it is clear that the variation of current density is linear for potential scan rate less than 10 mV s^{-1} and it is nonlinear beyond 10 mV s^{-1}. Also it is reported that the specific capacitance value reduces from 1840 to 890 Fg^{-1}, when the potential rises from 1 to 50 mV s^{-1}, which means that the synthesized material is useful for supercapacitor applications at low potential only.

Rakhi et al. synthesized the self-organized nano structured cobalt oxide electrodes for supercapacitor applications via substrate assisted solvothermal route. Their investigations revealed that cobalt oxide nanowires with brush like morphology exhibits the specific capacitance of 911 Fg^{-1}. Compared to nanowires, brush like and flower like structures shows long cyclic stability i.e. they retain almost 94% of capacitance after 5000 cycles. Furthermore, the energy and power densities for brush- and flower-like morphologies show 69.7 and 37 Wh/kg and 81 and 55.1 Wh/kg respectively. From all these results it is understood that nano wires with brush-like morphology is served as most prominent material for supercapacitors. Scanning electron microscope pictures of Co_3O_4 nanowires with distinct morphology were shown in Figure 17.1 of the published work [65]. Wang et al. [66] synthesized Ni1/3Co1/3Mn1/3(OH)2 composite material through chemical precipitation technique for supercapacitor electrode. Amorphous structure and high surface area of the prepared samples were confirmed by XRD and SEM. The specific capacitance of the prepared material is found to be 1840 and 1750 Fg^{-1} at 1.0 and 5.0 mV s^{-1}. They reported that the electrode materials show 85% of its initial capacity after 50 cycles and due to this these materials may not be suitable for real time applications.

Chen et al. fabricated nickel–cobalt layered double hydroxide (Ni-Co LDH) nanosheets for electrode materials for high-performance supercapacitors via one step hydrothermal co deposition route. They confirmed the structure and morphology with the help of XRD, SEM and TEM analysis. Figure 1 in the published paper [67] represents the typical SEM (1a, b & d) and TEM (1c) images of Ni-Co nanosheets. Figure 6 in the same work [67] shows the electrochemical studies of the prepared materials. Figure 6a exhibit the energy density vs power density curves of the nickel hydroxide, Ni-Co LDH hybrid-, RGO- electrodes, cobalt hydroxide- and the Ni-Co LDH//RGO asymmetric supercapacitor, 6b shows CV curves of the freeze-dried RGO electrode at distinct scan rates, 6c charge discharge curves of the same and 6d represents the specific capacity of reduced graphene oxide electrode at different current densities. They reported that the Ni-Co LDH materials achieved high specific capacity values as 2682 Fg^{-1} at 3.0 Ag^{-1}. Further, current density is 77.29 Wh kg^{-1} at 624 W kg^{-1} power density, and this value enhances to 187.9 Wh kg^{-1} at 1498 W kg^{-1} based on the active materials. Also it is mentioned that there is 82% of initial capacity after 5000 cycles. All these reports specify that the prepared materials are prominent aspirants for high performance supercapacitor electrodes. Similarly, Jagadale et al. [68] synthesized $Co(OH)_2$ thin films by using potentiodynamic electrodeposition technique, and they obtained a specific capacitance of 44 Fg^{-1} at the scan rate of 5 mV s^{-1} and also found that energy density of 3.96 Wh kg^{-1} and power density 42 W kg^{-1}.

Cao et al. [69] prepared $La_{0.7}Sr_{0.3}Co_{0.1}Mn_{0.9}O_{3-\delta}$ composite via electrospinning technique. The structural analysis (using XRD) and morphology (using TEM and SEM) of the samples reveals the formation of nanofibers in the prepared samples. Electrochemical analyzer used to observe the electrochemical properties revealed that LNF-0.7 sample exhibits maximum capacitance of 485 Fg^{-1} at a current density of 1 Ag^{-1}. Also it is observed that there is no variation of C_s after 2000 cycles. So, this composition shows long cyclic stability with good specific capacitance which may be appropriate for electrode material in supercapacitors. Shen et al. synthesized novel $NiCo_2S_4$ hollow spheres via solvothermal route and the obtained hallow spheres are represented in Figure 2 of reference [70]. These are the new type of cobalt-based materials for supercapacitor application. Electrochemical properties of the prepared samples are good when compared to samples reported in the literature and these spheres shows long charge discharge stability and only 13% of capacity will be discharged after 2000 charge discharge cycles. Highest capacitance obtained is 1036 Fg^{-1} at 1 Ag^{-1}. Liu et al. [71] prepared the different perovskite materials such as $Ba_{0.5}Sr_{0.5}Co_{0.8}Fe_{0.2}O_{3-\delta}$ (BSCF), Co_3O_4 and $SrCoO_{3-\delta}$ (SC) and compared their electrochemical properties. Figure 4 in the published work [71] shows the comparative studies of electrochemical performance of all the materials. From these studies they reported that, among these materials SC shows good electrochemical properties and is a much favorable candidate for supercapacitor electrode material in practical application. This material shows high energy density (27.51 Wh kg^{-1} at power density of 750 W kg^{-1} and long cyclic stability (retains 92% after 3500 cycles) compare to BSCF and Co_3O_4. In the similar manner Liu et al. [72] proposed that $La_{0.7}Sr_{0.3}CoO_{3-\delta}$ perovskite material also shows good electrochemical property. Xingwei et al. [73] synthesized Ag decorated amorphous cobalt carbon nanofibers via electrospinning route for high performance electrochemical storage devices. They reported that, the prepared composite shows improved conductivity due to the presence of Ag also exhibits long cyclic capability. Zhu et al. synthesized CoS/rGO for same application and they achieved 813 F^{-1} g at 0.5 Ag^{-1}. It is almost double the capacitance value of CoS. Figure 6 in the published work [74] show the comparison of CV curves of pure CoS and CoS/rGO. So, the prepared material shows better performance and it is one of the prominent materials in advancement of supercapacitor electrodes. Singh et al. [75] prepared cobalt oxide-manganese dioxide -nickel oxide (Co_3O_4)-MnO_2-NiO ternary 1D hybrid nanotubes for applications in high performance supercapacitors as electrodes. They found that the prepared nanotubes show the high specific capacitance value of the order of 2525 Fg^{-1} with long cyclic stability nearly 80% after 5700 cycles. The current vs. potential curves of (Co_3O_4)-MnO_2-NiO ternary nanotubes shown in Figure 3 [75], revealed that, the unique structure and combination of three extremely active redox materials established that the prepared materials are highly conducting and its fast charge transport aids to attain improved electrochemical properties. Due to this the prepared nanotubes are suitable for electrodes in supercapacitors. Likewise, Zhao et al. [76], synthesized Hierarchical ferric-cobalt-nickel ternary oxide Fe-Co-Ni ternary oxide (FCNO) nano wire arrays via sol-gel method for supercapacitor applications. They confirmed the structure and morphology by XRD and SEM, the investigations on electrochemical properties revealed that the nano wires exhibits high specific capacitance (61.58 mF·cm^{-2} at 0.1 mA cm^{-2}) and energy density (16.74 μWh·cm^{-2} at 69.94 μWh·cm^{-2} power density). Also reported that the prepared nano wires have extremely long cyclic stability, which retains almost 87.5% of initial capacitance after 8000 cycles and possess admirable capacitance retaining of 90.9% after 4000 cycles. In the advancement of electrodes of supercapacitors, Liu et al. [77] prepared Co_9S_8@$NiCo_2O_4$ hierarchically structured nano brushes for supercapacitor electrodes and reported that the prepared material shows remarkable specific capacitance when compared to individual materials which is around 1966 Fg^{-1} at 1 Ag^{-1}. In addition, this composite material shows excellent energy density of 86 Wh kg^{-1} at 792 Wkg^{-1} and long cyclic stability of 91.2% retention after 10,000 cycles. Research is going on to develop good electrode materials for high performance energy storage devices. In this process cobalt based materials have attracted much attention owing to its availability and performance.

REFERENCES

1. Conway B. E. 1999. *Electrochemical Supercapacitors: Scientific Fundamentals and Technological Applications*. Kluwer Academic: New York.
2. Wang G. P., Zhang L., Zhang J. J. 2012. A review of electrode materials for electrochemical supercapacitors, *Chem. Soc. Rev.* 41: 797–828.
3. Miller J. R., Burke A. F. 2008. Electrochemical capacitors: Challenges and opportunities for real-world applications, *ECS Interface* 17: 53.
4. Simon P., Gogotsi Y. 2008. Materials for electrochemical capacitors, *Nat. Mater.* 7: 845–854.
5. Wang G. P., Zhang L., Zhang J. J. 2012. A review of electrode materials for electrochemical supercapacitors, *Chem. Soc. Rev.* 41: 797–828.
6. Zhang L.; Zhao X. S. 2009. Carbon-based materials as supercapacitor electrodes, *Chem. Soc. Rev.* 38: 2520–2531.
7. Zhu Y., Murali S., Stoller M. D., et al. 2011. Carbon-based supercapacitors produced by activation of graphene, *Science* 332: 1537–1541.
8. Yang X. W., Cheng C., Wang Y. F., Qiu L., Li D. 2013. Liquid-mediated dense integration of graphene materials for compact capacitive energy storage, *Science* 341: 534–537.
9. Mai L. Q., Yang F., Zhao Y. L., Xu X., Xu L., Luo Y. Z. 2011. Hierarchical $MnMoO_4$/$CoMoO_4$ heterostructured nanowires with enhanced supercapacitor performance, *Nat. Commun.* 2: 381.
10. Winter M., Brodd R. J. 2004. What are batteries, fuel cells, and supercapacitors? *Chem. Rev.* 104: 4245.
11. Burke A. 2000. Ultracapacitors: Why, how, and where is the technology, *J. Power Sources* 91: 37.
12. Panasonic, Electric Double Layer Capacitor, www.thomasnet.com/articles/automation-electronics/super-capacitors.
13. Wu X., Han Z., Zheng X., et al. 2017. Core-shell structured Co_3O_4@ $NiCo_2O_4$ electrodes grown on flexible carbon fibers with superior electrochemical properties, *Nano Energy* 31: 410–417.
14. Lv W., Xue R., Chen S., Jiang M. 2018. Temperature stability of symmetric activated carbon supercapacitors assembled with in situ electrodeposited poly (vinyl alcohol) potassium borate hydrogel electrolyte, *Chin. Chem. Lett.* 29: 637–640.
15. Jiang W., Hu F., Yan Q., Xiang Wu. 2017. Investigation on electrochemical behaviors of $NiCo_2O_4$ battery-type supercapacitor electrodes: The role of an aqueous electrolyte, *Inorg. Chem. Front.* 4: 1642–1648.

16. El-Kady M.F., Strong V., Dubin S., Kaner R.B. 2012. Laser scribing of high-performance and flexible graphene-based electrochemical capacitors, *Science* 335: 1326–1330.
17. Shao Y., El-Kady M.F., Wang L.J., et al. 2015. Tailoring pores in graphene-based materials: From generation to applications, *Chem. Soc. Rev.* 44: 3639–3665.
18. Liu J., Jiang J., Cheng C., et al. 2011. Co_3O_4 nanowire@ MnO_2 ultrathin nanosheet core/shell arrays: A new class of high-performance pseudocapacitive materials, *Adv. Mater.* 23: 2076–2081.
19. Liu L.Y., Zhang X., Li H.X., et al. 2017. Synthesis of Co–Ni oxide microflowers as a superior anode for hybrid supercapacitors with ultra long cycle life, *Chin. Chem. Lett.* 28: 206–212.
20. Yan J., Fan Z., Sun W., et al. 2012. Advanced asymmetric supercapacitors based on Ni $(OH)_2$/graphene and porous graphene electrodes with high energy density, *Adv. Funct. Mater.* 22: 2632–2641.
21. Becker H. I. 1957. Low voltage electrolytic capacitor, *US Patent*, 2800616.
22. Ho J., Jow R., Boggs S. 2010. Historical introduction to capacitor technology, *IEEE Electrical Insulation Magazine.* 26: 20–25.
23. Pandolfo T., Ruiz V., Sivakkumar S., Nerkar J. 2013. General properties of electrochemical capacitors, in: F. Beguin, E. Frackowiak (Eds.), *Supercapacitors: Materials, Systems, and Applications*, Wiley-VCH, Weinheim, 69–109.
24. Schindall J. G. 2007. The Change of the Ultra-Capacitors, *IEEE Spectrum.*
25. Evans D. A. 2007. The Littlest Big Capacitor – An Evans Hybrid Technical Paper, Proceedings of Fifth International seminar on double layer capacitors and similar energy storage devices, 12/95, Deer field beach, Florida.
26. Wang G., Zhang L., Zhang J. 2012. A review of electrode materials for electrochemical supercapacitors, *Chem. Soc. Rev.* 41: 797–828.
27. Zhang L. L., Zhao X. S. 2009. Carbon-based materials as supercapacitor electrodes, *Chem. Soc. Rev.* 38: 2520–2531.
28. Chen X., Paul R., Dai L. 2017. Carbon-based supercapacitors for efficient energy storage, *Natl. Sci. Rev.* 4: 453–489.
29. Conway B.E. 2013. *Electrochemical Supercapacitors: Scientific Fundamentals and Technological Applications*, Springer Science & Business Media, New York.
30. Brousse T., B´elanger D., Long J. W. 2015. To be or not to be pseudocapacitive? *J. Electrochem. Soc.* 162: A5185–A5189.
31. Béguin F., Presser V., Balducci A., Frackowiak E. 2014. Carbons and electrolytes for advanced supercapacitors, *Adv. Mater.* 26: 2219–2251.
32. Zhang G. Q., Lou X. W. 2013. Controlled growth of $NiCo_2O_4$ nanorods and ultrathin nanosheets on carbon nanofibers for high-performance supercapacitors, *Sci. Rep.* 3: 1470–1476.
33. Lei Z., Zhang J., Zhao X. S. 2012. Ultrathin MnO_2 Nanofibers grown on graphitic carbon spheres as high-performance asymmetric supercapacitor electrodes, *J. Mater. Chem.* 22: 153–160.
34. Wang B., Park J., Su D.W., Wang C. Y., Ahn H., Wang G. X. 2012. Solvothermal synthesis of CoS_2–Graphene nanocomposite material for high-performance supercapacitors, *J. Mater. Chem.* 22: 15750–15756.
35. Miller J. R., Outlaw R. A., Holloway B. C. 2010. Graphene double-layer capacitor with ac line-filtering performance, *Science* 329: 1637–1639.
36. Jaidev, Ramaprabhu, S. 2012. Poly(p-phenylenediamine)/Graphene nanocomposites for supercapacitor application, *J. Mater. Chem.* 22: 18775–18783.
37. DeBlase, C. R. Silberstein, K. E. Truong, T-T. Abruña, H. D. Dichtel, W. R. 2013. ß-Ketoenamine-linked covalent organic frameworks capable of pseudocapacitive energy storage, *J. Am. Chem. Soc.* 135: 16821–16824.
38. Zhu J., Tang S., Wu J., Shi X., Zhu B., Meng X. 2017. Wearable high-performance supercapacitors based on silver-sputtered textiles with $FeCo_2S_4$–$NiCo_2S_4$ composite nanotube-built multitripod architectures as advanced flexible electrodes, *Adv. Energy Mater.* 7: 1601234.

39. Kulkarni P., Nataraj S. K., Balakrishna R. G., Nagaraju D. H., Reddy M. V. 2017. Nanostructured binary and ternary metal sulfides: Synthesis methods and their application in energy conversion and storage devices, *J. Mater. Chem. A* 5: 22040–22094.
40. Li X., Shen J., Li N., Ye M. 2015. Fabrication of γ-MnS/rGO composite by facile one-pot solvothermal approach for supercapacitor applications, *J. Power Sources* 282: 194–201.
41. Xiong D., Li X., Bai Z., Li J., Shan H., Fan L., Long C., Li D., Lu X. 2018. Rational design of hybrid Co_3O_4/graphene films: Free-standing flexible electrodes for high performance supercapacitors, *Electrochim. Acta* 259: 338–347.
42. Hou L., Shi Y., Zhu S., Rehan M., Pang G., Zhang X., Yuan C. 2017. Hollow mesoporous hetero-$NiCoS_4$/Co_9S_8 submicro-spindles: Unusual formation and excellent pseudocapacitance towards hybrid supercapacitors, *J. Mater. Chem. A* 5: 133–144.
43. He Y.B., Li G.R., Wang Z.L., Su C.Y., Tong Y.X. 2011. Single-crystal ZnO nanorod/amorphous and nanoporous metal oxide shell composites: Controllable electrochemical synthesis and enhanced supercapacitor performances, *Energy Environ. Sci.* 4: 1288.
44. Fan L.Q., Liu G.J., Zhang C.Y., Wu J.H., Wei Y.L. 2015. Facile one-step hydrothermal preparation of molybdenum disulfide/carbon composite for use in supercapacitor, *Int. J. Hydro. Energy* 40: 10150–10157.
45. Li X., Shen J., Li N., Ye M. 2015. Fabrication of γ-MnS/rGO composite by facile one-pot solvothermal approach for supercapacitor applications, *J. Power Sources* 282: 194–201.
46. Yuan C., Li J., Hou L., Zhang X., Shen L., Lou X. W. D. 2012. Ultrathin mesoporous $NiCo_2O_4$ nanosheets supported on Ni Foam as advanced electrodes for supercapacitors, *Adv. Funct. Mater.* 22: 4592–4597.
47. Yu X. Y., Yu L., Lou X. W. D. 2016. Carbon coated porous nickel phosphides nanoplates for highly efficient oxygen evolution reaction, *Adv. Energy Mater.* 6: 1502217.
48. Wu Z., Zhu Y., Ji X. 2014. $NiCo_2O_4$-based materials for electrochemical supercapacitors, *J. Mater. Chem. A* 2: 14759–14772.
49. Wu Z. S., Wang D. W., Ren W., Zhao J., Zhou G., Li F., Cheng H. M. 2010. Anchoring hydrous RuO_2 on graphene sheets for high-performance electrochemical capacitors, *Adv. Funct. Mater.* 20: 3595–3602.
50. Bi R. R., Wu X. L., Cao F. F., Jiang L. Y., Guo Y. G., Wan L. J. J. 2010. Highly dispersed RuO_2 nanoparticles on carbon nanotubes: Facile synthesis and enhanced super capacitance performance, *Phys. Chem. C* 114: 2448–2451.
51. Gao Y. Y., Chen S. L., Cao D. X., Wang G. L., Yin J. L. 2010. Electrochemical capacitance of Co_3O_4 nanowire arrays supported on nickel foam, *J. Power Sources* 195: 1757–1760.
52. Xia X. H., Tu, J. P., Zhang Y. Q., Mai Y. J., Wang X. L., Gu C. D., Zhao X. B. 2012. Metal oxide/hydroxide-based materials for supercapacitors, *RSC Adv.* 2: 1835–1841.
53. Xu J. A., Gao L., Cao J. Y., Wang W. C., Chen Z. D. 2010. Preparation and electrochemical capacitance of cobalt oxide (Co_3O_4) nanotubes as supercapacitor material, *Electrochim. Acta* 56: 732–736.
54. Xia X. H., Tu J. P., Mai Y. J., Wang X. L., Gu C. D., Zhao X. B. J. 2011. Self-supported hydrothermal synthesized hollow Co_3O_4 nanowire arrays with high supercapacitor capacitance, *Mater. Chem.* 21: 9319–9325.
55. Hosono E., Fujihara S., Honma I., Zhou H. S. J. 2005. Fabrication of morphology and crystal structure controlled nanorod and nanosheet cobalt hydroxide based on the difference of oxygen-solubility between water and methanol, and conversion into Co_3O_4, *Mater. Chem.* 15: 1938–1945.
56. Yuan Y. F., Xia X. H., Wu J. B., Huang X. H., Pei Y. B., Yang J. L., Guo, S. Y. 2011. Hierarchically porous Co_3O_4 film with mesoporous walls prepared via liquid crystalline template for supercapacitor application, *Electrochem. Commun.* 13: 1123–1126.

57. Meher S. K., Rao G. R. 2011. Ultralayered Co_3O_4 for high-performance supercapacitor applications, *J. Phys. Chem. C*, 115: 15646–15654.
58. Deng M. J., Huang F. L., Sun I. W., Tsai W. T., Chang J. K. 2009. An effective surface-enhanced Raman scattering template based on a Ag nanocluster–ZnO nanowire array, *Nanotechnology* 20: 17.
59. Tao F., Zhao Y.Q., Zhang G.Q., Li, H.L. 2007. Electrochemical characterization on cobalt sulfide for electrochemical supercapacitors, *Electrochemistry Communications* 9: 1282–1287.
60. Kandalkar S. G., Lee H. M., Seo S. H., Lee K., Kim C. K., 2011. Cobalt–nickel composite films synthesized by chemical bath deposition method as an electrode material for supercapacitors, *J Mater Sci.* 46: 2977–2981.
61. Wei T., Chen C., Chien H., Lu S., Hu, C. 2010). A cost-effective supercapacitor material of ultrahigh specific capacitances: Spinel nickel cobaltite aerogels from an epoxide-driven Sol–Gel process, *Adv. Mater.* 22: 347–351.
62. Chen J. S., Zhu T., Hu Q. H., Gao J., Su F., Qiao S. Z., Lou X. W. 2010. Shape-controlled synthesis of cobalt-based nanocubes, nanodiscs, and nanoflowers and their comparative lithium-storage properties, *ACS Appl. Mater. Interfaces* 2: 3628–3635
63. Wang H., Gao Q., Jiang, L. 2011. Facile approach to prepare nickel cobaltite nanowire materials for supercapacitors, *Small* 7: 2454–2459.
64. Wang G., Zhang L., Kimz J., Zhang J. 2012. Nickel and cobalt oxide composite as a possible electrode material for electrochemical supercapacitors, *J. Power Sources* 217: 554–561.
65. Rakhi R. B., Chen W., Cha D., Alshareef H. N. 2012. Substrate dependent self-organization of mesoporous cobalt oxide nanowires with remarkable pseudocapacitance, *Nano Lett.* 12: 2559–2567.
66. Wang G., Liu L., Zhang L., Zhang J. 2013. Nickel, cobalt, and manganese oxide composite as an electrode material for electrochemical supercapacitors, *Ionics* 19: 689–695
67. Chen H., Hu L., Chen M., Yan Y., Wu L. 2014. Nickel–cobalt layered double hydroxide nanosheets for high-performance supercapacitor electrode materials. *Adv. Funct. Mater* 24: 934–942.
68. Jagadale A.D., Kumbhar V.S., Dhawale D.S., Lokhande C.D. 2013. Performance evaluation of symmetric supercapacitor based on cobalt hydroxide $[Co(OH)_2]$ Thin film electrodes, *Electrochimica Acta* 98: 32–38.
69. Cao Y., Lin B., Sun Y., Yang H., Zhang X. 2014. Structure, morphology and electrochemical properties of $La_xSr_{1-x}Co_{0.1}Mn_{0.9}O_{3-\delta}$ perovskite nanofibers prepared by electrospinning method, *J. Alloy Compd.* doi: 10.1016/j.jallcom.2014.10.178.
70. Jin R., Liu C., Liu G. 2015. Hierarchical $NiCo_2S_4$ hollow spheres as a high performance anode for lithium ion batteries, *RSC Adv.* 5: 84711–84717.
71. Liu Y., Dinh J., Tade M. O., Shao Z. 2016. Design of perovskite oxides as anion-intercalation-type electrodes for supercapacitors: Cation leaching effect, *ACS Appl. Mater. Interfaces* 8: 23774–23783.
72. Liu P., Liu J., Cheng S., Cai W., Yu F., Zhang Y., Wu P., Liu M. 2017. A. High-performance, electrode for supercapacitors: Silver nanoparticles grown on a porous perovskite type material $La_{0.7}Sr_{0.3}CoO_{3-\delta}$ substrate, *Chem. Eng. J.* 328: 1–10.
73. Sun X., Li C., Bai J. 2019. Amorphous cobalt carbon nanofibers decorated with conductive Ag as free-standing flexible electrode material for high-performance supercapacitors, *J. Electron. Mater.* doi:10.1007/s11664-019-06971-8.
74. Zhu J., Zhou W., Zhou Y., et al. 2019. Cobalt sulfide/reduced graphene oxide nanocomposite with enhanced performance for supercapacitors, *J. Electron. Mater.* 48: 1531.
75. Singh A. K., Sarkar D., Karmakar K., Mandal K., Khan G. G. 2016. High-performance supercapacitor electrode based on cobalt oxide–manganese dioxide–nickel oxide ternary 1D hybrid nanotubes, *ACS App. Mater. Interfaces* 8: 20786–20792.

76. Zhao J., Li C., Zhang Q., et al. 2018. Hierarchical ferric-cobalt-nickel ternary oxide nanowire arrays supported on graphene fibers as high-performance electrodes for flexible asymmetric supercapacitors, *Nano Res.* 11: 1775.
77. Liu Q., Hong X., Zhang X., Wang W., Guo W., Liu X., Ye M. 2018. Hierarchically structured Co_9S_8@$NiCo_2O_4$ nanobrushes for high-performance flexible asymmetric supercapacitors, *Chem. Eng. J.* doi: 10.1016/j.cej.2018.09.095
78. Sharma R.K., Tomar A.K., Singh G. 2018. Fabrication of Mo-doped strontium cobaltite perovskite hybrid supercapacitor cell with high energy density and excellent cycling life, *Chem Sus Chem.* 11: 4123–4130. doi:10.1002/cssc.201801869.
79. Lang X., Zhang H., Xue X., Li C., Sun X., Liu Z., Nan H., Hu X., Tian H. 2018. Rational design of $La_{0.85}Sr_{0.15}MnO_3$@$NiCo_2O_4$ core–shell architecture supported on Ni foam for high performance supercapacitors, *J. Power Sources* 402: 213–220.

Index

Note: Page numbers in italic and bold refer to figures and tables, respectively.

A

acid-base property, *259*
activated carbon (AC), 63, 90
additive and removed template (A & RT) route, 166, *167*
additive and self-removed template (A & SRT) route, 166–167, *168*
air annealed titania nanotubes (A-TNT) arrays, 38–39
alcoholic solvent assisted (ASA), 164–165
α-Fe_2O_3, 44–45
amino acids, 308, *309*
  arginine, 312–314
  aspartic, 315–317
  cysteine, 314–315
  glutamic, 319
  glycine, 317–318
  guanidine, 318–319
  histidine, 311–312
  lysine, 308–311
amorphous phase, 240
annealing process, 8
  and metal oxide, 198
  titania nanotubes arrays, Ar, 36–38
anodized titanium oxide (ATO), 40
areal capacitance, 45
  vs. scan rates, H-TNT, 38, *39*
arginine amino acid, 312–314
ASA (alcoholic solvent assisted), 164–165
aspartic amino acid, 315–317
asymmetric supercapacitors, 258
autoclave, 175, *176*

B

Becker, H.I., 327
biomolecule, 160–161
biomolecules assisted (BA), 160–162
black phosphorous (BP), 142–143
Boos, Donald L., 326
Boost capacitors, 326
bottom-up method, 192–193

C

capacitance, 36, 40–41, 328
capacitor, 134, *135*
  charging, mechanism, 136
carbon aerogel (CA), 95–96
carbon-based electrodes, 124
carbon black, 62
carbon electrode, negative polarization, *281*
carbon fiber cloth (CFC), 96
carbon materials, 126
carbon nanofibers (CNFs), 62
carbon nanotubes (CNTs), 62–63, 124
CC/Fe_2O_3 electrode materials, *264*
charge-discharge process, 40
chemical vapors deposition (CVD), 178; *see also* ultrasonic nebulizer assisted chemical vapors deposition (UNA-CVD)
Co_3O_4 nanosheets, *42*, 42–43
cobalt based supercapacitor, 328–335, **330–332**
cobalt sulphide (CoS_x), 329
Co-based layered MOF (Co-LMOF), 193, *193*
colloidal synthesis (CS), 180
  advantages, 182
  capping agents role, 181
  disadvantages, 183
  nucleation process, 180
  Ostwald repining, 181
  precursor role in, 181
  single molecular precursors, 181
$CoMoO_4$, 45, *46*
conductive polymers (CPs)/conducting polymers, 96, 125–126, 205–206
conventional capacitor, 89
Conway, B.E., 326
copper tungstate ($CuWO_4$) nano-powder, 113
corundum (c)-V_2O_3, 44
CS, *see* colloidal synthesis (CS)
C-titania nanotubes arrays, 36
CuO sea urchin-like microcrystals, 170, *171*
CuS nanostructures, *162*
cyclic voltammetry (CV) curves, 36, 49
  VS_2 nanoplates, 53
  graphene-MoS_2 nanocomposites electrode, 245
  WS_2 quantum dots, *51*, 52, *52*
cysteine amino acid, *314*, 314–315

D

decomposition of single-source precursor (DSSP) route, 157–158
direct current (DC), 134

E

EAAJD (electrostatic aerosol assisted jet deposition), 179
Ebelmen, M., 176
ECs, *see* electrochemical capacitors (ECs)
EDLCs, *see* electrochemical double-layer capacitors (EDLCs); electrochemical double-layer capacitors (EDLCs)
EDLS (electrostatic double-layer supercapacitors), 2, 18
EG (ethylene glycol), 163–164
electrochemical double-layer capacitors (EDLCs), 79, 90–91, 279–282, 327, *327*
  hybrid supercapacitors, 123
  MOFs, supercapacitor/development strategies, 194
  pseudocapacitor, 123
  SCs, *216*
  storage mechanism, 122, *122*
electric double layer interface, 216–217
electric field assisted aerosol (EFAA) CVD, 179
electric voltage, 1
electrochemical capacitors (ECs), 2, 79, 120, 134, 137, 257
  and batteries, 188
  double-layer, *2*
electrochemical double-layer capacitors (EDLCs), 135, 145–146, 258
electrochemical reduction, 35–36
electrochemical SCs, *217*
electrode(s), 80
  configuration, 124
  double-layer capacitors, **81**
  materials, 138
  type and mechanism, **81**
electrolytes, 80
electrolytic capacitors, 90
Electrolytic-Hybrid Electrochemical capacitor, 326
electrostatic aerosol assisted jet deposition (EAAJD), 179
electrostatic double-layer supercapacitors (EDLS), 2, 18
electrostatic spray assisted vapor deposition (ESAVD), 179
energy storage capacity, Ragone plot, 120, *121*
equivalent series resistance (ESR), 80
ethylene glycol (EG), 163–164

F

facile hydrothermal method, 243, 248
faradaic redox process, 310
faradaic redox reactions, 282
faradaic supercapacitors (FS), 3, 18
Fe_2O_3@ACC electrode, *263*
FeOOH, 268
  graphene composite electrode, *260*
  nanoparticles, *261*
FeS_2/GNS anode, 113
flower-like zinc molybdate ($ZnMoO_4$), 25–26, *27*
Frenkel defect pair, 41

G

galvanostatic charge/discharge (GCD) curves, 49, *50*, 53, *54*, *70*
GF (graphene foam), 64, 245
glutamic amino acid, 319
glycine amino acid, 317–318
GNS/Co_3O_4 composite, 112
GO, *see* graphene oxide (GO)
Gold cap, 325
Gouy-Chapman model, 91
graphene, 96, 124, 127
  aerogel, 64, *64*
  T-Nb_2O_5 nanodots, *8*
graphene foam (GF), 64, 245
graphene oxide (GO), 62, 126, *206*, 290, 310, 316
green chemistry, 308
guanidine amino acid, 318–319

H

halide perovskite-supercapacitance performance, 145–146, *146*
hard-templated hydrothermal synthetic routes, 165–166
HDLC (hybrid double-layer supercapacitors), 3, 18
Helmholtz model, 90–91
hexagonal WO_3 (h-WO_3), 92, *94*
histidine amino acid, 311–312
H-TNT arrays, *see* hydrogenated titania nanotubes (H-TNT) arrays
hybrid aerogel, 64, *64*
hybrid capacitors, **81**
hybrid double-layer supercapacitors (HDLC), 3, 18
hybrid/electric vehicles, 83–84
hybrid supercapacitors, 3, 5–7, 123, 136, 264, 283–284, *284*, 328
  battery-type nature, 23, 26
  with graphene, 9
  with T-Nb_2O_5, 8
  tungsten disulfide-carbon, 96
  tungsten disulfide-conducting polymers, 96
  tungsten trioxide-carbon, 94–96
  tungsten trioxide-conducting polymers, 96
hydrogenated MnO_2, 40
hydrogenated titania nanotubes (H-TNT) arrays, 38–39, *39*
hydrogenation approach, 38–40
  microspheres, 41

hydrogen molybdenum bronze (HMB), 64
hydrogen plasma treatment, 39–40
hydrothermal synthesis, 156
classification, 156
exterior reaction environment adjustment, 169–170
hard-templated routes, 165–166
instrumentation for, 175–176
MA route, 170–171
MFA route, 171–172
soft-templated routes, 159–160, *160*
template-assisted, *166*
template-free, 156–159
template-mediated, 159–169
TiO_2 nanostructures, 158, *159*

I

ILA (ionic liquids assisted), 162–163, *163*
indirect-supply reaction source (ISRS) route, 157
industrial mobile equipment, 83
inner Helmholtz plane (IHP), 91
inorganic electrolytes, 82–83, **82–83**
use of, 84
interface-mediated growth, inorganic nanostructures, 173–176
ionic liquid electrolytes, 97
ionic liquids assisted (ILA), 162–163, *163*
iron-based electrodes, electrochemical behavior, 259–261
iron-based materials, 262–271

K

KF@MoS_2/rGO fabrication, *244*

L

$La_{0.7}Sr_{0.3}Co_{0.1}Mn_{0.9}O_{3-\delta}$ composite, 334
L-Arginine capped α-$Ni(OH)_2$ microstructures, 111
lead acid batteries, 120
Leyden jar, 89
Li-ion capacitors, 328
lithium-ion batteries, 120, 257, 318
load/current collector, 80
low temperature hydrothermal method, 241
Lu_2O_3 nanostructures, *165*
lysine amino acid, 308–311

M

magnetic field assisted (MFA) route, 171–172
MAHS (microwave-assisted hydro-thermal synthesis), 108–109, *109*
manganese oxide (MnO_2), 33, 40–41, 111, 113, 138–139
metal chalcogenides, 47
VS_2, 53–54
MoS_2, 47–51
WS_2, 51–53
metal-ions-cysteine complexes, 314
metal molybdates, 26, **26**
metal nitride (MNs), 141–142
metal organic frameworks (MOFs), 140, 189, *279*, 284–286
-based composites, 289–290
CNT, 291, *292*
conducting polymer, 292–293, *293*
coordination polymers, 190
derived metal oxides, 296
energy storage/conversion applications, 191
GO, 290, *291*
porous carbons, 293–296, *294*
pristine, 191–194, 286, 289, *289*
metal oxides, 33, 34, 125, 138–139
Co_3O_4, *42*, 42–44
$CoMoO_4$, 46, *46*
hollow spheres, 169, *169*
MnO_2, 33–34, 40
$NiCo_2O_4$, 46–47
$NiMoO_4$, 46
NiO, 45
TiO_2, 33–34
WO_3, 44
MFA (magnetic field assisted) route, 171–172
microwave(s)
dielectric heating, 104
energy storage devices, 103
generation and synthesis techniques, 103–104
heating, 102–104, 113
instruments, synthesis processes, 104–105, *105*
metals, 105
radiation, 102
microwave assisted (MA) method, 195, 170–171
microwave-assisted hydro-thermal synthesis (MAHS), 108–109, *109*
microwave-assisted inorganic materials
$CuWO_4$ nano-powder, 113
FeS_2/GNS anode, 113
GNS/Co_3O_4 composite, 112
hybrid polymer materials and composites, 113
L-Arginine capped α-$Ni(OH)_2$ microstructures, 111
MnO_2 nanostructures, 111
Ni-Co-Mn oxide nano-flakes, 112
stannous ferrite micro-cubes, 111
3D flower-like $NiMnO_3$ nano-balls, 110, *110*
3D flower-on-sheet nanostructure, NiCo LDHs, 110–111
3D hierarchical MnO_2 microspheres, 113
microwave-assisted solvothermal synthesis (MWSS), 108–109, *109*

microwave-assisted synthesis
  demonstration, 102
  MAHS/MWSS, 108–109, *109*
  methods and materials, 106, *106*
  single mode solid state, 107–108
  sol-gel/combustion, 109
  solid state microwave synthesis, 106
MOF-derived materials, 195, 293
  carbon composite, 202–203
  carbon nanomaterials, 204–205
  CPs, 205–206
  synthesis strategies, 195–197
  transition metal oxide, 197–199, *199*, 203–204
  transition metal phosphide/phosphate, 201–202
  transition metal selenide, 200–201, *201*
  transition metal sulfide, 199–200, *200*
MOFs, *see* metal organic frameworks (MOFs)
molybdenum sulfide, 252
MoS_2, 47, 242
  composites of, 248–249
  conducting polymer composite, 245–248
  defect-free/-rich, *50*
  electrochemical properties, 49, *50*
  MoS_2@3DG hybrid structure, *245*
  nanocarbon composite, 243–245
  nanosheets, 47, *48*
  nanostructures, *48*
  1T, 49
  PPy nanocomposites, *247*
  properties, 240–241
  specific capacitance, 48
  thiourea for, 49
MWSS (microwave-assisted solvothermal synthesis), 108–109
MXenes, 140–141

N

nanocrystals, TEM images, *175*, 176
nanoporous niobium pentoxide (Nb_2O_5), 4, 10
nanosheet
  Co_3O_4, *42*, 42–43
  MnO_2, 41
  MoS_2, 47, *48*
  Simonk, 22
  2D, 33
  ultrathin, 47
  $Zn_5(OH)_8C_{12}$, 21–23, *24*
nanowires (NWs)
  niobium oxide, 6–7
  T-Nb_2O_5, 1-D, 6–7, *7*
  ZnO, 20–21
nickel–cobalt layered double hydroxide (Ni-Co LDH) nanosheets, 333
nickel cobalt oxide ($Ni_{0.37}Co_{0.63}(OH)_2$), 329, 333
$NiCo_2O_4$, 46–47
Ni Co Mn doped SnS_2 graphene-based aerogel, 127
Ni-Co-Mn oxide nano-flakes, 112
NiF-G/ZHCNs electrode, *23*
$NiMnO_3$ nano-balls, 3D flowers, 110, *110*
$NiMoO_4$, 46
niobium oxide
  nanodots, 8–9
  nanorods, 7–8
  nanowires, 6–7
niobium pentoxide (Nb_2O_5), 4
  core-shell structure, 4–6, *6*
  nanoporous, 4, 10
  SEM image, *7*
  structural schemes, *5*
nitrogen-doped ordered mesoporous carbon, *311*
nMOFs supercapacitors, 193–194
non-faradaic supercapacitors, 327
non-hydrolytic sol–gel (NHSG) process, 177–178
non-renewable energy resources, 119
normal parallel plate capacitors, 1
nucleation process, 180
NWs, *see* nanowires (NWs)

O

O-Nb_2O_5 (orthorhombic), 4
1D hierarchical hollow Co_3O_4 nanotubes synthesis, *319*
organic acids assisted (OAA), 163–164, *164*
organic electrolytes, 82–83
Ostwald repining process, 181
outer Helmholtz plane (OHP), 91
oxides/hydroxides/polymer composites, 223, **224–228**, 229–231
oxygen deficient
  Co_3O_4, 43–44
  Fe_2O_3, 45
  SiO_2, 44
  WO_3, 44
oxygen vacancy (Ov), 33–34, 39–40
  Co_3O_4, 43, *43*
  rich MnO_2@MnO_2, 41, *41*

P

parallel plate capacitor, 1–2, 17, 281, 327
Pd incorporation, 43
performance assessment, 91, **92**
perovskite, 143
  materials, 217–218, **219–221**
  oxides, 144–145
  supercapacitors, fabrication, 143–144
phosphorus, 201
polarized capacitor, 90
polarized wave, magnetic/electric field orientation, *107*
polar materials, 104
polyaniline (PANI), 125–126, 129, 246

polymer-inorganic hybrid materials/composites, 113
polypyrrole (PPy), 126
porous carbon preparation, *295*
power density *vs.* energy density curve, *326*
PPy/GO/ZnO supercapacitor, *19*
pristine MOFs, 191
  3D-MOFs, 191–192
  2D-MOFs, 192–193
  0D and 1D-MOFs, *192*, 193–194
PRI Ultracapacitors, 326
proton intercalation, 34–35
pseudocapacitance, 282, 328
pseudocapacitive charge storage process, 65
pseudocapacitors, **81**, 136, 282–283, *283*, 328
  faradaic processes, 91
  storage mechanism, 123, *123*

Q

quantum dots, 51

R

Ragone plot, 215, *215*
  ASC, *265*
  energy/power density, 120, *121*
random-titania nanotubes arrays, 36
recrystallization of metastable precursors (RMP) route, 157
redox reaction rate, 91
reduced graphene oxide (rGO), 62–63, 91
  energy storage devices, 128
  rGO/WO_3 composites, 94–95
renewable energy, 239
  resources, 119
reshaping bulk materials (RBM) route, 157
rGO, *see* reduced graphene oxide (rGO)
Rietveld refinement techniques, 20, *20*, **21**
ruthenium oxide (RuO_2), 138, 283, 328–329

S

self additive and removed template (SA & RT) route, 169
self-additive and self-removed template (SA & SRT) route, 167, *168*, 169
self-doping process, 34–35
separators, 80
Simonkolleite (Simonk), 21
  nanosheets, 22
single-mode microwave apparatus, *104*
single-mode polarized microwave synthesis, 114
single mode solid state microwave synthesis, 107–108
single molecular precursors, 181
SiO_2, 44
$SnFe_2O_4$, 111
soft-templated hydrothermal synthesis routes, 159–160, *160*
sol gel synthesis, 176–178, *177*
solid state microwave synthesis, 106
solvothermal synthesis, 66, 108, 173
  instrumentation for, 175–176
sonochemical method, 196
sonoelectrochemistry, 285, *285*
  MOF-based materials as, **287–288**
specific surface area (SSA), 59, 61
spinels-based composites materials, 218, **222**, 223
Stern model, 91
Stober technique, 308
storage mechanism
  EDLCs, 122, *122*
  pseudocapacitor, 123, *123*
sulfur anions, 241
sulfurization, 315
supercapacitance device, *136*
supercapacitors (SCs), 2, 137, 188, 258, 278–280, *280*, 325
  advantages, 33, 199
  attributes, 215, **215**
  vs. batteries, 137
  battery/fuel cell/capacitor, 214, **214**
  classification, 135–136, 214, *214*
  cobalt based, 328–335, **330–332**
  conducting polymer composite, 245–248
  EDLC, 327, *327*
  vs. energy storage devices, 188
  energy storage mechanisms, 280–284
  Fe_2O_3-based materials, 262–266, *266*, **267**
  Fe_3O_4-based materials, 268–270
  FeOOH-based materials, 266, 268, *269*
  FeS/FeS_2-based materials, 270, **271**
  history, 326
  hybrid, 328
  inorganic materials, 217–231
  molybdenum based compound/composites, **250–251**
  nanocarbon composite for, 243–245
  PPy/GO/ZnO, *19*
  pseudocapacitors, 328
  on storage process, 189
  structural nanomaterials for, 20–26
  types, 326–328, *327*
  zinc-based materials, 18–20
surfactants assisted (SA), 160
synthesis routes, 241

T

template-free hydrothermal synthesis, 157
  DSSP route, 157–158
  growth mechanism, 158–159
  ISRS route, 157
  RBM route, 157

template-free hydrothermal synthesis (*Continued*)
RMP route, 157
states of, **158**
template-mediated hydrothermal synthesis, 159
ASA, 164–165
BA, 160–162
hard-templated, 165–166
ILA, 162–163, *163*
OAA, 163–164, *164*
A & RT route, 166, *167*
SA, 160
SA & RT route, 169
SA & SRT route, 167, *168*, 169
soft-templated, 159–160, *160*
A & SRT route, 166–167, *168*
thiourea, 49
3D-connected α-$Ni(OH)_2$ sheets, arginine amino acid, *313*
3D flower-like $NiMnO_3$ nano-balls, 110, *110*
3D flower-on-sheet nanostructure, NiCo LDHs, 110–111
3D hierarchical MnO_2 microspheres, 113
3D tubular MoS_2/PANI, *247*
tin-based materials
conducting polymers, 129
graphene (Gr), 127
manganese oxide, 127
metal oxides, 126–127
rGO, 128
SnO_2 graphene composite, 128–129
sulfides, 127–128
ternary oxide, 128
TiO_2, 33–34
anodic nanotubes, 39–40
nanotube array, 34, *35*, *37*
reduced, 35
titania nanotubes array, 34–36
C-axis and randomly oriented, *38*
titanium niobium oxide ($TiNb_2O_7$), 9
nanofiber, *9*, 9–10
nanotubes, 9
1-D nanofiber, 9–10, *10*
TMDs, *see* transition metal dichalcogenides (TMDs)
T-Nb_2O_5 (tetragonal), 4
fabrication process for, *8*
nanodots, 8
nanostructure synthesize, 7
1-D nanowire, 6–7, *7*
T-Nb_2O_5 NCs, 7–8, 11
top-down method, 192
transition metal dichalcogenides (TMDs), 34, 69, 240–241
transition metal nitrides, 68
transition metal oxides, 91, 125
tungsten disulfide (WS_2), 51–53, 92, 94
carbon, 96
conducting polymers, 96
quantum dots, *51*
tungsten trioxide (WO_3), 44, 92
CA, 95–96
carbon, 94–96
conducting polymers, 96
crystal structure, *h*-WO_3, 92, *94*
electrochemical properties, **93**
2D nanosheet, 33

U

ultracapacitors, 79, 134, 325–326
ultrasonic nebulizer assisted chemical vapors deposition (UNA-CVD), 178
advantages, 180
disadvantages, 180
precursors role, 178–179
solvent role, 179
temperature role, 179
variants of, 179
ultrathin nanosheets, 47

V

vanadium dioxide (VO_2), 60, 63
on graphene network, 64
lattice structure, 65, *65*
vanadium monoxide (VO), 68
vanadium nitrides (VN), 68–69
vanadium oxides, 59–60
with mixed valence states, 67–68
nanotubes, rolling mechanism, *161*
vanadium pentoxide (V_2O_5), 60
CNF nanocomposite, 62
CNT film, 62
conductivity, 61–62
gel structure, 61
nanoporous network, 61
rGO nanocomposite, 62–63
3D nanostructures, 60, *61*
vanadium trioxide (V_2O_3), *66*, 66–67
vanadyl phosphate ($VOPO_4$), 69–70, *70*
volumetric heating, 102–103
VS_2 nanoplates, *53*, 53–54, *54*

X

X-ray diffraction (XRD), 113, 241, 259–260, 329, 334

Z

zinc-based materials, 17–20
zinc hydroxychloride ($Zn_5(OH)_8C_{12}$) nanosheets, 21–23, *24*
zinc molybdate ($ZnMoO_4$), 25–26, *27*
zinc oxide (ZnO)
  colossal permittivity, *22*
  NWs, 20–21
  Rietveld refinements data, **21**
  SEM image, *22*
  wurtzite crystal structure, 18, *19*
zinc sulfide (ZnS) nanospheres, 23–24, *25*

9781032238166